Lecture Notes in Mathematics

Edited by A. Dold and B. Eckmann

924

Séminaire d'Algèbre Paul Dubreil et Marie-Paule Malliavin

Proceedings, Paris 1981
(34ème Année)

Edité par M.-P. Malliavin

Springer-Verlag
Berlin Heidelberg New York 1982

Editeur
Marie-Paule Malliavin
Université Pierre et Marie Curie, Mathématiques
10, rue Saint Louis en l'Ile, 75004 Paris, France

AMS Subject Classifications (1980): 12L10, 13C15, 13H15, 14F05, 14J05, 14L05, 14M15, 14M17, 16A08, 16A12, 16A15, 16A26, 16A27, 16A33, 16A46, 16A48, 16A54, 16A55, 16A62, 16A64, 16A68, 17B10, 17B35, 18G10, 18G40, 20F28, 55R40, 58G07.

ISBN 3-540-11496-3 Springer-Verlag Berlin Heidelberg New York
ISBN 0-387-11496-3 Springer-Verlag New York Heidelberg Berlin

CIP-Kurztitelaufnahme der Deutschen Bibliothek
Séminaire d'Algèbre Paul Dubreil et Marie-Paule Malliavin:
Proceedings/Séminaire d'Algèbre Paul Dubreil et Marie-Paule Malliavin. –
Berlin; Heidelberg; New York: Springer
34. Paris 1981: (34ème année). – 1982.
(Lecture notes in mathematics; Vol. 924)
ISBN 3-540-11496-3 (Berlin, Heidelberg, New York)
ISBN 0-387-11496-3 (New York, Heidelberg, Berlin)
NE: GT

Printed in Germany

Printing and binding: Beltz Offsetdruck, Hemsbach/Bergstr.
2141/3140-543210

Liste des auteurs

★

TABLE DES MATIERES

publié avec le concours de :

l'Université Pierre et Marie Curie

la Première Section de l'Ecole Pratique des Hautes Etudes

*

* *

PREVIOUS VOLUMES OF THE "SEMINAIRE PAUL DUBREIL" WERE PUBLISHED IN THE LECTURE NOTES, VOLUMES 586 (1976), 641 (1977), 740 (1978), 795 (1979) and 867 (1980).

VERMA MODULES AND SCHUBERT CELLS :

A DICTIONARY

by

Sergei GELFAND and Robert MacPHERSON

Preface

This paper was completed in June, 1980. Since that time there has been marked development in the subject. Most notably, a beautiful proof of the Kazhdan-Lusztig conjecture has been found by Beilinson and Bernstein [BB] and by Brylinski and Kashiwara [BK]. We have indicated some of these further developments by adding notes marked with an asterisk ★.

The proof of the Kazhdan-Lusztig conjecture has much in common with the ideas of this paper. It proceeds by establishing an equivalence of categories between an algebraic category including Verma modules as objects, and a topological category including intersection homology sheaves of Schubert varieties as objects. However, as explained in the note on page 36, the dictionary from topology to algebra arrising from this equivalence of categories is different from the one proposed in this paper.

Never the less, we believe there is still some merit in the point of view we present here. Several predictions we made on the basis of this dictionary have been substantiated. One is our conjecture 2.10 which has been proved by Deligne, Gabber, Beilinson, and Bernstein ([DBB], [GM4]).

Another is the fact that the decomposition of the coherent continuation of a projective module to the wall of a Weyl chamber parallels the decomposition of the projection of an intersection homology sheaf of a Schubert variety in G/B to G/P (See the note on page 35).

It would be very interesting to find a unification of the point of view presented here and that of [BB] and [BK].

Introduction

In [KLI] and [KL2], Kazhdan and Lusztig made a remarkable conjecture that relates properties of certain infinite dimensional representations of a semisimple Lie algebra with those of singularities of Schubert cells in a generalized flag manifold. This paper represents an attempt to understand the source of this relation. Being a preliminary draft, this paper contains almost no proofs.

Let G be a semi-simple Lie group with Lie algebra $\underline{g}$. There are two rather different categories associated to G . One is a subcategory of the category of $\mathcal{O}$ modules of highest weight of Bernstein, Gelfand, and Gelfand (see § 3.5). The other is a category of complexes of sheaves on a generalized flag manifold for G (see § 2.1). One relation between these two categories is the Kazhdan-Lusztig conjecture which asserts that the multiplicities of a simple module in the Jordan-Hölder series of a Verma module is the dimension of the stalk homology of the complex of sheaves that gives the middle intersection homology groups of Goresky-Mac Pherson. But it appears that these categories have much more in common. This paper contains a dictionary which puts some of these common features in a more or less organized form. We should mention two things we could not do. First we do not know how to construct a complex of sheaves from a g-module or vice versa. Second (which may well be implied by the first) we cannot prove the Kazhdan-Lusztig conjecture*. Although we could not find a direct relation between the category of sheaves and the category of G-modules we have found an indirect relation in the form of a functor from each of them to a third category (see § 2.12, § 2.18, § 3.13). This enables us to find a topological analogue (see § 2.8) for the method of "walking through the walls of a Weyl

* The Kazhdan-Lusztig conjectures have now been proved (see preface).

chamber (or coherent continuation) that was extensively used in representation theory ([BG], [Sch] , [V_1]). We have learned that D. Kazhdan has found several similar results.

In the other direction, we were able to refine the original Kazhdan-Lusztig conjecture by showing a Lie-algebra interpretation of the complete Poincaré polynomial of intersection homology sheaf (and not only of its value at the point $q = 1$, as in [KL1]). *

§1 contains notations and preliminary known results both from algebra and topology. §2 describes the topological side of the dictionary. §3 does the same for Lie algebras. We tried to arrange the material in more or less parallel form. For each result we use the notation (➡ §3.6) to indicate where its counterpart may be found on the other side. We have stated some results which are fairly trivial when the parallelism with the other side was interesting. §4 contains the final comparison of the two sides of the dictionary and some remarks. Some readers may wish to begin with this section. §5 contains some examples and tables.

Acknowledgements. The authors would like to thank J. Bernstein, P. Deligne, I. Gelfand, M. Goresky, D. Kazhdan, G. Lusztig, and D. Vogan for useful discussions on the subject of this paper. The second author would like to thank the Academy of Sciences of the U.S.S.R. and the American Academy of Sciences for support of this research through their exchange program.

---

* Gabber and Joseph also found this refined conjecture (see [GJ]).

§1. Notations, conventions and preliminary results.

1.1. Let G be a complex semi-simple Lie group with Lie algebra $\underline{g}$. The following data will be fixed throughout this paper.

$\underline{h}$ is a Cartan subalgebra of $\underline{g}$.

Δ_+ is a set of positive roots for the root system Δ of $\underline{h}$ in $\underline{g}$

Σ is the corresponding set of simple roots.

ρ is half the sum of all the positive roots.

$\underline{n}_+$ (resp. $\underline{n}_-$) is the nilpotent subalgebra of $\underline{g}$ generated by root vectors corresponding to positive (resp. negative) roots.

ι is the anti-involution (i.e. ι^2 = identity. $\iota[XY] = [\iota Y, \iota X]$ that is the identity on $\underline{h}$ and transforms $\underline{n}_+$ to $\underline{n}_-$.

W is the Weyl group of $(\underline{g},\underline{h})$. For any $\gamma \in \Delta_+$, $\sigma_\gamma \in W$ is the corresponding reflection. W acts from the left.

$\ell(w)$ is the length function on W corresponding to the set of generators σ_α , $\alpha \in \Sigma$.

w_o is the unique element of W of maximal length; $\ell(w_o) = \mathrm{card}(\Delta_+)$.

$\underline{h}^*$ is the dual vector space to $\underline{h}$

$\langle,\rangle$ is the bilinear form on $\underline{h}^*$ induced from the killing form on $\underline{h}$.

$\underline{h}^*_{\mathbb{Z}}$ is the lattice of integral weights ($\chi \in \underline{h}_{\mathbb{Z}}$ if and only if $\frac{2\langle\chi,\gamma\rangle}{\langle\gamma,\gamma\rangle} \in \mathbb{Z}$ for all $\gamma \in \Sigma$)

$\psi \lhd \chi$ for ψ and χ in $\underline{h}^*$ means $\chi - \psi = \sum_{\alpha\in\Sigma} n_\alpha \cdot \alpha$ for some non-negative integers n_α .

A weight $\chi \in \underline{h}^*$ is called dominant (resp. antidominant) if $\langle\chi,\alpha\rangle \geq 0$ (resp. $\langle\chi,\alpha\rangle \leq 0$) for all $\alpha \in \Sigma$.

1.2. Modules over $\underline{g}$

All the modules in this paper will belong to the category $\mathcal{O}$ (see [BGG2] or [B G]). For any $V \in \mathcal{O}$ and $\psi \in \underline{h}^*$ let V^{ψ} be the weight space of weight ψ in V. For any χ in $\underline{h}^*$ let $M(\chi)$ be the Verma module, and $f(\chi) \in M(\chi)$ be the generator (it has weight $\chi-\rho$). Let $L(\chi)$ and $P(\chi)$ be the unique simple module with highest weight $\chi-\rho$ and its projective cover.

Let Ω be an orbit of the Weyl group in $\underline{h}^*$. Denote by $\mathcal{O}(\Omega)$ the complete subcategory of $\mathcal{O}$, formed by those $V \in \mathcal{O}$, such that the only simple quotient in the Jordan-Hölder series for V are $L(\chi)$, $\chi \in \Omega$. It is well known, that $\mathcal{O}$ is a direct sum of the subcategories $\mathcal{O}(\Omega)$ for all orbits Ω.

For $V \in \mathcal{O}$, we will denote by $[V:L(\chi)]$ the multiplicity of $L(\chi)$ in the Jordan-Hölder series for V.

We will define a partial ordering in $\underline{h}^*$ by saying $\psi \leq \chi$ if $M(\psi)$ is a submodule of $M(\chi)$. It is easy to see that $\psi \leq \chi$ implies $\psi \lhd \chi$. We will write $\psi \to \chi$ if $\psi \leq \chi$, and there is no ϕ such that $\psi \leq \phi \leq \chi$ except ψ and χ. In this case $\psi = \sigma_{\gamma}\chi$ for some $\gamma \in \Delta_+$. Sometimes we will denote this by $\psi \overset{\gamma}{\to} \chi$.

1.3. Orbits of the Weyl group and subsets of Σ.

Let A be a subset of Σ. Denote by $W(A)$ the subgroup generated by all σ_{α}, $\alpha \in A$ and by $W^1(A)$ the set of elements $w \in W$ with the property $\ell(w_{\alpha}) = \ell(w)+1$ for all $\alpha \in A$.

Any coset $W/W(A)$ contains exactly one element of $W^1(A)$ (it is characterized as the unique element of minimal length in that coset.) Let Ω be an orbit of W in $\underline{h}^*_{\mathbb{Z}}$. Then Ω contains exactly one dominant weight χ_+ and exactly one antidominant weight χ_-. Denote by $A = \mathfrak{H}(\Omega) \subset \Sigma$ the set of all $\alpha \in \Sigma$ with $\sigma_\alpha \chi_- = \chi_-$. Then $W(A)$ is the stationary subgroup of χ_- in W so that we may (but don't) identify Ω with $W^1(A)$. The ordering $\leq$ on Ω determines an ordering $\leq$ on $W^1(A)$ and the relation $\chi \overset{\gamma}{\to} \psi$ determines a relation $w \overset{\gamma}{\to} v$ on $W^1(A)$. These depend only on A, not on Ω.

1.4. Schubert cells.

Let H be the Cartan subgroup of G, corresponding to $\underline{h}$. For any root let X_γ be the corresponding one-parameter subgroup. Also let B be the Borel subgroup generated by H and X_γ, $\gamma \in \Delta_+$.

For a subset $A \subset \Sigma$ we denote by $P(A)$ the parabolic subgroup of G, generated by B and all $X_{-\alpha}$, $\alpha \in A$, so that $P(\emptyset) = B$, $P(\Sigma) = G$. Denote by $\phi(A)$ the generalized flag manifold $\phi(A) = G/P(A)$. Let $d(A)$ be the dimension of $\phi(A)$ as a complex manifold ($d(A)$ is equal to $\mathrm{card}(\Delta_+ \setminus \Delta_+(A))$, where $\Delta_+(A)$ is the set of positive roots that are linear combinations of roots in A). For $A' \subset A$ denote by $\pi(A',A)$ the projection $\phi(A') \to \phi(A)$.

For each $w \in W$, define a Schubert cell $C(w,A) \subset \phi(A)$ by $C(w,A) = B\, w_0 w P(A)$. Summarize the properties of Schubert cells in the following proposition :

Proposition.

$C(w,A) = C(v,A)$ if $w^{-1}v$ lies in $W(A)$, and they are disjoint otherwise. (So there is a one to one correspondence between $W'(A)$ and the Schubert cells in $\phi(A)$).

The Schubert cells form a Whitney stratification of $\phi(A)$. In particular the closure of a Schubert cell is a union of Schubert cells (axiom of the frontier). If v and w are in $W^1(A)$, then $C(w,A) \subset \overline{C(v,A)}$ if and only if $w \geq v$.

Let $A' \subset A$, $w \in W^1(A)$, and $w' \in W^1(A')$. Suppose that $w^{-1}w' \in W(A)$. Then $\pi(A,A')\ \ C(w',A') = C(w,A)$. Moreover, the restriction of $\pi(A',A)$ to $C(w',A')$ is a fibration with a $2(d(A')-d(A)-\ell(w')+\ell(w))$ - (real) dimensional open ball as a fiber. In particular, taking $A = \Sigma$ we see that $C(w',A')$ is a cell in $\phi(A')$ of (real) codimension $2\ell(w')$.

Suppose $A' \subset A$ and $w \in W^1(A)$. Then $\pi(A',A)^{-1}\ \overline{C(w,A)} = \overline{C(w,A')}$.

1.5. Sheaves.

Let X be a topological space. We denote by $S(X)$ the category of sheaves of vector spaces over the complex numbers on X. The value of a sheaf $\underline{\underline{S}}$ on an open set $U \subset X$ is $\underline{\underline{S}}(U)$, the stalk of $\underline{\underline{S}}$ at $x \in X$ is $\underline{\underline{S}}_x$. A sheaf $\underline{\underline{S}}$ is called a constant sheaf with value V if $\underline{\underline{S}}$ is the sheaficiation of the presheaf that takes the value V on all open sets. Then $\underline{\underline{S}}(U) = V$ for all connected U. The constant sheaf with value $\mathbb{C}$ is called $\underline{\underline{\mathbb{C}}}(X)$.

We denote by $D^bS(X)$ the doubly bounded derived category of $S(X)$. (See [H], [Vel]). Thus an object $\underline{\underline{S}}^\cdot$ of $D^bS(X)$ is a sheaf of cochain complexes

$$\underline{\underline{S}}(U) = \{\cdots \to \underline{\underline{S}}^i(U) \to \underline{\underline{S}}^{i+1}(U) \to \cdots\}$$

where $\underline{\underline{S}}^i(U)$ is a complex vector space and $\underline{\underline{S}}^i(U) = 0$ for $i \gg 0$ and $i \ll 0$. A morphism from $\underline{\underline{S}}^\cdot$ to $\underline{\underline{T}}^\cdot$ in $D^bS(X)$ is determined by a diagram

of chain-maps

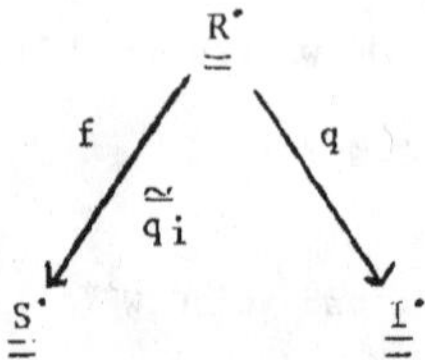

where $\underset{qi}{\simeq}$ indicates a quasi-isomorphism, i.e. f induces an isomorphism on all homology groups of all stalks. If q is also a quasi-isomorphism then the morphism is an isomorphism in $D^bS(x)$ from $\underline{\underline{S}}^{\cdot}$ to $\underline{\underline{T}}^{\cdot}$.

There are functors $\underline{\underline{I}}$, $\underline{\underline{H}}^i$, and $\underline{\underline{T}}$

$$S(X) \underset{\underline{\underline{H}}^i}{\overset{\underline{\underline{I}}}{\rightleftarrows}} D^bS(X) \circlearrowleft \underline{\underline{T}}$$

$\underline{\underline{I}}\ \underline{\underline{S}}\,(U)$ is the complex that is $\underline{\underline{S}}(U)$ in dimension zero and zero in all other dimensions. $\underline{\underline{H}}^i(\underline{\underline{S}}^{\cdot})$ is the sheafification of the presheaf $U \to H^i(\underline{\underline{S}},U)$. The translation functor T from $D^bS(X)$ to itself shifts the numbering of all chain complexes : $(\underline{\underline{T}}\ \underline{\underline{S}})^{i+1}(U) = \underline{\underline{S}}^i(U)$.

<u>Proposition.</u> $\underline{\underline{I}}(\underline{\underline{S}})$ is characterized up to equivalences in $D^bS(X)$ by the properties

$$\underline{\underline{H}}^i(\underline{\underline{I}}(\underline{\underline{S}})) = \begin{cases} \underline{\underline{S}} & \text{if } i = 0 \\ 0 & \text{if } i \neq 0 \end{cases}$$

The category $D^bS(X)$ is a triangulated category ([Ve1]) . The distinguished triangle denoted by

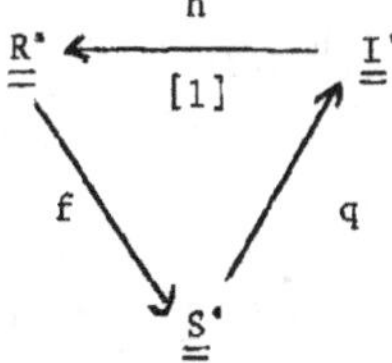

induces a long exact sequence

$$\cdots \xrightarrow{h_*} H^i(\underline{\underline{R}}') \xrightarrow{f_*} H^i(\underline{\underline{S}}') \xrightarrow{q_*} H^i(\underline{\underline{T}}') \xrightarrow{h_*} H^{i+1}(\underline{\underline{R}}') \xrightarrow{f_*} \cdots$$

Let $f : X \to Y$ be a continuous map. Then there are functors $f_!$ and f^*

$$S(X) \underset{f^*}{\overset{f_!}{\rightleftarrows}} S(Y)$$

The functor f^* , called the pullback is defined as follows : if $\tilde{\underline{\underline{S}}} \to Y$ is an étale map giving the sheaf $\underline{\underline{S}}$ on Y ([6], p. 110) then $f^*\underline{\underline{S}} = \underline{\underline{T}}$ is given by the étale map $\tilde{\underline{\underline{T}}} \to X$ such that the following square is a fiber square

$$\begin{array}{ccc} \tilde{\underline{\underline{T}}} & \longrightarrow & \tilde{\underline{\underline{S}}} \\ \downarrow & & \downarrow \\ X & \xrightarrow{f} & Y \end{array}$$

The functor $f_!$, called pushforward with proper supports, is defined as follows ([Ve2], p.3) : $f_!(\underline{\underline{S}})(U)$ is those sections in $\underline{\underline{S}}(f^{-1}U)$ whose support is proper over U .

The functors f^* and $f_!$ determine right derived functors Rf^* and $Rf_!$

$$D^b S(X) \underset{Rf^*}{\overset{Rf_!}{\rightleftarrows}} D^b\, S(Y)$$

$Rf_!(\underline{\underline{S}}^\cdot)$ is quasi-isomorphic to $\{\cdots \to f_!\, \underline{\underline{T}}^i \to f_!\, \underline{\underline{T}}^{i+1} \to \cdots\}$ where each $\underline{\underline{T}}^i$ is an injective sheaf and $\underline{\underline{T}}^\cdot$ is quasi-isomorphic to $\underline{\underline{S}}^\cdot$. Rf^* is similarly characterized.

Proposition.

a) $Rf^*\, \underline{\underline{I}} \underset{qi}{\simeq} \underline{\underline{I}}\, f^*$

b) If $\begin{array}{ccc} X' & \xrightarrow{g'} & X \\ f' \downarrow & & \downarrow f \\ Y' & \xrightarrow{g} & Y \end{array}$ is a fiber square, and f is a locally trivial topological fibration, then $f'_!\, g'^* = g^* f_!$ and $Rf'_!\; Rg'^* = Rg^*\, Rf_!$.

c) If $f : X \to Y$ is a fibration with fiber homeomorphic to $\mathbb{R}^n$, then $Rf_!\, Rf^* = \underline{\underline{T}}^n$.

One final functor, $R^i f_! : S(X) \to S(Y)$, is defined by commutativity of the diagram :

$$\begin{array}{ccc} S(X) & \xrightarrow{R^i f_!} & S(Y) \\ \underline{\underline{I}} \downarrow & & \uparrow \underline{\underline{H}}^i \\ D^b S(X) & \xrightarrow[Rf_!]{} & D^b S(Y) \end{array}$$

1.6. Intersection homology sheaves.

Let V be a complex analytic variety of complex dimension n. An element $\underline{S}^\cdot$ of $D^bS(V)$ is called constructible if there exists a filtration by closed subvarieties $\emptyset = V_{-1} \subset V_o \subset \cdots \subset V_n = V$ such that $\underline{H}^i\, S^\cdot$ restricted to $V_j - V_{j-1}$ is locally a constant sheaf for all i and j

Proposition ([GM3]) . Up to quasi-isomorphism there exists a unique constructible complex of sheaves $\underline{\underline{IC}}^\cdot(V)$ whose homology sheaves vanish in negative dimension such that :

1) For some open dense set U, $\underline{\underline{IC}}^\cdot(V)$ restricted to U is $\underline{\underline{I\mathbb{C}}}(U)$.

2) Support condition :

$$\dim_{\mathbb{C}}\{x \in V \mid (\underline{H}^i\ \underline{\underline{IC}}^\cdot(V))_x \neq 0\} < n-i$$

for all $i > 0$.

3) Dual support condition

$$\dim_{\mathbb{C}}\{x \in V \mid (\underline{H}_c^{2n-i}\ \underline{\underline{IC}}^\cdot(V))_x \neq 0\} < n-i$$

for all $i > 0$

Here, with respect to the directed system of neighborhoods U of x,

$$(\underline{H}^i\ \underline{\underline{IC}}^\cdot(V))_x = \lim_{\substack{\rightarrow \\ x\in U}} H^i(U)$$

$$(\underline{H}_c^i\ \underline{\underline{IC}}^\cdot(V))_x = \lim_{\substack{\leftarrow \\ x\in U}} H_c^i(U)$$

where H_c^i denotes cohomology with compact supports.

The sheaves of chain complexes $\underline{\underline{IC}}^{\cdot}(V)$ characterized by this proposition are called intersection homology sheaves. They were constructed first in geometric topology in [GM1] then in algebraic geometry in [D] (see [KL2]) . These constructions are proved to coincide in [GM3] . If V is compact, then the hypercohomology groups of $\underline{\underline{IC}}^{\cdot}(V)$, called intersection homology groups, satisfy Poincaré duality : the ith and the $(2n-i)^{th}$ Betti numbers are equal.

§2. Topology

2.1. Let A be a subset of Σ, the set of simple roots. Recall (§1.4) that $\phi(A)$ denotes the associated generalized flag manifold and $c(w) : C(w,A) \hookrightarrow \phi(A)$ are the inclusions of the Schubert cells of $\phi(A)$.

Definition.

The category Chains (A) is the full subcategory of $D^b S(\phi(A))$ (where $\phi(A)$ is considered with its classical topology) whose objects $\underline{\underline{S}}$ satisfy the following three conditions :

1. Finiteness. For all i, the stalks of $\underline{\underline{H}}^i \underline{\underline{S}}^\cdot$ are all finite dimensional over $\mathbb{C}$.

2. Evenness. $\underline{\underline{H}}^i \underline{\underline{S}}^\cdot$ is zero if i is odd.

3. Constructibility $c(w)^* \underline{\underline{H}}^i \underline{\underline{S}}^\cdot$ is a constant sheaf on $C(w,A)$ for all $w \in W^1(A)$ and all $i \in \mathbb{Z}$

Examples of objects in Chains (A) are the cell sheaves $\underline{\underline{C}}(w,A)$ of §2.2 and the intersection, homology sheaves $\underline{\underline{IC}}^\cdot(w,A)$ of §2.5 .

2.2. Definition.

For any $w \in W^1(A)$, the cell-sheaf $\underline{\underline{C}}(w,A)$ is $Rc(w)_! I[\mathbb{C}(C(w,A))]$.

The stalk at p of $H^i \underline{\underline{C}}(w,A)$ is

$\mathbb{C}$ if $p \in C(w,A)$ and $i = 0$

0 otherwise

Proposition.

A. (➡ 3.6 **A**) The cell sheaf $\underline{\underline{C}}(w,A)$ lies in Chains (A)

Б. (➡ 3.6 Б) The cell sheaves $\underline{\underline{C}}(w,A)$, $w \in W^1(A)$, have the property that for any exact triangle in Chains (A)

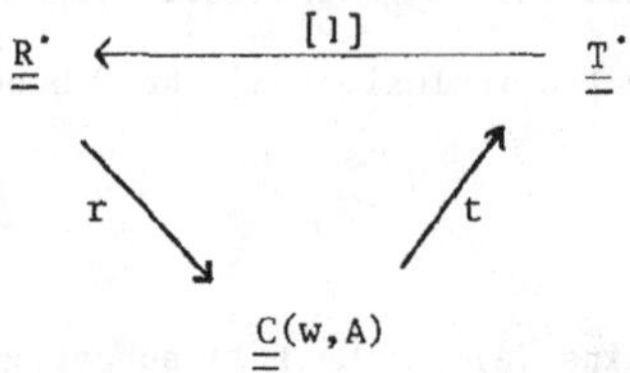

either $\underline{\underline{R}}^{\cdot} \underset{qi}{\cong} 0$ or $\underline{\underline{T}}^{\cdot} \underset{qi}{\cong} 0$.

Proof of Б . Cell-sheaves have this property because by constructibility $v_* H^0 \underline{\underline{R}}^{\cdot} \to H^0 \underline{\underline{C}}(w,A)$ is either zero or surjective. Then by evenness we find from the long exact sequence in cohomology that $\underline{\underline{R}}^{\cdot}$ or $\underline{\underline{T}}^{\cdot}$ has no cohomology and hence is quasi-isomorphic to zero.

2.3 (➡ 3.9) Let A be given. Let us choose a numbering $w_1,\dots,w_r$ of elements in $W^1(A)$ in such a way that $w_i \geq W_j$ implies $i \geq j$.

Proposition. Suppose $\underline{\underline{S}}^{\cdot}$ is in Chains (A). Then there exists a canonical filtration $\underline{\underline{S}} = \underline{\underline{S}}^{\cdot}_o \supset \underline{\underline{S}}^{\cdot}_1 \supset \dots \supset \underline{\underline{S}}^{\cdot}_n = 0$ such that $\underline{\underline{S}}^{\cdot}_i/\underline{\underline{S}}^{\cdot}_{i+1}$ in quasi-isomorphic to a direct sum of objects of the form $\underline{\underline{T}}^k\, \underline{\underline{C}}(w_i,A)$ for various k .

Proof. Let d(w) be the inclusion of the complement of the closure of C(w,A) into $\phi(A)$. Then

$$\underline{\underline{S}}^{\cdot}_i = d(w_i)_!\, d(w_i)^*\, \underline{\underline{S}}$$

satisfies the conditions of the proposition.

2.4 We define the Grothendieck group of the category Chains (A) , K(Chains (A)) , to be the Abelian group generated by quasi-isomorphism classes of objects in A subject to the relation $[\underline{\underline{S}}^{\cdot}] = [\underline{\underline{R}}^{\cdot}] + [\underline{\underline{T}}^{\cdot}]$ whenever we have a triangle

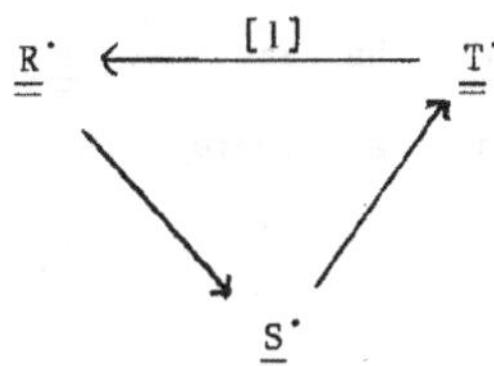

We denote by $[\underline{\underline{S}}^{\cdot}]$ the equivalence class of $\underline{\underline{S}}^{\cdot}$ in K(A) .

For any Abelian group J we denote by $J^{W^1(A)}$ the group of formal linear combinations $\Sigma j_i w_i$ of elements $w_i \in W^1(A)$ with coeficients $j_i \in J$. $\mathbb{Z}[q,q^{-1}]$ denotes the group of integral Laurent polynomials in q under addition .

Corollary (➡ 3.8)

A. The Grothendieck group K(Chains (A)) is the free Abelian group generated by all sheaves $T^{2n}\,\underline{\underline{C}}(w,A)$ for $n \in \mathbb{Z}$, $w \in W^1(A)$.

Б. There is an isomorphism

$$k : K(\text{Chains }(A)) \longrightarrow \mathbb{Z}[q,q^{-1}]^{W^1(A)}$$

which takes $[\underline{\underline{S}}^{\cdot}]$ to $\Sigma\, P_w \cdot w$ where P_w is the Poincaré polynomial in $q^{1/2}$ of the stalk cohomology of $\underline{\underline{S}}^{\cdot}$ at any point in C(w,A)

2.5. Let $\bar{c}(w) : \overline{C(w,A)} \subset \Phi(A)$ be the inclusion of the closure of the Schubert cell C(w,A).

Definition. (→ 3.6B) The intersection homology sheaf $\underline{\underline{IC}}^{\cdot}(w,A)$ is $\bar{c}(w)_! \; \underline{\underline{IC}}^{\cdot}(\overline{C(w,A)})$ (See §1.6)

$\underline{\underline{IC}}^{\cdot}(w,A)$ is an object in Chains (A) : finiteness is true of $\underline{\underline{IC}}^{\cdot}(V)$ for any algebraic variety V ; evenness was proved by Kazhdan and Lusztig [KL2] when $A = \phi$ and it follows in general by applying $\underline{\underline{F}}(A,\phi)$ of §2.8 ; and constructibility follows from the fact that $\underline{\underline{IC}}^{\cdot}(\overline{C(w,A)})$ is invariant under homeomorphisms ([GM3]) and the homeomorphism group of $\overline{C(w,A)}$ is transitive on the Schubert cells .

2.6. The Kazhdan-Lusztig polynomial, $P_{v,w}(q)$, depending on two elements v and w of $W^1(A)$, is defined by

$$[\underline{\underline{IC}}^{\cdot}(w,A)] = \sum_v P_{vw_o,ww_o} \cdot v$$

where $[\underline{\underline{IC}}^{\cdot}(w,A)]$ denotes the class of $\underline{\underline{IC}}^{\cdot}(w,A)$ in K(Chains (A)) . (See §2.4) . In other words, the coeficient of q^n in $P_{v,w}$ is the multiplicity of $T^{2n}\,\underline{\underline{C}}(vw_o,A)$ in the composition series of $\underline{\underline{IC}}^{\cdot}(ww_o,A)$.

We say an element e of $\mathbb{Z}[q,q^{-1}]^{W^1(A)}$ has leading term $P\cdot w$ if

$$e = P.w + \sum_{v \geq w} P_v \cdot v$$

The element $[\underline{\underline{IC}}^{\cdot}(w,A)]$ has leading term $1\cdot w$. This follows from §1.6 property 1 .

2.7. Proposition. (→ 3.6Γ). The intersection homology sheaf $\underline{\underline{IC}}^{\cdot}(w,A)$ is indecomposable.

Proof. Let $\underline{\underline{IC}}^{\cdot}(w,A) = \underline{\underline{R}}_1^{\cdot} \oplus \ldots \oplus \underline{\underline{R}}_m$ be a decomposition into indecomposable summands . Since $[\underline{\underline{IC}}^{\cdot}(w,A)]$ has leading term $1\cdot w$, for some i , $[\underline{\underline{R}}_i]$ also has leading term $1\cdot w$. Then this $\underline{\underline{R}}_i$ satisfies all the axioms of §1.6

for $\overline{c}(w)_! \, \underline{\underline{IC}}^{\cdot} \, (\overline{C(w,A)})$ and hence equals $\underline{\underline{IC}}^{\cdot}(w,A)$.

2.8. Let A and A' be two subsets of Σ . Then we have a fiber square of filtrations

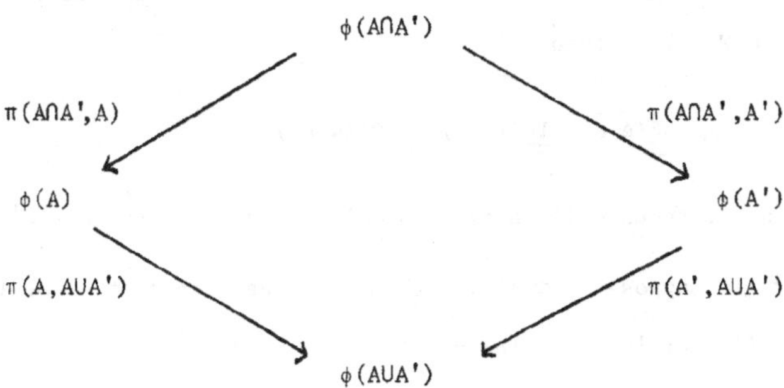

We have $R\pi(A\cap A',A')_! \, R\pi(A\cap A',A)^* = R\pi(A',A\cup A')^* \, R\pi(A,A\cup A')_!$ (see §1.5) .

Definition. (3.9) . The functor $\underline{\underline{F}}(A,A')$ from $D^bS(\phi(A))$ to $D^bS(\phi(A'))$ is given by

$$\underline{\underline{F}}(A,A') = R\pi(A\cap A',A')_! \, R\pi(A\cap A',A)^* = R\pi(A',A\cup A')^* \, R\pi(A,A\cup A')_!$$

2.9. Properties of $\underline{\underline{F}}(A,A')$ (➡ 3.10)

А. $\underline{\underline{F}}(A,A')$ defines a functor from Chains (A) to Chains(A')

Б. $\underline{\underline{F}}(A,A)$ is the identity functor .

В. Suppose $A\cap A' \subset A'' \subset A\cup A'$. Then $\underline{\underline{F}}(A'',A') \, \underline{\underline{F}} \, (A,A'') = \underline{\underline{F}}(A,A')$.

Г. Suppose $A \subset A'$. Then $\underline{\underline{F}}(A,A')$ transforms cell sheaves in Chains (A') to cell sheaves in Chains (A'). More precisely, if $w \in W^1(A)$,

$$\underline{\underline{F}}(A,A') \, \underline{\underline{C}}(w,A) = T^n \, \underline{\underline{C}}(v,A')$$

where $v = w(W(A) \cap W^1(A'))$ and $n = 2(d(A) - d(A') - \ell(w) + \ell(v))$. $\underline{\underline{F}}(A,A')$ does not in general transform indecomposable objects to indecomposable objects.

Д. Suppose $A \supset A'$. Then $\underline{\underline{F}}(A,A')$ transforms indecomposable objects in Chains (A) to indecomposable objects in Chains (A') . In particular if $w \in W^1(A)$, then

$$\underline{\underline{F}}(A\ A')\ \underline{\underline{IC}}^{\cdot}(w,A) = \underline{\underline{IC}}^{\cdot}(w,A')$$

$\underline{\underline{F}}(A,A')$ never takes cell sheaves to cell sheaves unless $A = A'$.

Е. Suppose $A \subset A'$. Let m_n be the number of elements of $W(A') \cap W^1(A)$ of length $d(A) - d(A') - n$. Then

$$\underline{\underline{F}}(A',A'')\circ \underline{\underline{F}}(A,A') = \Sigma\, m_n\, \underline{\underline{T}}^{2n}$$

Ж. $\underline{\underline{F}}(A,A')$ takes distinguished triangles to distinguished triangles .

2.10. Conjecture★ (⟶ 3.9). The functor $\underline{\underline{F}}(A,A')$ applied to an intersection homology sheaf $\underline{\underline{IC}}^{\cdot}(w,A)$ is quasi-isomorphic to a direct sum of translated intersection homology sheaves $\underline{\underline{T}}^{2n}\ \underline{\underline{IC}}^{\cdot}(v,A')$. (In other words, the sheaves of the form $\bigoplus_i \underline{\underline{T}}^{2n_i}\ \underline{\underline{IC}}^{\cdot}(w_i,A)$ are taken to sheaves of similar form by $\underline{\underline{F}}(A,A')$.) .

If $A \supset A'$ the conjecture is easy to prove, so the general case reduces to the case $A \subset A'$. This can be proved for several classes of examples. Two specific examples are illustrated in §5.4.

We do not know of any counterexample to the following conjecture : For any map $f : X \longrightarrow Y$ of algebraic varieties, $f_!\ \underline{\underline{IC}}(X)$ is

★ This conjecture is true. See note on next page.

quasi-isomorphic to a direct sum of translated intersection hormology groups of sub-varieties of Y with twisted coeficients*. The twisting of the coeficients can never happen in the case of Schubert varieties.

2.11. Resolutions.

Definition. For $A \subset \Sigma$ and $w \in W^1(A)$ we call a sequence $\{A_i\}$ of subsets of Σ

$$\phi = A_o , A_1 , \dots , A_m = A$$

Resolution data for the Schubert variety $\overline{C(w,A)}$ if the K(Chains (A)) element $[\underline{\underline{F}}(\{A_i\})\ \underline{\underline{C}}^{\cdot}(w_o,A_o)]$ where

$$\underline{\underline{F}}(\{A_i\})\ \underline{\underline{C}}^{\cdot}(w_o,A.) = \underline{\underline{F}}(A_{m-1},A_m) \cdots \underline{\underline{F}}(A_1,A_2)\ \underline{\underline{F}}(A_o,A_1)\ \underline{\underline{C}}^{\cdot}(w_o,A_o)$$

has leading term $1.w$.

Proposition. There exist resolution data for any $A \subset \Sigma$ and any $w \in W^1(A)$.

Resolution data for $w \in W^1(A)$ is not uniquely determined by A and w . Resolution data determines a diagram of spaces

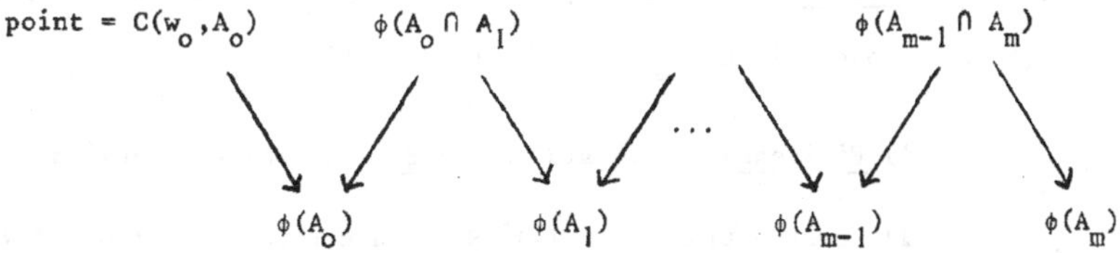

and hence the fiber product

$$C(w_o,A_o) \times_{\phi(A_o)} C(A_o \cap A_i) \times_{\phi(A_1)} \cdots \times_{\phi(A_{m-1})} \phi(A_{m-1} \cap A_m)$$

*This conjecture has been proved by Deligne, Gabber, Beilinson and Bernstein [DBB] (the map f must be proper). This implies conjecture 2.10. See [GM4] for some general consequences of this result.

which maps to $\overline{C(w,A)} \subset \phi(A_m)$. This mapping denoted by $\pi : X(\{A_i\}) \to \overline{C(w,A)}$ is a resolution of singularities of the Schubert variety $\overline{C(w,A)}$. It is called the <u>canonical resolution</u> associated to the resolution data . In case all sets $A_1 \ldots A_{m-1}$ have one element, it is the Demazure resolution [Dm].

We give two alternative characterizations of the sheaf $\underline{\underline{IC}}^{\cdot}(w,A)$ which would follow from conjecture 2.10.

A. (⟶ 3.10 a) . $\underline{\underline{IC}}^{\cdot}(w,A)$ is the unique indecomposable direct summand of $\underline{\underline{F}}(\{A_i\})\ \underline{\underline{C}}\ (w_o,A_o)$ whose class in K(<u>Chains (A)</u>) has leading term $1.w$.

B. $\underline{\underline{IC}}^{\cdot}(w,A)$ is the unique indecomposable direct summand of $R\,\pi_!\ \underline{\underline{\mathbb{C}}}(X(\{A_i\}))$ whose class in K(<u>Chains (A)</u>) has leading term $1\cdot w$.

(In fact $\underline{\underline{F}}(\{A_i\})\ \underline{\underline{C}}^{\cdot}\ (w_o,A_o) = R\,\pi_!\ \mathbb{C}(X(\{A_i\}))$.)

In many cases resolution data can be found for which $R\,\pi_!\ \underline{\underline{\mathbb{C}}}(X(\{A_i\}))$ is indecomposable. This happens when $X(\{A_i\})$ is a small resolution (see [GM3]) . An example is given in §5.4.

2.12. <u>Definition</u>. The category <u>Sheaves (A)</u> is the full sub-category of $S(\phi(A))$ whose objects $\underline{\underline{S}}$ satisfy the conditions :

1) <u>Finiteness</u>. All stalks of $\underline{\underline{S}}$ are finite dimensional over $\mathbb{C}$.

2) Constructibility $c(w)^* \underline{\underline{S}}$ is a constant sheaf on $C(w,A)$ for all $w \in W^1(A)$

<u>Definition</u>. (⟶ 3.13) . The total homology functor $\underline{\underline{H}}(A)$: <u>Chains (A)</u> ⟶ <u>Sheaves (A)</u> is defined by

$$\underline{\underline{H}}(A) = \sum_i \underline{\underline{H}}^i$$

2.13. Properties of the functor $\underline{\underline{H}}(A)$. (➡ 3.14)

A. $\underline{\underline{H}}(A)$ takes exact triangles in Chains (A)

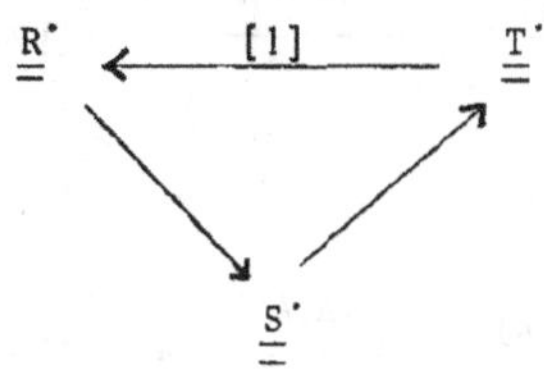

to short exact sequences in Sheaves (A)

$$0 \longrightarrow \underline{\underline{H}}(A)(R^{\cdot}) \longrightarrow \underline{\underline{H}}(A)\ (\underline{\underline{S}}^{\cdot}) \longrightarrow \underline{\underline{H}}(A)\ (\underline{\underline{T}}^{\cdot}) \longrightarrow 0\ .$$

Б. $\underline{\underline{H}}(A)$ induces a map h on Grothendieck groups

$$\mathbb{Z}[q,q^{-1}]^{W^1(A)} = K(\text{Chains (A)}) \xrightarrow{h} K(\text{Sheaves (A)}) = \mathbb{Z}^{W^1(A)}$$

which is given by specializing q to 1 .

B. $\underline{\underline{H}}(A)$ is not an equivalence of categories. In particular it does not preserve indecomposability .

2.14. We recall the diagram of 2.8

Definition. (➡ 2.19) . The functor $\underline{\underline{G}}(A,A')$ from Sheaves (A) to Sheaves (A') is given by

$$\underline{\underline{G}}(A,A') = \Sigma\, R^i\, \pi(A \cap A', A')_!\ \pi(A \cap A', A)^*$$

$$= \pi(A',A \cup A')^*\ \Sigma\, R^i\, \pi(A,A \cup A')_!$$

These functors satisfy properties similar to those of 2.9.

<u>Proposition</u>. (➡ 3.15) . The following diagram of categories and functors is commutative :

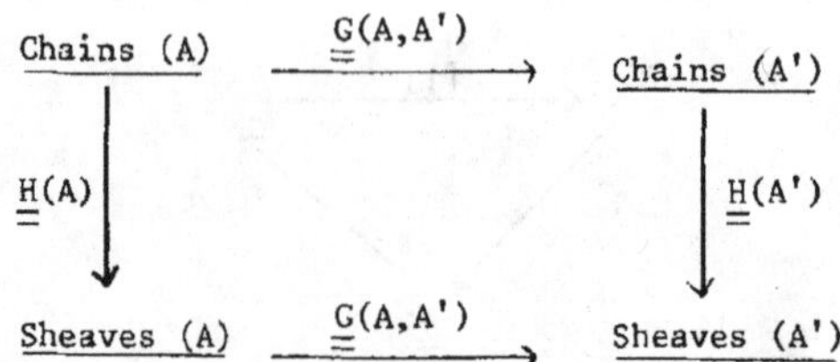

<u>Proof</u>. There is a spectral sequence $G(A,A') \circ \underline{\underline{H}}(A)\ \underline{\underline{S}}^{\cdot} \Longrightarrow \underline{\underline{H}}(A') \circ \underline{\underline{G}}(A,A')\underline{\underline{S}}^{\cdot}$. This spectral sequence degenerates because of the evenness condition of §2.1 and the fact that all the cells $C(w,A)$ are even dimensional.

2.15. Given an object $\underline{\underline{S}}$ in <u>Sheaves (A)</u> , let $\underline{\underline{S}}(w)$ be the stalk of $\underline{\underline{S}}$ at any point $p \in C(w,A)$ (Since $\underline{\underline{S}}$ is trivial on $C(w,A)$, $\underline{\underline{S}}(w)$ is canonically independent of the choice of the point p .)

<u>Definition 5</u>. (➡ 3.1) . If $W \longrightarrow V$, then for any $\underline{\underline{S}} \in$ <u>Sheaves (A)</u> the <u>generalization map</u> $y(v,w) : \underline{\underline{S}}(v) \longrightarrow \underline{\underline{S}}(w)$ is defined as follows : for any $p \in C(v,A)$ let $\mathcal{U}$ be a neighborhood such that $\underline{\underline{S}}(\mathcal{U}) \simeq \underline{\underline{S}}_p$. Then for any $q \in \mathcal{U} \cap C(w,A)$, $y(v,w)$ is defined so that the following diagram commutes :

$$
\begin{array}{ccccc}
\underline{\underline{S}}_p & \xleftarrow[\text{res}]{\simeq} & \underline{\underline{S}}(\mathcal{U}) & \xrightarrow[\text{res}]{} & \underline{\underline{S}}_q \\
\wr\!\!\!\shortparallel & & & & \wr\!\!\!\shortparallel \\
\underline{\underline{S}}(v) & & \xrightarrow{\quad y(v,w) \quad} & & \underline{\underline{S}}(w)
\end{array}
$$

2.16. Proposition. (⟶ 3.4) . Suppose $v \longrightarrow w$ and $\underline{\underline{S}}$ is an extension of $\underline{\underline{C}}(v,A)$ by $\underline{\underline{C}}(w,A)$

$$0 \longrightarrow \underline{\underline{C}}(w,A) \longrightarrow \underline{\underline{S}} \longrightarrow \underline{\underline{C}}(v,A) \longrightarrow 0 \quad .$$

Then this extension is nontrivial if and only if $y(v,w)\ \underline{\underline{S}}(v) = \{0\}$

2.17. Let A be a subset of Σ . The category $Att(A)$ of attaching schemes is defined as follows :

An object of A is a pair of data $\{E(w),e(v,w)\}$ where $E(w)$ is a complex vector space for each $w \in W^1(A)$ and $e(v,w)$ is a linear map $E(v) \longrightarrow E(w)$ for each $v,w \in W^I(A)$ such that $v \longrightarrow w$. These data are subject to the commutativity restriction : whenever $v = s_o \to s_1 \to \ldots \to s_n = W$ and $V = t_o \to t_1 \to \ldots \to t_n = W$, then
$e(s_{n-1},s_n) \ldots e(s_o,s_1) = e(t_{n-1},t_n) \ldots e(t_o,t_1)$.

A morphism $\{E(v),e(v,w)\} \to \{E'(v),e'(v,w)\}$ is a map $E(v) \to E'(v)$ for each $v \in W'(A)$ such that the obvious diagrams commute :

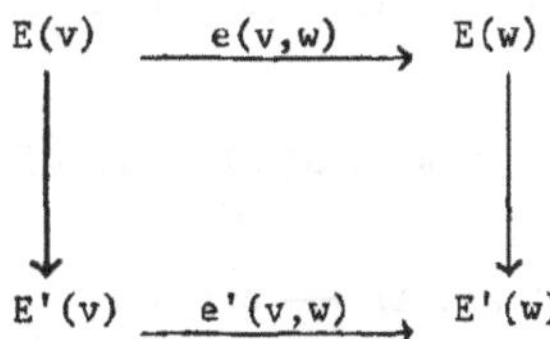

2.18. There is a contravariant functor $J(A)$: Sheaves (A)→Att(A) which assigns the data $E(w) = (\underline{\underline{S}}(w))^*$, $e(v,w) = (y(w,v))^t$ to the sheaf $\underline{\underline{S}}$. (Here $(S(w))^*$ is the vector space dual of $\underline{\underline{S}}(w)$ and $(y(w,v))^t$ is the adjoint of $y(w,v)$.)

Proposition. The contravariant functor $J(A)$ gives an equivalence of categories between Sheaves (A) and $Att(A)^{op}$.

Proof. We will construct the inverse functor $J(A)^{-1} : Att(A) \to \underline{Sheaves}\ (A)$.

Given an attaching scheme $\{E(w),e(v,w)\}$ and an open set $\mathcal{U} \subset \phi(A)$, we define

$$T(\mathcal{U}) = \varinjlim \{E(w) \mid C(w,A) \cap \mathcal{U} \neq \emptyset\}$$

That is, $T(\mathcal{U})$ comes equipped with maps $E(w) \to T(\mathcal{U})$ (whenever $C(w,A)$ meets $\mathcal{U}$) such that the diagrams

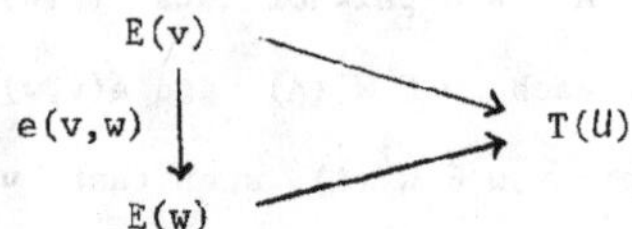

commute (whenever $v \to W$) and $T(\mathcal{U})$ is the universal vector space with this property. If $\mathcal{U}' \subset U$ then $\{E(w) \mid C(w,A) \cap \mathcal{U}' \neq \emptyset\} \subset \{E(w) \mid C(w,A) \cap \mathcal{U} \neq \emptyset\}$ so there is a unique map $T(\mathcal{U},\mathcal{U}') : T(\mathcal{U}) \longrightarrow T(\mathcal{U}')$ commuting with the maps from $E(w)$ by the universal property of $T(\mathcal{U})$. Now $J(A)^{-1}\ \{E(w),e(vw)\}$ is the sheafification of the presheaf $\underline{\underline{P}}$ whose value $\underline{\underline{P}}(\mathcal{U})$ is $T(\mathcal{U})^*$ and whose restriction map $P(\mathcal{U}',\mathcal{U})$ is $T(\mathcal{U},\mathcal{U}')^t$.

2.19. Let A and A' be subsets of Σ. We will define a functor

$$G(A,A') : Att(A) \longrightarrow Att(A')$$

Given an object $\{E(w), e(v,w)\}$ in $Att(A)$, we describe $G(A,A')\ \{E(w),e(v,w)\} = \{E'(w),e'(v,w)\}$ first in two special cases.

Case 1 : $A \supset A'$. Define $p : W^1(A') \longrightarrow W^1(A)$ by $p(w) = w\,W(A') \cap W^1(A)$ (i.e. $p(w)$ is the element of minimal length in the coset $w\,W^1(A')$.) .

Then

$$\text{if}\quad w \in W^1(A'),\quad E'(w) = E(p(w))$$

$$\text{if}\quad v \to w,\qquad e'(v,w) = \begin{cases} \text{the identity if } p(v) = p(w) \\ e(p(v),p(w)) \text{ otherwise} \end{cases}$$

Case 2 : $A \subset A'$: Define similarly $p : W^1(A) \to W^1(A')$ by $p(w) = w\, W(A) \cap W^1(A')$.

Then

$$\text{if}\quad w \in W^1(A'),\quad E'(w) = \bigoplus_{v \in p^{-1}(w)} E(v)$$

$$\text{if}\quad v \to w \qquad e'(v,w) = \sum_{\substack{v' \in p^{-1}(v) \\ w' \in p^{-1}(w) \\ v' \to w'}} e(v',w')$$

General case : for arbitrary $A, A' \subset \Sigma$, we define

$$G(A,A') = G(A \cap A', A')\; G\; (A, A \cap A')$$

There is an evident definition of $G(A,A')$ on morphisms of $Att(A)$

The functor $G(A,A')$ satisfies formal properties similar to §2.9.

2.20. Proposition. For any two subsets A and A' of Σ , the following diagram of categories and functors is commutative :

$$\begin{array}{ccc} \underline{\text{Sheaves}}\ (A) & \xrightarrow{\underline{G}(A,A')} & \underline{\text{Sheaves}}\ (A') \\ \downarrow{\scriptstyle H(A)} & & \downarrow{\scriptstyle H(A')} \\ Att(A) & \xrightarrow{G(A,A')} & Att(A') \end{array}$$

In the special case of $G(A,\Sigma)$, this proposition reduces to a formula for the total cohomology of a sheaf $\underline{\underline{S}}$ in Sheaves (A) .

$$\bigoplus_i H^i(\underline{\underline{S}}) = \bigoplus_{w \in W^I(A)} \underline{\underline{S}}(w)$$

If $\underline{\underline{S}}$ is a constant sheaf, this is the usual formula deduced by regarding $\cup C(w,A)$ as a C-W decomposition of $\phi(A)$ with even dimensional cells.

3. Algebra

3.1. Let Ω be an orbit of W in $\underline{\underline{h}}^*_{\mathbb{Z}}$. Let χ_+ (resp. χ_-) be the unique element of Ω lying in the closure of the positive (resp. negative) Weyl chamber. Let us fix for each $\chi \in \Omega$ an inclusion $M(\chi) \subset M(\chi_+)$. Then for any $\psi \longrightarrow \chi$ with $\psi, \chi \in \Omega$ we have $f(\psi) = X(\psi,\chi)\, f(\chi)$ for a unique $X(\psi,\chi) \in U(\underline{n}_-)$.

Definition. ($\Rightarrow$ 2.15). The characteristic elements $Y(\chi,\psi) \in U(\underline{n}_+)$ for $\psi \longrightarrow \chi$ are defined by

$$Y(\chi,\psi) = \iota(X(\psi,\chi))$$

ι being the anti-involution of §1.1.

3.2. Lemma. If $\chi_1 = \psi_1 \longrightarrow \psi_2 \longrightarrow \dots \longrightarrow \psi_k = \chi_2$ and $\chi_1 = \varphi_1 \longrightarrow \varphi_2 \longrightarrow \dots \longrightarrow \varphi_k = \chi_2$, then

$$Y(\psi_k,\psi_{k-1}) \dots Y(\psi_2,\psi_1) = Y(\varphi_k,\varphi_{k-1}) \dots Y(\varphi_2,\varphi_1)$$

The proof is obvious.

3.3. Proposition. Let $\psi \longrightarrow \chi$. Then up to a constant multiple, $Y(\psi,\chi)$ is the unique element in $U(\underline{n}_+)$ of weight $\chi-\psi$ with the property

$$Y(\psi,\chi)\ (M(\chi)^{\psi-\rho}) = \{0\}$$

Proof. Let M be the maximal proper submodule in $M(\chi)$ and let $(\ ,\ )$ be the Jantzen bilinear form on $M(\chi)$.

Then M coincides with the kernel of the bilinear form $(\ ,\ )$ (See [J],1.6). It is easy to see that the weight space M^ψ is one dimensional and generated by $f(\psi) = X(\psi,\chi)\, f(\chi)$. Now for $X \in U(\underline{n}_-)$ of the weight $\psi-\chi$ it is clear that $Xf(\chi)$ lies in the kernel of Jantzen form if and only if $\iota X(M(\chi)^{\psi}) = 0$

3.4. Proposition. (➡ 2.16). Suppose $\psi \longrightarrow \chi$ and M is an extension

$$0 \longrightarrow M(\chi) \longrightarrow M \longrightarrow M(\psi) \longrightarrow 0$$

Then this extension is non-trivial if and only if $Y(\chi,\psi)(M^{\psi}) \neq \{0\}$.

3.5. $U(\underline{n}_-)$ - free modules. We will consider $\underline{g}$ modules $V \in \mathcal{O}$ with the following property :

(F) V is free as a $U(\underline{n}_-)$ module.

Definition. $\mathcal{O}_-$ is the full subcategory of $\mathcal{O}$ whose objects satisfy (F)

Definition. For a W-orbit Ω in $\underline{h}^*_{\mathbb{Z}}$ we let $\mathcal{O}_-(\Omega)$ be $\mathcal{O}_- \cap \mathcal{O}(\Omega)$.

3.6. Examples.

(➡ 2.2 A) If $\chi \in \Omega$, the Verma module $M(\chi)$ lies in $\mathcal{O}_-(\Omega)$

(➡ 2.2 Б) The Verma modules $M(\chi)$, $\chi \in \Omega$, are characterized by the property that for any exact sequence in $\mathcal{O}_-(\Omega)$

$$0 \longrightarrow V \longrightarrow M(\chi) \longrightarrow V' \longrightarrow 0$$

either $V = 0$ or $V' = 0$.

(➡ 2.5) If $\chi \in \Omega$, the projective module $P(\chi)$ lies in $\mathcal{O}_-(\Omega)$.

(➡ 2.7) $P(\chi)$ is indecomposable in $\mathcal{O}_-(\Omega)$.

$L(\chi)$ is not in $\mathcal{O}_-$ unless χ is in the negative Weyl chamber .

3.7. (⟶ 2.3). Let Ω be given. Let us choose a numbering $\chi_1,\dots,\chi_n$ of elements in Ω in such a way that $\chi_i \rhd \chi_j$ implies $i \leq j$.

Proposition. Let $V \in \mathcal{O}_-(\Omega)$. Then there exists a unique filtration $\{0\} = V_o \subset V_1 \subset \dots \subset V_n = V$ such that V_i/V_{i-1} is isomorphic to a direct sum of several copies of $M(\chi_i)$.

Proof. Define V_i as the $\underline{g}$-submodule of V generated by the weight spaces $V^{\chi_1 - \rho},\dots,V^{\chi_i - \rho}$. It is easy to see that these V_i satisfy the conditions of the proposition.

3.8. Corollary. (⟶ 2.4). The Grothendieck group $K(\mathcal{O}_-(\Omega))$ of the category $\mathcal{O}_-(\Omega)$ is the free Abelian group generated by all $M(\chi)$ for $\chi \in \Omega$. So there is an isomorphism

$$K(\mathcal{O}_-(\Omega)) \longrightarrow \mathbb{Z}^{W^1(A)}$$

where $A = \mathcal{H}(\Omega)$ which takes $M(w\chi_-)$ to $1.\, ww_o$.

3.9. The functors $F(\Omega,\Omega')$. Let Ω and Ω' be two orbits of W in $\underline{h}^*_{\mathbb{Z}}$. Then $F(\Omega,\Omega')$ is the projective functor $\mathcal{O}(\Omega) \longrightarrow \mathcal{O}(\Omega')$ determined by the following property : Let χ_+ and ψ_+ be maximal elements of Ω and Ω' respectively, and let ψ be the (unique) minimal element in the set $W(\Omega)\psi_+$. Then $F(\Omega,\Omega')$ is the indecomposable projective functor such that $F(\Omega,\Omega')M(\chi_+) = P(\psi)$. (See [BG] for the definition and properties of projective functors, in particular :

Proposition. (⟶ 2.10). $F(\Omega,\Omega')$ takes projective modules in $\mathcal{O}(\Omega)$ to projective modules in $\mathcal{O}(\Omega')$.

3.10. Properties of $F(\Omega,\Omega')$ (➡ 2.9)

А. $F(\Omega,\Omega')$ defines a functor from $\mathcal{O}\_(\Omega)$ to $\mathcal{O}\_(\Omega')$

Б. If $Ж(\Omega) = Ж(\Omega')$, then $F(\Omega,\Omega')$ is an equivalence of categories

В. Suppose $Ж(\Omega) \cap Ж(\Omega') \subset Ж(\Omega'') \subset Ж(\Omega) \cup Ж(\Omega')$ then $F(\Omega'',\Omega')\ F(\Omega,\Omega'') = F(\Omega,\Omega')$.

Г. Suppose $Ж(\Omega) \subset Ж(\Omega')$. Then $F(\Omega,\Omega')$ transforms Verma modules in $\mathcal{O}\_(\Omega)$ to Verma modules in $\mathcal{O}\_(\Omega')$. More precisely, if $\chi\_$ and $\psi\_$ are minimal elements in Ω and Ω' respectively, and if $w \in W^1(Ж(\Omega))$, then

$$F(\Omega,\Omega')\ M(w\ \chi\_) = M(v\ \psi\_)$$

where $v = w\ W(Ж(\Omega)) \cap W^1(Ж(\Omega'))$. $F(\Omega,\Omega')$ does not in general transform indecomposable objects to indecomposable objects.

Д. Suppose $Ж(\Omega) \supset Ж(\Omega')$. Then $F(\Omega,\Omega')$ transforms indecomposable objects in $\mathcal{O}\_(\Omega)$ to indecomposable objects in $\mathcal{O}\_(\Omega')$. In particular, if $\chi\_$ and $\psi\_$ are minimal elements in Ω and Ω' respectively and $w \in W'(Ж(\Omega))$, then

$$F(\Omega,\Omega')\ P(w\ \chi\_) = P(w\ \psi\_)$$

$F(\Omega,\Omega')$ never takes Verma modules to Verma modules unless $Ж(\Omega) = Ж(\Omega')$.

Е. Let Ω and Ω be two orbits with $A = Ж(\Omega) \subset A' = Ж(\Omega')$ so that $W(A) \subset W(A')$. Then

$$F(\Omega',\Omega)\ F(\Omega,\Omega') = [W(A') : W(A)]\ \mathrm{Id}$$

where Id is the identity functor in $\mathcal{O}\_(\Omega')$.

Ж. $F(A,A')$ is an exact functor.

Proof. All of these properties (aside from the first one in **Д**) follow easily from [BG] , especially theorem 3.4.

3.10a. We give a well-known characterization of the indecomposable projective modules $P(\psi)$ in terms of the projective functors $F(\Omega,\Omega')$ (➡ 2.11A).

<u>Proposition</u>. Suppose $\psi \in \underline{h}^*_{\mathbb{Z}}$. Let Ω be the W orbit containing ψ , let $A = \mathcal{H}(\Omega)$, and let $w \in W^1(A)$ be the element that takes the antidominant weight in Ω to ψ . Choose a sequence of W orbits $\Omega_o,\ldots,\Omega_m$ so that $\mathcal{H}(\Omega_o),\ldots,\mathcal{H}(\Omega_m)$ is resolution data for $\overline{C(w,A)}$. Let χ^+ be the dominant weight in Ω_o . Then $P(\psi)$ is the unique indecomposable direct summand of

$$F(\Omega_{m-1},\Omega_m)\circ \ldots \circ F(\Omega_o,\Omega_1)\ M(\Omega_+)$$

whose class in $K(\mathcal{O}_-(\Omega))$ has leading term $1\cdot w$.

3.11. Let $V \in \mathcal{O}(\Omega)$ and $\chi \in h^*_{\mathbb{Z}}$. Define the vector space $V[\chi]$ by

$$V[\chi] = V^{\chi-p}/U(\underline{g})\{ \bigoplus_{\substack{\psi \rhd \chi \\ \psi \neq \chi}} V^{\psi-p}\} \cap V^{\chi-p}$$

It is easy to see that $V(\chi)$ may be different from zero only if $\chi \in \Omega$.

<u>Lemma</u>. Let $V \in \mathcal{O}_-(\Omega)$ and $0 = V_o \subset V_1 \subset \ldots \subset V_n = V$ be the filtration of proposition 3.7 so that $V_i/V_{i-1} = M(\chi_i) \oplus \ldots \oplus M(\chi_i)$ (n_i times) . Then $\dim V[\chi_i] = n_i$

3.12. <u>Proposition</u>. Let $V \in \mathcal{O}_-(\Omega)$ and χ , $\psi \in \Omega$ with $\psi \longrightarrow \chi$. Then the characteristic element $Y(\psi,\chi)$ defines a linear transformation $\tilde{Y}(\psi,\chi) : V(\psi) \longrightarrow V(\chi)$.

The proposition follows easily from propositions 3.3 and 3.7.

3.13. (➡ 2.12, 2.18) . Let Ω be a W orbit in $\underline{h}^*_{\mathbb{Z}}$ with $A = \mathcal{H}(\Omega)$ and let χ_- be the minimal element in Ω.

Let Att(A) be the corresponding attaching category (See §2.17).

<u>Proposition</u>. The map $V \rightsquigarrow \{E(w), e(v,w)\}$ where $E(w) = V[w\,\chi_-]$ for $w \in W^1(A)$ and $e(v,w) = \tilde{Y}(v\,\chi_-, w\,\chi_-)$ for $v, w \in W^1(A)$ with $v \longrightarrow w$, defines a functor $a(\Omega) : \mathcal{O}_-(\Omega) \longrightarrow \mathrm{Att}(A)$.

3.14. Properties of the functor $a(\Omega)$ (➡ 2.13)

А. $a(\Omega)$ is an exact functor.

Б. $a(\Omega)$ induces an isomorphism of Grothendieck groups.

В. $a(\Omega)$ is not an equivalence of categories.

In particular it does not preserve indecomposability.

3.15. Let Ω and Ω' be two W orbits with $A = Ж(\Omega)$ and $A' = Ж(\Omega')$.

<u>Proposition</u>. (➡ 2.13). The following diagram of categories and functors is commutative :

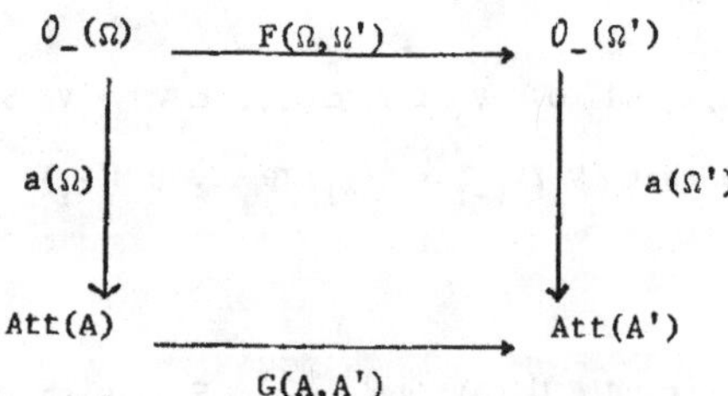

(where $G(\Omega,\Omega')$ is defined in §2.19)

It is enough to prove this for two cases : $A \supset A'$ and $A \supset A'$. In the first case it is easy. In the second case one has to use some properties of characteristic elements.

3.16. In [J],ch.5, Jantzen defined a filtration of a Verma module $M(\chi)$ by $\underline{g}$-submodules

$$M(\chi) = M_o \supset M_1 \supset M_2 \supset \ldots \supset M_k = \{0\} .$$

We will formulate conjectures about the relationship of this filtration with the Kazhdan - Lusztig polynomials of §2.6 and [KL1]

Let Ω be an W-orbit in $\underline{h}^*_{\mathbb{Z}}$ and χ_- be the minimal element in Ω . Let $A = \mathcal{H}(\Omega)$.

Definition. For w_1 , $w_2 \in W^1(A)$, let $m_i(w_1,w_2)$ be the multiplicity of the simple module $L(w_1\,\chi_-)$ in the quotient M_i/M_{i+1} of Jantzen's filtration of the Verma module $M(w_2\,\chi_-)$.

3.17. Proposition. (i) $m_i(w_1,w_2)$ does not depend on Ω , but only on A, w_1 ,w_2 and i .

(ii) $m_i(w_1,w_2) = 0$ unless $w_2 \geq w_1$

(iii) $m_o(w,w) = 1$; $m_i(w,w) = 0$ for $i > 0$

The proof of the first part of this proposition is rather complicated and relies on the behavior of Jantzen's filtration under projective functors $F(\Omega,\Omega')$ with $\mathcal{H}(\Omega) = \mathcal{H}(\Omega')$. Parts (ii) and (iii) are easy.

3.18. Definition . (of the Jantzen polynomial) . Let w_1 and w_2 be elements of $W^1(A)$. Define

$$J_{w_1,w_2}(q) = \sum_i m_i(w_1,w_2)q^{1/2(\ell(w_1)-\ell(w_2)-i)}$$

3.19. Let $P_{w_1,w_2}(q)$ be the Kazhdan - Lusztig polynomial for $W^1(A)$. Let w_o be the unique maximal element in $W^1(A)$ (under the ordering $\leq$).

Conjecture. (improved Kazhdan - Lusztig conjecture). $J_{w_1,w_2}(q)$ is a polynomial in q, and

$$(*) \qquad J_{w_1,w_2}(q) = P_{w_2 w_o, w_1 w_o}(q) .$$

Remark. For $q = 1$ (*) becomes usual Kazhdan - Lusztig conjecture, see [K-L1], (1.56).

3.20. We cannot prove, of course, conjecture 3.19. The strongest evidence for this conjecture is that it agrees with properties of Jantzen filtration from [J], Satz 5.3.

Namely, one can prove the following result.

Let $R_{w_1,w_2}(q) = q^{(\ell(w_2)-\ell(w_1))} P_{w_2 w_o, w_1 w_o}(q^{-2})$

Also for any $w \in W^1(A)$ let $\Gamma(w)$ be the set of all $w' \in W^1(A)$ with the properties :

(i) $w' < w$.

(ii) $w^{-1} w' \in \sigma_\gamma W(A)$ for some root γ.

Proposition* The polynomials R satisfy the following equality

$$\frac{d}{dq} R_{w_1,w_2}(q) \Big|_{q=1} = \sum_{w' \in \Gamma(w_2)} R_{w_1,w'}(1) .$$

* This proposition was also proved by Gabber and Joseph (see [GL], §4.10)

4. The dictionary.

We list corresponding concepts from the topology of Schubert varieties and from the algebra of $\underline{g}$-modules on opposite sides of the page.

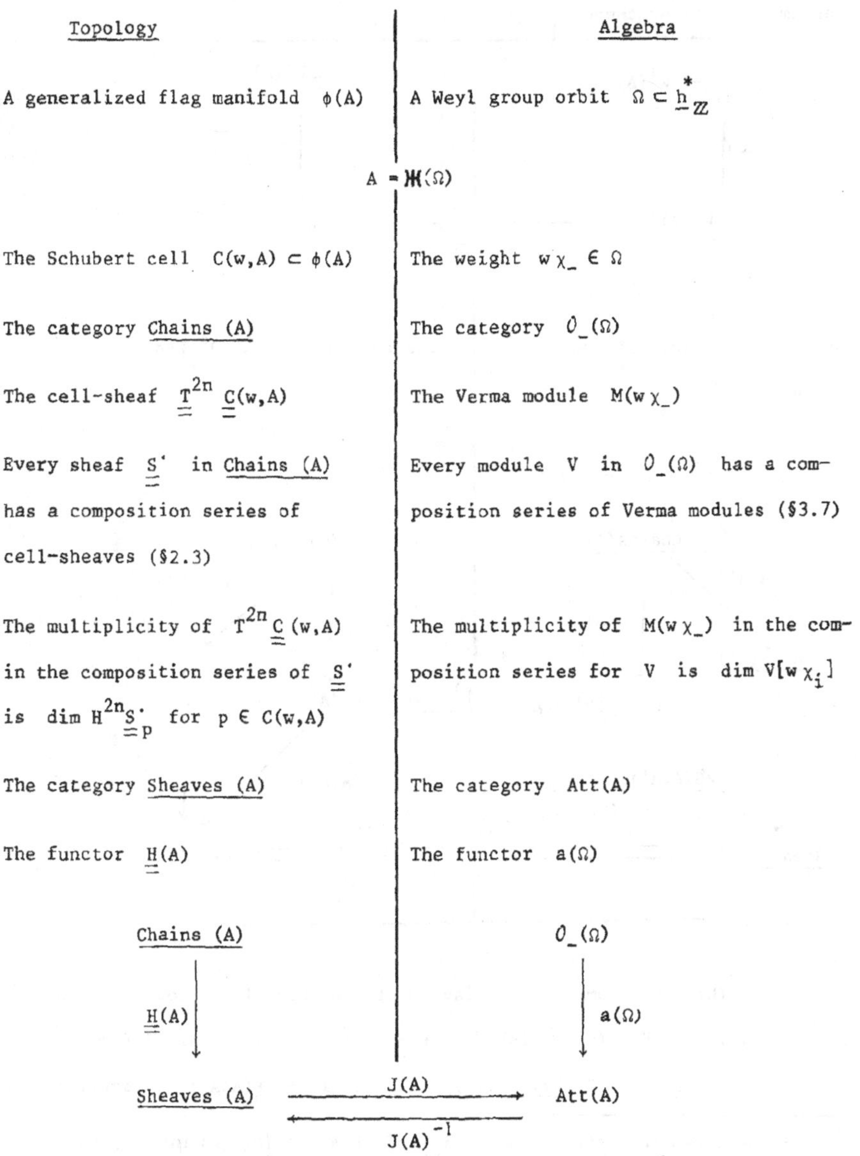

Topology	Algebra
A generalized flag manifold $\phi(A)$	A Weyl group orbit $\Omega \subset \underline{h}^*_{\mathbb{Z}}$
$A = \mathcal{H}(\Omega)$	
The Schubert cell $C(w,A) \subset \phi(A)$	The weight $w\chi_- \in \Omega$
The category Chains (A)	The category $\mathcal{O}_-(\Omega)$
The cell-sheaf $\underline{\underline{T}}^{2n}\,\underline{\underline{C}}(w,A)$	The Verma module $M(w\chi_-)$
Every sheaf $\underline{\underline{S}}^{\cdot}$ in Chains (A) has a composition series of cell-sheaves (§2.3)	Every module V in $\mathcal{O}_-(\Omega)$ has a composition series of Verma modules (§3.7)
The multiplicity of $T^{2n}\underline{\underline{C}}(w,A)$ in the composition series of $\underline{\underline{S}}^{\cdot}$ is $\dim H^{2n}\underline{\underline{S}}^{\cdot}_p$ for $p \in C(w,A)$	The multiplicity of $M(w\chi_-)$ in the composition series for V is $\dim V[w\chi_i]$
The category Sheaves (A)	The category Att(A)
The functor $\underline{\underline{H}}(A)$	The functor $a(\Omega)$

The functor $J(A)$ is contravariant. It gives an equivalence of categories between Sheaves (A) and $\mathrm{Att}(A)^{op}$.

This diagram of categories and functors gives rise to the following diagram of their Grothendieck groups :

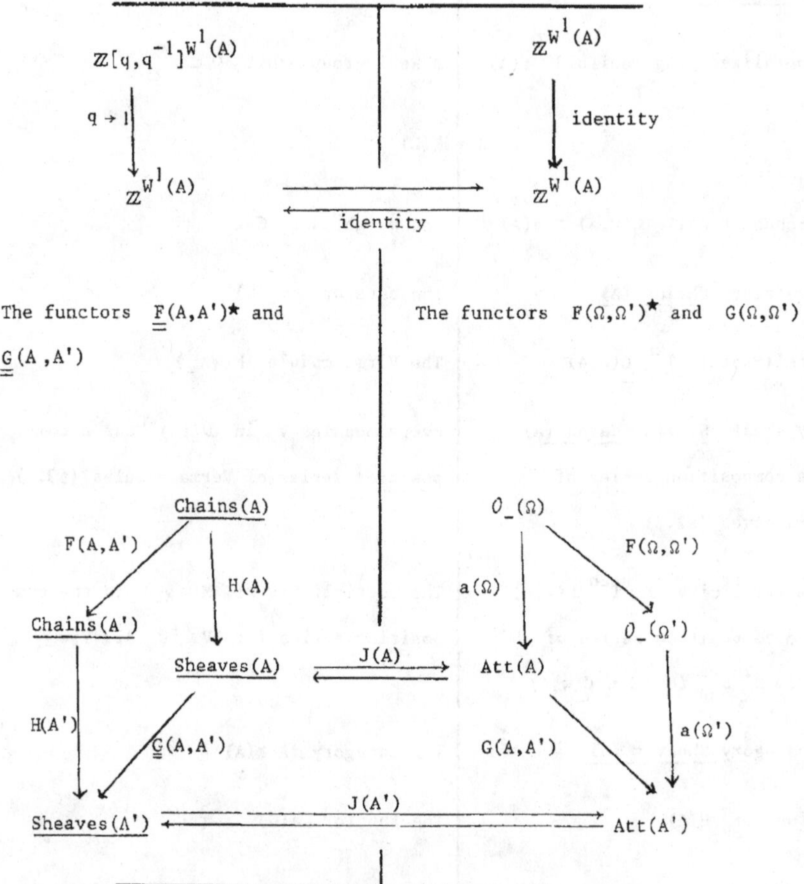

This is a commutative diagram of functors. It follows that the map $K(\mathcal{O}\_(\Omega)) \longrightarrow K(\mathcal{O}\_(\Omega'))$ induced from $F(\Omega,\Omega')$ is obtained from the map $K(\text{Chains }(A)) \longrightarrow K(\text{Chains }(A'))$ induced from $\underline{\underline{F}}(A,A')$ by setting $q = 1$

★It is now known that the decomposition of $\underline{\underline{F}}(A,A')$ $\underline{\underline{IC}}(w,A)$ into direct summands exactly parallels the decomposition of $F(\Omega,\Omega')$ $P(w)(\_)$ into direct summands. This can be seen using [DBB], [BB], and [BK].

Direct sums $\bigoplus_i \underline{\underline{T}}^{2n_i} \underline{\underline{IC}}^\cdot(w_i,A)$	Projective modules
The intersection homology sheaf $\underline{\underline{IC}}^\cdot(w,A)$	The indecomposable projective module $P(w\chi_-)$

The Kazhdan - Lusztig conjecture asserts that

for $p \in C(w,A)$ $\sum_n \dim H^n \underline{\underline{IC}}^\cdot(v,A)_p$	$= \dim P(v\chi_-)[w\chi_-]$

It is known that

$$\dim P(v\chi_-)[w\chi_-] = [M(w\chi_-) : L(v\chi_-)]$$

$$= \sum_n [M(w\chi_-)_{c+n}/M(w\chi_-)_{c+n+1} : L(v\chi_-)]$$

where [:] represents the multiplicity of the irreducible in the composition series, subscripts denote the Jantzen filtration, and $c = \ell(w)-\ell(v)$. We can state the improved Kazhdan - Lusztig conjecture.

for $p \in C(w,A), \dim H^n IC^\cdot(v,A)_p$	$= [M(w\chi_-)_{c+n}/M(w\chi_-)_{c+n+1} : L(v\chi_-)]$

★The proof of the Kazhdan-Lusztig conjectures in [BB], [BK] establishes an equivalence of categories between $\mathcal{O}(\Omega)$ and a full subcategory of $D^bS(\emptyset(A))$ called the category of perverse sheaves, at least when A is empty. (A perverse sheaf by definition satisfies conditions 1) and 3) of §.2.1 and conditions 2) and 3) of §.1.6 with $\leq$ replacing $<$; see [DBB]). This yields a dictionary between topology and algebra which is different from the one above. In this dictionary $\underline{\underline{IC}}^\cdot(w)$ corresponds to an irreducible module rather than a projective module, and the association of weights in Ω to Schubert cells differs from the one above by multiplication by w_o.

5. <u>Examples</u>.

In this section we display $W^1(A)$ for several Lie groups G and several subsets $A \subset \Sigma$

5.1. Let $G = SL(n)$ be the $n \times n$ matrices of determinant one with entries $E_{i,j}$. We choose $\underline{h}$ to be the diagonal matrices of trace zero and Σ to be $\{\gamma_1, \ldots, \gamma_{n-1}\}$ where $\gamma_i(E_{j,j} - E_{j+1,j+1}) = 2\,\delta_{i,j}$.

In figures 1-5 of $W^1(A)$, the elements w are represented by vertices $\boxed{w}$. The relation $v \xrightarrow{\gamma_i} w$ is symbolized by marked edges (w is above v) :

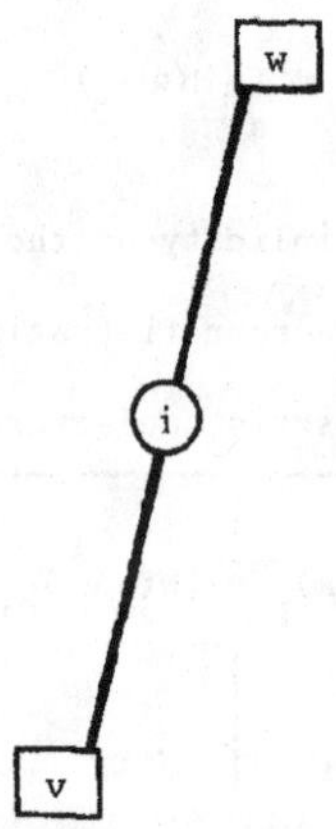

Elements w and v satisfy $w \geq v$ if and only if there is a string of continuously rising edges (marked or not) from $\boxed{v}$ to $\boxed{w}$. The identity is at the bottom of the diagram, and elements on the same row have the same length, given by the number at the right. In figure 2, elements labeled n and n' form a coset $w\,W(\{\gamma_2\})$ and in figure 4, similarly labeled elements form a coset $w\,W(\{\gamma_3\})$.

The entry in the column for w and the row for v of the tables 1-5 is the Kazhdan - Lusztig polynomial $P_{v w_0, w w_0}$.

We recall briefly the topological and the Lie-algebraic significance of this data.

5.2. Topology.

The vertices of figures 1-5 label the Schubert cells in the flag manifold $\phi(A)$. The edges give the incidence relations : The cell $\boxed{w}$ is in the closure of the cell $\boxed{v}$ if and only if there is an increasing path along edges from $\boxed{v}$ to $\boxed{w}$. Cells of the same dimension are on a horizontal row : the number at the right is the codimension. The flag manifold of figure 2 fibers over that of figure 1 by $\pi(\emptyset,\{\gamma_2\})$; the map is cellular with cells $\boxed{n}$ and $\boxed{n'}$ mapping to $\boxed{n}$. A similar remark applies to figure 4 and 3.

The entry in the column for w and the row for v in the tables 1-5 is the Poincaré polynomical in $q^{1/2}$ of the stalk at a point of $\boxed{v}$ of the intersection homology sheaf of the closure of $\boxed{w}$.

5.3. Lie algebra modules.

A vertex $\boxed{\chi}$ of figures 1-5 corresponds to a weight χ in $\Omega \subset \underline{h}^*_{\mathbb{Z}}$. One such correspondance is indicated by the first column of tables 1-5 which gives $\chi(E_{1,1} - E_{2,2})$, $\chi(E_{2,2} - E_{3,3})$, ...
The Verma module $M(\chi)$ is contained in $M(\psi)$ if and only if there is an increasing string of edges from $\boxed{\chi}$ to $\boxed{\psi}$.

The functor $F(\emptyset,\{\gamma_2\})$ takes $M(\chi)$ and $M(\chi')$ in figure 2 to $M(\chi)$ in figure 1 : similarly for figures 4 and 3.

If p is the polynomial in the row of χ and the column of ψ in tables 1-5 , then $p(1)$ is $\dim P(\chi)\,[\psi]$.

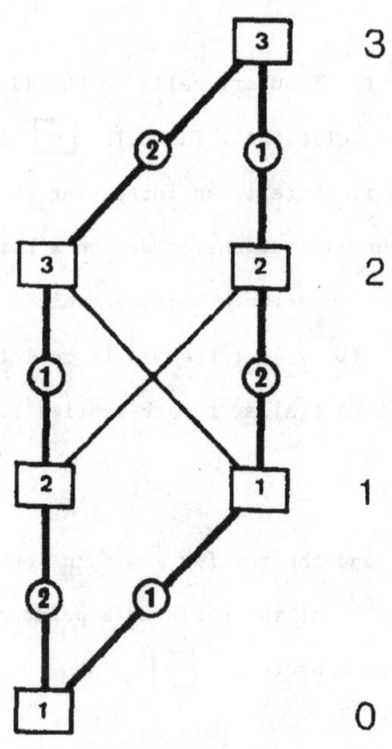

Figure 2 : $W^1(\emptyset)$ for SL(3)

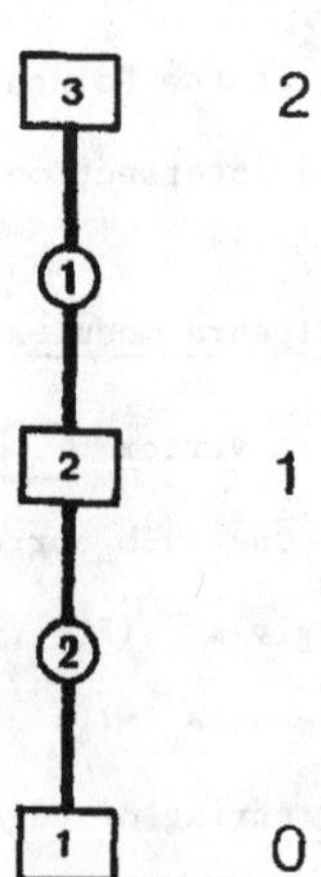

Figure 1 : $W^1(\{\gamma_2\})$ for SL(3)

	weight		1	2	3
3	1	0	1	1	1
2	-1	1	1	1	
1	0	-1	1		

Table 1

	weight		1	2	1'	3	2'	3'
3'	1	1	1	1	1	1	1	1
2'	-1	2	1	1	1		1	
3	2	-1	1	1	1	1		
1'	1	-2	1		1			
2	-2	1	1	1				
1	-1	-1	1					

Table 2

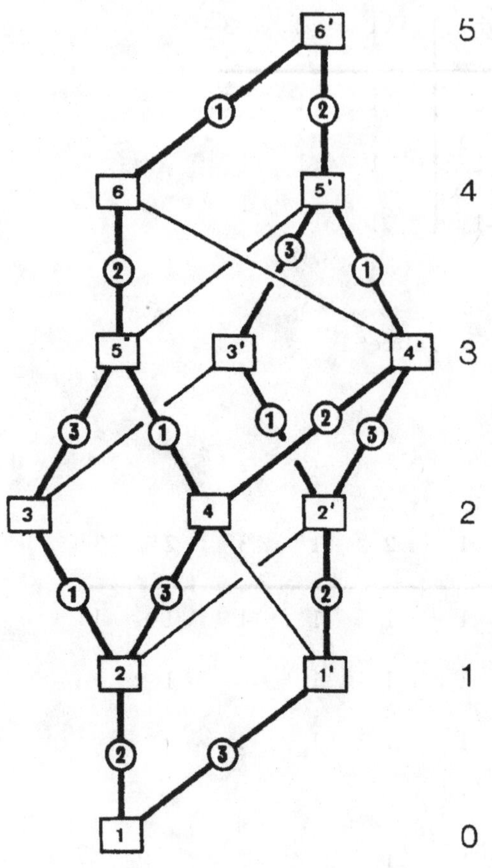

Figure 4 : $W^1(\{\gamma_3\})$ for SL(4)

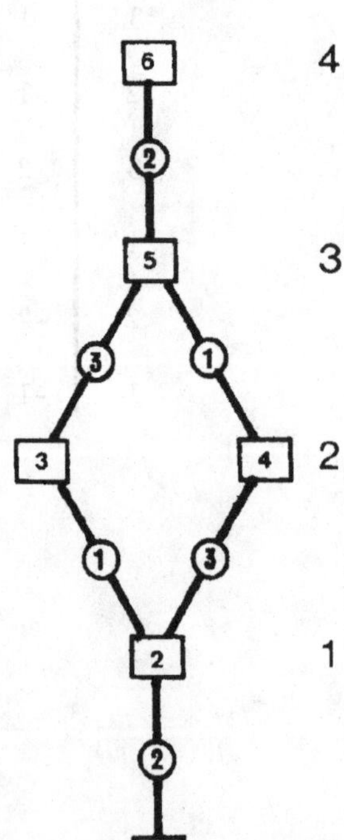

Figure 3 :
$W^1(\{\gamma_1,\gamma_3\})$ for SL(4)

	weight	1	2	3	4	5	6
6	0 1 0	1	1+q	1	1	1	1
5	-1 1 -1	1	1	1	1	1	
4	-1 0 1	1	1		1		
3	1 0 -1	1	1	1			
2	1 -1 1	1	1				
1	0 -1 0	1					

Table 3

	weight	1	2	1'	3	4	2'	5	3'	4'	6	5'	6'
6'	1 1 0	1	1+q	1	1	1	1	1	1	1	1	1	1
5'	2 -1 1	1	1	1	1	1	1	1	1	1		1	
6	-1 2 0	1	1+q	1	1	1	1	1			1		
4'	-2 1 1	1	1	1		1	1			1			
3'	2 0 -1	1	1	1	1		1		1				
5	1 -2 2	1	1	1	1	1		1					
2'	-2 2 -1	1	1	1			1						
4	-1 -1 2	1	1	1		1							
3	1 0 -2	1	1		1								
1'	0 -2 1	1		1									
2	-1 1 -2	1	1										
1	0 -1 -1	1											

Table 4

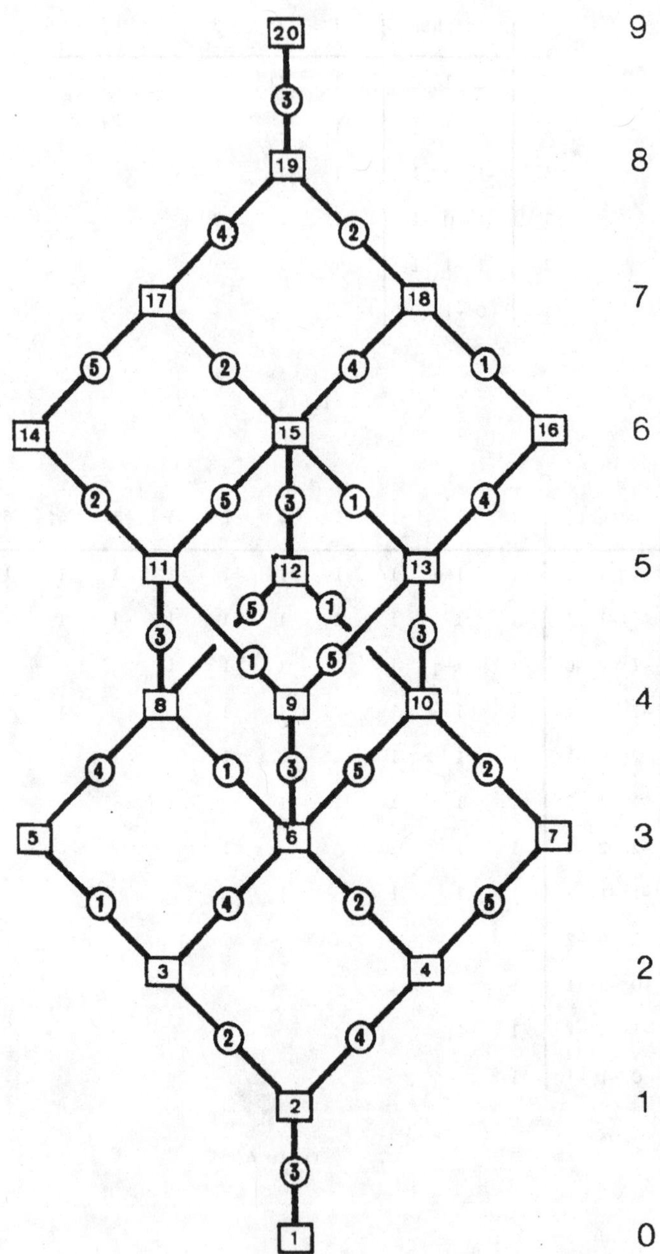

Figure 5 : $W^1(\{\gamma_1,\gamma_2,\gamma_4,\gamma_5\})$ for SL(6)

	1	2	3	4	5	6	7	8	9	10	11	12	13	14	15	16	17	18	19	20
20	1	$1+q+q^2$	1+q	1+q	1	$1+2q+q^2$	1	1+q	$1+q+q^2$	1+q	1+q	1	1+q	1	1+q	1	1	1	1	1
19	1	1+q	1+q	1+q	1	$1+2q+q^2$	1	1+q	1	1+q	1	1	1	1	1	1	1	1	1	
18	1	1+q	1	1+q	1	1+q	1	1+q	1	1	1	1	1		1	1		1		
17	1	1+q	1+q	1	1	1+q	1	1	1	1+q	1	1	1	1	1		1			
16	1	1+q	1	1+q		1+q	1		1	1			1			1				
15	1	1+q	1	1	1	1	1	1	1	1	1	1	1		1					
14	1	1+q	1+q	1	1	1+q		1	1		1			1						
13	1	1+q	1	1		1	1		1	1			1							
12	1	1	1	1	1	1	1	1		1		1								
11	1	1+q	1	1	1	1		1	1		1									
10	1	1	1	1		1	1			1										
9	1	1+q	1	1		1			1											
8	1	1	1	1	1	1		1												
7	1	1		1			1													
6	1	1	1	1		1														
5	1	1	1		1															
4	1	1		1																
3	1	1	1																	
2	1	1																		
1	1																			

Table 5

5.4. Remarks.

Consider figures 1 and 2. The map

$$\pi(\emptyset,\{\gamma_2\}) : \overline{C(1',\emptyset)} \longrightarrow \overline{C(1,\{\gamma_2\})} \qquad *$$

is the projection of the complex projective plane $\mathbb{C}P^2$ blown up at a point $(C(3,\{\gamma_2\}))$ on to $\mathbb{C}P^2$ (so the fiber over $C(3,\{\gamma_3\}$ is $\mathbb{C}P^1$) .

$$R\pi(\emptyset,\{\gamma_2\})_! \underline{\underline{IC}}^\cdot(1',\emptyset) = IC^\cdot(1,\{\gamma_2\}) \oplus \underline{\underline{T}}^2 \underline{\underline{IC}}(3,\{\gamma_2\}$$

(Compare §2.10). Likewise,

$$F(\phi,\{\gamma_2\})\ P(1') = P(1) \oplus P(3)$$

Consider figures 3 and 4. $\overline{C(2,\{\gamma_1,\gamma_3\}}$ is a complex 3-fold with an ordinary quadratic singularity (locally like $w^2+x^2+y^2+z^2 = 0$ in $\mathbb{C}^4$) at $C(6,\{\gamma_1,\gamma_2\})$.)

The map

$$\pi(\{\gamma_1\},\{\gamma_1,\gamma_3\}) : \overline{C(2',\{\gamma_1\})} \longrightarrow \overline{C(2,\{\gamma_1,\gamma_3\}} \qquad **$$

is one of the two usual resolutions of it . Here again the exceptional fiber is a $\mathbb{C}P^1$. But now

$$R\pi(\{\gamma_1\},\{\gamma_1,\gamma_3\})_! \underline{\underline{IC}}^\cdot(2',\{\gamma_1\}) = \underline{\underline{IC}}^\cdot(2,\{\gamma_1,\gamma_3\})$$

which is indecomposable (since it is a small resolution, see [GM3]) . Likewise

$$F(\{\gamma_1\},\{\gamma_1,\gamma_3\})\ P(2') = P(2)$$

is indecomposable.

The map * is the same as the canonical resolution for the resolution data ϕ , $\{\gamma_1\}$, $\{\gamma_2\}$. (Compare §2.11). More efficient resolution data are ϕ , $\{\gamma_1,\gamma_2\}$, $\{\gamma_2\}$. The map ** is the resolution for ϕ , $\{\gamma_1,\gamma_3\}$, $\{\gamma_1,\gamma_2\}$, $\{\gamma_1,\gamma_3\}$.

5.5. We give here an algorithm due to Kazhdan and Lusztig for computing the polynomials in tables 1-5 from the data in figures 1-5.

We denote by $\tilde{P}_{v,w}$ the entry in the row of v and the column of w . ($\tilde{P}_{v,w} = P_{v w_o, w w_o}$) . For the w at the top of the diagram, we have

$$\tilde{P}_{v,w} = \begin{cases} 1 & \text{if } v = w \\ 0 & \text{otherwise} \end{cases}$$

Suppose $\tilde{P}_{v,w'}$ has been constructed for all $w' > w$. We want to construct $\tilde{P}_{v,w}$. First choose w' so that

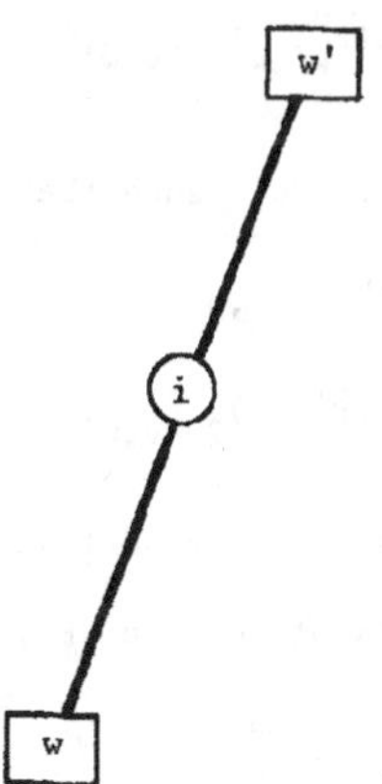

Now for each

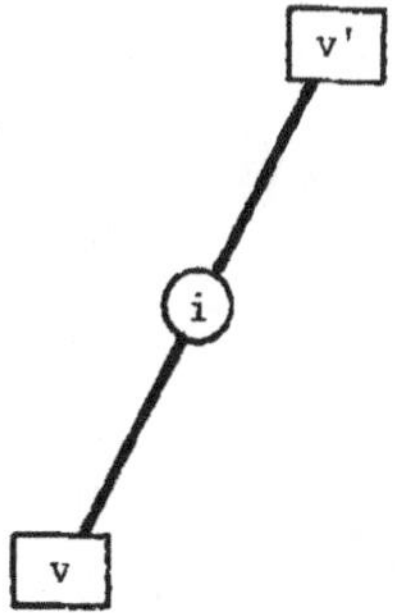

set $\tilde{P}^{o}_{v,w} = \tilde{P}^{o}_{v',w} = \tilde{P}_{v,w'} + q\tilde{P}_{v',w'}$

For each v which is not a vertex of an edge

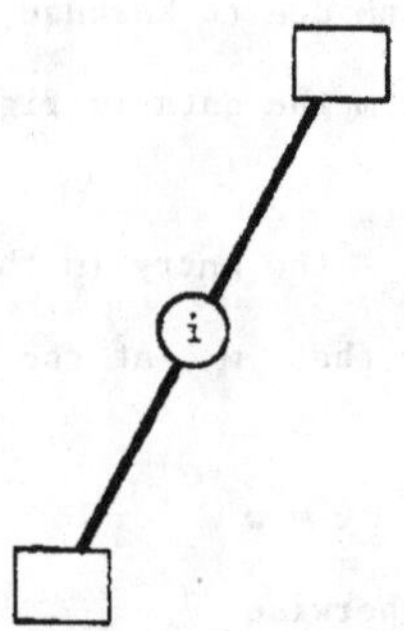

set $P^{o}_{v,w} = \tilde{P}_{v,w'} + q\tilde{P}_{v,w'} = (1+q)\,\tilde{P}_{v,w'}$

Next $$\tilde{P}^{1}_{v,w} = \tilde{P}^{o}_{v,w} - \Sigma(\tilde{P}^{o}_{w',w})_1\,\tilde{P}_{v,w'}$$

where the sum is taken over all w' such that $\ell(w') = \ell(w) + 2$ and $(\tilde{P}^{o}_{w',w})_1$ means the coeficient of q in $\tilde{P}^{o}_{w',w}$.

Next $$\tilde{P}^{2}_{v,w} = \tilde{P}^{1}_{v,w} - \Sigma(\tilde{P}^{1}_{w',w})_2\,\tilde{P}_{v,w'}$$

where the sum is taken over all w' such that $\ell(w') = \ell(w) + 4$ and $(P^{1}_{w',w})_2$ means the coeficients of q^2 in $\tilde{P}^{1}_{w',w}$. Proceeding until the process stabilizes at $\tilde{P}^{n}_{v,w}$, we have $P_{v,w} = \tilde{P}^{n}_{v,w}$.

BIBLIOGRAPHY

[BG] J.N. Bernstein, S.I. Gelfand : Tensor products of finite and infinite-dimensional representations of semisimple Lie algebras, Compositio Math., 1980 (in press).

[BGG1] J.N. Bernstein, I.M. Gelfand, S.I. Gelfand : The structure of representations generated by vectors of highest weight. Funk. Anal. Appl. 5(1971), N° 1, 1-9 (in Russian).

[BGG2] J.N. Bernstein, I.M. Gelfand, S.I. Gelfand : On a category of -modules, Funk. Anal. Appl., 10(1976), N° 2, 1-8 (in Russian).

[D] P. Deligne : Letter to G. Lusztig and D. Kazhdan, 20 April 1979.

[De] M. Demazure : Desingularisation des variétés de Schubert generalisé, Annales Scientifiques de l'E.N.S. t.7 fasc.1(1974)

[Dl] J. Dixmier : Algèbres envellopantes, Paris, Gauthier - Villars, 1974.

[GM1] M. Goresky and R. MacPherson : La dualité de Poincaré pour les espaces singulières, C.R. Acad. Sci. t. 284, Serie A(1977), 1549-1551.

[GM2] M. Goresky and R. MacPherson : Intersection Homology Theory, Topology 19(1980) pp. 135-162.

[GM3] M. Goresky and R. MacPherson : Intersection Homology II - to appear.

[G] R. Godement : Théorie des faisceaux, Act. Sci. et Ind 1252, Hermann, Paris, 1958.

[H] R. Hartshorne : Residues and Duality, Springer Lect. Notes Math. N° 20, 1966.

[J] J. Jantzen : Moduln mit einem höchsten Gewicht, Lect. Notes Math., ... (1980), Springer-Verlag, Berlin.

[KL1] D. Kazhdan, G. Lusztig : Representations of Coxeter groups and Hecke algebras, Inventiones Math., 53(1979), 165-184.

[KL2] D. Kazhdan, G. Lusztig : Schubert varieties and Poincaré duality, Geometry of the Laplace Operator, Proc. Sympos. Pure Math. XXXVI, Amer. Math. Soc. (1980), pp. 185-203.

[Sch] W. Schmid : Two character identities for semisimple Lie groups in Lect. Notes Math., 587(1977) Springer-Verlag, Berlin.

[Ve1] J-L. Verdier : Categories Dérivées, SGA $4\frac{1}{2}$, Springer Lect. Notes Math. N° 569(1977), pp. 262-312

[Ve2] J-L. Verdier : Dualité dans la cohomologie des espaces localement compacts, Sem. Bourbaki (1965) n° 300.

[V1] D. Vogan : Irreductible characters of semisimple Lie groups, I Duke Math. J., 46 (1979), 61-108.

[V2] D. Vogan : Irreductible characters of semisimple Lie groups, II : The Kazhdan - Lusztig conjectures, Duke Math. J., 46 (1979), 805-859.

[GJ] O. Gabber and A. Joseph : Towards the Kazhdan - Lusztig conjecture, preprint.

[BB] A.A. Beilinson and J.N. Bernstein : C.R. Acad. Sc., 1981

[BK] J.L. Brylinski and M. Kashiwara : Kazhdan - Lusztig conjecture and holonomic systems, preprint Ecole Polytechnique, 1980

[DBB] P. Deligne, A.A. Beilinson, J.N. Bernstein : Luminy Conference on analysis and topology on singular varieties, to appear in Asterisque.

[GM4] M. Goresky and R. Mac Pherson : On the topology of complex algebraic maps, La Rabida Conference on singularities, Springer Lect. Notes, to appear.

Institut des Hautes Etudes Scientifiques
35, route de Chartres
91440 Bures-sur-Yvette (France)

November 1980

IHES/M/80/45

Invariant Dimension and Restricted Extension of Noetherian Rings

by Walter Borho

Summary: This paper is an alternative and complement to [1]. First we take up the idea of an axiomatic notion of dimension generalizing and unifying Gelfand-Kirillov- and Gabriel-Rentschler-dimension. We introduce an "axiom of invariance", generalizing an idea of Stafford. Next we apply this to re-prove the main results of [1] on "good behaviour" of prime ideals in certain extension-rings of noncommutative rings, including an additivity principle for Goldie-ranks. Finally we discuss the extent to which our "restriction" on extensions is also necessary in order to have results of this type.

Address: FB 7 - Mathematik,
Gaußstraße 20
5600 Wuppertal 1
West-Germany

Preprint, Wuppertal,
June 1981

Introduction

We consider right noetherian rings A and B. A right restricted extension of A is a ring B containing A as a subring, such that the A-bimodule AbA is right noetherian for all $b \in B$. For example, B may be an enveloping algebra of a finite-dimensional Lie-algebra $\underline{b}$, and A the enveloping algebra of a subalgebra of $\underline{b}$, see 2.3.

A dimension for left A-modules is a function d from left A-modules into some totally ordered set, satisfying certain axioms (see 1.1). Examples are provided by the well-known dimension-concepts of Gelfand-Kirillov (GK-dimension) resp. of Gabriel-Rentschler (GR-dimension). A dimension d is called invariant (for B), if tensoring by an A-bimodule (in B) never increases the value of the dimension (see 1.4). GK-dimension for example is invariant for all restricted extensions (1.5). We note that J.T. Stafford in [11] introduced the phrase " A is ideal-invariant" to say that, in our terminology, GR-dimension is invariant for A itself. Whether or not all bi-noetherian rings A have this property, seems to be a difficult open problem, see [12]. Stafford's concept of ideal-invariance was extensively studied in [4], [5], [9] see also [6], and it inspired the present generalization.

In the present paper, we use an invariant dimension d for a restricted extension B of A, as a tool to study prime ideals. We prove the properties of "good behaviour" of a prime P of B, the same properties as formulated in [1], 7.2. These properties include e.g. "equidimensionality" of the primes $P_1,\ldots,P_n$ minimal over $P \cap A$, that is to say that $d(A/P_i)=d(A/P_j)$ for all $i,j=1,\ldots,n$, and also an "additivity-principle" for Goldie-ranks, generalizing work of Joseph-Small [8].

Let us now point out some major differences of the present paper, if compared to [1]. A crucial difference is that the "axiom of symmetry", on which [1] was based, is replaced here by the "axiom of invariance". This has the advantage to make the complicated interplay between right and left noetherian assumptions (see [1]) much clearer here. The results obtained here are slightly more general than in [1], if applied to GK-dimension. For GR-dimension, however, the results are incomparable, in view of certain open problems about invariance as well as symmetry for GR-dimension (cf. [12]). - Moreover, in contrast to [1], we concentrate here just on a stream-lined proof of the main theorem.

In the last section, we deal with the question, whether our restrictedness-condition on an extension is in some sense also necessary for good behaviour of primes. For certain obvious reasons, this can only be expected after modifying the notion of restrictedness into a slightly weaker condition: B is weakly right restricted over A if AbA has, for all $b \in B$, finite reduced right length in the sense of Goldie (see 4.1). For a detailed statement of some results towards the goal of a necessary and sufficient condition for good behaviour of primes, we refer to 4.2 and 4.7 in the text.

Acknowledgement. The material in this paper was presented in two lectures at Paris in December 1980 resp. January 1981, and I want to thank M.P. Malliavin resp. J. Dixmier for providing the opportunity for doing this in their seminars. I am grateful to T. Lenagan and R. Rentschler for useful hints to the literature.

§1. Axiomatics of dimension

In this paper, we study an extension $B \supset A$ of two right noetherian (r-noetherian) rings A an B. "Rings" are understood to be associative with a unit element 1, and the subring A has the same unit element as the big ring B. We denote by k a commutative field. Sometimes, B will be assumed to be a k-algebra, and in this case it is understood that A is a k-subalgebra of B.

1.1. Definition of d

Let us fix some totally ordered set Ω admitting suprema, and a function d assigning to each left A-module M a value $d(M)$ in Ω, such that the following four properties hold:

(d1) $d(M) > d(0)$ if $M \neq 0$.

(d2) $d(M) \geq \sup(d(N), d(M/N))$ for all sumodules $N \subset M$, with equality for a direct summand N.

(d3) $d(\bigcup_i N_i) = \sup_i d(N_i)$ for all inductive systems $(N_i)_i$ of submodules of M.

(d4) $d(M/Ms) < d(M)$ for all monomorphisms $s: M \to M$ of a cyclic module M.

Such a function d is called a <u>dimension</u>, or -in more detail- <u>an Ω-valued dimension for left A-modules.</u>

1.2 Examples

a) Let A be a finitely generated k-algebra of finite Gelfand-Kirillov-dimension γ (GK-dimension), [3]. Then GK-dimension is an Ω-valued dimension for left A-modules, where Ω is the set of real numbers $\leq \gamma$.

b) Let A be a ring, such that its left GR-dimension (Krull-dimension in the sense of Gabriel-Rentschler) exists, and is the ordinal number ω (see [7]). Then GR-dimension is an Ω-valued dimension for left A-modules, where Ω is the set of ordinals $\leq \omega$.

c) Let d be a dimension in the sense of [1], 1.2. Then d is an $\mathbb{R}\cup\{\pm\infty\}$- valued dimension for left A-modules in the present sense.

1.3. Comments:

The verification of axioms (d1)-(d4)in 1.1 for the above examples is left to the reader. If necessary, the more detailed discussion in [1],§1 may be consulted for hints. - The main difference of our present definition of a dimension in comparison with [1],1.2 is that it allows to include ordinal-valued GR-dimension here. The other minor differences are essentially only matters of taste.

We also leave to the reader the elaboration of a list of elementary formal properties of d as in [1],1.5. We only state explicitly that axiom (d4) implies the following basic fact about our dimension (which is proved as in [1], 1.6):

Lemma: $d(A/Q) < d(A/P)$ for any two primes $Q \supsetneq P$ of A.

1.4.

On our three data A,B,d considered throughout the paper, we shall impose two additional basic assumptions:

a) The extension B is called <u>r-restricted</u> over A , if the A-bimodule AbA is r-noetherian for all $b \in B$.

b) The dimension d is called <u>B-invariant</u>, if $d(AbA \otimes_A M) \le d(M)$ for all left A-modules M and all $b \in B$.

<u>Remark:</u> Note that in both definitions a), b), we could equivalently ask those properties for all finitely generated A-bisubmodules of B , instead of only the cyclic ones (of the form AbA).

<u>Comment:</u> If d is B-invariant, it is in particular A-invariant. Instead of "A-invariant", we also use the term <u>"ideal-invariant".</u> The notion of ideal-invariance was introduced by Stafford [11] in the case of GR-dimension, and was extensively studied for this case in [4],[5],[9]. We refer to these papers for many examples of rings with ideal-invariant GR-dimension. It seems to be a difficult open problem, whether all bi-noetherian (left and right noetherian) rings have this property.

1.5 <u>Invariance of GK-dimension</u>

<u>Lemma: Let</u> A <u>be a finitely generated k-algebra,</u> <u>and</u> d <u>the GK-dimension. If</u> B <u>is r-restricted, then</u> d <u>is B-invariant.</u>

Proof: Let $b \in B$ be given. By r-restrictedness, we can find $E \subset B$ finite-dimensional such that $AbA=EA$. We fix a finite-dimensional k-subspace V generating A with $1 \in V$. Then $VE \subset EV^m$ for some $m \in \mathbb{N}$, since the subspaces EV^j form an ascending chain exhausting EA. By induction, $V^nE \subset EV^{mn}$ for all n. Now let M be a left A-module, and W any finite-dimensional submodule of $AbA \otimes_A M$. By enlarging E if necessary, we may assume $W \subset E \otimes F$ for some finite-dimensional subspace F of M. Then

$$V^nW \subset V^nE \otimes F \subset EV^{nm} \otimes F = E \otimes V^{nm}F .$$

Since the latter is a k-linear image of $E \otimes_k V^{nm}F$, we see

$$\dim(V^nW) \leq \dim(E)\dim(V^{mn}F) \qquad \text{for all } n \in \mathbb{N}.$$

This implies that

$$d(AW) \leq d(AF) \leq d(M).$$

Passing to the supremum over all W, this gives $d(AbA \otimes_A M) \leq d(M)$, as desired. Q.e.d.

§2 Main theorem on good behaviour of primes

2.1 Fix the three data A,B,d as in §1. A prime ideal P of B is called well-behaved over A, if the five properties listed below hold. The following notation is needed: $\bar{B}=B/P$ resp. $\bar{A}=A/P\cap A$ are the residue-algebras; $P_1,\dots,P_n$ are the distinct prime ideals of A minimal over $P\cap A$, and S is the "Small-set" of $\bar{A}$, that is to say the set of those elements of $\bar{A}$, which become nonzerodivisors modulo the nilradical of $\bar{A}$.

Now we are ready to list the "five rules of good behaviour":

(1) homogeneity: $d(\bar{A}x)=d(\bar{A})$ for all $0\neq x\in\bar{B}$

(2) regularity: S consists of nonzerodivisors in $\bar{B}$.

(3) localizability: S satisfies the right Ore-condition for $\bar{B}$.

(4) equidimensionality: $d(A/P_i)=d(\bar{A})$ for all $i=1,\dots,n$.

(5) additivity: $\operatorname{rk}\bar{B}=\sum_{i=1}^{n} z_i \operatorname{rk} A/P_i$ for suitable positive integers $z_1,\dots,z_n$. (Here rk denotes Goldie-rank.)

2.2 Theorem: Let A and B be r-noetherian rings. We assume that B is a r-restricted extension of A, and that d is a B-invariant dimension for left A-modules. Then all primes P of B are well-behaved over A.

The proof will be given in §3.

2.3 Corollary (A,B r-noetherian): Let A be a k-algebra finitely generated by $V\subset A$, such that B is locally ad V-finite (this notion is defined in the proof). Let d be

GK-dimension, and assume $d(A) < \infty$. Then all primes of B are well-behaved over A .

Proof: We write $(\mathrm{ad}\ v)(a) = [v,a] = va-av$ for the commutator of $v,a \in B$, and $(\mathrm{ad}\ V)(W) = [V,W]$ for the k-span of all commutators $vw-wv$, $v \in V$, $w \in W$. Let $b \in B$ be arbitrary and let $E = \Sigma_j (\mathrm{ad}\ V)^j(b) = kb+[V,b]+[V,[V,b]]+\ldots$ be the ad V-stable k-subspace generated by b in B . The assumption "B locally ad V-finite" means just that E is finite-dimensional. Hence EA is r-noetherian. Now observe that $AbA \subset AEA = EA$ by ad V-stability of E. Hence AbA is r-noetherian. We have proved that B is r-restricted over A. By Lemma 1.5, d is B-invariant. Now the theorem applies and gives the desired result. Q.e.d.

2.4 Examples

Here is a typical example, where Corollary 2.3 applies: $B = U(\underline{b})$ is an enveloping algebra of a finite-dimensional k-Lie-algebra $\underline{b}$, and $A = U(\underline{a})$ is an enveloping algebra of a subalgebra $\underline{a} \subset \underline{b}$ (see [1], 8.3).

In this example, A may be replaced by any homomorphic image of $U(\underline{a})$, and B by any locally ad $\underline{a}$-finite extension, e.g. by the ad $\underline{a}$-finite k-endomorphisms of some $\underline{a}$-module M. This includes the results of Joseph-Small [8] as a special case.

§ 3 Proof of the main theorem

Let A,B,d be as in the theorem, and let P be any prime ideal of B . By replacing B by $\bar{B}=B/P$, A by $\bar{A}=A/A\cap P$, d by $d_{\bar{A}}$ with $d_{\bar{A}}(M)=d(M)$ for all $\bar{A}$-modules M , and observing that the assumptions of the theorem pass over to homomorphic images, we reduce the proof to the case P=O . So from now on, we assume that $\bar{B}=B$ is prime, and P=O .

3.1 Homogeneity:

Let J be the set of those $b \in B$ such that $d(Ab)<d(A)$. We have to prove J=O . We first prove that J is an ideal of B : Trivially, J is a right ideal (use that $d(Abx)\leq d(Ab)$ for all $x \in B$, since Abx is a homomorphic image of Ab). In order to prove $xb \in J$ for $x \in B$, $b \in J$, we use B-invariance of d : Since $AxA \otimes_A Ab$ maps k-linearly onto $AxAAb \supset Axb$, we see that $d(Axb) \leq d(AxA \otimes_A Ab) \leq d(Ab) < d(A)$. Hence J is actually a two-sided ideal of B . Since B is prime noetherian, $J\neq O$ would imply that J contains a non-zerodivisor y of B , and hence a left A-submodule $Ay \cong A$. But $d(Ay)=d(A)$ contradicts $y \in J$. Hence J=O . Q.e.d.

3.2 Regularity:

An element $s \in A$ is called <u>regular</u> for an A-module M , if it acts by an injection on M . A subset of A is called regular for M , if all its elements are. The <u>Small-set</u> S of A (cf. 2.1) is defined as the set of elements l(eft)-regular for $A/\sqrt{O}$, where $\sqrt{O}$ denotes the nilradical of A . We shall prove here that S is l-regular for B. Then S is also r-regular by ([2], 2.7b), since B is prime r-noetherian.

Putting $B_i=\{b \in B \mid \sqrt{O}^i b=0\}$, we obtain a finite filtration $0=B_0 \subsetneqq B_1 \subsetneqq B_2 \subsetneqq \dots \subsetneqq B_r=B$ by A-bisubmodules. By lemma 3.3 below, each factor $F_i=B_i/B_{i-1}$, as a left A-module, is homogeneous of dimension $d(F_i)=d(A)$. We are going to prove that S is 1-regular for each F_i and hence for B too. In fact, let $f \in F_i$ and $s \in S$ such that $sf=0$. Since also $\sqrt{O}f=0$, we see that Af is a homomorphic image of the left A-module $A/(As+\sqrt{O})$. Hence $d(Af) \le d(A/As+\sqrt{O})) < d(A/\sqrt{O}) \le d(A)$, where the strict inequality comes from axiom (d4), since s is 1-regular for $A/\sqrt{O}$ by definition of the Small-set. Now homogeneity of F_i (3.1) implies $Af=0$, hence $f=0$. This was left to be proved. Q.e.d.

<u>3.3</u> With notations as in 3.2, we want to prove the

<u>Lemma:</u> $F_i=B_i/B_{i-1}$ is homogeneous of dimension $d(A)$ as a left A-module, for all $i=1,\dots,r$.

<u>Proof:</u> Let $\bar{L}$ be a left A-submodule $\neq 0$ in F_i, and $L \supsetneqq B_{i-1}$ its pre-image in B_i. Then $\sqrt{O}^{i-1}L\neq 0$, and therefor $d(\sqrt{O}^{i-1}L)=d(A)$ by homogeneity of B (3.1) On the other hand, the obvious surjection $\sqrt{O}^{i-1} \otimes_A \bar{L} \to \sqrt{O}^{i-1}L$ is a homomorphism of left A-modules. By ideal-invariance of d, we conclude that $d(A)=d(\sqrt{O}^{i-1}L) \le d(\sqrt{O}^{i-1} \otimes_A \bar{L}) \le d(\bar{L})$. This proves the lemma. Q.e.d.

3.4 Localizability: Given any $b \in B$, the A-bimodule $M = AbA$ is r-noetherian - by r-restrictedness of B. Since also S is l-regular for M (3.2), the general localization theorem in [1], 5.1 applies. It gives that S is right orean for M, and hence for B.

Remark: An alternative proof of this step is given in 4.3.

3.5 Additivity: For the case that A is r-artinian, we refer to [8], 3.8 for a proof. The general case is reduced to the artinian case by localization with respect to S as follows: The quotient ring AS^{-1} exists (by 3.4), contains A as a subring (by 3.2,3.5) and is r-artinian [10], 2.11. If $P_1,\dots,P_n$ are the minimal primes of A, then $P_1S^{-1},\dots,P_nS^{-1}$ are those of AS^{-1} ([2],2.10), and the Goldie-ranks are preserved under localization (loc. cit.). Finally, BS^{-1} also exists (3.4). Hence

$$\text{rk } B/P = \text{rk } BS^{-1}/PS^{-1} = \sum_{i=1}^{n} z_i \text{rk } AS^{-1}/P_iS^{-1} = \sum_{i=1}^{n} z_i \text{rk } A/P_i .$$

Q.e.d.

3.6 Equidimensionality:

Let us pick a prime Q_1 of A such that the ideal $I_1 = \{a \in A \mid Q_1 a = 0\}$ of A is $\neq 0$. The existence of such a prime is trivial, since A is r-noetherian. We claim that A/I_1

is homogeneous of dimension $d(A/I_1)=d(A)$. In fact, let $\bar{L}$ be a left ideal $\neq 0$ in A/I_1 , and $L \supsetneq I_1$ its pre-image in A . Now $Q_1 L \neq 0$, and therefor $d(Q_1 L)=d(A)$ by homogeneity of A (3.1). On the other hand, $Q_1 L$ is a homomorphic image of $Q_1 \otimes_A \bar{L}$, and hence $d(Q_1 L) \leq d(Q_1 \otimes_A \bar{L}) \leq d(\bar{L})$ by ideal-invariance of d . We conclude that $d(\bar{L}) \geq d(A)$, as was to be shown.

Repeating this chain of arguments, we can produce a chain of ideals $0=I_o \subset I_1 \subset \ldots . \subset I_m=A$ such that each factor $F_j=I_j/I_{j-1}$ ($=1,\ldots,m$) is annihilated by some prime Q_j of A , and homogeneous of dimension $d(A)$. Now we have $d(A/Q_j) \geq d(F_j)=d(A)$, and this implies that Q_j must be a <u>minimal</u> prime of A (Lemma 1.3). But conversely, each of the minimal primes $P_1,\ldots,P_n$ of A occurs as a Q_j : In fact, since $I=Q_1 \cap \ldots \cap Q_m$ satisfies $I^m A=0$, we conclude that $I^m=0$, $I \subset \sqrt{0}=P_1 \cap \ldots \cap P_n$. So for each i , we have $Q_1 \cap \ldots \cap Q_m \subset P_i$, and hence $Q_j \subset P_i$ for some j , by primality of P_i . Finally, $Q_j=P_i$ by minimality of P_i . We conclude that $d(A/P_i)=d(A)$ for all i . Q.e.d.

§ 4 A necessary and sufficient condition for localizability

4.1 Let A be a r-noetherian ring with Small-set S bi-regular for A . By Small's theorem [10], 2.11 A has a r-artinian quotientring AS^{-1} . For any right A-module M , the AS^{-1}-module $MS^{-1}=M \otimes_A AS^{-1}$ is defined. Its length will be called the reduced (r-) length of M . Evidently, this is a finite number, whenever M is r-noetherian. (We avoid the term "reduced rank" introduced by Goldie, see [13], in order to avoid confusion with the notion of Goldie-rank.)

We shall use the phrase " A is finitely generated by a k-Lie-algebra $\underline{a}$ " to express that $\underline{a}$ is a finite dimensional k-subspace generating A as a ring such that $[\underline{a},\underline{a}] \subset \underline{a}$. Equivalently, this means that A is a homomorphic image of an enveloping algebra of a finite-dimensional Lie-algebra.

Lemma: Let A be finitely generated by a k-Lie-algebra. Then the following statements are equivalent for an A-bimodule B :

(i) AbA has finite reduced r-length for all $b \in B$.

(ii) $\sum_{\nu=0}^{\infty} a^{\nu}bA$ has finite reduced r-length for all $b \in B$, $a \in A$.

We shall say that a bimodule B is weakly r-restricted over A , if (ii) is satisfied. Clearly, r-restricted implies weakly r-restricted.

<u>Proof</u> of the lemma: The non-trivial implication is (ii) ⇒ (i). Let $x_1,\dots,x_n$ be a k-basis of the Lie-algebra generating A. Given $b\in B$, set $M_o=bA$, and $M_i=k[x_i]M_{i-1}=\sum_{\nu=0}^{\infty} x_i^{\nu} M_{i-1}$ for $i=1,\dots,n$.
We claim that M_iS^{-1} is finitely generated as right AS^{-1}-module: In fact, if $M_{i-1}S^{-1}$ is finitely generated by $e_1,\dots,e_m$ as a right AS^{-1}-module, then we have that

$$M_iS^{-1} = \sum_{j=1}^{m} k[x_i]e_jAS^{-1} ,$$

where each of the m summands is a finitely generated right AS^{-1}-module by (ii). Hence M_iS^{-1} is also finitely generated, completing the induction argument. Now for $i=n$, we have $M_n=AbA$ by the Poincaré-Birkhoff-Witt-theorem. Hence $AbAS^{-1}$ is finitely generated as a right module over the artinian ring AS^{-1}, so has finite length. This proves (i). Q.e.d

<u>4.2</u> <u>Theorem:</u> <u>Let</u> B <u>be a finitely generated k-algebra over an uncountable field</u> k . <u>Let</u> A <u>be a subalgebra finitely generated by a k-Lie-algebra. Assume the Small set</u> S <u>of</u> A <u>bi-regular for</u> B . <u>Then a necessary and sufficient condition for</u> S <u>to be r-orean for</u> B <u>is that</u> B <u>is weakly r-restricted over</u> A .

4.3 Proof of sufficiency:

Let $s\in S$ and $b\in B$ be given. The quotientring AS^{-1} exists [10], as does the localized module NS^{-1} of the right A-module $N = \sum_{\nu=0}^{\infty} s^{\nu}bA$. The assumption on weak restrictedness means that NS^{-1} has finite length as a right AS^{-1}-module. Left multiplication by s is an endomorphism of N, and extends uniquely to an endomorphism φ of NS^{-1}. From 1-regularity of s for B, we conclude that φ is injective. Since NS^{-1} has finite length, φ must also be surjective. Hence $\varphi(NS^{-1})=NS^{-1}$ contains b, say $b=\varphi(ct^{-1})$ for some $t\in S$, $c\in N\subset B$. Now we conclude that $bt=\varphi(ct^{-1})t=\varphi(ct^{-1}t)=\varphi(c)=sc$. This proves the right Ore-condition. Q.e.d.

Comment: The uncountability assumption on k will be used in the proof of necessity (4.4). I could not find a proof avoiding this assumption, although the theorem as stated above should certainly hold for countable k too. On the other hand, the arguments below do not use that A is finitely generated by a Lie-algebra.
The main point, which I want to make here is not so much to give the most general result possible, but to point out that there is some intrinsic relation between our "restrictedness condition" on an extension -in a weakenedform at least- on one hand, and "good behaviour" or localizability as studied in 2.2, on the other hand.

4.4 Proof of necessity (case A semiprime).

We assume that S is r-orean for B, but that for some $s \in A$ and $b \in B$ the right A-module $N = \sum_{\nu=0}^{\infty} s^{\nu} bA$ has finite reduced r-length. We shall derive a contradiction, and thus prove weak r-restrictedness of B over A. Let $Q=AS^{-1}$ denote the artinian quotientring of A [10].

For the sake of clearity of ideas, we first give the proof in the special case where A is semiprime, so Q is semisimple. By assumption, the right quotientring $BS^{-1}=BQ$ exists, and is a bi-semisimple Q-bimodule, and $NQ=k[s]bQ$ is a right Q-submodule of infinite length. Without loss of generality, we may assume b to be chosen such that $E=bQ$ is a simple right Q-module. Then the infinity of the length of the sum $E+sE+s^2E \ldots =NQ$ implies that this sum is direct. Hence $NQ \cong k[s] \otimes_k E$ as a $(k[s],Q)$-bisubmodule of BQ. Inparticular s is transcendental over k and NQ is free as a left $k[s]$-module.
Now consider the subring $k(s)'$ of Q generated by $k[s]$ and by all elements $(s-\alpha)^{-1}$ with $\alpha \in k$ such that $s-\alpha$ is invertible in Q. Since those elements are k-independent, $k(s)'$ is a k-algebra of uncountable k-dimension (see 4.5). The Q-bimodule BQ contains $k(s)'NQ \cong k(s)' \otimes_k E$, which is a right Q-module of uncoutable rank. But on the other hand, B has countable k-dimension, because it is a finitely generated k-algebra. Hence BQ can have only coutable rank as a right Q-module. This is the desired contradiction. Q.e.d.

4.5 Lemma: *Given any element* s *of an artinian k-algebra* Q *, the elements* $s-\alpha$ *are invertible in* Q *for all but finitely many* $\alpha \in k$.

Proof: For any $\alpha \in k$, consider the right ideal R_α of those elements $q \in Q$ such that $(s-\alpha)q=0$. Their sum, as α ranges over k , is evidently direct. Since Q is artinian, the sum must be finite. Hence $R_\alpha=0$ for all α up to at most a finite number of exceptions. But $R_\alpha=0$ means $s-\alpha$ 1-regular for Q , and in an artinian ring, 1-regular implies invertible. Q.e.d.

4.6 Proof of necessity (general case):

In the general case, we chose $s \in S$ and $b \in B$ as in 4.4, with bQ of the minimum possible length as a right Q-module, such that $N= \sum_{\nu=0}^{\infty} s^\nu bA$ has infinite reduced r-length.

Now consider the radical series of B as a right A-module. This is the filtration $B=B_0 \supset B_1 \supset B_2 \supset \dots \supset B_m=0$ of B by the A-bisubmodules $B_i=B\sqrt{0}^i$ for $i=1,\dots,m$, where $\sqrt{0}$ is the nilradical of A and $\sqrt{0}^m=0$. We note that $\sqrt{0}Q$ is the nilradical of $Q=AS^{-1}$, and $\sqrt{0}^iQ$ is its i-th power for all i (cf. [10], Theorem 1.9). Since we are assuming S r-orean for B , we conclude that BQ and all B_iQ $(0 \le i \le m)$ are Q-bimodules. Now let $\bar{N}Q$ resp. $\bar{E}$ denote the image of NQ resp. bQ in $\bar{B}Q=BQ/B_iQ$, where i is minimal such that $b \notin B_i$. By construction, we can conclude that $\bar{E}$ is a simple right Q-module, and that $\bar{N}Q=k[s]\bar{E}$ has infinite length as a right Q-module. Now we can proceed as in 4.4 to derive a contradiction. Q.e.d.

4.7 Of the assumptions in theorm 4.2, the one on regularity of the Small-set will in general be difficult to check. Therefor, we point out the

Corollary: Assume k uncountable. Let B be a prime finitely generated k-algebra, and A a subalgebra finitely generated by a Lie-algebra. Let S denote the Small-set of A, then the following conditions (i), (ii) are equivalent:

(i) S is r-orean for B.

(ii) S is bi-regular for B, and B is weakly r-restricted over A.

Proof: Since B is prime, (i) implies bi-regularity of S by [2], 2.11. Now theorem 4.2 applies and gives (ii) ⇒ (i). Q.e.d.

BIBLIOGRAPHY

[1] Borho, W.: On the Joseph-Small additivity principle for Goldie-ranks, IHES preprint No.4, Jan. 1981.

[2] Borho, W.- Gabriel, P.- Rentschler, R.: Primideale in Einhüllenden auflösbarer Lie-Algebren, Springer Lecture Notes in Math. 357 (1973).

[3] Borho, W.- Kraft, H.: Über die Gelfand-Kirillov-Dimension, Math. Ann. 220 (1976), 1-24.

[4] Brown, K.A.- Lenagan,, T.H.- Stafford, J.T.: Weak ideal invariance and localization, J. London Math. Soc. 21 (1980), 53-61.

[5] Brown, K.A.- Lenagan, T.H.- Stafford, J.T.: K-Theory and stable structure of Noetherian group rings, Proc. London Math. Soc., to appear.

[6] Goldie, A.- Krause, G.: Artinian quotient rings of ideal invariant Notherian rings, J. of Algebra 63 (1980), 374-388.

[7] Gordon, R.- Robson, J.C.: Krull dimension, Mem. AMS 133 (1973).

[8] Joseph,A.- Small, L.W.: An additivity principle for Goldie rank, Israel J. Math. 31 (1978)105-114.

[9] Krause, G.- Lenagan, T.H.- Stafford, J.T.: Ideal invariance and Artinian quotient rings, J. Algebra 55 (1978),145-154.

[10] Small, L.W.: Orders in Artinian rings, J. Algebra 4 (1966),13-41.

[11] Stafford, J.T.: Stable structure of non-commutative Noetherian rings, J. Algebra 47 (1977),244-267.

[12] Stafford, J.T.: On the regular elements of Noetherian rings, Proc. 1978 Atwerp Conference, Marcel Dekker Co., New York.

[13] Chatters, A.W.- Goldie, A.W.- Hajarnavis, C.R.- Lenagan, T.H.: Reduced Rank in Noetherian Rings, J. Algebra 61 11969), 582-589).

GENERATING MODULES EFFICIENTLY OVER NONCOMMUTATIVE RINGS

BY

J.T. STAFFORD

If R is a commutative Noetherian ring then the Forster-Swan Theorem and the stable version of that result due to Eisenbud and Evans give information concerning the number of generators of a module in terms of local data. In this article we will describe versions of these results for noncommutative Noetherian rings. As an application, we give noncommutative versions of Serre's Theorem and Bass's Cancellation Theorem which, over a commutative ring, show that a'big' projective module has a free direct summand and that the complementary direct summand is unique up to isomorphism.

Section 1 describes and gives the motivation for the results that we are interested in. Sections 2 and 3 are devoted to proving noncommutative versions of the Forster-Swan Theorem and its K-theoretic applications, respectively. Most of the results of this article come from [St 3] and the reader is referred to that paper for the full generality of the results given here. In particular, although these theorems are proved in [St 3] for an arbitrary Noetherian ring, we will only give them here under the additional assumption of weak ideal invariance since this enables one to give considerably easier proofs. For a similar reason, in the K-theoretic applications we will only be concerned with the problem of finding free direct summands of projective modules. The results in [St 3] consider conditions under which an arbitrary module has a direct summand isomorphic to a given projective module.

1) THE MAIN THEOREMS

Throughout this article all rings will contain an identity and all modules will be unitary. If M is a finitely generated right module over a ring R, then $g_R(M)$ - or just $g(M)$ if there can be no confusion over the ring - will denote the minimal number of generators of M as an R-module.

Let R be a commutative Noetherian ring and M a finitely generated R-module. If P is a prime ideal of R then the local number of generators of M at P , $g_{R_P}(M_P)$, is certainly a lower bound on g(M) and, in some sense, is trivial to determine. For, by Nakayama's lemma, $g(M_P) = g(M_P/MP_P)$ and this is just the dimension of the vector space M_P/MP_P over the field R_P/P_P . The Forster-Swan Theorem gives an upper bound on g(M) in terms of this local information :

THEOREM 1.1. [F,Sw1] *Let M be a finitely generated module over a commutative Noetherian ring R. Then*

$$\begin{aligned} g(M) &\leqslant \max\{g_{R_P}(M_P) + \text{Kdim } R/P : P \text{ a prime ideal}\} \\ &\leqslant \max\{g_{R_P}(M_P) : P \text{ a maximal ideal}\} + \text{Kdim } R . \end{aligned} \tag{1}$$

REMARK As stated, this is the result that was proved by Forster. Swan then gave a refined version which, in particular, replaced "Kdim R/P" by "dim(max Spec R/P)" and" Noetherian" by the condition that "max Spec R be Noetherian". However, it is the weaker version of Theorem 1.1 that is pertinent to the present discussion.

We are interested in generalising this theorem to noncommutative Noetherian rings. Before formulating such a generalisation there are two points to be considered. First, one needs an interpretation of the number of generators of a module that makes sense when one cannot localise. Secondly, and modulo the first point, one has to ask what kind of generalisation is reasonable in an arbitrary Noetherian ring, where there may be relatively few maximal ideals and therefore relatively little local data that is available.

The first problem has been answered by Warfield. Let P a prime ideal of a commutative Noetherian ring R and M a finitely generated R-module. Then :

$$g(M_P) = g(M_P/MP_P) = g_Q(M/MP \otimes Q) ,$$

where Q is the quotient field of R/P , and this is an interpretation of $g(M_P)$ which does generalise to a noncommutative setting. For, suppose that P is now a prime ideal of a right Noetherian ring R . Then by Goldie's Theorem R/P has a simple Artinian, right quotient ring, say Q(R/P) . So, for a finitely generated right R-module M , write :

$$g(M,P) = g_{Q(R/P)}(M/MP \otimes_{R/P} Q(R/P)). \tag{2}$$

As in the commutative case, this is a number that is 'trivial' to determine and, indeed, is a natural invariant of M . For, if N is a finitely generated Q = Q(R/P) -module, then $N \cong S \oplus \dots \oplus S$, where S is the unique simple Q-module. Thus, if :

$$\hat{g}_Q(N) = \text{length }(N)/\text{length }(Q),$$

then $g_Q(N)$ is the least integer greater than or equal to $\hat{g}_Q(N)$. We will write $\hat{g}(M,P) = \hat{g}_Q(M/MP \otimes Q)$. Note that this number can be defined without (explicit) reference to Q. For, let N be a finitely generated right R/P-module, with torsion submodule $T(N)$. Then the reduced rank of N, written $\rho_{R/P}(N)$, is defined to be uniform dimension of $N/T(N)$. It is readily checked that :

$$\hat{g}(M,P) = \rho_{R/P}(M/MP)/\rho_{R/P}(R/P) .$$

The definition (2) is due to Warfield [W] and he used it in that paper to generalise the Forster-Swan Theorem to fully bounded rings. (A prime ring is bounded if every essential right ideal contains a non-zero two-sided ideal and a ring is fully bounded if every prime factor ring is bounded).

THEOREM 1.2 [W, Theorems 2 and 5] *Let M be a finitely generated right module over a right Noetherian, fully bounded ring R. Then*

$$\begin{aligned} g(M) &\leq \max \{ g(M,P) + \text{Kdim } R/P : P \text{ a prime ideal} \} \\ &\leq \max \{ g(M,P) : P \text{ a maximal ideal} \} + \text{Kdim } R . \end{aligned}$$

In Theorem 1.2, Krull dimension is defined, as in the commutative case, in terms of chains of prime ideals. The comments given above therefore show that there is a reasonable interpretation of the number of generators of a module and, if the ring has "enough" two-sided ideals, then this can be used to generalise Theorem 1.1 . This leaves the question of finding the appropriate formulation of the Forster-Swan Theorem for rings that have relatively few ideals. We will illustrate this in the extreme case, when R is a simple, Noetherian, non-Artinian ring and M is a finitely generated, torsion right R-module. Now, (0) is the only prime ideal of R and $M \otimes Q(R) = 0$. Thus $g(M,0) = 0$. In other words, if Equ. 1 holds then the number of generators of M depends only upon the dimension of R . Fortunately this is true, as is illustrated by the following special case of [St2, Theorem 5.1]. We will give a proof of this result since, although it is easy, it does illustrate a technique that will be used several times in the next section.

PROPOSITION 1.3 *Let R be a simple, Noetherian, non-Artinian ring and $M = \Sigma_1^s a_i R$ be a finitely generated, Artinian right R-module. Let $f \neq 0 \in R$. Then M is cyclic. Indeed, there exist $\lambda_i \in R$ such that*

$$M = (a_1 + \Sigma_2^s a_i \lambda_i f) R .$$

<u>Proof</u> By induction, it suffices to prove the result when $s = 2$. As M is Artinian, there exists $b \in R$ such that a_2bR is a simple module. By induction on the length of a_2R, there exists $\mu \in R$ such that $M = (a_1 + a_2\mu f)R + a_2bR$. So, replacing a_1 by $a_1+a_2\mu f$ and a_2 by a_2b reduces the problem to the case when a_2R is a simple module. As M is Artinian but R is not, M is also a torsion R-module. So there exists c regular in R such that $a_1c = 0$. Now, $fc \neq 0$, so $a_2RfcR = a_2R \neq 0$. Thus $a_2\lambda fc \neq 0$ for some $\lambda \in R$. However :

$$(a_1 + a_2\lambda f)R \supseteq a_2\lambda fcR = a_2R ;$$

as a_2R is simple. Thus $M = a_1R + a_2R = (a_1 + a_2\lambda f)R$; as required.

<u>REMARK</u> The obvious choice of the element f in the above proposition is $f = 1$. However, the present formulation is both natural and useful. In particular, if $\text{Kdim } R = 1$ (see below for the definition) and f is regular, then it follows from Proposition 1.3 applied to R/fR with $a_1 = 0$ and $a_2 = 1$, that $R = fR + \lambda fR$. In other words, this is a rather strong form of the statement that $R = RfR$.

Proposition 1.3 shows that is reasonable to expect that Theorem 1.1 can be generalised to noncommutative rings in the manner of Theorem 1.2 . The dimension on the ring R that seems to be most appropriate, since it generalises the notion of an Artinian module, is the Krull dimension of Rentschler and Gabriel [RG] . This is defined as follows. Let M be a module over a ring R . Then $\text{Kdim } M = -1$ if $M = 0$ and, inductively, for any ordinal α , $\text{Kdim } M = \alpha$ if $\text{Kdim } M \nless \alpha$ but every descending chain $M = M_0 \supset M_1 \supset \ldots$, with $\text{Kdim } M_i/M_{i+1} \nless \alpha$ for each i, is of finite length. The Krull dimension of R is defined by $\text{Kdim } R = \text{Kdim } R_R$. This always exists if R is Noetherian. Further, it coincides with the usual commutative definition in terms of chains of prime ideals when R is a commutative (or even a fully bounded) Noetherian ring, provided that one of the two dimensions is finite.

We are now able to state our first main result from [St3] :

<u>THEOREM 1.4</u> <u>Let R be a Noetherian ring and M a finitely generated right R-module. Then</u> :

$$g(M) \leqslant \max \{ g(M,P) + \text{Kdim } R/P \mid P \text{ a prime ideal} \} .$$

Theorem 1.4 will be proved in the next section, though under the additional assumption of weak ideal invariance. The reader should note that the second inequality of Theorem 1.2 does not hold in general. For example, let R be a Noetherian domain, of Krull dimension one, such that R has exactly one non-zero ideal, say Q. Let M be the direct sum of m copies of Q. Then $g(M) \geqslant m$. However, since $MQ = M, g(M,Q) = 0$ and so :

$$\max \{g(M,P) \mid P \text{ a maximal ideal}\} + \operatorname{Kdim} R = 1 .$$

Of course, such an example is not surprising since R/Q provides very little information about R. However, if maximal is replaced by primitive then a generalisation is possible :

COROLLARY 1.5 [St3, Corollary 4.6] *Let M be a finitely generated right module over a Noetherian ring R. Then* :

$$g(M) \leqslant \max \{g(M,P) \mid P \text{ right primitive}\} + \operatorname{Kdim} R .$$

Having obtained an upper bound for the number of generators of a module M, one would also like to obtain a corresponding set of generators. Note that this did indeed happen in Proposition 1.3, since in that result an arbitrary set of generators was refined, by suitable elementary transformations, to one with the required number of generators. This can be generalised as follows. Let M be a finitely generated right R-module. Then define the stable number of generators of M by :

$s(M) \leqslant s$ if, whenever $M = \Sigma_1^{r+1} a_i R$ with $r \geqslant s$, then there exist

$$\lambda_i \in R \text{ such that } M = \Sigma_1^r (a_i + a_{r+1}\lambda_i) R .$$

THEOREM 1.6 [St3, Theorem 3.1] *Let M be a finitely generated right module over a Noetherian ring R. Then* :

$$\begin{aligned} S(M) &\leqslant \max \{g(M,P) + \operatorname{Kdim} R/P : P \text{ a prime ideal}\} \\ &\leqslant \max \{g(M,P) \mid P \text{ a right primitive ideal}\} + \operatorname{Kdim} R. \end{aligned}$$

The commutative case of Theorem 1.6 is due to Eisenbud and Evans [EE, Theorem B]. Theorem 1.6 will also be proved in Section 2 ; indeed, if one assumes weak ideal invariance then the natural proof of Theorem 1.4 leads to the formulation given in Theorem 1.6 .

A special case of Theorem 1.6 is that $s(R) \leqslant 1 + \operatorname{Kdim} R$ which, in the commutative case, is known as the Stable Range Theorem. This results is important in classical algebraic K-theory as it provides information about $K_1(R)$. We will mention some of these corollaries here.

Let S be a ring and write $E_n(S)$ for the subgroup of $Gl_n(S)$ generated by the elementary matrices ; ie, those of the form $I_n + e_{ij}a$ for $i \neq j$ and $a \in S$. $Gl_n(S)$ can be regarded as a subgroup of $Gl_{n+1}(S)$ in the natural way. Define $Gl(S) = \lim\limits_{n \to \infty} Gl_n(S)$ and $E(S) = \lim\limits_{n \to \infty} E_n(S)$. It can be shown that $E(S)$ is isomorphic to the commutator subgroup $Gl(S)' = (Gl(S),GL(S))$. Thus $K_1(S) = Gl(S)/E(S)$ is an abelian group. The application of the Stable Range Theorem lies in the following result.

<u>THEOREM</u> 1.7 [V] <u>Suppose that</u> S <u>is a ring and that</u> $s(S) \leqslant n$ <u>for some integer</u> $n \geqslant 2$. <u>Then</u> $Gl_m(S)/E_m(S) \cong Gl(S)/E(S) = K_1(S)$ <u>for all</u> $m \geqslant n+1$. <u>Furthermore,</u> $E_m(S) = Gl_m(S)'$.

There are a number of interesting rings S for which $K_1(S)$ has been calculated. So, for example, we have :

<u>COROLLARY</u> 1.8 <u>Let</u> k <u>be a field and suppose that either</u>

i) $R = U(g)$, <u>the universal enveloping algebra of a finite dimensional Lie algebra</u> g <u>over</u> k , <u>or</u>

ii) $R = kG$, <u>the group ring of a poly (infinite cyclic) group</u> G .

<u>Let</u> $n = \dim_k g$ <u>or</u> $n = h(G)$, <u>the Hirsch number of</u> G , <u>in the respective cases.</u> <u>Then for</u> $m \geqslant n+2$, $E_m(R) = Gl_m(R)'$ <u>and</u> $Gl_m(R)/E_m(R) \cong k^\star$, <u>the multiplicative group of non-zero elements of</u> k .

<u>Proof.</u> By [Q, Theorem 7] and [FH, Theorem 29] , respectively, $K_1(R) \cong k^\star$. By [RG] and [Sm], respectively, Kdim $R \leqslant n$. The corollary now follows from Theorems 1.6 and 1.7 .

The Stable Range Theorem can also be used to show that certain projective modules are free. For example, let R be one of the rings described in Corollary 1.8. Then by [Q] or [FH], again, every finitely generated, projective R-module P is <u>stably free</u> ; that is, $P \oplus F \cong G$ for some finitely generated free modules F and G . It is another consequence of the Stable Range Theorem (use for example the proof of [Sw2, Proposition 12.1]) that P is actually free whenever $g(P,0) \geqslant n + 1$.

In the case of a commutative Noetherian ring, there are two further results, Serre's Theorem and Bass's Cancellation Theorem, which are related to the Stable Range Theorem and give further information about projective modules. We will remind the reader of these results. Let M be a finitely generated, projective module over a commutative, Noetherian ring R. Then for any prime ideal P of R, M_P is a free R_P-module. As with the Forster-Swan Theorem, one can obtain global information about the structure of M from this local data :

THEOREM 1.9 Let M be a finitely generated, projective module over a commutative Noetherian ring R. Suppose that, for every maximal ideal P of R, there exists an integer $s \geqslant 1 + \text{Kdim}\, R$ such that $M_P \cong R_P^{(s)}$. Then

i) (Serre's Theorem) $M \cong M' \oplus R$ for some module M',

ii) (Bass's Cancellation Theorem) if $M \oplus R \cong N \oplus R$ for some module N, then $M \cong N$.

We are interested in generalising this to noncommutative rings. Note that the initial condition on M can easily be rephrased so that it makes sense in a noncommutative setting, as it is equivalent to saying that $g(M,P) \geqslant 1 + \text{Kdim}\, R$ for every maximal (or indeed prime) ideal P of R. In this form the result does generalise to noncommutative rings :

THEOREM 1.10 [St3, Corollaries 5.10 and 5.11] Let M be a finitely generated, projective right module over a Noetherian ring R, such that $\hat{g}(M,P) \geqslant 1 + \text{Kdim}\, R$ for every prime ideal P. Then :

i) $M \cong M' \oplus R$ for some module M',

ii) if $M \oplus R \cong N \oplus R$ for some module N, then $M \cong N$.

This result will be proved in Section 3 although, as it depends upon Theorem 1.6, it will only be proved here under the assumption of weak ideal invariance.

2 - PROOF OF THE FORSTER-SWAN THEOREM

In this section we will prove Theorem 1.6 for weakly ideal invariant, Noetherian rings. The proof in the general case is considerably more difficult and can be found in [St3]. The reader is reminded that a right Noetherian ring is weakly ideal invariant if $\text{Kdim}\, M \otimes T < \text{Kdim}\, R/T$, whenever T is an ideal and M is a finitely generated right R-module such that $\text{Kdim}\, M < \text{Kdim}\, R/T$. We do not intend to enter into a discussion about weak ideal invariance here, but will content ourselves with the observation that, although there do exist right Noetherian rings that are not weakly ideal invariant, there is no known Noetherian counterexample. Further, many of the standard examples of Noetherian rings are known to be weakly ideal invariant. The usefulness of weak ideal invariance in the present circumstances comes from the following observation. Let $A \subset B$ be finitely generated right modules over a right Noetherian, weakly ideal invariant ring R and P be a prime ideal of R, such that $\mathrm{Kdim}\, B/A < \mathrm{Kdim}\, R/P$. Then $g(A,P) = g(B,P)$.

Given a finitely generated right module M over a Noetherian ring R, and a prime ideal P of R, set :

$$b(M,P) = \begin{cases} g(M,P) + \mathrm{Kdim}\, R/P & \text{if } g(M,P) \neq 0 \\ 0 & \text{if } g(M,P) = 0 . \end{cases}$$

Let $b(M) = \max\ \{ b(M,P) : P \text{ a prime ideal} \}$. The idea behind the proof of the theorem is to use induction on the number $\max\ \{ b(M), \mathrm{Kdim}\, M + 1 \}$. We begin with some lemmas which show how to reduce this number in certain special cases.

LEMMA 2.1 Let M be a finitely generated right module over a right Noetherian ring R, with $\mathrm{Kdim}\, M = \alpha$. Then :

i) there are only finitely many prime ideals P of R such that $\mathrm{Kdim}\, M/MP = \mathrm{Kdim}\, R/P = \alpha$,

ii) suppose that $P_1,\dots,P_m$ are prime ideals such that $\mathrm{Kdim}\, R/P_i \leqslant \alpha$ but $\mathrm{Kdim}\, M/MP_i < \alpha$ for each i. Then $\mathrm{Kdim}\, M/M(\cap P_i) < \alpha$.

Proof This follows by a routine induction and is similar to the proof of [St2, Lemma 3.3].

LEMMA 2.2. Let A be a finitely generated module over a prime, right Noetherian ring R and let $x,y \in A$. Then there exists $f \in R$ such that either $(x+yf)R \cong R$ or $xR + yR/(x+yf)R$ is a torsion R-module. Equivalently :

$$g(xR + yR/(x+yf)R\ ,\ 0) = \max\ \{g(xR + yR, 0) - 1, 0\}.$$

Proof We may assume that the torsion submodule of A equals zero. Further, by induction, we may assume that yR is uniform. If $I = \text{r-ann}(x) = 0$, then $xR \cong R$ and we may take $f = 0$. So, suppose that $I \neq 0$ and that $y \neq 0$. Then $yRI \neq 0$ and there exist $f \in R$ and $b \in I$ such that $yfb \neq 0$. This element will do the trick. For, $(x+yf)R \ni yfb \neq 0$. Since yR is uniform, yfb generates an essential submodule of yR . Thus $(x+yf)R + yR/(x+yf)R$ is a torsion module ; as required.

<u>LEMMA</u> 2.3 <u>Let $M = \Sigma_1^s m_i R$ be a finitely generated right module over a right Noetherian ring R . Let $\{P_i : 1 \leq i \leq n\}$ be a finite collection of prime ideals of R such that $g(M,P_i) > 0$ for each i . Then there exist $f_i \in R$ such that, if $c = m_1 + \Sigma_2^s m_i f_i$, then</u> :

$$g(M/cR , P_i) = g(M,P_i) - 1 \quad \text{for} \quad 1 \leq i \leq n .$$

<u>Proof</u> This is a special case of [W, Theorem 1] . Note that, for any $c \in M$, $g(M/cR , P_i) \geq g(M , P_i) - 1$, so one only has to prove the inverse inequality. Order the prime ideals $\{P_i\}$ in such a way that, if $P_i \subsetneqq P_j$, then $i > j$. Let $0 \leq r \leq n-1$ and suppose that there exist $g_i \in R$ such that, if $d = m_1 + \Sigma_2^s m_i g_i$, then

$$g(M/dR,P_i) = g(M,P_i) - 1 \quad \text{for} \quad 1 \leq i \leq r .$$

Set $T = P_1 \cap \ldots \cap P_r$ (or $T = R$ if $r = 0$, in which case d can be taken to be m_1). By the ordering of the P_i's, $T \not\subseteq P_{r+1}$ and so there exists $t \in T$ which is regular mod P_{r+1} . Note that this implies that :

$$g(M,P_{r+1}) = g((dR + \Sigma_2^s m_i tR + MP_{r+1})/MP_{r+1} , P_{r+1}) .$$

By Lemma 2.2 and induction, there exist $f_i \in R$ such that, if $c = d + \Sigma m_i t f_i$, then $g(M/cR,P_{r+1}) \leq g(M,P_{r+1}) - 1$. However, since $t \in T$, $cR + MP_i = dR + MP_i$ for $1 \leq i \leq r+1$. Thus $g(M/cR,P_i) = g(m,P_i) - 1$ for $1 \leq i \leq r+1$. Induction completes the argument.

For a fully bounded Noetherian ring, the Forster-Swan Theorem follows fairly easily from Lemma 2.3 . For a general Noetherian ring one also requires that in certain circumstances one also has $\mathrm{Kdim}\, M/cR < \mathrm{Kdim}\, M$. This is achieved in the next proposition. A Noetherian module M is said to be <u>critical</u> if $\mathrm{Kdim}\, M' < \mathrm{Kdim}\, M$ for every proper factor module M' of M .

<u>PROPOSITION</u> 2.4 <u>Let $M = \Sigma_1^s m_i R$ be a finitely generated right module over a right Noetherian, weakly ideal invariant ring R . Let $\mathrm{Kdim}\, M = \alpha$ and suppose that $g(M,Q) \leq 1$ for every prime ideal Q with $\mathrm{Kdim}\, R/Q = \alpha$. Then, for any set of prime ideals $\Lambda = \{P_i \mid 1 \leq i \leq n\}$ there exist $f_i \in R$ such that, if $c = m_1 + \Sigma_2^s m_i f_i$, then</u> :

a) $\mathrm{Kdim}\, M/cR < \alpha$ and

b) $g(M/cR,P_i) = \max\ \{0, g(M,P_i) - 1\}$ for $1 \leqslant i \leqslant n$.

Proof We may clearly suppose that $g(M,P_i) > 0$ and hence that $\mathrm{Kdim}\, R/P_i \leqslant \alpha$ for each $P_i \in \Lambda$. Further, after possibly expanding the set Λ, we may suppose by Lemma 2.1 i) that, if Q is a prime ideal such that $g(M,Q) > 0$ but $\mathrm{Kdim}\, R/Q = \alpha$, then $Q = P_i$ for some i.

By Lemma 2.3 there exist $g_i \in R$ such that, if $d = m_1 + \Sigma_2^s m_i g_i$, then

$$g(M/dR,P_i) = g(M,P_i) - 1 \quad \text{for} \quad 1 \leqslant i \leqslant n. \tag{3}$$

Note that this implies that $\mathrm{Kdim}\, M/dR + MP_i < \alpha$ for each i. So, if $T = P_1 \cap \ldots \cap P_n$ then, by Lemma 2.1 ii), $\mathrm{Kdim}\, M/dR + MT < \alpha$. Thus, if $N = \Sigma_2^s m_i T$, then $\mathrm{Kdim}\, M/dR + N < \alpha$. Notice that, if d is replaced by $d + n$ for some $n \in N$, then Equ.3 will still hold. So it remains to find $n \in N$ such that $\mathrm{Kdim}\, M/(d+n)R < \alpha$. The proof of this is similar to that of [St1, Proposition 2.1] and uses the idea behind the proof of Proposition 1.3.

Choose a non-zero submodule L of N such that $Q = \text{r-ann}\ L$ is as large as possible. So Q is a prime ideal. Since every Noetherian module has a critical submodule [GR, Theorem 2.1], we may also assume that L is critical and cyclic, say $L = aR$. By a Noetherian induction there exists $n \in N$ such that $\mathrm{Kdim}\, M/(d+n)R + L < \alpha$. Replace d by $d+n$. If $\mathrm{Kdim}\, L < \alpha$ then $\mathrm{Kdim}\, M/dR < \alpha$ and the proof is complete. So, assume that $\mathrm{Kdim}\, L = \alpha$. Similarly, if $dR \cap L \neq 0$, then as L is critical, $\mathrm{Kdim}\, L + dR/dR = \mathrm{Kdim}\, L/L \cap dR < \alpha$ and again $\mathrm{Kdim}\, M/dR < \alpha$. Thus we are left with the possibility that $dR \cap L = 0$.

Let $K = \text{r-ann}(d)$. We will first show that $K \not\subseteq Q$. If $K \subseteq Q$ then

$$\mathrm{Kdim}\, R/Q \leqslant \mathrm{Kdim}\, dR \leqslant \alpha = \mathrm{Kdim}\, L \leqslant \mathrm{Kdim}\, R/Q.$$

So $\mathrm{Kdim}\, R/Q = \alpha$. Further, $dR/dQ \cong R/Q$. Now, by weak ideal invariance,

$$\mathrm{Kdim}\, MQ/(dr + L)Q \leqslant \mathrm{Kdim}\, (M/dR + L) \otimes Q < \alpha.$$

Thus $g(M,Q) \geqslant g(dR + L + MQ/MQ,Q) = g(dR + L,Q) > 1$, since $dR/dQ \cong R/Q$, $g(L,Q) > 0$ and $dR \cap L = 0$. This contradicts our initial hypothesis on M. So $K \not\subseteq Q$. Pick $k \in K \setminus Q$. Then there exists $r \in R$ such that $ark \neq 0$. Thus

$$(d + ar)R \supseteq (d + ar)kR = arkR \neq 0.$$

As in the last paragraph, this implies that $\mathrm{Kdim}\, M/(d+ar)R < \alpha$, and the proof is complete.

Proposition 2.4 is the only place in this section where ideal invariance is used. The Forster-Swan Theorem follows easily from Lemma 2.3 and Proposition 2.4, provided that one can show that there are only finitely many prime ideals P for which $b(M,P) = \max\{b(M), \text{Kdim } M+1\}$. This is the point of the next result.

PROPOSITION 2.5. Let M be a finitely generated right module over a right Noetherian, weakly ideal invariant ring R, with $b(M) < \infty$. If $b(M) = b(M,P)$ for infinitely many prime ideals P, then $b(M) \leq \text{Kdim } M$.

REMARK If R is commutative or fully bounded, then $b(M)$ can be attained by only finitely many prime ideals [W, Lemma 5]. This is however not the case in general, as is shown [St3, Section 7].

Proof Suppose that the result is false. Then certainly $\alpha = \text{Kdim } M < \infty$. So there exist integers r and s such that

a) $b(M,P_i) \geq s \geq \alpha + 1$ for infinitely many prime ideals P_i with $\text{Kdim } R/P_i = r$, and

b) for every prime ideal Q with $\text{Kdim } R/Q > r$, $b(M,Q) \leq s$ and $b(M,Q) < s$ for all but finitely many of them.

Given such a counterexample, Lemma 2.1 i) implies that $\alpha > r$. Choose among all possible modules M that satisfy a) and b) one for which $\alpha - r$ is as small as possible and, modulo this, such that s is as small as possible. Let $Q_1,\ldots,Q_m$ be the prime ideals such that $\text{Kdim } R/Q_i > r$ but $b(M,Q_i) = s$ (if any such ideals exist). Suppose first that $s > \alpha + 1$. Then, by Lemma 2.3, there exists $c \in M$ such that $b(M/cR,Q_i) = b(M,Q_i) - 1 = s - 1$ for $1 \leq i \leq m$. Write $M' = M/cR$. So certainly $b(M',P_i) \geq s - 1$ for each i and $b(M',Q) \leq s - 1$ for each prime ideal Q with $\text{Kdim } R/Q > r$. Of course, there may now exist infinitely many prime ideals Q with $b(M,Q) = s - 1$ but $\text{Kdim } R/Q > r$. However, there certainly exists an integer $r' \geq r$ such that:

a') $b(M',P_i) \geq s - 1$ for infinitely many prime ideals P_i with $\text{Kdim } R/P_i = r'$,

b') $b(M',Q) \leq s - 1$ for all prime ideals Q with $\text{Kdim } R/Q > r'$ and $b(M',Q) < s$ for all but finitely many of them.

This contradicts the minimality of either $\alpha - r$ or s.

This leaves the case $s = \alpha + 1$. Notice that the proof of the last paragraph will still work in this case except that M' could still have Krull dimension α, violating the condition $s \geq \alpha + 1$. However, we now have

$g(M,Q) \leqslant 1$ for every prime ideal Q with $\text{Kdim } R/Q = \alpha$. So the proof can be completed by repeating the argument of the last paragraph, but with Lemma 2.3 replaced by Proposition 2.4 .

Our generalisation of the Forster-Swan Theorem is now easy to prove.

<u>THEOREM</u> 2.6. <u>Let</u> M <u>be a finitely generated right module over a weakly ideal invariant, right Noetherian ring</u> R . <u>Then</u>

$$s(M) \leqslant \max\{\text{Kdim } M + 1,\ \sup\{b(M,P) \mid P \text{ a prime ideal}\}\}$$
$$\leqslant \max\{g(M,P) + \text{Kdim } R/P \mid P \text{ a prime ideal}\}.$$

<u>REMARK</u> 2.7 By Nakayama's Lemma, it is enough to find elements of the required form that generate $M/MJ(R)$. So $\text{Kdim } M$ can be replaced by $\text{Kdim } M/MJ(R)$. For a similar reason, the second inequality is a triviality. For, we may assume that $J(R) = 0$. Now, if $\text{Kdim } M = \text{Kdim } R$ then, by Lemma 2.1 ii), there exists a prime ideal P of R such that $\text{Kdim } R/P = \text{Kdim } R$ yet $g(M,P) \geqslant 0$, from which the second inequality follows.

<u>Proof</u> It remains to prove the first inequality. Suppose that $\text{Kdim } M = \alpha$ and $M = \sum_1^s m_i R$ for some $s > \max\{\alpha + 1, b(M)\}$. Set $X = \{\text{prime ideals } P \mid b(M,P) = \max(b(M), 1 + \text{Kdim } M)\}$. Of course, X may be empty but, by Proposition 2.5, X is a finite set, say $X = \{P_1,\dots,P_n\}$.

Suppose first that $b(M) > \alpha + 1$. Then X is non-empty and by Lemma 2.3 there exist $f_i \in R$ such that, if $c = m_1 + \sum_2^s m_i f_i$, then :

$$g(M/cR,P_i) = g(M,P_i) - 1 \qquad \text{for} \qquad 1 \leqslant i \leqslant n .$$

Let $\bar{M} = M/cR$. Then $b(\bar{M}) \leqslant b(M) - 1$ and $\bar{M} = \sum_2^s \bar{m}_i R$, where $\bar{m}_i$ denotes the image of m_i in $\bar{M}$. So $s - 1 > \max\{\text{Kdim } \bar{M} + 1, b(\bar{M})\}$. Thus, by induction on $b(M)$, there exist $g_i \in R$ that $\bar{M} = \sum_2^{s-1} (\bar{m}_i + \bar{m}_s g_i)R$. Hence :

$$M = \sum_1^{s-1} (m_i + m_s g_i)R \qquad \text{where} \qquad g_1 = f_s - \sum_2^{s-1} g_i f_i \ ;$$

as required.

Suppose, finally, that $b(M) \leqslant \alpha + 1$. Then Proposition 2.4 can be applied to find elements $f_i \in R$ such that, if $c = m_1 + \sum_2^s m_i f_i$, then $\text{Kdim } M/cR < \alpha$ and $g(M/cR,P_i) = g(M,P_i) - 1$ for all $P_i \in X$. Thus $\max\{b(M/cR), \text{Kdim}/cR + 1\} \leqslant \alpha$.

As in the last paragraph, induction can be used to complete the proof of the theorem.

If R is a fully bounded ring, then the term $\text{Kdim } M + 1$ can be ignored as it is bounded above by $b(M)$. In general, however, it is necessary as is shown by considering a torsion module over a simple ring.

3 - PROOF OF THE K-THEORETIC APPLICATIONS

In this section we show how the generalisation of the Forster-Swan Theorem can be applied to give noncommutative versions of Serre's Theorem and Bass's Cancellation Theorem. Since we will be using the results of the last section, these results will only be proved under the assumption of weak ideal invariance. The reader is referred to [St 3] for the general result.

We first need a module to which Theorem 1.6 can be applied and for this some notation is required. Let A and B be modules. Then $A^{(n)}$ will denote the direct sum of n copies of A and elements of $A^{(n)}$ will frequently be written as n-tuples. Since $\text{Hom}(B,A^{(n)}) \cong \text{Hom}(B,A)^{(n)}$, an element $\theta \in \text{Hom}(B,A^{(n)})$ can also be written as an n-tuple, say $\theta = (\theta_1,\dots,\theta_n)$ for some $\theta_i \in \text{Hom}(B,A)$.

LEMMA 3.1 *Let M be a finitely generated projective right module and A,B arbitrary right modules over a right Noetherian ring R. Let X be a finite set of prime ideals of R and $\alpha \in \text{Hom}(M,A)$, $\beta \in \text{Hom}(B,A)$ be homomorphisms. Finally, suppose that $\hat{g}(M,P) \geq \hat{g}(A,P)$ for all $P \in X$. Then there exists $\gamma \in \text{Hom}(M,B)$ such that :*

$$\hat{g}(A/(\alpha+\beta\gamma)(M),P) = \hat{g}(A/\alpha(M) + \beta(B),P) \quad \textit{for all} \quad P \in X .$$

Proof This is a special case of [W, Theorem 1] . In outline, if X consists of just one prime ideal, say P, then elements of $\text{Hom}(M/MP, B/BP)$ can be lifted to elements of $\text{Hom}(M,B)$ and so one may, by factoring out P, assume that $P=0$. The result is now a fairly easy exercise in uniform dimension, similar to the proof of Lemma 2.2 . The general case follows by "patching together" the individual homomorphisms obtained in this manner along the lines of Lemma 2.3 .

PROPOSITION 3.2 *Let M be a finitely generated, projective right module over a right Noetherian, weakly ideal invariant ring R. Suppose that $\text{Kdim } R = n < \infty$, and $\hat{g}(M,P) \geq n+1$ for all prime ideals P of R. Let $\theta_1 \in \text{Hom}(M,R)$ be such that $b(R/\theta_1(M)) \leq n$. Then there exist $\theta_2,\dots,\theta_{n+1} \in \text{Hom}(M,R)$ such that, if $\theta = (\theta_1,\dots,\theta_{n+1}) \in \text{Hom}(M,R^{(n+1)})$, then $b(R^{(n+1)}/\theta(M)) \leq n$.*

REMARK 3.3 $\hat{g}$ is defined in Section 1 and $b(N)$ in Section 2 . Note also that $\mathrm{Kdim}\, R^{(n+1)}/\theta(M) < n$. For, let L be a finitely generated right R-module with $b(L) \leqslant n$. Then $g(L,P) = 0$ for all prime ideals P with $\mathrm{Kdim}\, R/P = n$. Thus, by Lemma 2.1 ii), $\mathrm{Kdim}\, L/LN < n$, where $N = N(R)$ is the nilradical. By weak ideal invariance, $\mathrm{Kdim}\, LN^r/LN^{r+1} < n$ for every r . So $\mathrm{Kdim}\, L < n$.

Proof Fix $1 \leqslant r \leqslant n$ and suppose that there exist $\theta_2,\dots\theta_r \in \mathrm{Hom}(M,R)$ such that, if $\phi = (\theta_1,\dots,\theta_r)$, then $b(R^{(r)}/\phi(M)) \leqslant n$. By Remark 3.3 and Proposition 2.5 the set $X = \{\text{prime ideals } P \mid b(R^{(r)}/\phi(M),P) = n\}$ is finite. Expand X to include all those prime ideals Q for which $\mathrm{Kdim}\, R/Q = n$. We will identify $R^{(r)}$ with the first r coordinates of $R^{(r+1)}$ and let R_1 be the submodule $(0,\dots,0,R)$ of $R^{(r+1)}$.

Since $\hat{g}(M,P) \geqslant r+1$ for each $P \in X$, we may apply Lemma 3.1 with $A = R^{(r+1)}$, $B = R_1$, $\alpha = \phi$ and β the inclusion map. This provides $\theta_{r+1} \in \mathrm{Hom}(M,R_1)$ such that, for all $P \in X$,

$$\begin{aligned}\hat{g}(R^{(r+1)}/(\phi + \beta\theta_{r+1})(M),P) &= \hat{g}(R^{(r+1)}/\phi(M) + \beta(R_1),P) \\ &= \hat{g}(R^{(r)}/\phi(M),P) .\end{aligned}$$

Identify θ_{r+1} with the corresponding element of $\mathrm{Hom}(M,R)$ and write $\psi = (\theta_1,\dots,\theta_{r+1})$. Then $b(R^{(r+1)}/\psi(M),P) \leqslant n$ for all $P \in X$.

In fact this in turn implies that $b(R^{(r+1)}/\psi(M)) \leqslant n$. For, suppose that Q is a prime ideal of R such that $Q \notin X$. Then :

$$\mathrm{Kdim}\, R/Q < n \qquad \text{and} \qquad b(R^{(r)}/\phi(M),Q) < n . \tag{4}$$

Consider the short exact sequence, with the obvious homomorphisms :

$$0 \to R_1/R_1 \cap (\psi(M) + R^{(r+1)}Q) \to R^{(r+1)}/\psi(M) + R^{(r+1)}Q \to R^{(r)}/\phi(M) + R^{(r)}Q \to 0.$$

This implies that $g(R^{(r+1)}/\psi(M),Q) \leqslant g(R^{(r)}/\phi(M),Q) + 1$. Together with Equ. 4 this implies that $b(R^{r+1)}/\psi(M),Q) \leqslant n$, as required. The proof is completed by induction.

The next result can be regarded as a generalisation of the result dual to [B, Theorem 9.1] or [Sw 2, Theorem 12.5]. Both Serre's Theorem and Bass's Cancellation Theorem are easy consequences of it. If $a \in R$ then $\tilde{a}$ will denote the endomorphism of R given by left multiplication by a.

THEOREM 3.4 *Let M be a finitely generated, projective right module over a weakly ideal invariant, right Noetherian ring R, with $\mathrm{Kdim}\, R = n < \infty$ and $\hat{g}(M,P) \geqslant n+1$ for all prime ideals P of R. Suppose that $a \in R$ and $\phi \in \mathrm{Hom}(M,R)$ are such that $R = aR + \phi(M)$. Then there exists $\phi_1 \in \mathrm{Hom}(M,R)$ such that*

$$R = (\phi + \tilde{a}\phi_1)(M).$$

Proof Let $P_1, \ldots, P_m$ be the prime ideals of R such that $\mathrm{Kdim}\, R/P_i = n$. Since $\hat{g}(M,P_i) \geqslant 1$ for each i, there exists by Lemma 3.1 some $\psi \in \mathrm{Hom}(M,R)$ such that

$$g(R/(\phi + \tilde{a}\psi)(M), P_i) = 0 \quad \text{for} \quad 1 \leqslant i \leqslant m.$$

By Proposition 3.2, there exist $\theta_1 = \phi + \tilde{a}\psi, \theta_2, \ldots, \theta_{n+1} \in \mathrm{Hom}(M,R)$ such that, if $\theta = (\theta_1, \ldots, \theta_{n+1}) \in \mathrm{Hom}(M, R^{(n+1)})$, then $b(R^{n+1)}/\theta(M)) \leqslant n$. Let η_i be the element of $R^{(n+1)}$ with 1 in the i-th coordinate and zeros elsewhere. Since $R = \theta_1(M) + aR$, certainly :

$$R^{(n+1)} = \theta(M) + \textstyle\sum_2^{n+1} \eta_i R + \eta_1 aR.$$

So, by Remark 3.3 and Theorem 2.6 applied to $R^{(n+1)}/\theta(M)$, there exist $f_i \in R$ such that :

$$R^{(n+1)} = \theta(M) + \textstyle\sum_2^{n+1} (\eta_i + \eta_1 af_i)R.$$

Thus there exist $m \in M$ and $r_i \in R$ such that :

$$\eta_1 = \theta(m) + \textstyle\sum_2^{n+1} (\eta_i + \eta_1 af_i) r_i.$$

Identifying coordinates of this equation gives $0 = \theta_i(m) + r_i$ for $2 \leqslant i \leqslant n+1$ and (for $i = 1$)

$$1 = \theta_1(m) + \textstyle\sum_2^{n+1} af_i r_i = \theta_1(m) - \sum_2^{n+1} af_i\theta_i(m)$$
$$= (\phi + \tilde{a}\psi - \textstyle\sum_2^{n+1} \tilde{a}\tilde{f}_i\theta_i)(m).$$

So set $\phi_1 = \psi - \sum_2^{n+1} \tilde{f}_i\theta_i$.

We can now give the generali ations of Serre's Theorem and the Cancellation Theorem.

COROLLARY 3.5 Let M be a finitely generated, projective right module over a weakly ideal invariant, right Noetherian ring R. Suppose that $\mathrm{Kdim}\, R = n < \infty$ and that $\hat{g}(M,P) \geqslant n+1$ for every prime ideal P of R. Then $M \cong M' \oplus R$ for some module M'.

Proof Take $\phi = 0$ and $a = 1$ in the theorem.

COROLLARY 3.6 Let R and M be as in Corollary 3.5. If $M \oplus R \cong N \oplus R$ for some module N, then $M \cong N$.

Proof This is dual to the proof of [Sw2, Corollary 12.6]. Let $\sigma: M \oplus R \to N \oplus R$ be the given isomorphism. We will modify σ so as to give an isomorphism between M and N. If π is the projection from $N \oplus R$ to R, then $\pi\sigma = (\phi, d)$ for some $\phi \in \mathrm{Hom}(M,R)$ and $d \in R$. By Theorem 3.4, there exists $\phi_1 \in \mathrm{Hom}(M,R)$ such that $R = (\phi + \tilde{d}\phi_1)(M)$. So $(1-d) = (\phi + d\phi_1)(x)$ for some $x \in M$. Let $\alpha, \beta \in \mathrm{Aut}(M \oplus R)$ be defined by

$$\alpha(m \oplus f) = (m+xf) \oplus f \quad \text{and} \quad \beta(m \oplus f) = m \oplus (f + \phi_1(m)).$$

Then
$$\begin{aligned} \pi\sigma\beta\alpha(0 \oplus f) &= \pi\sigma\beta(xf \oplus f) = \pi\sigma(xf \oplus (f + \phi_1(xf)) \\ &= \phi(xf) + df + d\phi_1(xf) \\ &= f. \end{aligned}$$

Thus $\sigma\beta\alpha(0 \oplus f) = \nu(f) \oplus f$ for some $\nu \in \mathrm{Hom}(R,N)$. Let $\gamma \in \mathrm{Aut}(N \oplus R)$ be defined by $\gamma(n \oplus f) = (n - \nu(f)) \oplus f$; so $\gamma\sigma\beta\alpha(0 \oplus f) = 0 \oplus f$. Thus

$$M \cong \frac{M \oplus R}{0 \oplus R} \cong \frac{\gamma\sigma\beta\alpha(M \oplus R)}{\gamma\sigma\beta\alpha(0 \oplus R)} \cong \frac{N \oplus R}{0 \oplus R} \cong N.$$

REFERENCES

[B] H. Bass, K-theory and stable structure, Publ. Math. I.H.E.S. 22 (1964), 5-60.

[EE] D. Eisenbud and E.G. Evans, Jr., Generating modules efficiently : theorems from algebraic K-theory, J. Algebra 27 (1973), 278-305 .

[FH] F.T. Farrell and W.C. Hsiang, A formula for $K_1R_\alpha[t]$, Proc. Symp. Pure Math. 17 (1970), 192-208 .

[F] O. Forster, Über die Anzahl der Erzeugenden eines ideals in einem Noetherschen Ring, Math. Z. 84 (1964), 80-87 .

[GR] R. Gordon and J.C. Robson, Krull dimension, Mem. Amer. Math. Soc., 133 (1973) .

[Q] D. Quillen, Higher algebraic K-theory I, in Algebraic K-theory I : Higher K-theories, Lecture Notes in Math., No 341, Springer-Verlag, Berlin/New York, 1973 .

[RG] R. Rentschler and P. Gabriel, Sur la dimension des anneaux et ensembles ordonnés, C.R. Acad. Sci. Paris Sér. A 265 (1967), 712-715 .

[Sm] P.F. Smith, On the dimension of group rings, Proc. London Math. Soc., 25 (1972), 288-302 ; Corrigendum, ibid, 27 (1973), 766-768 .

[St1] J.T. Stafford, Stable structure of noncommutative Noetherian rings, J. Algebra 47 (1977), 244-267 .

[St2] J.T. Stafford, Stable structure of noncommutative Noetherian rings, II, J. Algebra 52 (1978), 218-235 .

[St3] J.T. Stafford, Generating modules efficiently : algebraic K-theory for noncommutative Noetherian rings, J. Algebra, 69(1981), 312 - 346.

[Sw1] R.G. Swan, The number of generators of a module, Math. Z. 102 (1967); 318-322 .

[Sw2] R.G. Swan, Algebraic K-theory, Lecture Notes in Math., No 76, Springer-Verlag, Berlin/New York, 1968 .

[V] L.N. Vaseršteĭn, On the stabilisation of the general linear group over a ring, Math USSR-Sb. 8 (1969), 383-400 .

[W] R.B. Warfield, Jr., The number of generators of a module over a fully bounded ring, J. Algebra 66 (1980), 425-447 .

Gonville and Caius College,
Cambridge CB2 1TA .

ALGEBRES SIMPLES CENTRALES SUR LES CORPS

ULTRA-PRODUITS DE CORPS p-ADIQUES

par

Faleh TAHA

Le théorème suivant joue un rôle important dans la Théorie Algébrique des Nombres et plus particulièrement dans l'édification de la Théorie du Corps de Classes Global :

(A) *Soit* K *un corps de nombres* (i.e. une extension algébrique de degré fini du corps $\mathbb{Q}$ des rationnels). *Si* A *est une* K-*algèbre simple centrale de dimension finie sur* K, *pour presque tous les complétés* $\mathfrak{p}$-*adiques* $\hat{K}_{\mathfrak{p}}$ *de* K (i.e. sauf au plus pour un nombre fini d'entre eux) *l'algèbre* $A \otimes_K \hat{K}_{\mathfrak{p}}$ *est isomorphe à une* $\hat{K}_{\mathfrak{p}}$-*algèbre de matrices* $M_n(\hat{K}_{\mathfrak{p}})$.

Ce théorème est souvent démontré par recours aux méthodes de la théorie analytique des nombres (cf.[6]). Mais, dans un travail célèbre (cf.[1]et[2]),Ax et Kochen ont proposé une méthode de démonstration nouvelle et entièrement algébrique s'appuyant sur les ultraproduits. En fait Ax et Kochen abordent le problème de la résolution locale des équations diophantiennes ; mais on peut facilement déduire (A) de leurs résultats en passant par l'intermédiaire de la norme réduite.

Le but du présent travail est d'obtenir une démonstration directe de l'assertion (A) *en détaillant la structure des algèbres ultraproduits.*

Le théorème principal que nous obtenons peut s'énoncer ainsi :

THEOREME - Soit A un anneau de Dedekind de corps de fractions K. On suppose que A a une infinité d'idéaux premiers et qu'il satisfait à la condition suivante :

(*) A est de caractéristique 0 et, pour tout nombre premier p, l'ensemble des idéaux premiers $\mathfrak{p}$ de A tels que $A/\mathfrak{p}$ soit de caractéristique p est fini.

Si B est une K-algèbre simple centrale (de dimension finie) sur K, alors à l'exception peut-être d'un nombre fini d'idéaux premiers $\mathfrak{p}$, l'algèbre $B \otimes_K \hat{K}_{\mathfrak{p}}$, obtenue par extension des scalaires de K au complété $\mathfrak{p}$-adique $\hat{K}_{\mathfrak{p}}$, est décomposée si et seulement si la réduite $B(\mathfrak{p})$ de B en $\mathfrak{p}$ l'est.

Nous reviendrons au §.5 n°1 sur la notion de réduction en $\mathfrak{p}$. Notons seulement que l'utilisation systématique des ultraproduits permet de démontrer rapidement certains résultats délicats de cette réduction modulo $\mathfrak{p}$. Ainsi par utilisation des ultraproduits nous démontrerons aisément que toutes les réduites de l'algèbre simple centrale B sont des algèbres simples centrales sur les corps résiduels, à l'exception peut-être d'un nombre fini d'entre elles.

§.1 ULTRAPRODUITS D'ENSEMBLES.

Nous regroupons dans ce paragraphe un certain nombre de résultats plus ou moins bien connus sur les ultraproduits d'ensembles. D'une façon générale on pourra consulter l'ouvrage de Bell et Slomson [7].

n°1. DEFINITION DES ULTRAPRODUITS.

Soient I un ensemble, $\mathcal{U}$ un ultrafiltre sur I, $(X_i)_{i\in I}$ une famille d'ensembles indexée par I.

LEMME 1 - On définit une relation d'équivalence R sur le produit ensembliste $\prod_{i\in I} X_i$ par : $(x_i) \equiv (y_i) \pmod R$ si et seulement si $\{i\in I \mid x_i = y_i\} \in \mathcal{U}$.

Preuve : Comme I est dans $\mathcal{U}$, on a $\{i\in I \mid x_i = x_i\} \in \mathcal{U}$ pour tout (x_i) de $\prod X_i$; c'est dire que $(x_i) \equiv (x_i)$. La symétrie est évidente; reste la transitivité. Si $(x_i) \equiv (y_i)$ et $(y_i) \equiv (z_i)$, les ensembles $\{i \mid x_i = y_i\}$ et $\{i \mid y_i = z_i\}$ sont des ensembles de $\mathcal{U}$; il en est de même de leur intersection qui est incluse dans $\{i \mid x_i = z_i\}$ ∎

Remarque : On dira souvent que la relation d'équivalence R précédente est la relation d'égalité $\mathcal{U}$-presque partout. Et on écrira $(x_i) = (y_i)$ $(\mathcal{U})$.

DEFINITION 1 - On appelle ultraproduit des X_i sur $\mathcal{U}$ l'ensemble quotient du produit ensembliste $\prod X_i$ par la relation d'équivalence qu'est l'égalité $\mathcal{U}$-presque partout.

Notation : L'ultraproduit des $(X_i)_{i\in I}$ se notera $\prod_{\mathcal{U}} X_i$. Si $\eta \in \prod_{\mathcal{U}} X_i$, on appelle système de représentants de η, un élément $(x_i) \in \prod X_i$ dont la classe est η. Si (x_i) est un élément de $\prod X_i$ nous noterons souvent $\widetilde{(x_i)}$ la classe de cet élément dans l'ultraproduit.

LEMME 2 - Soit $\mathcal{U}$ l'ultrafiltre des parties de I contenant l'élément $j \in I$ (on dit que $\mathcal{U}$ est l'ultrafiltre principal associé à j). On a $(x_i) = (y_i)$ $(\mathcal{U})$ si et seulement si $x_j = y_j$. L'application de $j^{\text{ième}}$-projection $pr_j : \prod X_i \longrightarrow X_j$ passe au quotient et définit une bijection de $\prod_{\mathcal{U}} X_i$ sur X_j.

Preuve : Il est immédiat que pr_j est compatible avec l'égalité $\mathcal{U}$-presque partout qui se réduit à l'égalité en j, puisque $\{j\} \in \mathcal{U}$. Comme pr_j est une surjection, pr_j définit ainsi une surjection p de $\prod_{\mathcal{U}} X_i$ sur X_j. Mais p est aussi une injection : deux éléments (x_i) et (y_i) de $\prod X_i$ sont $\mathcal{U}$-égaux si et seulement si $x_j = y_j$, ce que l'on peut aussi écrire $p(\widetilde{(x_i)}) = p(\widetilde{(y_i)})$.∎

La construction d'un ultraproduit sur un ultrafiltre principal n'apporte donc rien. Aussi dans ce qui suit supposerons-nous toujours que les ultrafiltres considérés sont non principaux. Nous devrons donc nous placer dans une théorie des ensembles satisfaisant au Théorème des Ultrafiltres. Mais la construction des ultraproduits fait intervenir les produits infinis ; nous admettrons donc l'Axiome de Choix.

Si tous les X_i sont égaux à un même ensemble X, on parle plutôt de l'ultrapuissance $X^{\mathcal{U}}$ de X. L'ultrapuissance $X^{\mathcal{U}}$ apparaît comme le quotient de l'ensemble $\mathcal{F}(I,X)$ des applications de I dans X par la relation d'équivalence R' : $f \equiv g \pmod{R'}$ si et seulement si $\{i \in I \mid f(i) = g(i)\} \in \mathcal{U}$.

LEMME 3 - Soit X un ensemble fini. On définit une bijection j de X sur X en associant à $x \in X$ la classe dans $X^{\mathcal{U}}$ de l'application constante $i \longmapsto x$.

<u>Preuve</u> : L'application est injective : si $x,y \in X$ on a $j(x) = j(y)$ si et seulement si les applications $i \longmapsto x$ et $i \longmapsto y$ coïncident sur un ensemble de $\mathfrak{U}$; mais alors $x = y$. L'application j est surjective : soit f une application de I dans X. Comme $I = \bigcup_{x \in X} \{i | f(i) = x\}$ et comme I, ensemble de l'ultrafiltre $\mathfrak{U}$, apparait ainsi comme une réunion finie d'ensembles deux à deux disjoints, on a $\{i \in I | f(i) = x\} \in \mathfrak{U}$ pour un x et un seul. Mais f et $i \longmapsto x$ ont alors la même classe dans l'ultrapuissance. ∎

n°2. PROPRIETES FONCTORIELLES DES ULTRAPRODUITS.

Soient I un ensemble, $\mathfrak{U}$ un ultrafiltre sur I, $(X_i)_{i \in I}$ et $(Y_i)_{i \in I}$ deux familles d'ensembles indexées par I, $(f_i)_{i \in I}$ une famille d'applications $f_i : X_i \longrightarrow Y_i$. A cette famille d'applications correspond une application unique $\prod f_i : \prod_{i \in I} X_i \longrightarrow \prod_{i \in I} Y_i$ telle que $pr_i \circ (\prod f_i)((x_j)) = f_i(x_i)$.

LEMME 4 - <u>Il existe une application f et une seule de $\prod_{\mathfrak{U}} X_i$ dans $\prod_{\mathfrak{U}} Y_i$ telle que pour tout $(x_i) \in \prod X_i$ on ait</u>

$$f(\widetilde{(x_i)}) = \widetilde{(\prod_{i \in I} f_i)((x_i))} = \widetilde{(f_i(x_i))} .$$

<u>Preuve</u> : L'existence et l'unicité de f sont assurées par la compatibilité de l'application $\prod f_i$ avec les relations d'égalité $\mathfrak{U}$-presque partout sur $\prod X_i$ et $\prod Y_i$. En effet si $(x_i) = (y_i)$ $(\mathfrak{U})$, on a $\{i \in I | x_i = y_i\} \in \mathfrak{U}$. Mais cet ensemble est inclus dans $\{i \in I | f_i(x_i) = f_i(y_i)\}$. Il en résulte que $(f_i(x_i)) = (f_i(y_i))$ $(\mathfrak{U})$. D'où le lemme. ∎

DEFINITION 2 - L'application f construite dans le lemme précédent est appelée <u>l'application associée à la famille des applications</u> f_i. Elle sera notée $(f_i)^a$.

PROPOSITION 1 - (Propriétés fonctorielles de l'utraproduit)

i) <u>Si $f_i : X_i \longrightarrow Y_i$ est l'application identique pour tout i, l'application associée à $\prod_{\mathfrak{U}} X_i$ dans $\prod_{\mathfrak{U}} Y_i$ est l'application identique.</u>

ii) <u>Soient $f_i : X_i \longrightarrow Y_i$ et $g_i : Y_i \longrightarrow Z_i$ deux familles d'applications. L'application associée à la famille des $g_i \circ f_i$ s'obtient en composant $(f_i)^a$ et $(g_i)^a$.</u>

<u>Preuve</u> : i) Pour tout $(x_i) \in \Pi X_i$ on a :

$$(f_i)^a(\widetilde{(x_i)}) = \widetilde{(f_i(x_i))} = \widetilde{(x_i)} = id(\widetilde{(x_i)}).$$

ii) Soit $(x_i) \in \Pi X_i$. On a :

$$(g_i \circ f_i)^a(\widetilde{(x_i)}) = \widetilde{(g_i \circ f_i(x_i))} = \widetilde{(g_i(f_i(x_i)))} = (g_i)^a(\widetilde{f_i(x_i)})$$

$$= (g_i)^a \circ (f_i)^a \widetilde{(x_i)} . \blacksquare$$

COROLLAIRE - <u>Si pour tout</u> i <u>l'application</u> $f_i : X_i \longrightarrow Y_i$ <u>est une bijection, l'application</u> $(f_i)^a$ <u>est aussi une bijection</u>.

<u>Preuve</u> : Une f_i est une bijection si et seulement s'il existe une inverse $g_i : Y_i \longrightarrow X_i$. La proposition 1 montre que l'application $(g_i)^a$ associée à la famille des inverses des f_i est l'inverse de $(f_i)^a$.$\blacksquare$

En fait on a un résultat beaucoup plus fort.

LEMME 5 - <u>Soit</u> $f_i : X_i \longrightarrow X_i$ <u>une famille d'applications. Pour que</u> $(f_i)^a$ <u>soit l'application identique de</u> $\prod_{\mathcal{U}} X_i$ <u>dans lui-même il suffit que</u> $f_i = id_{X_i}$ <u>pour</u> $\mathcal{U}$<u>-presque tous les i, i.e. pour les</u> i <u>d'un ensemble de l'ultrafiltre</u>.

<u>Preuve</u> : Comme $\{i \in I | f_i = id_{X_i}\} \in \mathcal{U}$, pour tout $(x_i) \in \Pi X_i$ on a $\{i \in I | f_i(x_i) = x_i\} \in \mathcal{U}$. On en déduit que $(f_i)^a(\widetilde{(x_i)}) = \widetilde{(x_i)}$. $\blacksquare$

PROPOSITION 2 - <u>Soit</u> $f_i : X_i \longrightarrow Y_i$ <u>une famille d'applications</u>.

i) <u>Si les applications</u> f_i <u>sont</u> $\mathcal{U}$<u>-presque toutes des injections</u>, $(f_i)^a$ <u>est une injection</u>.

ii) <u>Si les applications</u> f_i <u>sont</u> $\mathcal{U}$<u>-presque toutes des surjections</u>, $(f_i)^a$ <u>est une surjection</u>.

iii) <u>Si les applications</u> f_i <u>sont</u> $\mathcal{U}$<u>-presque toutes des bijections</u>, $(f_i)^a$ <u>est une bijection</u>.

<u>Preuve</u> : i) Dire que f_i est une injection, c'est aussi dire qu'il existe une application $g_i : Y_i \longrightarrow X_i$ telle que $g_i \circ f_i = id_{X_i}$. Comme l'ensemble des i tels que f_i soit une injection est un ensemble de $\mathcal{U}$, on peut construire une famille g_i d'applications de Y_i dans X_i telle que $\{i \in I | g_i \circ f_i = id_{X_i}\} \in \mathcal{U}$. D'après la proposition 1 et le lemme 5 l'application $(g_i)^a$ est telle que $(g_i)^a \circ (f_i)^a = id_{\prod_{\mathcal{U}} X_i}$. C'est dire que $(f_i)^a$ est injective.

ii) Démonstration analogue en considérant des applications $h_i : Y_i \longrightarrow X_i$ telles que $f_i \circ h_i = id_{Y_i}$.

iii) Il suffit de composer i) et ii).■

COROLLAIRE 1 - Soient $(X_i)_{i\in I}$ et $(Y_i)_{i\in I}$ deux familles d'ensembles telles que $\{i|X_i = Y_i\}\in\mathcal{U}$. On définit une bijection de $\prod_{\mathcal{U}} X_i$ sur $\prod_{\mathcal{U}} Y_i$ en prenant l'application associée à la famille $f_i : X_i \longrightarrow Y_i$ où $f_i = id_{X_i}$ si $i\in\{j\in I|X_j = X_j\}$, f_i est une application quelconque de X_i dans Y_i si $X_i \neq Y_i$.

Preuve : Les f_i sont alors $\mathcal{U}$-presque toutes des bijections.■

COROLLAIRE 2 - Soient $(X_i)_{i\in I}$ une famille d'ensembles, $(R_i)_{i\in I}$ une famille de relations d'équivalence, R_i étant une relation d'équivalence sur X_i. On définit une relation d'équivalence R sur $\prod_{\mathcal{U}} X_i$ par : $\widetilde{(x_i)} \equiv \widetilde{(y_i)}$ mod R si et seulement si $\{i\in I|x_i \equiv y_i \text{ mod } R_i\}\in\mathcal{U}$. Si p_i est la projection canonique de X_i sur X_i/R_i, l'application associée $(p_i)^a$ est une application de $\prod_{\mathcal{U}} X_i$ sur $\prod_{\mathcal{U}}(X_i/R_i)$ qui est compatible avec la relation d'équivalence R et qui, par passage au quotient, définit une bijection de $\prod_{\mathcal{U}} X_i/R$ sur $\prod_{\mathcal{U}}(X_i/R_i)$.

Preuve : Nous laissons au lecteur le soin de vérifier que R est une relation d'équivalence. La proposition 2 ii) montre que l'application associée aux p_i, est une surjection. Reste à vérifier que $(p_i)^a(\widetilde{(x_i)}) = (p_i)^a(\widetilde{(y_i)})$ si et seulement si $\widetilde{(x_i)} \equiv \widetilde{(y_i)}$ mod R. Or $(p_i)^a(\widetilde{(x_i)}) = (p_i)^a(\widetilde{(y_i)})$ si et seulement si $\{i\in I|p_i(x_i) = p_i(y_i)\}\in\mathcal{U}$; et $\{i\in I|p_i(x_i) = p_i(y_i)\} = \{i\in I|x_i \equiv y_i (R_i)\}$.■

COROLLAIRE 3 - Soient $(Y_i)_{i\in I}$ une famille d'ensembles, $(X_i)_{i\in I}$ une famille de sous-ensembles $X_i \subset Y_i$. L'application associée aux injections canoniques $X_i \longrightarrow Y_i$ est une injection de $\prod_{\mathcal{U}} X_i$ dans $\prod_{\mathcal{U}} Y_i$.

Preuve : C'est une conséquence immédiate de la proposition 2 i).■

DEFINITION 3 - Soit (X_i) une famille d'ensembles. On dit qu'un sous-ensemble W de $\prod_{\mathcal{U}} X_i$ est un sous-ultraproduit s'il existe une famille de sous-ensembles $W_i \subset X_i$ telle que W soit l'image de $\prod_{\mathcal{U}} W_i$ par l'application associée à la famille d'injections $W_i \longrightarrow X_i$.

PROPOSITION 3 - Soient $(X_i)_{i\in I}$ et $(Y_i)_{i\in I}$ deux familles d'ensembles. Pour tout $i\in I$ on désigne par $p_i : X_i \times Y_i \longrightarrow X_i$ et $q_i : X_i \times Y_i \longrightarrow Y_i$ la première et la deuxième projection. L'application $\Phi : \prod_{\mathcal{U}}(X_i \times Y_i) \longrightarrow \prod_{\mathcal{U}} X_i \times \prod_{\mathcal{U}} Y_i$

définie par $\Phi((\widetilde{x_i,y_i})) = ((\widetilde{x_i}),(\widetilde{y_i}))$ <u>ou</u> $\Phi = (p_i)^a \times (q_i)^a$ <u>est une bijection.</u>

<u>Preuve</u> : La démonstration est très voisine de celle de la proposition 2, et les détails sont laissés au lecteur.■

n°3. <u>APPLICATIONS D'UN ENSEMBLE DANS UN ULTRAPRODUIT.</u>

Soient $(X_i)_{i\in I}$ une famille d'ensembles, W un ensemble, $(f_i)_{i\in I}$ une famille d'applications de W dans X_i. A cette famille d'applications correspond une application h de W dans $\prod_{i\in I} X_i$ telle que $pr_i(h(x)) = f_i(x)$, pour tout $x\in W$ et par passage au quotient une application f de W dans $\prod_{\mathcal{U}} X_i$; si p est la projection canonique de $\prod_{i\in I} X_i$ sur $\prod_{\mathcal{U}} X_i$, $f = p \circ h$. On dit que <u>l'application</u> f <u>est déduite de la famille</u> (f_i), et on la note $(f_i)^d$.

PROPOSITION 4 - <u>Soient</u> W <u>un ensemble</u>, (X_i) <u>et</u> (Y_i) <u>deux familles d'ensembles</u> (f_i) <u>une famille d'applications</u> $f_i : X_i \longrightarrow Y_i$, (g_i) <u>une famille d'applications</u> $g_i : W \longrightarrow X_i$. <u>On a</u> $(f_i \circ g_i)^d = (f_i)^d \circ (g_i)^d$.

<u>Preuve</u> : La démonstration est immédiate.■

PROPOSITION 5 - <u>Soit</u> (f_i) <u>une famille d'applications</u> $f_i : W \longrightarrow X_i$ <u>qui sont</u> $\mathcal{U}$<u>-presque toutes injectives. L'application déduite</u> $(f_i)^d : W \longrightarrow X_i$ <u>est injective.</u>

<u>Preuve</u> : Soient $x,y\in W$ avec $x \neq y$. Comme les (f_i) sont $\mathcal{U}$-presque toutes injectives, $\{i\in I | f_i(x) \neq f_i(y)\} \in \mathcal{U}$. Il en résulte que $(\widetilde{f_i(x)})$ est différent de $(\widetilde{f_i(y)})$, ce que l'on peut aussi écrire $(f_i)^d(x) \neq (f_i)^d(y)$.■

<u>Remarque</u> : L'énoncé analogue à celui de la proposition 5 mais portant sur des surjections est faux. Prenons par exemple $I = \mathbb{N}$ et $X_i = \mathbb{N}$ pour tout i. Si $\mathcal{U}$ est un ultrafiltre non principal sur $\mathbb{N}$, l'application $f : \mathbb{N} \longrightarrow \mathbb{N}$ déduite des applications identiques $\mathbb{N} \longrightarrow \mathbb{N} = X_i$ est injective mais non surjective : la classe dans $\mathbb{N}^{\mathcal{U}}$ de l'application $h(n) = n$ n'appartient pas à l'image de f. Cette remarque conduit aux modèles non standard de Skolem en Arithmétique.

La proposition 5 admet le corollaire suivant :

COROLLAIRE - <u>Soit</u> $(X_i)_{i\in I}$ <u>une famille d'ensembles</u> $\mathcal{U}$<u>-presque tous infinis. L'ultraproduit</u> $\prod_{\mathcal{U}} X_i$ <u>est infini.</u>

Preuve : Comme nous disposons de l'Axiome de Choix, dire que X_i est infini c'est dire qu'il existe une injection h_i de $\mathbb{N}$ dans X_i. D'après la proposition 5, l'application déduite des h_i est une injection de $\mathbb{N}$ dans l'ultraproduit $\prod_{\mathcal{U}} X_i$.∎

n°4. UN CRITERE DE FINITUDE.

Pour achever les généralités sur les ultraproduits, nous donnerons le résultat suivant qui ne nous sera pas utile directement, mais qui illustre assez bien les méthodes de raisonnement sur les ultraproduits que nous emploierons.

PROPOSITION 6 - *Soit $(X_i)_{i\in I}$ une famille d'ensembles. On suppose que pour tout ultrafiltre non principal $\mathcal{U}$ sur I l'ultraproduit $\prod_{\mathcal{U}} X_i$ est fini. Alors il existe un entier n tel que, à l'exception peut être d'un nombre fini d'entre eux, tous les X_i aient au plus n éléments.*

Preuve : Montrons d'abord que l'ensemble E des $i\in I$ tels que X_i soit infini est fini, en procédant par l'absurde. Si E est infini, il existe au moins un ultrafiltre non principal sur E, d'après le Théorème des ultrafiltres. Un tel ultrafiltre se prolonge en un ultrafiltre $\mathcal{U}$ sur I. Pour cet ultrafiltre $\mathcal{U}$ presque tous les X_i sont infinis ; le corollaire de la proposition 5 du numéro précédent montre que l'ultraproduit $\prod_{\mathcal{U}} X_i$ est infini, ce qui contredit l'hypothèse. Nous pouvons supposer que l'ensemble E est vide : en effet si tel n'est pas le cas nous introduirons la famille des ensembles X'_i tels que $X'_i = X_i$ si $i\notin E$ et $X_i = \{\emptyset\}$ si $i\in E$. Le corollaire 1 de la proposition 2 du n°2 montre que $\prod_{\mathcal{U}} X'_i$ et $\prod_{\mathcal{U}} X_i$ sont en bijection. Dans ce qui suit nous supposerons donc que $E = \emptyset$ c'est-à-dire que tous les X_i sont finis. Pour achever la démonstration de la proposition, nous procéderons à nouveau par l'absurde, en supposant donc que pour tout entier n il existe des i tels que Card $X_i \geq n$. Plus précisément soit $E_n = \{i\in I \mid \text{Card } X_i \geq n\}$. D'après notre hypothèse les E_n sont tous non vides et $E_m \cap E_n \supset E_{\sup(m,n)}$; il existe donc un ultrafiltre non principal $\mathcal{U}$ sur I contenant tous les E_n. La contradiction cherchée s'obtiendra à partir du lemme 6 suivant.∎

LEMME 6 - *Soient $(X_i)_{i\in I}$ une famille d'ensembles finis, $\mathcal{U}$ un ultrafiltre non principal sur I. On suppose que pour tout entier n les X_i aient $\mathcal{U}$-presque tous plus de n éléments. Alors $\prod_{\mathcal{U}} X_i$ est infini.*

Preuve : Raisonnons par l'absurde. Si $\prod_{\mathcal{U}} X_i$ est fini avec p éléments, il existe une bijection ω de $\{1,2,\ldots,p\}$ sur $\prod_{\mathcal{U}} X_i$. Comme $\{i\in I \mid \text{Card } X_i \geq p+1\} \in \mathcal{U}$, on définit une famille d'applications $j_i : \{1,2,\ldots,p+1\} \longrightarrow X_i$ qui sont $\mathcal{U}$-presque toutes des injection de la façon suivante :

$$j_i \text{ est } \begin{cases} \text{une application quelconque de } \{1,2,\ldots,p+1\} \text{ dans } X_i \text{ si } X_i \\ \text{a moins de } p+1 \text{ éléments} \\ \text{une injection de } \{1,2,\ldots,p+1\} \text{ dans } X_i \text{ si } X_i \text{ a plus de} \\ p+1 \text{ éléments.} \end{cases}$$

D'après la proposition 5 du n°3 l'application déduite $(j_i)^d$ est une injection de $\{1,2,\ldots,p+1\}$ dans $\prod_{\mathcal{U}} X_i$, qui est un ensemble à p éléments, ce qui est absurde. ■

§.2 ULTRAPRODUITS D'ANNEAUX ET ALGEBRE LINEAIRE SUR LES CORPS ULTRAPRODUITS.

n°1. ANNEAUX ULTRAPRODUITS.

Soit $(A_i)_{i\in I}$ une famille d'anneaux unitaires. On munit le produit ensembliste $\prod_{i\in I} A_i$ de la structure d'anneau produit. On considère enfin un ultrafiltre $\mathcal{U}$ sur I.

LEMME 7 - Si J est le sous-ensemble de $\prod_{i\in I} A_i$ des éléments (a_i) tels que $\{i\in I \mid a_i = 0\} \in \mathcal{U}$, l'ensemble J est un idéal bilatère de l'anneau produit $\prod_{i\in I} A_i$. La relation d'équivalence définie par J n'est rien d'autre que la relation d'égalité $\mathcal{U}$-presque partout.

Preuve : La démonstration est immédiate. ■

Le lemme 7 revient à dire que l'ensemble ultraproduit $\prod_{\mathcal{U}} A_i$ peut aussi se décrire comme l'ensemble sous-jacent à l'anneau quotient $\prod_{i\in I} A_i/J$. Par suite $\prod_{\mathcal{U}} A_i$ hérite de la structure d'anneau quotient de $\prod_{i\in I} A_i/J$.

DEFINITION 4 - On appelle anneau ultraproduit des A_i l'anneau quotient $\prod_{i\in I} A_i/J$, dont l'ensemble sous-jacent est l'ultraproduit $\prod_{\mathcal{U}} A_i$.

PROPOSITION 7 - Soit $(A_i)_{i\in I}$ une famille d'anneaux unitaires $\mathcal{U}$-presque tous commutatifs. L'ultraproduit $\prod_{\mathcal{U}} A_i$ est commutatif.

Preuve : Soient $\widetilde{(x_i)}$ et $\widetilde{(y_i)} \in \prod_{\mathcal{U}} A_i$. Comme $E = \{i\in I \mid A_i$ est commutatif$\}$ est un ensemble de $\mathcal{U}$, on a $x_i y_i = y_i x_i (\mathcal{U})$.■

PROPOSITION 8 - Soient $(A_i)_{i\in I}$ et $(B_i)_{i\in I}$ deux familles d'anneaux unitaires, $(f_i)_{i\in I}$ une famille d'homomorphismes d'anneaux $f_i : A_i \longrightarrow B_i$. L'application associée $(f_i)^a$ est un homomorphisme d'anneaux de $\prod_{\mathcal{U}} A_i$ dans $\prod_{\mathcal{U}} B_i$.

Preuve : En effet $(f_i)^a$ s'obtient par passage au quotient suivant la $\mathcal{U}$-égalité à partir de l'homomorphisme $\prod_{i\in I} f_i$ du produit $\prod_{i\in I} A_i$ dans le produit $\prod_{i\in I} B_i$.■

PROPOSITION 9 - Soient $(A_i)_{i\in I}$ une famille d'anneaux unitaires et $(J_i)_{i\in I}$ une famille d'idéaux à gauche (resp. à droite, resp. bilatères). L'ensemble J des $\eta \in \prod_{\mathcal{U}} A_i$ qui admettent un système de représentants $(x_i)_{i\in I}$ tel que $\mathcal{U}$-presque partout $x_i \in J_i$ est un idéal à gauche (resp. à droite, resp. bilatère).

Preuve : Nous ne ferons la démonstration que pour les idéaux à gauche, les autres démonstrations étant très voisines. Si ξ et η sont des éléments de l'ensemble J, nous choisirons des systèmes de représentants (x_i) de ξ et (y_i) de η tels que $x_i \in J_i$ et $y_i \in J_i$ pour tout i. Nous aurons $\xi + \eta = \widetilde{(x_i + y_i)} \in J$. De même si $\widetilde{(a_i)} = \alpha$ est un élément de $\prod_{\mathcal{U}} A_i$, $\alpha\xi = \widetilde{(a_i x_i)}$; or $a_i x_i \in J_i$ pour tout i .■

On dira que J est l'idéal ultraproduit des J_i, avec les notations de la proposition 9.

n°2. MODULES ULTRAPRODUITS.

Soient $(A_i)_{i\in I}$ une famille d'anneaux, $(M_i)_{i\in I}$ une famille de modules, M_i étant un A_i-module à gauche. Le produit $\prod_{i\in I} M_i$ est de façon naturelle un $\prod_{i\in I} A_i$-module. Pour un ultrafiltre $\mathcal{U}$ sur I, soit N le sous-ensemble de $\prod_{i\in I} M_i$ des (m_i) tels que $m_i = 0 \ (\mathcal{U})$.

LEMME 8 - L'ensemble N est un sous-$\prod_i A_i$-module de M. La relation d'équivalence associée est l'égalité $\mathcal{U}$-presque partout. Si J est l'idéal bilatère de $\prod_i A_i$ des

(a_i) tels que $a_i = 0$ ($\mathcal{U}$), $JN = 0$.

Preuve : La démonstration est immédiate.∎

Puisque N décrit la relation d'égalité $\mathcal{U}$-presque partout, le module quotient M/N a comme ensemble sous-jacent l'ultraproduit $\prod_{\mathcal{U}} M_i$; on peut donc équiper l'ultraproduit $\prod_{\mathcal{U}} M_i$ de la structure de module M/N. D'autre part ce module est annulé par l'idéal bilatère J de $\prod_i A_i$. C'est dire que l'ultraproduit $\prod_{\mathcal{U}} M_i$ peut être considéré comme un module sur l'anneau $\prod_i A_i/J$, autrement dit sur l'anneau $\prod_{\mathcal{U}} A_i$.

DEFINITION 5 - Le $\prod_{\mathcal{U}} A_i$-module d'ensemble sous-jacent $\prod_{\mathcal{U}} M_i$ décrit dans l'alinéa précédent est appelé le module ultraproduit des M_i.

PROPOSITION 10 - Soient (A_i) une famille d'anneaux unitaires, $(M_i)_{i\in I}$ et $(N_i)_{i\in I}$ deux familles de modules, M_i et N_i étant des A_i-modules. Pour tout i, on considère une application A_i-linéaire f_i de M_i dans N_i. L'application associée $(f_i)^a$ est une application $\prod_{\mathcal{U}} A_i$-linéaire de $\prod_{\mathcal{U}} M_i$ dans $\prod_{\mathcal{U}} N_i$.

Preuve : Comme $\prod f_i : \prod_i M_i \longrightarrow \prod_i N_i$ est une application $\prod A_i$-linéaire, on en déduit que l'application $(f_i)^a$ obtenue à partir de $\prod f_i$ par passage au quotient suivant des sous-modules est $\prod_i A_i$-linéaire. Mais $\prod M_i$ et $\prod N_i$ sont en fait des $\prod_{\mathcal{U}} A_i$-modules. La $\prod A_i$-linéarité de $(f_i)^a$ est donc aussi une $\prod_{\mathcal{U}} A_i$-linéarité. ∎

Remarque : Dans la proposition 10 on peut seulement supposer que $\mathcal{U}$-presque toutes les f_i sont A_i-linéaires.

Soient $(A_i)_{i\in I}$ une famille d'anneaux unitaires et $(M_i)_{i\in I}$ une famille de modules, M_i étant un A_i-module. On considère pour tout $i\in I$ un sous-A_i-module N_i de M_i. Et on introduit le sous-ensemble N de $\prod_{\mathcal{U}} M_i$ des éléments qui ont un système de représentants $(x_i) \in \prod_i M_i$ tel que $x_i \in N_i$ pour tout i.

PROPOSITION 11 - Le sous-ensemble N est un sous- $\prod_{\mathcal{U}} A_i$-module de $\prod_{\mathcal{U}} M_i$. Si p_i est l'homomorphisme canonique de M_i sur M_i/N_i, le module N est le

<u>noyau de l'homomorphisme surjectif</u> $(p_i)^a$.

<u>Preuve</u> : Il est immédiat de vérifier que N est un sous-$\prod_{\mathcal{U}} A_i$-module de $\prod_{\mathcal{U}} M_i$. Quant à la deuxième partie de la proposition elle résulte aussitôt du Corollaire 2 de la Proposition 2 du n°2 du §.1 : la relation d'équivalence définie par N est en fait la relation d'équivalence associée par ultraproduit à la famille des relations définies par les N_i, puisque $\widetilde{(x_i)} \in N$ si et seulement si $x_i \in N_i$ pour $\mathcal{U}$-presque tous les i.

<u>Remarque</u> : On dira souvent que N <u>est le sous-module ultraproduit des</u> N_i, terminologie à rapprocher de celle des idéaux ultraproduits.

Il sera commode de reformuler la proposition précédente en termes de suites exactes :

PROPOSITION 12 - <u>Soient</u> $(A_i)_{i\in I}$ <u>une famille d'anneaux</u>, $(M_i)_{i\in I}$, $(N_i)_{i\in I}$ <u>et</u> $(P_i)_{i\in I}$ <u>trois familles de modules sur les</u> A_i. <u>On suppose que pour tout</u> i <u>les modules</u> M_i, N_i <u>et</u> P_i <u>entrent dans une suite exacte de</u> A_i<u>-modules</u> :

$$M_i \xrightarrow{f_i} N_i \xrightarrow{g_i} P_i .$$

<u>La suite de</u> A_i<u>-modules</u>

$$\prod_{\mathcal{U}} M_i \xrightarrow{(f_i)^a} \prod_{\mathcal{U}} N_i \xrightarrow{(g_i)^a} \prod_{\mathcal{U}} P_i$$ <u>est exacte.</u>

(Autrement dit le passage à l'ultraproduit est un foncteur exact).

<u>Preuve</u> : Comme $g_i \circ f_i = 0$ pour tout i et comme $(g_i)^a \circ (f_i)^a = (g_i \circ f_i)^a$, l'image de $(f_i)^a$ est contenue dans le noyau de $(g_i)^a$. Soit maintenant (x_i) un élément du noyau de $(g_i)^a$; on a $g_i(x_i) = 0$ pour $\mathcal{U}$-presque tous les i. Plus précisément soit E l'ensemble des i tels que $g_i(x_i) = 0$. Par exactitude des suites $M_i \longrightarrow N_i \longrightarrow P_i$, on peut construire pour tout $i \in E$ un $t_i \in M_i$ tel que $x_i = f_i(t_i)$. Complétons la définition de la famille t_i en prenant $t_j = 0$ pour $j \notin E$. On a alors $f_i(t_i) = x_i$ pour $\mathcal{U}$-presque tous les i. C'est dire que $(x_i) = (f_i)^a((t_i))$. □

PROPOSITION 13 - <u>Soient</u> $(A_i)_{i\in I}$ <u>une famille d'anneaux</u>, $(S_i)_{i\in I}$ <u>une famille de modules</u>, S_i <u>étant un</u> A_i<u>-module pour tout</u> $i\in I$. <u>On suppose que pour presque tous les</u> i, <u>au sens de l'ultrafiltre</u> $\mathcal{U}$, <u>les</u> S_i <u>sont des</u> A_i<u>-modules simples.</u>

<u>Alors</u> $\prod_{\mathcal{U}} S_i$ <u>est un</u> $\prod_{\mathcal{U}} A_i$<u>-module simple</u>.

<u>Preuve</u> : On sait qu'un module T sur un anneau B est simple si pour tout $t \neq 0$ de T le sous-module Bt de T est égal à T. C'est ce critère de simplicité que nous allons utiliser. Soit $s = \widetilde{(s_i)}$ un élément non nul de l'ultraproduit $\prod_{\mathcal{U}} S_i$. Considérons un $t = \widetilde{(t_i)}$ quelconque dans cet ultraproduit ; il s'agit de montrer qu'il existe $a = \widetilde{(a_i)}$ dans $\prod_{\mathcal{U}} A_i$ tel que $t = as$. Comme s est non nul, s_i est $\neq 0$ pour $\mathcal{U}$-presque tous les i ; on peut donc trouver pour $\mathcal{U}$-presque tous les i, un $a_i \in A_i$ tel que $t_i = a_i s_i$. Prolongeons cette famille de a_i en une famille $(a_j)_{j \in I}$ en prenant $a_i = 1$ pour les i en lesquels $s_i = 0$. On a encore $t_i = a_i s_i$ pour $\mathcal{U}$-presque tous les i, soit $t = as$ dans $\prod_{\mathcal{U}} S_i$.■

COROLLAIRE - <u>Soient</u> $(A_i)_{i \in I}$ <u>une famille d'anneaux</u>, $\mathcal{U}$ <u>un ultrafiltre sur</u> I. <u>Considérons une famille</u> (J_i) <u>d'idéaux à gauche</u>, J_i <u>étant un idéal à gauche de</u> A_i, <u>telle que pour</u> $\mathcal{U}$<u>-presque tous les</u> i <u>l'idéal</u> J_i <u>soit un idéal à gauche maximal de</u> A_i. <u>L'idéal</u> J, <u>ultraproduit des</u> J_i, <u>est un idéal à gauche maximal de</u> $\prod_{\mathcal{U}} A_i$.

<u>Preuve</u> : Pour $\mathcal{U}$-presque tout i, le A_i-module A_i/J_i est simple . La proposition précédente montre que l'ultraproduit des A_i/J_i est un $\prod_{\mathcal{U}} A_i$-module simple. Mais d'après la proposition 11, l'ultraproduit des A_i/J_i est isomorphe à $\prod_{\mathcal{U}} A_i/J$.■

n°3. <u>CORPS ULTRAPRODUIT</u>.

Soit $(K_i)_{i \in I}$ une famille d'anneaux unitaires.

PROPOSITION 14 - <u>Soit</u> $\mathcal{U}$ <u>un ultrafiltre sur</u> I <u>tel que pour</u> $\mathcal{U}$<u>-presque tous les</u> i <u>l'anneau</u> K_i <u>soit un corps</u>. <u>L'ultraproduit</u> $\prod_{\mathcal{U}} K_i$ <u>est un corps</u>.

<u>Preuve</u> : Soit E l'ensemble des i tels que K_i soit un corps ; c'est un ensemble de $\mathcal{U}$. Soit $x = \widetilde{(x_i)}$ un élément non nul de l'ultraproduit $\prod_{\mathcal{U}} K_i$: dire que x est non nul c'est dire que $\{i \mid x_i = 0\}$ n'appartient pas à $\mathcal{U}$; comme $\mathcal{U}$ est un ultrafiltre, l'ensemble complémentaire $F = \{i \mid x_i \neq 0\}$ est dans $\mathcal{U}$. Posons $y_i = x_i^{-1}$ pour $i \in E \cap F$, $y_i = 1$ pour $i \notin E \cap F$. Pour $i \in E \cap F \in \mathcal{U}$ on a $y_i x_i = x_i y_i = 1$, soit $y_i x_i = x_i y_i = 1$ pour $\mathcal{U}$-presque tous les i. C'est dire que $\widetilde{(y_i)}$ est l'inverse de x dans l'ultraproduit $\prod_{\mathcal{U}} K_i$.■

DEFINITION 6 - L'ensemble $\prod_{\mathcal{U}} K_i$ muni de la structure de corps décrite dans la proposition 14 est appelé <u>le corps ultraproduit des</u> K_i.

PROPOSITION 15 - <u>Soient</u> $(A_i)_{i\in I}$ <u>une famille d'anneaux unitaires,</u> $\mathcal{U}$ <u>un ultrafiltre sur</u> I. <u>On suppose que pour</u> $\mathcal{U}$<u>-presque tous les</u> i,<u>l'anneau</u> A_i <u>est un anneau commutatif intègre. Alors l'ultra-produit</u> $\prod_{\mathcal{U}} A_i$ <u>est un anneau commutatif intègre. Si on note</u> K_i <u>le corps des fractions de</u> A_i <u>pour</u> $i\in\{j|A_j$ <u>est commutatif est intègre</u>$\}$ <u>et</u> $K_i = A_i$ <u>pour</u> $i\notin\{j|A_j$ <u>est commutatif intègre</u>$\}$ <u>et si</u> j_i <u>est l'injection canonique de</u> A_i <u>dans</u> K_i, <u>alors l'application</u> $(j_i)^a$ <u>est une injection de</u> $\prod_{\mathcal{U}} A_i$ <u>dans le corps</u> $\prod_{\mathcal{U}} K_i$ <u>qui réalise</u> $\prod_{\mathcal{U}} K_i$ <u>comme le corps des fractions de</u> $\prod_{\mathcal{U}} A_i$.

<u>Preuve</u> : D'après la proposition 7 du n°1 l'anneau $\prod_{\mathcal{U}} A_i$ est commutatif. Soit $a = \widetilde{(a_i)}$ un élément non nul de $\prod_{\mathcal{U}} A_i$ et soit $b = \widetilde{(b_i)}$ tel que $\widetilde{(a_i b_i)} = 0$. Comme $\mathcal{U}$ est un ultrafiltre, on voit, comme dans la démonstration de la proposition 15, que $\{i|a_i \neq 0\}$ est un ensemble E de l'ultrafiltre $\mathcal{U}$. D'autre part on suppose que $F = \{i|b_i a_i = 0\}$ et $G = \{i|A_i$ est commutatif intègre$\}$ sont aussi des ensembles de $\mathcal{U}$. Il en résulte que pour $i\in E\cap F\cap G$ on a $b_i = 0$; autrement dit b_i est nul pour $\mathcal{U}$-presque tous les i, et $\widetilde{(b_i)} = 0$ dans $\prod_{\mathcal{U}} A_i$. On a ainsi établi que $\prod_{\mathcal{U}} A_i$ est un anneau intègre.

Comme application associée à une famille d'injections, l'application $(j_i)^a$ est une injection. Reste à vérifier que tout élément de $\prod_{\mathcal{U}} K_i$ peut se représenter comme une fraction dont les deux termes sont dans $\prod_{\mathcal{U}} A_i$. Soit $x \in \prod_{\mathcal{U}} K_i$ et considérons un système de représentants $\widetilde{(x_i)}$ de x. Pour tout $i\in G = \{j|A_j$ est commutatif intègre$\}$ on peut trouver a_i et b_i dans A_i tels que $x_i = a_i/b_i$. Posons $y_i = a_i$ pour $i\in G$, $y_i = x_i$ pour $i\notin G$, $z_i = b_i$ pour $i\in G$ et $z_i = 1$ pour $i\notin G$; on a $y_i = x_i z_i$ pour tout i soit $\widetilde{(y_i)} = x\widetilde{(z_i)}$ dans $\prod_{\mathcal{U}} K_i$. ∎

PROPOSITION 16 - <u>Soient</u> $(K_i)_{i\in I}$ <u>une famille d'anneaux commutatifs unitaires,</u> $\mathcal{U}$ <u>un ultrafiltre sur</u> I. <u>On suppose que</u> $\mathcal{U}$<u>-presque tous les</u> K_i <u>sont des corps. S'il existe un nombre premier</u> p <u>tel que</u> $\mathcal{U}$<u>-presque tous les</u> K_i <u>soient de caractéristique</u> p, $\prod_{\mathcal{U}} K_i$ <u>est de caractéristique</u> p. <u>Sinon</u> $\prod_{\mathcal{U}} K_i$ <u>est de caractéristique</u> 0.

Preuve : Supposons que $\mathfrak{U}$-presque tous les K_i soient de caractéristique p. On a alors $p.1 = 0$ dans $\mathfrak{U}$-presque tous les K_i ; c'est dire que $p.1 = 0$ dans $\prod_{\mathfrak{U}} K_i$. Réciproquement, supposons que $\prod_{\mathfrak{U}} K_i$ soit de caractéristique $p \neq 0$. Comme $p.1 = 0$ dans $\prod_{\mathfrak{U}} K_i$ on a, par définition de l'ultraproduit, $\{i | p.1 = 0 \text{ dans } K_i\} \in \mathfrak{U}$.∎

Remarque : Dans ce numéro nous venons d'utiliser pleinement pour la première fois le fait que $\mathfrak{U}$ est un ultrafiltre et non pas seulement un filtre.

n°4. ALGEBRE LINEAIRE SUR LES CORPS ULTRAPRODUITS.

Soient $(K_i)_{i\in I}$ une famille de corps commutatifs, $(V_i)_{i\in I}$ une famille d'espaces vectoriels, V_i étant un K_i-espace vectoriel. On considère un ultrafiltre $\mathfrak{U}$ sur I. Les constructions du n°2 montre que $\prod_{\mathfrak{U}} V_i$ est de façon naturelle un $\prod_{\mathfrak{U}} K_i$-espace vectoriel.

PROPOSITION 17 - Les notations étant celles de l'alinéa précédent, soit $(a_1,\ldots,a_n)$ une famille finie d'éléments de $\prod_{\mathfrak{U}} V_i$. Pour tout a_j nous choisirons un système de représentants $a_j(i)$ de a_j.

a) La famille des $(a_j)_{j=1,\ldots,n}$ est libre dans $\prod_{\mathfrak{U}} V_i$ si et seulement si pour $\mathfrak{U}$-presque tous les i la famille $(a_j(i))$ est libre dans V_i.

b) La famille des $(a_j)_{j=1,\ldots,n}$ est un système de générateurs dans $\prod_{\mathfrak{U}} V_i$ si et seulement si pour $\mathfrak{U}$-presque tous les i la famille des $a_j(i)$ est un système de générateurs de V_i.

c) La famille des $(a_j)_{j=1,\ldots,n}$ est une base de $\prod_{\mathfrak{U}} V_i$ si et seulement si pour $\mathfrak{U}$-presque tous les i la famille des $a_j(i)$ est une base de l'espace vectoriel V_i.

Preuve : c) s'obtient évidemment en combinant a) et b). Démontrons d'abord a). Si pour presque tous les i la famille des $a_j(i)$ est libre dans V_i et si $\sum \lambda_j a_j = 0$ dans $\prod_{\mathfrak{U}} V_i$, on a, en introduisant pour tout j un système de représentants $(\lambda_j(i))$ de λ_j,

$$\sum \lambda_j(i)\, a_j(i) = 0 \text{ pour } \mathfrak{U}\text{-presque tous les } i.$$

Comme les $(a_j(i))$ sont $\mathfrak{U}$-presque toutes des familles libres, on a $\lambda_j(i) = 0$

pour tout j dans un ensemble de i appartenant à l'ultrafiltre $\mathcal{U}$. La condition énoncée dans a) est donc suffisante. Il est plus délicat de montrer qu'elle est nécessaire : nous procéderons par l'absurde en considérant une famille (a_j) qui est libre dans $\prod_{\mathcal{U}} V_i$ et telle que $(a_j(i))$ soit liée pour les i d'un $F \in \mathcal{U}$. Pour tout $i \in F$ on peut donc trouver des $\lambda_j(i) \in K_i$ non tous nuls tels que $\Sigma \lambda_j(i)\, a_j(i) = 0$. Posons $\lambda_j(i) = 0$ pour tout $i \notin F$ et tout j ; et soit λ_j l'élément de $\prod_{\mathcal{U}} K_i$ associé à la famille des $\lambda_j(i)$ ainsi complétée. Si pour tout j, on pose $F_j = \{i \in I | \lambda_j(i) \neq 0\}$, on a $F = F_1 \cup \ldots \cup F_n$; comme $\mathcal{U}$ est un ultrafiltre, l'un au moins des F_j est dans $\mathcal{U}$, ce qui revient à dire que l'un au moins des λ_j est non nul. Mais en passant à l'ultraproduit à partir des identités: $\Sigma \lambda_j(i)\, a_j(i) = 0$, valables pour tout i (on a fait ce qu'il faut pour, en choisissant $\lambda_j(i) = 0$ pour $i \notin F$), on obtient $\Sigma \lambda_j\, a_j = 0$; ce qui est contradictoire puisque (a_j) est une famille libre et que les λ_j ne sont pas tous nuls. Pour démontrer b) nous noterons que la partie suffisante de l'assertion est immédiate. Pour démontrer que la condition est nécessaire nous procéderons par l'absurde en utilisant le critère suivant : "Si $v_1, \ldots, v_r$ sont des éléments de l'espace vectoriel V qui n'engendrent pas V, il existe une forme linéaire h sur V qui est non nulle et vérifie $h(v_i) = 0$ pour tout i". Soit (a_j) une famille d'éléments de $\prod_{\mathcal{U}} V_i$ telle que pour les i d'un ensemble $E \in \mathcal{U}$ la famille $(a_j(i))$ ne soit pas un système de générateurs de V_i. Pour $i \in E$ on peut trouver un $x_i \neq 0$ dans V_i et une forme K_i-linéaire h_i sur V_i telle que $h_i(x_i) \neq 0$ et $h_i(a_j(i)) = 0$ pour tout j. Pour $i \notin E$ posons $x_i = 0$ et $h_i = 0$. L'application $(h_i)^a$ associée à la famille des h_i est une application $\prod_{\mathcal{U}} K_i$-linéaire de $\prod_{\mathcal{U}} V_i$ dans $\prod_{\mathcal{U}} K_i$: comme $h_i(x_i) \neq 0$ pour $\mathcal{U}$-presque tous les i, $(h_i)^a((x_i)) \neq 0$, ce qui entraîne que $(h_i)^a$ n'est pas identiquement nulle. D'autre part $h_i(a_j(i)) = 0$ pour tout i et tout j ; ce qui entraîne $(h_i)^a\, (a_j) = 0$ pour tout j. C'est dire que la famille des (a_j) n'engendre pas $\prod_{\mathcal{U}} V_i$.■

COROLLAIRE - Soit V un $\prod_{\mathcal{U}} K_i$-espace vectoriel ultraproduit des K_i-espaces vectoriels V_i. S'il existe un entier n tel que $\{i \in I | \dim_{K_i} V_i \leq n\} \in \mathcal{U}$, V est de dimension finie $\leq n$. Plus précisément, il existe un entier m unique tel que les V_i sont $\mathcal{U}$-presque tous de dimension m ; et on a alors $\dim \prod_{\mathcal{U}} V_i = m$.

<u>Preuve</u> : Soit E l'ensemble des $i\in I$ tels que $\dim_{K_i} V_i \leq n$. Pour tout $i\in E$ choisissons un système de générateurs à n éléments de V_i, soit $(a_1(i),\dots,a_n(i))$. Posons $a_j(i) = 0$ pour tout $i\notin E$ et tout j compris entre 1 et n, a_j = "classe de $(a_j(i))$ dans $\prod_{\mathcal{U}} V_i$". Presque tous les systèmes $(a_j(i))$ engendrent V_i ; l'assertion b) de la proposition précédente montre que (a_j) est un système de générateurs de $\prod_{\mathcal{U}} V_i$. C'est dire que $\dim \prod_{\mathcal{U}} V_i \leq n$.
Posons $E_p = \{i\in I \mid \dim_{K_i} V_i = p\}$. L'ensemble E de $\mathcal{U}$ est réunion de $E_1,\dots,E_n$. Ces E_p sont des ensembles deux à deux disjoints ; comme $\mathcal{U}$ est un ultrafiltre, il existe un p unique tel que $E_p\in\mathcal{U}$. Reprenons alors la démonstration de l'alinéa précédent en remplaçant E par E_p et le système de générateurs $(a_j(i))$ par une base $(e_j(i))$ de V_i pour $i\in E_p$. Des $e_j(i)$ on déduit une base e_j de $\prod_{\mathcal{U}} V_i$. D'où le corollaire.■

PROPOSITION 18 - <u>Soient $(V_i)_{i\in I}$ une famille de K_i-espaces vectoriels. On suppose que pour tout ultrafiltre (non principal) $\mathcal{U}$ l'ultraproduit $\prod_{\mathcal{U}} V_i$ est un espace vectoriel de dimension finie. Alors il existe un entier n tel que $\dim_{K_i} V_i \leq n$ sauf peut-être pour un nombre fini de i.</u>

<u>Preuve</u> : La démonstration est dans ses grandes lignes voisine de celle de la proposition 6 du n°4 du §.1. Montrons d'abord que l'ensemble $E = \{i\in I \mid \dim_{K_i} V_i \text{ est infinie}\}$ est fini. Si E était infini, il serait élément d'un ultrafiltre non principal $\mathcal{U}$ sur I. Définissons pour tout $i\in E$ une famille libre de n éléments $(a_j(i))$ de V_i et posons $a_j(i) = 0$ pour $i\notin E$. Si $a_j = \widetilde{(a_j(i))} \in \prod_{\mathcal{U}} V_i$, la proposition 17 a) montre que la famille des a_j est une famille libre de n éléments dans $\prod_{\mathcal{U}} V_i$. Comme n est arbitraire, $\prod_{\mathcal{U}} V_i$ est de dimension infinie, contrairement à l'hypothèse. Posons pour tout entier n, $F_n = \{i\in I \mid i\notin E \text{ et } \dim_{K_i} V_i \geq n\}$. Nous allons montrer que les F_n sont vides pour n assez grand, ce qui achèvera la démonstration de la proposition. Supposer que les F_n sont tous non vides, c'est supposer, la suite F_n étant décroissante, qu'il existe un ultrafiltre non principal $\mathcal{U}$ sur I contenant tous les F_n. Comme $F_n\in\mathcal{U}$, on peut reprendre la construction de l'alinéa précédent avec F_n remplaçant E ; cette construction montre que $\dim \prod_{\mathcal{U}} V_i$ est $\geq n$. Comme n est à nouveau arbitraire, $\dim \prod_{\mathcal{U}} V_i$ est infinie, ce qui est encore en contradiction avec l'hypothèse. Donc l'un au moins des F_n est vide.■

La proposition 19 reprend un argument déjà utilisé plus haut mais qu'il est commode de préciser.

PROPOSITION 19 - <u>Soient</u> $(V_i)_{i\in I}$ <u>une famille de</u> K_i<u>-espaces vectoriels telle que</u> $\sup \dim_{K_i} V_i < \infty$, $\mathfrak{U}$ <u>un ultrafiltre sur</u> I. <u>Il existe un entier</u> n <u>et un seul tel que les</u> V_i <u>soient</u> $\mathfrak{U}$<u>-presque tous de dimension</u> n. <u>Et l'espace vectoriel ultraproduit</u> $\prod_{\mathfrak{U}} V_i$ <u>est de dimension</u> n.

<u>Preuve</u> : Soit $r = \sup \dim_{K_i} V_i$. Posons $F_m = \{i\in I \mid \dim_{K_i} V_i = m\}$. Par hypothèse I, ensemble de l'ultrafiltre $\mathfrak{U}$, est réunion des F_m pour m compris entre 1 et r. Ces F_m sont deux à deux disjoints ; il existe donc un m unique tel que F_m appartienne à $\mathfrak{U}$. La dernière partie de la proposition est une conséquence du corollaire de la proposition 17.∎

PROPOSITION 20 - <u>Soient</u> $(V_i)_{i\in I}$ <u>et</u> $(W_i)_{i\in I}$ <u>deux familles d'espaces vectoriels,</u> V_i <u>et</u> W_i <u>étant des</u> K_i<u>-espaces vectoriels. On suppose que</u> $\prod_{\mathfrak{U}} V_i$ <u>et</u> $\prod_{\mathfrak{U}} W_i$ <u>sont de dimension finie. Pour toute application</u> $\prod_{\mathfrak{U}} K_i$<u>-linéaire</u> f <u>de</u> $\prod_{\mathfrak{U}} V_i$ <u>dans</u> $\prod_{\mathfrak{U}} W_i$, <u>il existe au moins une famille</u> (f_i) <u>d'applications</u> K_i<u>-linéaires</u> $f_i : V_i \longrightarrow W_i$ <u>telle que</u> $f = (f_i)^a$. <u>L'application</u> f <u>est injective (resp. surjective, resp. bijective) si et seulement si</u> f_i <u>est</u> $\mathfrak{U}$<u>-presque partout injective (resp. surjective, resp. bijective)</u>.

<u>Preuve</u> : Soient $(e_j)_{j=1,\dots,m}$ et $(\varepsilon_k)_{k=1,\dots,n}$ des bases de $\prod_{\mathfrak{U}} V_i$ et $\prod_{\mathfrak{U}} W_i$ respectivement. Introduisons des systèmes de représentants $(e_j(i))$ et $(\varepsilon_k(i))$. D'après la proposition 17 on peut trouver un ensemble $E\in\mathfrak{U}$ tel que pour tout $i\in E$ la famille des $(e_j(i))$ soit une base de V_i et la famille des $(\varepsilon_k(i))$ une base de W_i. L'application f est entièrement déterminée par sa matrice dans les bases (e_j) et (ε_k) :

$$f(e_j) = \sum u_j^k \varepsilon_k .$$

Les u_j^k sont des éléments de $\prod_{\mathfrak{U}} K_i$; nous choisirons des systèmes de représentants $u_j^k(i)\in K_i$. Et nous définirons pour $i\in E$ l'application linéaire f_i de V_i dans W_i par :

$$f_i(e_j(i)) = \sum u_j^k(i)\, \varepsilon_k(i).$$

Pour $i\notin E$ nous poserons $f_i = 0$. Il est alors immédiat de vérifier que $f(e_j) = (f_i)^a(e_j)$ pour tout j, soit $f = (f_i)^a$. Pour que f soit injective, il faut et il suffit que la famille des $f(e_j)$ soit libre. Comme $f(e_j) = (f_i)^a(e_j) = (\widetilde{f_i(e_j(i))})$, la proposition 17 montre que $(f(e_j))$ est libre si et seulement si les $(f_i(e_j(i)))$ sont $\mathfrak{U}$-presque toutes libres. D'où le critère d'injectivité pour f. On démontre de la même façon les critères de surjectivité et de bijectivité en utilisant les assertions b) et c) de la proposition 17.∎

§.3 ULTRAPRODUITS D'ALGEBRES.

Nous abordons dans ce paragraphe des résultats qui sont essentiels pour la démonstration du théorème cité dans l'introduction.

n°1. ALGEBRES ULTRAPRODUITS.

Soient $(K_i)_{i\in I}$ une famille de corps commutatifs, $(A_i)_{i\in I}$ une famille d'algèbres, A_i étant une K_i-algèbre unitaire. D'après les résultats du §.2 l'ultraproduit $\prod_{\mathfrak{U}} A_i$ est de façon naturelle un anneau unitaire et un $\prod_{\mathfrak{U}} K_i$-espace vectoriel.

LEMME 9 - *L'ultraproduit est une* $\prod_{\mathfrak{U}} K_i$-*algèbre pour les structures d'anneau et d'espace vectoriel décrites ci-dessus.*

Preuve : Il s'agit de montrer que dans $\prod_{\mathfrak{U}} A_i$ on a $a(\lambda b) = \lambda(ab)$ pour tous $a,b \in \prod_{\mathfrak{U}} A_i$ et tout $\lambda \in \prod_{\mathfrak{U}} K_i$. En prenant des systèmes de représentants on vérifie aussitôt cette identité.∎

DEFINITION 7 - L'ultraproduit des A_i muni de la structure précédente d'algèbre est appelée *l'algèbre ultraproduit des* A_i.

La proposition suivante est essentielle pour la suite.

PROPOSITION 21 - *Soient* $(K_i)_{i\in I}$ *une famille de corps commutatifs,* $\mathfrak{U}$ *un ultrafiltre sur* I, K *le corps ultraproduit des* K_i, A *une* K-*algèbre de dimension finie* n. *Il existe une famille de* K_i-*algèbres* A_i, $\mathfrak{U}$-*presque toutes de dimension* n, *telle que* A *soit* K-*isomorphe à l'algèbre ultraproduit des* A_i.

Preuve : Soit $(e_k)_{k=1,\dots,n}$ une K-base de l'algèbre A. La structure d'algèbre de A est entièrement caractérisée par les constantes de structure (c^t_{rs}) définie par :

$$e_r . e_s = \sum c^t_{rs} e_t \quad , \quad c^t_{rs} \in K.$$

Ces constantes sont liées par les identités traduisant l'associativité :

$$(**) \qquad \sum c^w_{rs} c^v_{wt} = \sum c^v_{rw} c^w_{st} \quad ,$$

les sommations portant sur tous les couples d'indices apparaissant à la fois en bas et en haut. Remarquons que les identités (**) sont en nombre fini. Puisque des c^t_{rs} sont des éléments de l'ultraproduit K des K_i, nous pouvons choisir des systèmes de représentants $(c^t_{rs}(i))$. Les identités (**), qui sont, rappelons-le, en nombre fini, entraînent la validité $\mathfrak{U}$-presque partout des identités

$$(**i) \qquad \sum c^w_{rs}(i)\, c^v_{wt}(i) = \sum c^v_{rw}(i)\, c^w_{st}(i).$$

Nous désignerons par E l'ensemble de l'ultrafiltre $\mathfrak{U}$ des i où toutes les (**i) sont vérifiées. Pour $i \in E$ soit A_i l'espace vectoriel sur K_i des combinaisons linéaires à coefficients dans K_i des entiers 1,2,...,n. On fait de A_i une K_i-algèbre en adoptant la table de multiplication

$$r.s = \sum c^t_{rs}(i)\, t.$$

Pour $i \notin E$ nous poserons $A_i = K_i$. Soit alors A^* la $K = (\prod_{\mathfrak{U}} K_i)$-algèbre ultraproduit des A_i. D'après la proposition 17 du n°4 du §.2 on obtient une une K-base de l'algèbre A^* en posant : $\varepsilon_k = \widetilde{(x_k)}$ où $x_k = \sum \lambda_r r$ est l'élément de A_k combinaison linéaire des r à coefficients λ_r nuls pour $r \neq k$, $\lambda_k = 1$ (ce que l'on peut écrire $x_k = k$). On définit une application linéaire F de A^* dans A en posant $F(\varepsilon_k) = e_k$ pour tout k ; comme cette application transforme une base en une base, elle est bijective. En passant à l'ultraproduit et en notant que $r.s - \sum c^t_{rs}(i)\, t = 0$ pour $\mathfrak{U}$-presque tous les i, on trouve $\varepsilon_r . \varepsilon_s = \sum c^t_{rs} \varepsilon_t$. Il en résulte que F est aussi un isomorphisme de K-algèbres.■

COROLLAIRE - _Les notations étant celles de la proposition_ 21, _si la K-algèbre_ A _est_ K-_isomorphe à une algèbre ultraproduit_ $\prod_{\mathfrak{U}} B_i$, _pour_ $\mathfrak{U}$-_presque tout_ i _les algèbres_ A_i _et_ B_i _sont_ K_i-_isomorphes_ (unicité de la représentation d'une K-algèbre de dimension finie comme ultraproduit).

Ce corollaire est cité ici, bien que sa démonstration utilise la proposition 24, pour ne pas rompre le cours de l'exposé.

<u>Preuve du corollaire</u> : Etant toutes deux isomorphes à l'algèbre A, les K-algèbres $\prod_{\mathcal{U}} A_i$ et $\prod_{\mathcal{U}} B_i$ sont K-isomorphes. Un tel isomorphisme $F : \prod_{\mathcal{U}} A_i \longrightarrow \prod_{\mathcal{U}} B_i$ est, d'après la proposition 20, de la forme $(f_i)^a$, où les f_i sont des applications K_i-linéaires de A_i dans B_i. La proposition 24 montrera qu'en fait $\mathcal{U}$-presque toutes les f_i sont des homomorphismes d'algèbres.∎

PROPOSITION 22 - <u>Soient</u> $(K_i)_{i\in I}$ <u>une famille de corps commutatifs</u>, $(A_i)_{i\in I}$ <u>une famille d'algèbres</u> (A_i <u>étant une</u> K_i<u>-algèbre</u>) <u>telle que</u> sup dim $A_i < \infty$, $\mathcal{U}$ <u>un ultrafiltre sur</u> I, K <u>le corps ultraproduit des</u> K_i, A <u>l'algèbre ultraproduit des</u> A_i. <u>Tout idéal à gauche</u> (<u>resp. à droite, resp. bilatère</u>) <u>de</u> A <u>est un idéal ultraproduit</u>.

<u>Preuve</u> : Compte-tenu de la définition des idéaux ultraproduits (cf. Proposition 9, n°1, §.2) il s'agit de montrer qu'il existe une famille d'idéaux (J_i), J_i idéal de A_i, telle que $x \in J$ si et seulement s'il existe un système de représentants (x_i) de x tel $x_i \in J_i$ pour tout i. Nous traiterons seulement le cas des idéaux à gauche, les deux autres cas se traitant de façon similaire. Nous introduirons une K-base (e_k) de A et une K-base (ε_j) de J, considéré comme sous-espace vectoriel de A. Choisissons des systèmes de représentants $(e_k(i))$ de e_k, $(\varepsilon_j(i))$ de ε_j. Dire que J est un idéal à gauche de A c'est dire que la condition suivante est satisfaite :

(*) pour tout j et tout k, l'élément $e_k.\varepsilon_j$ est combinaison linéaire à coefficients dans K des ε_r.

En passant aux systèmes de représentants et en notant que les conditions (*) sont en nombre fini on voit que pour $\mathcal{U}$-presque tous les i les conditions

(*i) pour tout j et k, l'élément $e_k(i).\,\varepsilon_j(i)$ est combinaison linéaire à coefficients dans K_i des $\varepsilon_r(i)$,

sont satisfaites. Soit E l'ensemble des $i \in I$ tels que (*i) soit vérifié. Nous définirons J_i par :

$$J_i = \begin{cases} \text{le sous-espace vectoriel de } A_i \text{ engendré par les } \varepsilon_j(i) \text{ pour } i \in E, \\ 0 \text{ pour } i \notin E. \end{cases}$$

Pour tout i le sous-espace vectoriel J_i de A_i est un idéal à gauche de A_i : c'est évident si $i \notin E$ et, si $i \in E$, c'est une conséquence de la définition de E comme ensemble des i en lesquels (*i) est vérifié. Soit $\widetilde{(x_i)}$ un élément de $\prod_{\mathcal{U}} A_i$; pour que $\widetilde{(x_i)}$ appartienne à J il faut et il suffit que

(x_i) soit combinaison linéaire des (ε_k) à coefficients dans $K = \prod_{\mathcal{U}} K_i$. On peut traduire la condition précédente en disant aussi que pour $\mathcal{U}$-presque tous les i, x_i est combinaison linéaire à coefficients dans K_i des $\varepsilon_k(i)$, autrement dit que pour $\mathcal{U}$-presque tout i on a $x_i \in J_i$. Si (x_i') est définie par $x_i' = x_i$ si $x_i \in J_i$ et $x_i' = 0$ dans le cas contraire, (x_i) et (x_i') sont $\mathcal{U}$-égaux et $\widetilde{(x_i)} = \widetilde{(x_i')}$; d'autre part $x_i' \in J_i$ pour tout i.∎

COROLLAIRE - <u>Les notations étant celles de la proposition 22, l'idéal J est un idéal à gauche maximal si et seulement si $\mathcal{U}$-presque tous les J_i sont des idéaux à gauche maximaux de A_i.</u>

<u>Preuve</u> : La condition est suffisante d'après le Corollaire de la Proposition 13 du n°2, §.2. Soit J un idéal à gauche maximal de A, que l'on représentera d'après la proposition 22 comme ultraproduit d'idéaux (J_i). Si J_i' est pour tout i un idéal contenant J_i, l'idéal J', ultraproduit des J_i', contient J. Comme J est maximal, on a soit $J' = J$, soit $A = J'$. Si $J' = J$ on a $\dim_{K_i} J_i' = \dim_{K_i} J_i$ pour $\mathcal{U}$-presque tous les i, d'où $J_i' = J_i$ pour $\mathcal{U}$-presque tous les i. De même si $J' = A$ on a pour $\mathcal{U}$-presque tout i l'égalité $\dim J_i' = \dim A_i$ qui entraîne $A_i = J_i'$ pour $\mathcal{U}$-presque tous les i.∎

PROPOSITION 23 - <u>Soient $(K_i)_{i \in I}$ une famille de corps commutatifs, $\mathcal{U}$ un ultrafiltre sur I, $K = \prod_{\mathcal{U}} K_i$ le corps ultraproduit des K_i, A une K-algèbre de dimension finie identifiée à un ultraproduit $\prod_{\mathcal{U}} A_i$. Si V est un A-module de dimension finie sur K, il existe une famille de A_i-modules V_i telle que V soit A-isomorphe à l'ultraproduit $\prod_{\mathcal{U}} V_i$.</u>

<u>Preuve</u> : La démonstration suit les grandes lignes de celle de la proposition 22, à quelques complications près. Introduisons une base (e_k) de l'algèbre A, les constantes de structure c_{rs}^t associées à cette base, et enfin une base (v_j) de V (Il s'agit des bases pour la structure de K-espace vectoriel de A et de V). La structure de A-module de V est entièrement caractérisée par la donnée des constantes (d_{rj}^l) telles que :

$$e_r . v_j = \sum d_{rj}^l \, v_l \, .$$

Ces constantes ne sont pas quelconques ; elles sont liées par les relations :

$$(**) \qquad \sum_t c_{rs}^t \, d_{tj}^l = \sum_p d_{sj}^p \, d_{rp}^l$$

qui traduisent les identités d'associativité $(e_r.e_s).v_j = e_r.(e_s.v_j)$. Choisissons comme d'habitude des systèmes de représentants pour les e, les v, les c et les d. Comme les relations de (**) sont en nombre fini, on a pour $\mathfrak{U}$-presque tous les i,

$$(**i)\qquad \sum_t c^t_{rs}(i)\, d^l_{tj}(i) = \sum_p d^p_{sj}(i)\, f^l_{rp}(i)\ .$$

Soit E l'ensemble de l'ultrafiltre $\mathfrak{U}$ des i tels que (**i) soit vérifié. Pour $i\in E$ soit V_i le K_i-espace vectoriel des combinaisons linéaires formelles à coefficients dans K_i des entiers $1,2,\dots,w$ (avec $w = \dim_K V$). On fait de V_i un A_i-module en posant :

$$c_r(i).j = \sum_{\ell} d^l_{rj}(i)\,\ell\qquad ;$$

en effet, comme (**i) est vérifiée, les formules écrites plus haut réalisent les $e_r(i)$ comme des applications linéaires de V_i dans lui-même satisfaisant aux conditions d'associativité :

$$e_r(i)(e_s(i).v) = (e_r(i).e_s(i)).v$$

Pour $i\notin E$ on posera $V_i = 0$. Il reste alors à vérifier que l'application linéaire de V dans $\prod_{\mathfrak{U}} V_i$ qui à v_k associe la classe de (x_i) avec $x_i = k$ pour tout i, est un isomorphisme de modules. Mais le A_i-module V_i a précisément été construit à l'aide des constantes $d^l_{rj}(i)$ pour qu'il en soit bien ainsi.

PROPOSITION 24 - _Soient A et B deux K-algèbres de dimension finie, K étant un corps commutatif ultraproduit $\prod_{\mathfrak{U}} K_i$ de corps K_i. Compte-tenu de la proposition 21 on identifie A à un ultraproduit d'algèbres A_i, B à un ultraproduit d'algèbres B_i. Tout homomorphisme f de A dans B est de la forme $(f_i)^a$ où les f_i sont $\mathfrak{U}$-presque toutes des homomorphismes d'algèbres de A_i dans B_i._

Preuve : D'après la proposition 20 du n°4, §.2 l'application linéaire f est de la forme $(f_i)^a$, les f_i étant K_i-linéaires. Il reste à voir que l'ensemble $\{i \mid f_i$ est un homomorphisme d'algèbres$\}$ est un ensemble de $\mathfrak{U}$; soit E le complémentaire de cet ensemble. Pour tout $i\in E$ on peut trouver $x_i, y_i\in A_i$ tels que $f_i(x_i)f_i(y_i) \neq f_i(x_iy_i)$; complétons les familles x_i et y_i en posant $x_i = 1$ et $y_i = 1$ pour $i\notin E$, et posons $x = \widetilde{(x_i)}$, $y = \widetilde{(y_i)}$. Comme f est un homomorphisme d'algèbres, $f(xy) = f(x)f(y)$, soit $f_i(x_iy_i) = f_i(x_i)f_i(y_i)$ pour $\mathfrak{U}$-presque tous les i. Il en résulte que E n'appartient pas à l'ultrafiltre $\mathcal{U}$: sinon $\emptyset = E \cap \{i \in I \mid f_i(x_i y_i) = f_i(x_i) f_i(y_i)\}$ serait un ensemble de $\mathcal{U}$. ■

Soient $(A_i)_{i \in I}$ et $(B_i)_{i \in I}$ deux familles de K_i-algèbres. On définit une application :

$$\Phi : \prod_{i \in I} A_i \times \prod_{i \in I} B_i \longrightarrow \prod_{\mathcal{U}} (A_i \otimes_{K_i} B_i)$$

en posant :

$$\Phi((a_i),(b_i)) = \widetilde{((a_i \otimes b_i))}.$$

LEMME 10 - Si $(a_i) = (a_i')$ $(\mathcal{U})$ et $(b_i) = (b_i')(\mathcal{U})$, on a :

$$\Phi((a_i),(b_i)) = \Phi((a_i'),(b_i')).$$

L'application Φ, qui est ainsi compatible avec la $\mathcal{U}$-égalité, définit une application $K = (\prod_{\mathcal{U}} K_i)$-bilinéaire de $\prod_{\mathcal{U}} A_i \times \prod_{\mathcal{U}} B_i$ dans $\prod_{\mathcal{U}} (A_i \otimes B_i)$, puis une application linéaire de $\prod_{\mathcal{U}} A_i \otimes_K \prod B_i$ dans $\prod_{\mathcal{U}} (A_i \otimes_{K_i} B_i)$.

Preuve : Immédiat. ■

La propriété suivante sera constamment utilisée.

PROPOSITION 25 - Supposons que $\prod_{\mathcal{U}} A_i$ et $\prod_{\mathcal{U}} B_i$ sont des algèbres de dimension finie. L'application linéaire de $\prod_{\mathcal{U}} A_i \otimes \prod_{\mathcal{U}} B_i$ dans $\prod_{\mathcal{U}} (A_i \otimes B_i)$ construite dans le lemme précédent est un isomorphisme.

Preuve : On va montrer que Φ transforme une base en une base. Soient (e_k) une base $\prod_{\mathcal{U}} A_i$, (ε_l) une base de $\prod B_i$. Si on passe à des systèmes de représentants, $(e_k(i))$ (resp. $\varepsilon_l(i)$), resp $(e_k(i) \otimes \varepsilon_l(i)))$ est pour $\mathcal{U}$-presque tout i une base de A_i (resp. de B_i, resp. de $A_i \otimes_{K_i} B_i$). Enfin $(e_k \otimes \varepsilon_l)$ est une base de $\prod_{\mathcal{U}} A_i \otimes_K \prod_{\mathcal{U}} B_i$. Or :

$$\Phi(e_k \otimes \varepsilon_l) = \Phi(\widetilde{(e_k(i))} \otimes \widetilde{(\varepsilon_l(i))}) = \Phi(\widetilde{(e_k(i))}, \widetilde{(\varepsilon_l(i))}) = \Phi((e_k(i)), (\varepsilon_l(i))) = \widetilde{= (e_k(i) \otimes \varepsilon_l(i))}. ■$$

PROPOSITION 26 - Soit $A = \prod_{\mathcal{U}} A_i$ une $K = \prod_{\mathcal{U}} K_i$-algèbre de dimension finie. Le centre $Z(A)$ de A est l'ultraproduit des centres $Z(A_i)$ des A_i, i.e. $x \in Z(A)$ si et seulement s'il possède un système de représentants (x_i) tel que $x_i \in Z(A_i)$ pour tout i.

Preuve : Soit (e_k) une K-base de A. Un élément x de A appartient à Z(A) si et seulement si $xe_k = e_kx$ pour tout k. En termes de systèmes de représentants il revient au même de dire que : $E = \{i \in I \mid$ le système des $e_k(i)$ est une base de A_i et $x(i)e_k(i) = e_k(i)x(i)$ pour tout $k\}$ est un ensemble de $\mathfrak{U}$. Si x a un système de représentants $(x(i))$ avec $x(i) \in Z(A_i)$ pour tout i, la condition ci-dessus est évidemment satisfaite. Si x appartient au centre de A, soit E l'ensemble introduit ci-dessus. Posons $x'(i) = x(i)$ si $i \in E$, $x'(i) = 1$ si $i \notin E$. Comme E est dans $\mathfrak{U}$, $(x(i))$ et $(x'(i))$ sont $(\mathfrak{U})$-égaux. Et on a évidemment $x'(i) \in Z(A_i)$ pour tout i.■

n°2 . RADICAL ET SEMI-SIMPLICITE D'UNE ALGEBRE ULTRAPRODUIT.

LEMME 11 - Soit A un anneau ultraproduit d'anneau A_i. Si (a_i) est une famille dont $\mathfrak{U}$-presque tous les termes sont inversibles, $\widetilde{(a_i)}$ est inversible dans A.

Preuve : Immédiat.■

COROLLAIRE - Soit A un anneau ultraproduit d'une famille d'anneaux unitaires A_i. L'idéal de A ultraproduit des radicaux de Jacobson des A_i est inclus dans le radical de Jacobson de A.

Preuve : L'idéal J ultraproduit des radicaux $R(A_i)$ est un idéal bilatère de A (cf. §.2 n°1 proposition 9). Soit $x = \widetilde{(x_i)}$ un élément de J. Comme par définition d'un idéal ultraproduit $x_i \in J_i$ pour $\mathfrak{U}$-p.t. i, $1-x_i$ est inversible dans A_i pour $\mathfrak{U}$-p.t. i ; le lemme montre que $(1-x)$ est inversible dans A. Or R(A) est le plus grand des idéaux bilatères de A tels que $1-r$ soit inversible pour tout $r \in R(A)$.

LEMME 12 - Soit $A = \prod_{\mathfrak{U}} A_i$ une $K = \prod_{\mathfrak{U}} K_i$-algèbre de dimension finie et soit J un idéal de A représenté comme un ultraproduit d'idéaux J_i de A_i (cf. §.3 n°1 proposition 22). Si J est un idéal nilpotent, pour $\mathfrak{U}$-presque tout i l'idéal J_i est nilpotent.

Preuve : Dire que J est nilpotent, c'est dire qu'il existe n entier>0 tel que $J^n = 0$ ou que, pour toute suite $x_1,\dots,x_n$ d'éléments de J, on a : $x_1 \dots x_n = 0$. Par linéarité on peut supposer que les x_i sont pris parmi les éléments d'une K-base $a_1,\dots,a_p$ de J : ce qui revient à écrire $x_i = a_{j(i)}$ pour une application convenable j de $\{1,2,\dots,n\}$ dans $\{1,2,\dots,p\}$. Dire que J est nilpotent c'est finalement dire qu'il existe un entier n positif tel que, pour toute application j de $\{1,2,\dots,n\}$ dans $\{1,2,\dots,p\}$, on a

$a_{j(1)}\ldots a_{j(n)} = 0$. Mais ces applications j forment <u>un ensemble fini</u>. En passant aux systèmes de représentants $a_k(i)$ des a_k on voit qu'il existe un ensemble E de l'ultrafiltre $\mathcal{U}$ tel que pour toute application j de $\{1,2,\ldots,n\}$ dans $\{1,2,\ldots,p\}$ et tout $i \in E$ on ait $a_{j(1)}(i)\ldots a_{j(n)}(i) = 0$. Le sous-ensemble E' des i de E tels que $a_1(i),\ldots,a_p(i)$ soit une base de J est encore dans $\mathcal{U}$. Par suite, en reprenant le raisonnement fait au début de la démonstration, on voit que, pour tout $i \in E'$, toute suite de n éléments de J_i a un produit nul ; c'est à dire que $J_i^n = 0$ pour tout $i \in E' \in \mathcal{U}$.■

PROPOSITION 27 - <u>Soit</u> $A = \prod_{\mathcal{U}} A_i$ <u>une</u> $K = (\prod_{\mathcal{U}} K_i)$<u>-algèbre de dimension finie. Le radical de Jacobson de</u> A <u>s'identifie à l'idéal ultraproduit des radicaux de Jacobson des</u> A_i.

<u>Preuve</u> : D'après le corollaire du lemme 11, l'idéal J, ultraproduit des $R(A_i)$, est inclus dans le radical $R(A)$. D'après la proposition 22 du §.3 n°1, le radical $R(A)$ peut se représenter comme un idéal ultraproduit d'idéaux bilatères J_i. Comme A est de dimension finie, le radical $R(A)$ est nilpotent ; et le lemme 12 montre que $J_i \subset R(A_i)$ pour $\mathcal{U}$-presque tout i, puisque $R(A_i)$ est le plus grand idéal nilpotent de A_i. On en déduit que $R(A_i) = J_i$ pour $\mathcal{U}$-presque tous les i, puis que $R(A)$ est l'idéal ultraproduit des $R(A_i)$, par un calcul de dimension ($R(A)$ contient l'idéal ultraproduit des $R(A_i)$ et a même dimension que lui sur K d'après la proposition 17 du §.2 n°4).■

COROLLAIRE 1 - <u>Pour que la</u> $K = (\prod_{\mathcal{U}} K_i)$<u>-algèbre de dimension finie</u> $A = \prod_{\mathcal{U}} A_i$ <u>soit semi-simple il faut et il suffit que</u> $\mathcal{U}$<u>-presque toutes les</u> A_i <u>le soient.</u>

<u>Preuve</u> : En effet une algèbre de dimension finie est semi-simple si et seulement si son radical est nul.■

Le corollaire qui suit a une grande importance.

COROLLAIRE 2 - <u>Soit</u> $(A_i)_{i \in I}$ <u>une famille de</u> K_i<u>-algèbres de dimension finie telle que</u> $\sup \dim_{K_i} A_i < \infty$. <u>Si pour tout ultrafiltre non principal</u> $\mathcal{U}$ <u>sur</u> I <u>l'algèbre ultraproduit</u> $\prod_{\mathcal{U}} A_i$ <u>est semi-simple, les</u> A_i <u>sont semi-simples à l'exception peut-être d'un nombre fini d'entre elles.</u>

<u>Preuve</u> : Supposons que $E = \{i \in I \mid A_i$ n'est pas semi-simple$\}$ soit infini. Il existe alors un ultrafiltre non principal $\mathcal{U}$ sur I contenant E. Les A_i sont alors non-semi-simples pour $\mathcal{U}$-presque tous les i. Et le corollaire 1 montre

que $\prod_{\mathcal{U}} A_i$ n'est pas semi-simple.∎

PROPOSITION 28 - <u>Soit</u> $(A_i)_{i\in I}$ <u>une famille de</u> K_i<u>-algèbres telle que</u> $A = \prod_{\mathcal{U}} A_i$ <u>soit de dimension finie sur</u> $K = \prod_{\mathcal{U}} K_i$. <u>Il y a équivalence entre</u> :

a) <u>Pour</u> $\mathcal{U}$<u>-presque tout</u> i <u>les</u> K_i<u>-algèbres</u> A_i <u>sont simples</u>,

b) <u>la</u> K<u>-algèbre</u> A <u>est simple</u>.

<u>Preuve</u> : Une algèbre est simple si et seulement si elle est semi-simple et si son centre est un corps. La proposition est alors une conséquence immédiate du Corollaire 1 de la proposition 27 et de la proposition 26 du n°1.∎

COROLLAIRE 1 - <u>Les hypothèses étant celles de la proposition</u> 28, <u>il y a équivalence entre</u> :

a) <u>Les</u> A_i <u>sont</u> $\mathcal{U}$<u>-presque toutes des</u> K_i<u>-algèbres simples centrales</u>,

b) <u>l'algèbre</u> $A = \prod_{\mathcal{U}} A_i$ <u>est une</u> K<u>-algèbre simple centrale</u>.

<u>Preuve</u> : Par rapport à la proposition 28 il faut traduire la condition supplémentaire $Z(A) = K$. Or d'après la proposition 26 du n°1, $Z(A)$ est l'ultraproduit des $Z(A_i)$. Le corollaire de la proposition 21 montre que $Z(A)$ n'est égal à K que si $Z(A_i) = K_i$ pour $\mathcal{U}$-presque tous les i.∎

COROLLAIRE 2 - <u>Soit</u> $(A_i)_{i\in I}$ <u>une famille de</u> K_i<u>-algèbres</u> A_i <u>telles que</u> $\sup \dim_{K_i} A_i < \infty$. <u>Il y a équivalence entre</u> :

a) <u>les</u> K_i<u>-algèbres</u> A_i <u>sont simples centrales à l'exception d'un nombre fini d'entre elles</u>,

b) <u>pour tout ultrafiltre non principal</u> $\mathcal{U}$ <u>sur</u> I, <u>la</u> $\prod_{\mathcal{U}} K_i$<u>-algèbre</u> $\prod_{\mathcal{U}} A_i$ <u>est simple centrale</u>.

a) $\Longrightarrow$ b) : il suffit de noter que l'ensemble fini des i tels que A_i ne soit pas simple centrale n'appartient à aucun ultrafiltre non principal $\mathcal{U}$, ce qui permet d'appliquer le corollaire 1.

b) $\Longrightarrow$ a) : procédons par l'absurde en supposant que l'ensemble E des i tels que A_i ne soit pas simple centrale est infini. Cet ensemble appartient alors à un ultrafiltre non principal $\mathcal{U}$. Et la proposition 28 montre que $\prod_{\mathcal{U}} A_i$ n'est pas une algèbre simple centrale.∎

Soient $(A_i)_{i\in I}$ une famille de K_i-algèbres simples centrales, S_i un A_i-module simple pour tout i, D_i le corps commutant de S_i pour tout i. On considère un ultrafiltre $\mathcal{U}$ sur I. Aux injections canoniques $D_i \longrightarrow \mathcal{L}_{K_i}(S_i)$ est associé à un homomorphisme injectif de $\prod_{\mathcal{U}} D_i$ dans $\prod_{\mathcal{U}} \mathcal{L}_{K_i}(S_i)$. D'autre part la proposition 8 du §.2 n°1 montre que l'on a un isomorphisme de $\prod_{\mathcal{U}} \mathcal{L}_{K_i}(S_i)$ sur $\mathcal{L}_K(\prod_{\mathcal{U}} S_i)$ -avec $K = \prod_{\mathcal{U}} K_i$- en associant à toute famille (u_i) l'application $(u_i)^a$. Par composition on obtient ainsi un homomorphisme injectif de $\prod_{\mathcal{U}} D_i$ dans $\mathcal{L}_K(\prod_{\mathcal{U}} S_i)$.■

PROPOSITION 29 - <u>Les hypothèses étant celles de l'alinéa précédent, l'homomorphisme</u> $\prod_{\mathcal{U}} D_i \longrightarrow \mathcal{L}_K(\prod_{\mathcal{U}} S_i)$ <u>est un isomorphisme, si</u> $\sup \dim_{K_i} A_i < \infty$.

<u>Preuve</u> : Nous devons montrer que tout élément f du commutant du module simple $\prod_{\mathcal{U}} S_i$ appartient à l'image de $\prod_{\mathcal{U}} D_i$, autrement dit qu'il existe une famille f_i d'endomorphismes A_i-linéaires de S_i telle que $f = (f_i)^a$. Comme f est K-linéaire et comme $\prod_{\mathcal{U}} S_i$ est de dimension finie, la proposition 20 du §.2 n°4 montre qu'il existe des applications K_i-linéaires f_i de S_i telles que $f = (f_i)^a$. Nous allons montrer que les f_i sont $\mathcal{U}$-presque toutes A_i-linéaires, ce qui achèvera la démonstration. Soit $E = \{i \in I | f_i$ n'est pas A_i-linéaire$\}$. Pour tout $i \in I$ il existe un $a_i \in A_i$ et un $s_i \in S_i$ tels que $f_i(a_i x_i) \neq a_i f(x_i)$. Prolongeons ces familles (a_i) et (s_i) en posant $a_i = 1$ et $s_i = 0$ pour $i \notin E$; ce qui permet d'introduire les classes a et s de (a_i) et de (s_i) dans les ultraproduits $\prod_{\mathcal{U}} A_i$ et $\prod_{\mathcal{U}} S_i$. Comme $f = (f_i)^a$ est A-linéaire, l'ensemble $\{i \in I | f_i(a_i s_i) = a_i f_i(s_i)\}$ est un ensemble de $\mathcal{U}$. Alors E ne peut être un ensemble de $\mathcal{U}$: en effet, on a par construction $f_i(a_i s_i) \neq a_i f_i(s_i)$ pour tout i dans E et $E \cap \{i \in I | f_i(a_i s_i) = a_i f_i(s_i)\} = \emptyset$. Si E n'est pas dans $\mathcal{U}$, son complémentaire est dans $\mathcal{U}$. C'est dire que $\mathcal{U}$-presque toutes les f_i sont A_i-linéaires.■

COROLLAIRE 1 - <u>Soit</u> $(A_i)_{i \in I}$ <u>une famille de</u> K_i<u>-algèbres, avec</u> $\sup \dim_{K_i} A_i < \infty$. <u>Il y a équivalence entre</u> :

a) <u>pour</u> $\mathcal{U}$<u>-presque tous les</u> i, <u>les</u> A_i <u>sont des algèbres de matrices sur</u> K_i,

b) <u>l'algèbre</u> $\prod_{\mathcal{U}} A_i$ <u>est une algèbre de matrices sur le corps</u> $K = \prod_{\mathcal{U}} K_i$.

<u>Si ces conditions sont satisfaites, il existe un entier</u> n <u>tel que</u> $A_i = M_n(K_i)$ <u>pour</u> $\mathcal{U}$<u>-presque toutes les</u> K_i.

<u>Preuve</u> : Pour que $\prod_{\mathcal{U}} A_i$ soit une algèbre de matrices sur K il faut et il suffit qu'elle soit simple centrale et que le commutant d'un $\prod_{\mathcal{U}} A_i$-module simple soit K. L'équivalence des assertions a) et b) résulte alors du corollaire 1 de

la proposition 28 (condition pour que $\prod_{\mathcal{U}} A_i$ soit simple centrale), de la proposition 29 (identification du commutant d'un $\prod_{\mathcal{U}} A_i$-module simple) et du corollaire de la proposition 21 du §.3 n°1 (condition pour qu'une algèbre ultraproduit soit isomorphe à K) ou de la proposition 17 du §.3 n°4 (condition pour qu'un espace vectoriel ultraproduit soit de dimension 1 sur K).■

COROLLAIRE 2 - *Soit* $(A_i)_{i\in I}$ *une famille de* K_i*-algèbre* A_i, *avec* $\sup \dim_{K_i} A_i < \infty$. *Il y a équivalence entre :*

a) *à l'exception d'un nombre fini d'entre elles, les* K_i*-algèbres* A_i *sont des algèbres de matrices* $M_{n_i}(K_i)$,

b) *pour tout ultrafiltre non principal* $\mathcal{U}$ *sur* I, *l'algèbre* $\prod_{\mathcal{U}} A_i$ *est une algèbre de matrices* $M_n(K)$ *sur le corps* $\prod_{\mathcal{U}} K_i$.

Preuve : On déduit ce corollaire du corollaire 1 en utilisant la même méthode que dans la déduction du corollaire 2 de la proposition 28 à partir du corollaire 1 de la même proposition.■

LEMME 13 - *Soit* $(A_i)_{i\in I}$ *une famille de* K_i*-algèbres. Pour que l'élément* $e = (\widetilde{e_i})$ *de l'ultraproduit* $\prod_{\mathcal{U}} A_i$ *soit un idempotent, il faut et il suffit que* $\mathcal{U}$*-presque tous les* e_i *soient des idempotents.*

Preuve : Immédiat : il suffit d'écrire que $e_i^2 = e_i$ $(\mathcal{U})$.

PROPOSITION 30 - *Soit* $(D_i)_{i\in I}$ *une famille de* K_i*-algèbres telle que* $\dim_K \prod_{\mathcal{U}} D_i < \infty$, *avec* $K = \prod_{\mathcal{U}} K_i$. *Il y a équivalence entre :*

a) *pour* $\mathcal{U}$*-presque tous les* i *les* D_i *sont des corps,*

b) *l'ultraproduit* $\prod_{\mathcal{U}} D_i$ *est un corps.*

Preuve : Il suffit de remarquer qu'une algèbre de dimension finie sur un corps est un corps si et seulement si elle est simple et si elle ne contient aucun idempotent non nul autre que l'identité : c'est ce qui résulte du Théorème de Wedderburn représentant toute algèbre simple comme une algèbre de matrices sur un corps. La proposition 30 est donc une conséquence immédiate de la proposition 28 et du lemme 13.■

COROLLAIRE 1 - *Soit* D *un corps de centre* $K = \prod_{\mathcal{U}} K_i$, *de dimension finie* n *sur* K. *Il existe une famille* $(D_i)_{i\in I}$ *de corps tels que* D_i *soit* $\mathcal{U}$*-presque partout un corps de centre* K_i *et de dimension* n *sur* K_i, *le corps* D *étant isomorphe au corps ultraproduit* $\prod_{\mathcal{U}} D_i$.

Preuve : D'après la proposition 21 du n°1 de ce §.3 l'algèbre D est K-isomorphe à un ultraproduit de K_i-algèbres D_i. On peut alors appliquer la proposition 30. ■

COROLLAIRE 2 - Soit $(K_k)_{i \in I}$ une famille de corps commutatifs. On considère un ultrafiltre $\mathcal{U}$ sur I tel que les K_i aient $\mathcal{U}$-presque partout un groupe de Brauer nul. Le corps ultraproduit $\prod_{\mathcal{U}} K_i$ a un groupe de Brauer nul.

Preuve : Soit D un corps de centre $K = \prod_{\mathcal{U}} K_i$. D'après le corollaire 1 le corps D est isomorphe à un ultraproduit de corps D_i qui sont des K_i-algèbres simples centrales. Il en résulte que $\mathcal{U}$-presque partout les corps D_i sont réduits à leur centre K_i. Par suite l'ultraproduit des D_i est de dimension 1 sur K, et $D = K$. ■

COROLLAIRE 3 - Tout corps ultraproduit de corps finis a un groupe de Brauer nul.

Preuve : En effet tout corps fini a un groupe de Brauer nul. ■

Remarque : En fait on a un résultat beaucoup plus précis que le corollaire 3. D'après Ax[2], tout corps ultraproduit de corps finis est quasi-fini.

COROLLAIRE 4 - Soit K un corps ultraproduit de corps commutatifs K_i. Pour toute K-algèbre simple centrale A il existe une famille $(D_i)_{i \in I}$ de corps de centre K_i et de dimension finie sur K_i et un entier n tel que A soit isomorphe à l'ultraproduit des algèbres $M_n(D_i)$.

Preuve : D'après le Théorème de Wedderburn, toute algèbre simple centrale A sur K est isomorphe à une algèbre de matrices $M_n(D)$ sur un corps D de centre K. Mais $M_n(D) \cong M_n(K) \otimes D$. Le corollaire 1 de la proposition 29 montre que $M_n(K)$ est isomorphe à l'algèbre ultraproduit des algèbres $M_n(K_i)$. D'après le corollaire 1 de la proposition 30 il existe une famille $(D_i')_{i \in I}$ de K_i-algèbres simples centrales qui sont $\mathcal{U}$-presque toutes des corps telle que D soit isomorphe à l'algèbre ultraproduit des D_i'. Si l'on pose $D_i = D_i'$ si D_i' est un corps de centre K_i, $D_i = K_i$ dans le cas contraire, le corps D est encore ultraproduit des D_i qui sont tous des corps de centre K_i. La proposition 25 du n°1 montre que $M_n(K \otimes D)$ est isomorphe à l'ultraproduit des $M_n(K_i) \otimes D_i$, et $M_n(K_i) \otimes D_i$ est K_i-isomorphe à $M_n(K_i)$.

§.4 ULTRAPRODUIT D'ANNEAUX LOCAUX.

n°1. ULTRAPRODUIT D'ANNEAUX LOCAUX.

Soit $(A_i)_{i \in I}$ une famille d'anneaux commutatifs locaux, l'idéal maximal de A_i étant noté $\mathfrak{m}_i$.

PROPOSITION 31 - *L'anneau ultraproduit* $\prod_{\mathcal{U}} A_i$ *est un anneau commutatif local dont l'idéal maximal* $\mathfrak{m}$ *est l'idéal ultraproduit des* $\mathfrak{m}_i$. *L'idéal* $\mathfrak{m}$ *est aussi l'ensemble des éléments* x *qui admettent des systèmes de représentants* (x_i) *dans lesquels les* x_i *sont* $\mathcal{U}$*-presque tous non inversibles.*

Preuve : Un élément x de $\prod_{\mathcal{U}} A_i$ qui n'appartient pas à l'idéal ultraproduit des $\mathfrak{m}_i$ a un système de représentants (x_i) satisfaisant à $\{i | x_i \notin \mathfrak{m}_i\} \in \mathcal{U}$, c'est-à-dire $\{i | x_i$ est inversible$\} \in \mathcal{U}$. Le lemme 11 du §.3 n°2 montre que x est un élément inversible. Et $\mathfrak{m}$, qui est ainsi l'ensemble des éléments non inversibles dans $\prod_{\mathcal{U}} A_i$, est maximal.■

COROLLAIRE 1 - *Le corps résiduel* $A/\mathfrak{m}$ *de l'anneau local ultraproduit des anneaux locaux* A_i *est isomorphe à l'ultraproduit des corps résiduels* $k_i = A_i/\mathfrak{m}_i$.

Preuve : La proposition 11 du §.2 n°2 montre que l'homomorphisme $\prod_{\mathcal{U}} A_i \longrightarrow \prod_{\mathcal{U}} k_i$ associé à la famille d'homomorphismes canoniques $p_i : A_i \longrightarrow k_i$ a pour noyau l'idéal ultraproduit des idéaux noyaux des p_i, c'est-à-dire l'idéal maximal de A.■

COROLLAIRE 2 - *Supposons que, pour tout nombre premier* p, $\{i \in I | k_i$ *est de caractéristique* $p\}$ *n'appartient pas à* $\mathcal{U}$. *L'anneau local* A *ultraproduit des* A_i *est un anneau local d'égale caractéristique* 0.

Preuve : D'après la proposition 16 du §.2 n°3 le corps ultraproduit des k_i est de caractéristique 0. Mais ce corps est isomorphe au corps résiduel de l'anneau ultraproduit des A_i.

n°2. ANNEAUX LOCAUX HENSELIENS.

Soit A un anneau local d'idéal maximal $\mathfrak{m}$ et de corps résiduel k. Si P est un polynôme de $A[X]$ nous noterons $\bar{P}$ le polynôme de $k[X]$ obtenu en réduisant les coefficients de P modulo $\mathfrak{m}$: si

$P = a_nX^n + a_{n-1}X^{n-1}+\ldots+a_1X+a_o$ et si pour tout i on note $\bar{a}_i$ la classe de a_i dans k, on a $\bar{P} = \bar{a}_nX^n+\ldots+\bar{a}_o$.

DEFINITION 8 - Soit A un anneau local d'idéal maximal $\mathfrak{m}$ et de corps résiduel k. On dit que A est <u>hensélien</u> si pour tout polynôme unitaire P de $A[X]$ et toute décomposition $\bar{P} = qr$ du polynôme réduit $\bar{P} \in k[X]$ en produit de deux polynômes premiers entre eux q et r de $k[X]$, il existe deux polynômes Q et R de $A[X]$ tels que $P = QR$, $q = \bar{Q}$ et $r = \bar{R}$.

La proposition suivante est bien connue :

PROPOSITION 32 - <u>Soient A un anneau local hensélien, P un polynôme unitaire de $A[X]$. Si le polynôme réduit $\bar{P} \in k[X]$ a une racine simple dans k, le polynôme P a une racine dans A.</u>

<u>Preuve</u> : On écrit qu'il existe un $\bar{a} \in k$ et un $q \in k[X]$ tels que $\bar{P} = (X-\bar{a})q$, $(X-\bar{a})$ et q étant premiers entre eux puisque $\bar{a}$ est racine simple. On remarque enfin que tout polynôme R tel que $\bar{R} = X-\bar{a}$ est de la forme $a_pX^p+\ldots+a_1X+a_o$ avec $a_i \in \mathfrak{m}$ pour $i \geq 2$. Comme on doit avoir $P = QR$, où Q est tel que $\bar{Q} = q$, la comparaison des degrés montre qu'en fait R est de degré 1, i.e. que les a_i sont nuls pour $i \geq 2$.■

PROPOSITION 33 - <u>Soient $(A_i)_{i \in I}$ une famille d'anneaux locaux qui sont $\mathcal{U}$-presque tous henséliens, pour un ultrafiltre $\mathcal{U}$ sur I. L'anneau ultraproduit des A_i est un anneau hensélien.</u>

<u>Preuve</u> : Soit $P = X^n+a_{n-1}X^{n-1}+\ldots+a_1X+a_o \in A[X]$ un polynôme unitaire sur l'anneau ultraproduit A des A_i. On suppose que sur le corps résiduel k on a une décomposition $\bar{P} = qr$ en polynômes premiers entre eux. Choisissons un système de représentants $(a_j(i))$ de chaque a_j et posons $P_i = X^n+a_{n-1}(i)X^{n-1}+\ldots+a_o(i) \in A_i[X]$. Nous introduirons de même, en notant que le corps résiduel k de A est isomorphe à l'ultraproduit des corps résiduels k_i des A_i, les polynômes $q_i \in k_i[X]$ et $r_i \in k_i[X]$. De l'identité $\bar{P} = qr$ on déduit que $\bar{P}_i = q_ir_i$ pour $\mathcal{U}$-presque tous les i. Dire que q et r sont premiers entre eux c'est dire, d'après l'identité de Bezout, qu'il existe des polynômes s et t de $k[X]$ tels que $sq + tr = 1$. En faisant intervenir les systèmes de représentants des coefficients on obtient que $s_iq_i + t_ir_i = 1$ pour $\mathcal{U}$-presque tous les i, autrement dit que q_i et r_i sont premiers entre eux pour $\mathcal{U}$-presque tous les i. Comme A_i est hensélien pour $\mathcal{U}$-presque tout i, on peut construire, pour $\mathcal{U}$-presque tout i, deux polynômes Q_i et R_i dans $A_i[X]$ tels que $P_i = Q_iR_i$, $\bar{Q}_i = q_i$ et $\bar{R}_i = r_i$. Comme chaque P_i est

de degré n, les Q_i et les R_i sont tous de degré $\leq n$. Soit E l'ensemble de l'ultrafiltre $\mathcal{U}$ sur lequel nous avons construit les Q_i et les R_i. Si $Q_i = \sum_{j \leq n} b_j(i)X^j$ et $R_i = \sum_{j \leq n} c_j(i)X^j$, nous définirons les $b_j(i)$ et les $c_j(i)$ pour $i \notin E$ par :

$$b_j(i) = a_j(i) \quad \text{(coefficients de } P_i)$$

$$c_j(i) = 0 \quad \text{pour } j \neq 0$$

$$c_o(i) = 1.$$

Nous avons alors $P_i = Q_i R_i$ pour tout i. Si $b_j = \widetilde{(b_j(i))}$ et $c_j = \widetilde{(c_j(i))}$ dans l'anneau ultraproduit des A_i, si $Q = \sum b_j X^j$ et $R = \sum c_j X^j$, on a donc $P = QR$. Par définition de l'ensemble E on a $\bar{Q}_i = q_i$ et $\bar{R}_i = r_i$ pour tout $i \in E$. Par passage à l'ultraproduit on obtient $\bar{Q} = q$ et $\bar{R} = r$, puisque E est un ensemble de $\mathcal{U}$. On a donc remonté la décomposition $\bar{P} = qr$ en une décomposition $P = QR$, ce qui achève la démonstration de la proposition.∎

n°3. CORPS DE COHEN.

DEFINITION 9 - Soit A un anneau local d'idéal maximal $\mathfrak{m}$ et de corps résiduel k. On dit qu'un sous-anneau unitaire S de A est un <u>corps de Cohen</u> pour A si S est un corps et si l'image de S par la projection canonique $p : A \longrightarrow k$ est k.

Tout homomorphisme unitaire d'un corps dans un anneau étant injectif la projection canonique p réalise donc un isomorphisme du corps de Cohen S sur le corps résiduel. On a $S \cap \mathfrak{m} = 0$, ce qui conduit à la décomposition "additive" de A sur la forme $A = S + \mathfrak{m}$.

PROPOSITION 34 - <u>Soit</u> A <u>un anneau commutatif unitaire qui est un anneau local hensélien d'égale caractéristique</u> 0. <u>Tout sous-corps de</u> A <u>est contenu dans un corps de Cohen.</u>

<u>Preuve</u> : (cf. [5] Chapitre II §.4 proposition 6). Soit N un sous-corps commutatif de A. D'après le Théorème de Zorn, la famille des sous-corps de A contenant N a au moins un élément maximal S. Nous allons montrer que cet S est un corps de Cohen. Ce corps S a une image par la projection canonique de A sur son corps résiduel qui est un sous-corps R du corps résiduel k. Le corps k est une extension algébrique de R : sinon il existerait au moins un élément a de A dont l'image $\bar{a}$ dans k serait transcendante sur R. Soit $P \in S[X]$;

l'image par la projection canonique $A \longrightarrow k$ de $P(a)$ dans k est évidemment $\bar{P}(\bar{a})$, si $\bar{P}$ est le polynôme obtenu en faisant la réduction des coefficients de P. Mais $\bar{P}$ est un polynôme de $R[X]$; et, puisque $\bar{a}$ est transcendant sur R, $\bar{P}(\bar{a}) \neq 0$ d'où, a fortiori, $P(a) \notin \mathfrak{m}$. Il en résulte que pour tout polynôme P de $S[X]$, l'élément $P(a)$ est inversible dans A et que l'homomorphisme de $S[X]$ dans A qui à un polynôme P fait correspondre $P(a)$ se prolonge en un homomorphisme du corps $S(X)$ des fractions rationnelles dans A. L'image de ce corps $S(X)$ est un corps qui contient strictement S (puisqu'elle contient a), en contradiction avec la définition de S. Il reste à montrer que k coïncide avec l'image R de S. Soit $\bar{a}$ un élément quelconque de k et soit $q(X) \in R[X]$ le polynôme minimal de $\bar{a}$ sur R. Comme A est d'égale caractéristique 0, R est de caractéristique 0 et $\bar{a}$ est racine simple de son polynôme minimal q. Si $q = \sum r_j X^j$, on peut trouver que pour tout j un $s_j \in S$ unique tel que r_j soit l'image de s_j par la projection $A \longrightarrow k$. Soit $Q(X) = \sum s_j X^j$, polynôme unitaire de $S[X]$, donc de $A[X]$, dont l'image par réduction des coefficients est q. La proposition 32 montre alors qu'il existe au moins une racine a pour le polynôme Q dans A. Le plus petit sous-anneau de A contenant S et a est un corps, extension algébrique de S. Par définition de S ce corps coïncide avec S, i.e. $a \in S$ et $\bar{a} \in R$. Or $\bar{a}$ était un élément quelconque du corps résiduel k de A.∎

Nous utiliserons le corollaire suivant :

COROLLAIRE - <u>Soient</u> $(A_i)_{i \in I}$ <u>une famille d'anneaux locaux noethériens d'idéaux maximaux</u> $\mathfrak{m}_i$. <u>On suppose que, pour un ultrafiltre sur</u> I, <u>les</u> A_i <u>soient</u> $\mathcal{U}$<u>-presque tous complets pour la topologie</u> $\mathfrak{m}_i$<u>-adique et que, pour tout nombre premier</u> p, $\mathcal{U}$<u>-presque tous les corps résiduels</u> k_i <u>ne sont pas de caractéristique</u> p ($\mathcal{U}$-incompatibilité des caractéristiques des corps résiduels). <u>Tout sous-corps de l'anneau ultraproduit des</u> A_i <u>sur</u> $\mathcal{U}$ <u>est contenu dans un corps de Cohen.</u>

<u>Preuve</u> : Tout anneau local noethérien complet pour la topologie de son idéal maximal est hensélien. Le corollaire est alors une conséquence de la proposition 34, de la proposition 32 et du corollaire 2 de la proposition 31.∎

§.5 DECOMPOSITION LOCALE DES ALGEBRES SIMPLES.

n°1. REDUCTION MODULO $\mathfrak{p}$ D'UNE ALGEBRE.

Soit A un anneau commutatif unitaire. Un idéal premier $\mathfrak{p}$ de A est minimal s'il est non nul et s'il n'existe aucun idéal premier non nul inclus dans $\mathfrak{p}$ et distinct de $\mathfrak{p}$.

Soit A un anneau commutatif unitaire. Rappelons (cf. N. Bourbaki Algèbre Commutative ch. VII §.1 n°7 Théorème 4) que A est un anneau de Krull si 1°) pour tout idéal premier minimal $\mathfrak{p}$ de A, le localisé A est un anneau de valuation discrète ; 2°) si A et ses localisés $A_{\mathfrak{p}}$ sont considérés comme des sous-ensembles du corps des fractions K de A, l'anneau A est l'intersection des $A_{\mathfrak{p}}$, pour $\mathfrak{p}$ parcourant l'ensemble des idéaux premiers minimaux de A ; 3°) pour tout $x \neq 0$ de A il n'existe qu'un nombre fini d'idéaux premiers minimaux $\mathfrak{p}$ de A tels que $x \in \mathfrak{p}$.

Soit A un anneau de Krull. Nous désignerons par $\mathfrak{M}$ l'ensemble des idéaux premiers minimaux de A, et nous supposerons $\mathfrak{M}$ infini. Pour tout $\mathfrak{p} \in \mathfrak{M}$ nous noterons $k(\mathfrak{p})$ le corps résiduel de l'anneau localisé $A_{\mathfrak{p}}$, $h_{\mathfrak{p}}$ l'homomorphisme de A dans $k(\mathfrak{p})$ qui à a associe la classe de a, considéré comme élément de $A_{\mathfrak{p}}$, dans $k(\mathfrak{p})$. Le noyau de $h_{\mathfrak{p}}$ est, comme on le vérifie aussitôt, $\mathfrak{p}$.

LEMME 14 - Soit $\mathfrak{U}$ un ultrafiltre (non principal) sur l'ensemble $\mathfrak{M}$ des idéaux premiers minimaux de A. L'homomorphisme h de A dans $\prod_{\mathfrak{U}} k(\mathfrak{p})$ déduit de la famille des $h_{\mathfrak{p}}$ est injectif.

Preuve : Soit $a \in A$. On a $h(a) = 0$ si et seulement si $a \in \mathfrak{p}$ pour $\mathfrak{U}$-presque tous les $\mathfrak{p}$. Comme $\mathfrak{U}$ est non principal et ne contient donc aucun ensemble fini, $\{\mathfrak{p} \in \mathfrak{M} \mid a \in \mathfrak{p}\}$ est infini, ce qui entraîne $a = 0$.■

COROLLAIRE 1 - Soit A un anneau de Krull de corps des fractions K. L'homomorphisme $h : A \longrightarrow \prod_{\mathfrak{U}} k(\mathfrak{p})$ se prolonge en un isomorphisme de K dans $\prod_{\mathfrak{U}} k(\mathfrak{p})$.

Preuve : Ceci résulte de la propriété universelle du corps des fractions.■

COROLLAIRE 2 - Soit A un anneau de Krull. Pour tout ultrafiltre non principal $\mathfrak{U}$ sur l'ensemble $\mathfrak{M}$ des idéaux premiers minimaux de A, l'anneau ultraproduit des localisés $A_{\mathfrak{p}}$ de A en les $\mathfrak{p} \in \mathfrak{M}$ est un anneau local d'égale caractéristique (égale à celle de A).

<u>Preuve</u> : En effet A admet un plongement dans $\prod_{\mathcal{U}} k(\mathfrak{p})$, corps résiduel de $\prod_{\mathcal{U}} A_{\mathfrak{p}}$. ∎

LEMME 15 - <u>L'homomorphisme de A dans l'ultraproduit des $A_{\mathfrak{p}}$ ($\mathfrak{p}\in\mathfrak{M}$), sur un ultrafiltre non principal déduit des injections $A \longrightarrow A_{\mathfrak{p}}$ se prolonge en un homomorphisme du corps K des fractions de A dans $\prod_{\mathcal{U}} A_{\mathfrak{p}}$.</u>

<u>Preuve</u> : Soit a un élément non nul de A pour $\mathcal{U}$-presque tout $\mathfrak{p}$ on a $a\notin\mathfrak{p}$, $a\notin\mathfrak{p}A_{\mathfrak{p}}$ (idéal maximal de l'anneau local $A_{\mathfrak{p}}$) ; et a est donc $\mathcal{U}$-presque partout inversible dans les $A_{\mathfrak{p}}$. ∎

Soient donc A un anneau de Krull de corps des fractions K et dont l'ensemble des idéaux premiers minimaux $\mathfrak{M}$ est infini. Pour tout ultrafiltre non principal $\mathcal{U}$ sur $\mathfrak{M}$, on fait de $\prod_{\mathcal{U}} k(\mathfrak{p})$ et de $\prod_{\mathcal{U}} A_{\mathfrak{p}}$ des K-algèbres en utilisant le corollaire 1 du lemme 14 et le lemme 15.

Soit B une K-algèbre de dimension finie n ; choisissons une base $e_1,\dots,e_n$ de B, à laquelle correspondent des constantes de structure $c_{ij}^{k}\in K$ par $e_i.e_j = \sum c_{ij}^{k}\, e_k$. Le lemme suivant est bien connu :

LEMME 16 - <u>Les localisés $A_{\mathfrak{p}}$ étant réalisés comme des sous-anneaux du corps K, pour tout $x\in K$ on a $x\in A_{\mathfrak{p}}$ à l'exception d'un nombre fini de $\mathfrak{p}\in\mathfrak{M}$.</u>

<u>Preuve</u> : On a $x = \frac{a}{b}$ avec $a, b\in A$ et $b\neq 0$. L'élément b n'appartient qu'à un nombre fini d'idéaux premiers minimaux ; et, si $b\notin\mathfrak{p}$, b est inversible dans $A_{\mathfrak{p}}$. ∎

Le lemme 16 entraîne qu'il existe un ensemble fini P d'idéaux premiers minimaux de A , ensemble dépendant de B et du choix de la base (e_i), tel que $c_{ij}^{k}\in A_{\mathfrak{p}}$ pour tout $\mathfrak{p}\notin P$. Pour $\mathfrak{p}\notin P$ soit $B_{\mathfrak{p}}^{*}$ le $A_{\mathfrak{p}}$-sous-module de B des combinaisons linéaires des e_i à coefficients dans $A_{\mathfrak{p}}$; c'est un $A_{\mathfrak{p}}$-module libre. Comme $c_{ij}^{k}\in A_{\mathfrak{p}}$ pour tout $\mathfrak{p}\notin P$, $B_{\mathfrak{p}}^{*}$ est aussi un sous-anneau de B et en fait une sous-$A_{\mathfrak{p}}$-algèbre de B. Comme chaque $k(\mathfrak{p})$ est aussi un $A_{\mathfrak{p}}$-module par la projection canonique, nous pouvons définir la réduite modulo $\mathfrak{p}$ par :

$$B(\mathfrak{p}) = B^{*}\otimes_{A_{\mathfrak{p}}} k(\mathfrak{p}) .$$

PROPOSITION 35 - <u>Dans $B(\mathfrak{p})$ les éléments $\varepsilon_i = e_i\otimes 1$ forment une base sur $k(\mathfrak{p})$. Et les constantes de structure de $B(\mathfrak{p})$ dans cette base sont les réduites c_{ij}^{k*} des c_{ij}^{k} modulo $\mathfrak{p}$.</u>

Preuve : Comme $B^*_{\mathfrak{p}}$ est un $A_{\mathfrak{p}}$-module libre de base (e_i), $B(\mathfrak{p})$ est un $k(\mathfrak{p})$-module de base (ε_i). Comme $e_i.e_j = \sum c^k_{ij}\, e_k$ la réduction modulo $\mathfrak{p}$ entraîne $\varepsilon_i.\varepsilon_j = \sum c^{k*}_{ij}\, \varepsilon_k$. A priori les réduites $B(\mathfrak{p})$ dépendent du choix de la base (e_i) de B. Mais :

PROPOSITION 36 - Soient $B(\mathfrak{p})$ et $B'(\mathfrak{p})$ deux systèmes de réduites modulo $\mathfrak{p}$ correspondant au choix des bases (e_i) et (e'_i) de B sur K. Il existe un ensemble fini Q d'idéaux premiers minimaux $\mathfrak{p}$ de A tels que pour tout $\mathfrak{p} \notin Q$ les algèbres $B(\mathfrak{p})$ et $B'(\mathfrak{p})$ soient $k(\mathfrak{p})$-isomorphes.

Preuve : Introduisons la matrice de passage $e'_i = \sum u^j_i\, e_j$. Il existe un ensemble fini Q d'idéaux premiers minimaux tel que $u^j_i \in A_{\mathfrak{p}}$ pour tous i,j et tout $\mathfrak{p} \notin Q$. La matrice des (u^j_i) est alors inversible pour chaque $A_{\mathfrak{p}}$; la matrice des réduites (u^{j*}_i) modulo $\mathfrak{p}$ est aussi inversible. Et elle définit alors l'isomorphisme cherché entre $B(\mathfrak{p})$ et $B'(\mathfrak{p})$.∎

Le système des réduites est donc caractérisé à un ensemble fini d'idéaux premiers minimaux près. C'est ce que nous allons retrouver par le biais des ultraproduits.

PROPOSITION 37 - Soit $B(\mathfrak{p})$ un système de réduites de B, définies sauf pour les $\mathfrak{p}$ d'un ensemble fini P (pour $\mathfrak{p} \in P$ on posera $B(\mathfrak{p}) = k(\mathfrak{p})$). Pour tout ultrafiltre (non principal) $\mathfrak{U}$ sur $\mathfrak{M}$ la $\prod_{\mathfrak{U}} k(\mathfrak{p})$-algèbre ultraproduit $\prod_{\mathfrak{U}} B(\mathfrak{p})$ est $\prod_{\mathfrak{U}} k(\mathfrak{p})$-isomorphe à l'algèbre $B \otimes_K \prod_{\mathfrak{U}} k(\mathfrak{p})$.

Preuve : Rappelons que d'après le corollaire 2 du lemme 14 le corps ultraproduit des $k(\mathfrak{p})$ est une extension de K. Soit (e_i) une base de B sur K. Pour $\mathfrak{p} \notin P$ on obtient une base de la réduite $B(\mathfrak{p})$ par $\varepsilon_i(\mathfrak{p}) = (e_i \otimes 1)$. Si ε_i est l'élément de $\prod_{\mathfrak{U}} B(\mathfrak{p})$ classe de $(\varepsilon_i(\mathfrak{p}))$, les ε_i forment une base de $\prod_{\mathfrak{U}} B(\mathfrak{p})$ sur $\prod_{\mathfrak{U}} k(\mathfrak{p})$, d'après la proposition 17 du §.2 n°4, puisque l'ensemble fini P n'appartient pas à l'ultrafiltre $\mathfrak{U}$. On définit un isomorphisme $\prod_{\mathfrak{U}} k(\mathfrak{p})$-linéaire F de $B \otimes_K \prod_{\mathfrak{U}} k(\mathfrak{p})$ dans $\prod_{\mathfrak{U}} B(\mathfrak{p})$, par $F(e_i \otimes 1) = \varepsilon_i$. Reste à voir que F est un isomorphisme d'algèbres, ce qui se démontre par le biais des constantes de structure. On a :

$$(e_i \otimes 1)(e_j \otimes 1) = \sum c^k_{ij}(e_k \otimes 1) \quad \text{et}$$

$$\varepsilon_i.\varepsilon_j = \sum \widetilde{(c^{k*}_{ij})}\, \varepsilon_k \quad \text{(cf. la proposition 35).}$$

Mais $\widetilde{(c^{k*}_{ij})}$ est l'élément de $\prod_{\mathfrak{U}} k(\mathfrak{p})$ image de c^k_{ij} par l'homomorphisme

$K \longrightarrow \prod_{\mathfrak{U}} k(\mathfrak{p})$. ∎

Le corollaire suivant est un des résultats les plus importants de ce travail :

COROLLAIRE - <u>Soit</u> B <u>une</u> <u>K-algèbre</u> <u>simple</u> <u>centrale</u>. <u>On</u> <u>considère</u> <u>un</u> <u>système</u> <u>de</u> <u>réduites</u> $(B(\mathfrak{p}))$ <u>de</u> B <u>en les idéaux premiers minimaux de</u> A. <u>A l'exception peut-être d'un nombre fini d'entre elles les réduites</u> $B(\mathfrak{p})$ <u>sont des</u> $k(\mathfrak{p})$<u>-algèbres simples centrales</u>.

<u>Preuve</u> : Comme B est une K-algèbre simple centrale, l'algèbre $B \otimes_K \prod_{\mathfrak{U}} k(\mathfrak{p})$ est une $\prod_{\mathfrak{U}} k(\mathfrak{p})$-algèbre simple centrale pour tout ultrafiltre $\mathfrak{U}$ non principal sur l'ensemble $\mathfrak{M}$ des idéaux premiers minimaux de A. D'après la proposition 37 l'algèbre $B \otimes_K \prod_{\mathfrak{U}} k(\mathfrak{p})$ est isomorphe à l'ultraproduit des réduites $B(\mathfrak{p})$. Il en résulte que tout ultraproduit de réduites sur un ultrafiltre non principal est une algèbre simple centrale. Le corollaire est alors une conséquence du corollaire 2 de la proposition 28 du §.3 n°2. ∎

n°2. <u>LE THEOREME PRINCIPAL</u>.

<u>Dans tout ce numéro</u> A <u>est un anneau de Krull de caractéristique</u> 0, <u>de corps des fractions</u> K, <u>d'ensemble des idéaux premiers minimaux</u> $\mathfrak{M}$ <u>infini</u>.

Pour tout $\mathfrak{p} \in \mathfrak{M}$ on notera $\widehat{K}_{\mathfrak{p}}$ le complété de K pour la valuation discrète définie par $A_{\mathfrak{p}}$, $\widehat{A}_{\mathfrak{p}}$ l'anneau de valuation de $\widehat{K}_{\mathfrak{p}}$ qui est aussi le complété de $A_{\mathfrak{p}}$. Le corps résiduel de $\widehat{A}_{\mathfrak{p}}$ est ainsi isomorphe à $k(\mathfrak{p})$. Pour tout $\mathfrak{p} \in \mathfrak{M}$, on a une injection canonique de K dans $\widehat{K}_{\mathfrak{p}}$, d'où l'on déduit un homomorphisme injectif de K dans $\prod_{\mathfrak{U}} \widehat{K}_{\mathfrak{p}}$. Compte-tenu du lemme 15 on a :

LEMME 17 - <u>L'image de</u> K <u>est contenue dans le sous-anneau</u> $\prod_{\mathfrak{U}} \widehat{A}_{\mathfrak{p}}$ <u>de</u> $\prod_{\mathfrak{U}} \widehat{K}_{\mathfrak{p}}$ <u>pour tout ultrafiltre non principal sur</u> $\mathfrak{M}$. ∎

L'anneau ultraproduit $\prod_{\mathfrak{U}} \widehat{A}_{\mathfrak{p}}$ est un anneau ultraproduit d'anneaux de valuation discrètes complets et est donc un anneau hensélien (cf. Proposition 33 du §.4 n°1) d'égale caractéristique 0 (cf. Corollaire 2 du lemme 14 du n°1). D'après la proposition 34 du §.4 n°2 le sous-corps K de $\prod_{\mathfrak{U}} \widehat{A}_{\mathfrak{p}}$ (cf. lemme 17) est contenu dans un sous-corps de Cohen S de l'anneau local $\prod_{\mathfrak{U}} A_{\mathfrak{p}}$. Comme le corps résiduel de $\widehat{A}_{\mathfrak{p}}$ est isomorphe à $k(\mathfrak{p})$, S est isomorphe à l'ultraproduit des $k(\mathfrak{p})$.

Soit B une K-algèbre simple centrale. Pour tout ultrafiltre non principal $\mathfrak{U}$ sur $\mathfrak{M}$ on a les isomorphismes :

$$B \otimes_K \prod_{\mathfrak{U}} \widehat{K}_{\mathfrak{p}} \simeq \prod_{\mathfrak{U}} B \otimes_K \widehat{K}_{\mathfrak{p}} \quad \text{et} \quad B \otimes_K \prod_{\mathfrak{U}} \widehat{K}_{\mathfrak{p}} \simeq (B \otimes_K S) \otimes_S \prod_{\mathfrak{U}} \widehat{K}_{\mathfrak{p}} .$$

La proposition 37 montre que $B \otimes_K S$ est isomorphe à l'ultraproduit des réduites $\prod_{\mathfrak{U}} B(\mathfrak{p})$. D'où la proposition :

PROPOSITION 38 - Soient B une K-algèbre simple centrale, $\mathfrak{U}$ un ultrafiltre non principal sur l'ensemble $\mathfrak{M}$ des idéaux premiers minimaux de A. Si $\mathfrak{U}$-presque toutes les réduites modulo $\mathfrak{p}$ de B sont des algèbres de matrices, l'algèbre $\prod_{\mathfrak{U}} B \otimes_K \widehat{K}_{\mathfrak{p}}$ est une algèbre de matrices sur le corps ultraproduit des $\widehat{K}_{\mathfrak{p}}$.

La proposition 38 jointe au corollaire 2 de la proposition 29 du §.3 n°2 entraîne le théorème suivant, qui est le résultat principal de cet article :

THEOREME - Soient A un anneau de Krull de caractéristique 0, de corps des fractions K, d'ensemble des idéaux premiers minimaux $\mathfrak{M}$ infini. On considère une K-algèbre simple centrale dont les réduites $B(\mathfrak{p})$ en les $\mathfrak{p} \in \mathfrak{M}$ sont des algèbres de matrices, à l'exception peut-être d'un nombre fini d'entre elles. Alors les algèbres $B_{\mathfrak{p}} = B \otimes_K \widehat{K}_{\mathfrak{p}}$, obtenues par extension des scalaires de K aux complétés $\widehat{K}_{\mathfrak{p}}$ de K en les $\mathfrak{p} \in \mathfrak{M}$, sont, à l'exception peut-être d'un nombre fini d'entre elles, des algèbres de matrices.

En prenant pour A un anneau de Dedekind ou l'anneau de Dedekind des entiers algébriques dans un corps de nombres on retrouve le Théorème et l'assertion (A) de l'introduction.

Les raisonnements qui conduisent au Théorème précédent sont utilisables dans d'autres situations. Citons seulement la :

PROPOSITION 39 - Les hypothèses étant celles du Théorème, soit D un corps de centre K et de dimension finie sur K. On suppose que sauf pour un nombre fini de $\mathfrak{p}$ les $D_{\mathfrak{p}} = D \otimes_K \widehat{K}_{\mathfrak{p}}$ soient encore des corps. Alors à l'exception peut-être d'un nombre fini de $\mathfrak{p}$ les réduites $D(\mathfrak{p})$ sont des corps.

Preuve : Procédons par l'absurde. Il existe alors un ensemble infini W d'idéaux premiers minimaux de A tels que les réduites $D(\mathfrak{p})$ ne soient pas des corps ; on peut trouver un ultrafiltre non principal $\mathfrak{U}$ sur $\mathfrak{M}$ contenant W.

D'après la proposition 30 du §.3 n°2 l'utraproduit des $D(\mathfrak{p})$ sur $\mathfrak{U}$ n'est pas un corps : c'est une algèbre simple centrale qui contient des idempotents non triviaux. Si S est un corps de Cohen de $\prod_{\mathfrak{U}} \hat{A}_{\mathfrak{p}}$ contenant K, on a les isomorphisme $\prod_{\mathfrak{U}} D(\mathfrak{p}) \simeq D \otimes_K S$ et $\prod_{\mathfrak{U}} D \otimes_K \hat{K}_{\mathfrak{p}} \simeq (D \otimes_K S) \otimes_S \prod \hat{K}_{\mathfrak{p}}$. Il en résulte que l'algèbre ultraproduit des $D_{\mathfrak{p}}$ contient des idempotents non triviaux ; elle ne peut pas être un corps.∎

BIBLIOGRAPHIE

[1] J. AX et S. KOCHEN : Diophantine problems over local fields. I. Amer. J. of Maths., 87, 1965, p.605-630.

[2] J. AX : A Metamathematical approach to some problems in number theory. Symposia in Pure Mathematics 20, 1969, A.M.S., 1971.

[3] O.F.G. SCHILLING : The theory of valuations, A.M.S., Math. Surveys, n°IV, New-York, 1950.

[4] N. BOURBAKI : Algèbre, chapitre VIII, Hermann, Paris, 1958.

[5] J.P. SERRE : Corps locaux, Hermann, Paris, 1968.

[6] A. WEIL : Basic number theory, Springer, 1967.

[7] J.L. BELL et A.B. SLOMSON : Models and ultraproducts : an introduction, North-Holland, Amesterdam, 1969.

CARACTERISATION DES ALGEBRES DE REPRESENTATION FINIE SUR DES CORPS ALGEBRIQUEMENT CLOS

par Maher ZAYED

INTRODUCTION

Les algèbres de représentation finie jouent un rôle important dans la théorie des représentations d'algèbres. On connaît de nombreux exemples d'algèbres de représentation finie (non-commutatives) : voir par exemple [5].

Plusieurs caractéristions de ces algèbres ont été données, par exemple [1],[4],[9].

Le but du présent travail est d'obtenir une caractérisation simple des algèbres de représentation finie, s'appuyant sur les ultraproduits.

Toutes les algèbres sont supposées de dimension finie sur le corps de base. De même tous les modules sur une K-algèbre sont supposés de dimension finie sur K. Tous les ultrafiltres sont supposés non-principaux.

§.1 ULTRAPRODUITS

Nous regroupons dans ce paragraphe un certain nombre de résultats plus ou moins bien connus sur la cardinalité des ultraproduits d'ensembles. D'une façon générale on pourra consulter l'ouvrage de Bell et Slomson [(2).ch.6 §.3].

Cardinalité des Ultraproduits.

Soient I un ensemble, $\mathcal{U}$ un ultrafiltre sur I, $(X_i)_{i\in I}$ une famille d'ensembles indexée par I. On désigne par $\prod_{\mathcal{U}} X_i$, l'ultraproduit de la famille $(X_i)_{i\in I}$. Si pour tout $i\in I$, $X_i = X$, l'ultraproduit sera noté $X^{\mathcal{U}}$ et sera appelé l'ultrapuissance de X.

Lemme 1.1 - Soient α un cardinal fini et $\mathcal{U}$ un ultrafiltre sur I. Alors : $\mathrm{card}(\alpha^{\mathcal{U}}) = \alpha$.

Preuve : Comme α est fini, l'ultrafiltre $\mathcal{U}$ est α-complet et on a donc $\alpha^{\mathcal{U}} \simeq \alpha$ ∎

Corollaire 1.2 - Soient $\{\alpha_i \mid i\in I\}$ une famille de cardinaux finis, et $\mathcal{U}$ un ultrafiltre sur I. S'il existe un entier $n\in\mathbb{N}$ tel que $\{i\in I \mid \alpha_i = n\}\in\mathcal{U}$, alors $\mathrm{card}\,(\prod_{\mathcal{U}} \alpha_i) = n$.

Preuve : Pour tout entier $p\in\mathbb{N}$, soit $X_p = \{i\in I \mid i = p\}$. Comme les X_p sont deux à deux disjoints, il existe un et un seul des ensembles X_p dans $\mathcal{U}$: c'est X_n. On a alors un bijection de $\prod_{\mathcal{U}} \alpha_i$ sur n.

Corollaire 1.3 - Soient $\{\alpha_i \mid i\in I\}$ une suite bornée de cardinaux finis et $\mathcal{U}$ un ultrafiltre sur I, alors $\mathrm{card}(\prod_{\mathcal{U}} \alpha_i)$ est fini.

Preuve : Supposons que $\alpha_i \leq n$ pour tout $i\in I$. Si $I_p = \{i\in I \mid \alpha_i = p\}$, on a $I = I_1\cup\ldots\cup I_n$, d'où $I_p\in\mathcal{U}$ pour un p.

Lemme 1.4 - Soient $\{\alpha_n \mid n\in\mathbb{N}\}$ une suite de cardinaux finis et $\mathcal{U}$ un ultrafiltre sur $\mathbb{N}$ tel que $\mathcal{U}$ ne contienne aucun ensemble de la forme $X_n = \{i\in I \mid \alpha_i = n\}$. Alors $\mathrm{Card}\,(\prod_{\mathcal{U}} \alpha_i) = 2^{\aleph_0}$.

Preuve : Voir [2], (ch.6 §.3 Th.3.12).

Corollaire 1.5 - Soient $(D_n)_{n\in\mathbb{N}}$ une suite d'ensembles finis ou infinis dénombrables et $\mathcal{U}$ un ultrafiltre non-principal sur $\mathbb{N}$. On suppose que pour tout entier p, l'ensemble $\{n \mid D_n$ a au moins p éléments$\}$ appartient à $\mathcal{U}$. Alors $\prod_{\mathcal{U}} D_n$ a la puissance du continu.

Preuve : Comme $\prod_{\mathcal{U}} D_n$ est un quotient de $\prod_{n\in\mathbb{N}} D$, $\prod_{\mathcal{U}} D_n$ a au plus la puissance du continu. Posons :

$$D'_n = \begin{cases} D_n & \text{si } D_n \text{ est fini} \\ \text{un sous-ensemble à } n \text{ éléments de } D_n, & \text{si } D_n \text{ est infini.} \end{cases}$$

Les D'_n sont finis et pour tout p, $\{n|D'_n$ a p éléments$\} \notin \mathcal{U}$; en effet si $\{n|D'_n$ a p éléments$\} \in \mathcal{U}$, $\{n|D'_n$ a p éléments$\}$ serait infini (car $\mathcal{U}$ est non-principal). Par définition de D'_n, si D'_n a p éléments et si $n \neq p$, $D'_n = D_n$. D'où, $\mathcal{U}$ n'étant pas principal, $\{n|D_n$ a p éléments$\} \in \mathcal{U}$, en contradiction avec l'hypothèse. Donc $\{n|D'_n$ a p éléments$\} \notin \mathcal{U}$ pour tout p. D'après le lemme 2, $\prod_{\mathcal{U}} D'_n$ a la puissance du continu. D'où

$$\text{Card}(\prod_{\mathcal{U}} D_n) \geq \text{card}(\prod_{\mathcal{U}} D'_n) \geq 2^{\aleph_0} .$$

<u>Lemme</u> 1.6 - <u>Si</u> X <u>est un ensemble infini, pour tout cardinal</u> α <u>il existe une ultrapuissance</u> $X^{\mathcal{U}}$ <u>de</u> X <u>telle que</u> $\text{Card}(X^{\mathcal{U}}) \geq \alpha$.

<u>Preuve</u> : Voir [2] (ch.6 §.3 Th.3.21 et §.1 lemme.1.14).

<u>Applications aux Corps algébriquement clos</u>.

Soient K un corps et L une extension de K. On notera $tr_K L$ le degré de transcendance de l'extension L/K, c'est-à-dire le cardinal fini ou non d'une base de transcendance de L sur K.

Le théorème suivant est bien connu :

<u>Théorème de Steinitz</u> - <u>Soient</u> K <u>un corps</u>, L <u>et</u> Ω <u>deux extensions de</u> K, Ω <u>étant algébriquement clos. Si</u> $tr_K L \leq tr_K \Omega$ <u>il existe un</u> K-<u>isomorphisme de</u> L <u>dans</u> Ω . <u>Si</u> L <u>est algébriquement clos et si</u> $tr_K L = tr_K \Omega$ <u>il existe un</u> K-<u>isomorphisme de</u> L <u>sur</u> Ω .

On a d'autre part les résultats suivants :

<u>Lemme de Cantor</u> - <u>Si</u> K <u>est un corps infini</u>, K <u>et sa clôture algébrique</u> $\bar{K}$ <u>ont le même cardinal</u>.

<u>Lemme</u> 1.7 - <u>Soient</u> L <u>une extension de</u> K, B <u>une base de transcendance de</u> L <u>sur</u> K; L <u>est équipotent à</u> K $\times$ B <u>si l'un des ensembles</u> K, B <u>est infini</u> ; <u>c'est-à-dire que</u> $\text{card}(K(B)) = \text{card}(K) \times \text{card}(B)$.

<u>Preuve</u> : Voir [3], (ch.V §.5 ex.2) et [6] (ch.IV).

Corollaire 1.8 - Soit L une extension d'un corps infini K. On suppose que $\mathrm{card}(L) > \mathrm{card}(K)$. Alors $\mathrm{tr}_K L = \mathrm{card}(L)$.

Preuve : Si B est une base de transcendance de L sur K, on a :

$$\begin{aligned}\mathrm{card}(L) &= \mathrm{card}(K(B)) \quad \text{(lemme de (Cantor))}\\ &= \mathrm{card}(K) \times \mathrm{card}(B)\end{aligned}$$

Comme $\mathrm{card}(K) < \mathrm{card}(L)$, on a $\mathrm{card}(B) > \mathrm{card}(K)$. D'où $\mathrm{card}(L) = \mathrm{card}(B)$.

Proposition 1.9 - Soit $(K_n)_{n\in\mathbb{N}}$ une suite de corps algébriquement clos dénombrables de caractéristique 0. Si $\mathcal{U}$ un ultrafiltre sur $\mathbb{N}$ (non-principal), $\prod_{\mathcal{U}} K_n$ est isomorphe à $\mathbb{C}$.

Preuve : Comme $\prod_{\mathcal{U}} K_n$ est de caractéristique 0, le sous-corps premier P de $\prod_{\mathcal{U}} K_n$ est isomorphe à $\mathbb{Q}$, corps de nombres rationnels. D'après corollaire 1.5, $\prod_{\mathcal{U}} K_n$ a la puissance du continu. Donc le degré de transcendance de $\prod_{\mathcal{U}} K_n$ sur P est $2^{\aleph_0}$, P étant dénombrable. On peut alors appliquer le théorème de Steinitz, en remarquant que $\prod_{\mathcal{U}} K_n$ est algébriquement clos et que $\mathrm{tr}_P \prod_{\mathcal{U}} K_n = \mathrm{tr}_{\mathbb{Q}}\mathbb{C}$.

Proposition 1.10 - Soit K un corps algébriquement clos et soit L une extension de K, il existe un K-isomorphisme de L dans une ultrapuissance $K^{\mathcal{U}}$ de K.

Preuve : $K^{\mathcal{U}}$ est algébriquement clos. Soit α le degré de transcendance de L sur K et soit $\beta = \sup(\alpha, \mathrm{card}\, K)$. On peut trouver une ultrapuissance de K telle que $\mathrm{card}\, K^{\mathcal{U}} \geq 2^{\beta}$ (cf. lemme 1.6). Alors $\mathrm{tr}_K K^{\mathcal{U}} = \mathrm{card}\, K^{\mathcal{U}} \geq 2^{\beta} \geq \alpha$. On applique alors le théorème de Steinitz.

Associativité des Produits Tensoriels.

Soient $j : K \longrightarrow L$ un isomorphisme de corps et M un K-espace vectoriel. Nous noterons $M \otimes_{K,j} L$ le produit tensoriel de M et L considéré comme K-espace vectoriel à l'aide de j ; $M \otimes_{K,j} L$ est un L-espace vectoriel. Par conséquent si M est un K-espace vectoriel, $i : K \longrightarrow L$ et $j : L \longrightarrow \Omega$ deux isomorphismes de corps commutatifs, il existe un isomorphisme de Ω-espaces vectoriels entre $(M \otimes_{K,i} L) \otimes_{L,j} \Omega$ et $M \otimes_{K,j.i} \Omega$. (voir [3], ch. II §.5 n°1, Prop. 2).

Soit L une extension algébriquement close du corps K algébriquement clos. D'après la proposition 1.10, il existe un K-isomorphisme j de L dans une ultrapuissance $K^{\mathcal{U}}$ de K.

Proposition 1.11 - Il existe un isomorphisme ϕ de $K^{\mathcal{U}}$ dans une ultrapuissance L^{ϑ} de L tel que le diagramme suivant soit commutatif :

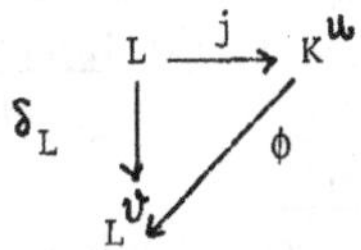

où δ_L, l'isomorphisme diagonal de L dans L^{ϑ}, est défini par $\delta_L(\lambda) = (\lambda)$ pour tout $\lambda \in L$.

Preuve : En effet, soit L' l'image de j dans $K^{\mathcal{U}}$, $L' \subset K^{\mathcal{U}}$. Il existe une ultrapuissance L'^{ϑ} de L' et un L'-isomorphisme ψ de $K^{\mathcal{U}}$ dans L'^{ϑ} (corollaire du Th. de Steinitz). D'autre part $(j^{-1})^{\vartheta}$ est un isomorphisme de L'^{ϑ} sur L^{ϑ}, où $(j^{-1})^{\vartheta}(\widetilde{x_i}) = \widetilde{(j^{-1}(x_i))}$, pour tout $\widetilde{(x_i)} \in L'^{\vartheta}$.

Posons $\phi = (j^{-1})^{\vartheta} \circ \psi$. Si $\lambda \in L$, $j(\lambda) \in K^{\mathcal{U}}$ et :

$$\phi . j(\lambda) = (j^{-1})^{\vartheta} \circ (j(\lambda)).$$

Comme $j(\lambda) \in L'$ et comme ψ est un L'-isomorphisme, on a: $\psi . j(\lambda) = j(\lambda) . \psi(1) = \delta_{L'}(j(\lambda))$. Alors $(j^{-1})^{\vartheta} . \psi(j(\lambda)) = (j^{-1})^{\vartheta} . \delta_{L'}(j(\lambda)) = \delta_L(\lambda)$ pour tout $\lambda \in L$, c'est-à-dire que $\phi \circ j = \delta_L$.

Corollaire 1.12 - Soient L une extension algébriquement close du corps K algébriquement clos et P un L-espace vectoriel. Il existe un isomorphisme de L^{ϑ}-espaces vectoriels entre :

$$(P \underset{L,j}{\otimes} K^{\mathcal{U}}) \underset{K^{\mathcal{U}},\phi}{\otimes} L^{\vartheta} \quad \text{et} \quad P \underset{L,\delta_L}{\otimes} L^{\vartheta}$$

§.2 ALGEBRE DE REPRESENTATION FINIE

Nous démontrons dans ce paragraphe des résultats qui sont essentiels pour la démonstration du théorème principal.

Modules sur l'Algèbre $A^{\mathcal{U}} \cong A \underset{K}{\otimes} K^{\mathcal{U}}$:

Soient $(K_i)_{i \in I}$ une famille de corps commutatifs, K le corps ultraproduit de K_i, $\mathcal{U}$ un ultrafiltre non-principal sur I et A une K-algèbre de dimension finie, identifiée à $\prod_{\mathcal{U}} A_i$ (prop. 21 [8]). La première partie de la proposition suivante est démontrée en ([8] prop. 23).

<u>Proposition</u> 2.1 - <u>Soient</u> M <u>et</u> N <u>deux</u> A-<u>modules</u> (<u>de dimension finie sur</u> K), M <u>identifié à</u> $\prod_{\mathcal{U}} M_i$ <u>et</u> N <u>identifié à</u> $\prod_{\mathcal{U}} N_i$. <u>Tout homomorphisme</u> f <u>de</u> M <u>dans</u> N <u>est de la forme</u> $\widetilde{(f_i)}$ <u>où les</u> f_i <u>sont</u> $\mathcal{U}$-<u>presque tous des homomorphismes du module</u> M_i <u>dans le module</u> N_i ; f <u>est injective</u> (<u>resp. surjective, resp. bijective</u>) <u>si et seulement si les</u> f_i <u>sont</u> $\mathcal{U}$-<u>presque toutes injectives</u> (<u>resp. surjective, resp. bijective</u>).

<u>Preuve</u> : Soient $(e_j)_{j=1,\dots,m}$ et $(\epsilon_k)_{k=1,\dots,n}$ des bases de $\prod_{\mathcal{U}} M_i$ et $\prod_{\mathcal{U}} N_i$ respectivement (sur $\prod_{\mathcal{U}} K_i$). Introduisons des systèmes de représentants $(e_j(i))$ et $(\epsilon_k(i))$. On peut trouver un ensemble $E \in \mathcal{U}$ tel que pour tout $i \in E$ la famille $(e_j(i))$ soit une base de M_i et la famille $(\epsilon_k(i))$ une base de N_i (sur K_i). L'application f est entièrement déterminée par sa matrice dans les bases (e_j) et (ϵ_k) :

$$f(e_j) = \sum u_j^k \epsilon_k .$$

Les u_j^k étant dans $\prod_{\mathcal{U}} K_i$, nous choisirons des systèmes de représentants $u_j^k(i) \in K_i$, et nous définirons pour $i \in E$, l'application linéaire f_i de M_i dans N_i par :

$$f_i(e_j(i)) = \sum u_j^k(i)\, \epsilon_k(i) .$$

Pour $i \notin E$, nous poserons $f_i = 0$, il est alors immédiat de vérifier que $f(e_j) = \widetilde{(f_i)}(e_j)$.

Il reste à voir que les f_i sont des homomorphismes de $(\prod_{\mathcal{U}} A_i)$-modules. Soit $\Omega = \{i \in I \mid f_i$ est un A_i-homomorphisme$\}$; pour tout $i \notin \Omega$ on peut trouver $c_i \in A_i$, $x_i \in M_i$ tels que $f_i(c_i x_i) \neq c_i f_i(x_i)$. Supposons que $\Omega \notin \mathcal{U}$; alors $(I-\Omega) \in \mathcal{U}$. On pose, $c = \widetilde{(c_i)}$, $x = \widetilde{(x_i)}$ et on a $f(x) \neq cf(x)$, avec $c \in \prod_{\mathcal{U}} A_i$, $x \in M = \prod_{\mathcal{U}} M_i$. D'où $(I-\Omega) \notin \mathcal{U}$ et, comme $\mathcal{U}$ est un ultrafiltre, $\Omega \in \mathcal{U}$. Pour que f soit injective, il faut et il suffit que la famille $f(e_j)$ soit libre. Comme $f(e_j) = \widetilde{(f_i)}(e_j) = \widetilde{(f_i(e_j(i)))}$, $f(e_j)$ est libre si et seulement si les $(f_i(e_j)(i)))$ sont $\mathcal{U}$-presque toutes libres D'où le critère d'injectivité pour f. On démontre de la même façon les critères de surjectivité et de bijectivité.

<u>Corollaire</u> 2.2 - <u>Soient</u> A <u>une</u> K-<u>algèbre</u>, M <u>et</u> N <u>deux</u> A-<u>modules</u>. M <u>est isomorphe à</u> N <u>si et seulement si</u> $M^{\mathcal{U}}$ <u>est isomorphe à</u> $N^{\mathcal{U}}$ (<u>comme</u> $A^{\mathcal{U}}$-<u>module</u>).

<u>Preuve</u> : Appliquons la proposition précédente aux familles $M_i = M$ et $N_i = N$ pour tout $i \in I$.

Lemme 2.3 - Soit $(M_i)_{i\in I}$ une famille d'espaces vectoriels sur K avec $\sup \dim_K M_i < \infty$. On considère un ultrafiltre $\mathcal{U}$ et une application $K^{\mathcal{U}}$-linéaire p de $\prod_{\mathcal{U}} M_i$ dans lui-même. Alors p est un projecteur si et seulement si $\mathcal{U}$-presque tous les p_i sont des projecteurs.

Preuve : Vérifions que $p^2 = p$ sur une base de $\prod_{\mathcal{U}} M_i$ qui est finie. D'après [8] lemme 6, il existe un ensemble $\mathcal{R}\in\mathcal{U}$ tel que, pour tout $i\in\mathcal{R}$, p_i. est une application K-linéaire de M_i dans lui-même et $p = \widetilde{(p_i)}$. De $p^2 = p$ on déduit $\widetilde{(p_i^2)} = \widetilde{(p_i)}$ c'est-à-dire, les espaces étant de dimension finie, $p_i^2 = p_i$ pour $\mathcal{U}$-presque tous les i.

Proposition 2.4 - Soit $(M_i)_{i\in I}$ une famille de A_i-modules telles que $\sup \dim_{K_i}(M_i) < \infty$. Alors $\prod_{\mathcal{U}} M_i$ est un $\prod_{\mathcal{U}} A_i$-module indécomposables si et seulement si $\mathcal{U}$-presque tous les M_i sont indécomposables.

Preuve : Le $\prod_{\mathcal{U}} A_i$-module $\prod_{\mathcal{U}} M_i$ admet une décomposition si et seulement s'il existe un projecteur p de $\prod_{\mathcal{U}} M_i$ dans lui même tel que $p \neq o$, $p \neq 1$. D'après le lemme 2.3 , il existe une famille (p_i) d'applications linéaires de M_i dans lui-même tel que $p_i^2 = p_i$ pour $\mathcal{U}$-presque tous les i. Supposons que $p \neq 0$, c'est-à-dire $p_i \neq 0$ pour $\mathcal{U}$-presque tous les i. Si $\prod_{\mathcal{U}} M_i$ est indécomposable , p = id, i.e. p_i = id pour $\mathcal{U}$-presque tous les i, et M_i est indécomposable pour $\mathcal{U}$-presque tout i. Réciproquement si les M_i sont $\mathcal{U}$-presque tous indécomposables p = id pour $\mathcal{U}$-presque tous les i, alors p_i = id.

Corollaire 2.5 - Soit M un A-module ; M est indécomposable si et seulement si $M^{\mathcal{U}}$ est un $A^{\mathcal{U}}$-module indécomposable.

Preuve : On applique la proposition précédente à la famille $M_i = M$ pour tout $i\in I$.

Proposition 2.6 - Soit A une algèbre (de dimension finie) sur le corps K. L'ultrapuissance $A^{\mathcal{U}}$ est un $K^{\mathcal{U}}$-algèbre qui est isomorphe à $A \otimes K^{\mathcal{U}}$.

Preuve : Soient $(e_k)_{k=1,\dots,n}$ une K-base de l'algèbre A et (c_{rs}^t) les constantes de structure associée à cette base. Si on considère l'injection canonique $d : K \longrightarrow K^{\mathcal{U}}$, $(e_k \otimes 1)$ est une $K^{\mathcal{U}}$-base de l'algèbre $A \otimes_K K^{\mathcal{U}}$ et $\alpha_{rs}^t = d(c_{rs}^t) = \widetilde{(c_{rs}^t)}$ sont les constantes de structure associée à cette base. Appliquons le lemme 8 de [8] et remarquons que $c_{rs}^t(i) = c_{rs}^t$ $\mathcal{U}$-presque partout. Alors on a $A \otimes_K K^{\mathcal{U}} \simeq \prod_{\mathcal{U}} A_i$, où $A_i = A$ et c'est dire que $A \otimes_K K^{\mathcal{U}} \simeq A^{\mathcal{U}}$.

Proposition 2.7 - Soit M un A-module (de dimension finie sur K). L'ultrapuissance $M^{\mathcal{U}}$ est un $A^{\mathcal{U}}$-module qui est isomorphe à $M \otimes K^{\mathcal{U}}$.

Preuve : la démonstration est analogue à celle de 2.6.

Proposition 2.8 - *Soit A une K-algèbre. On suppose qu'il existe un entier n tel qu'il y ait une infinité de A-modules indécomposables de dimension $\leq n$, deux à deux non isomorphes. Alors il existe une ultrapuissance $K^{\mathcal{U}}$ et un module indécomposable sur $A^{\mathcal{U}}$ de dimension $\leq n$ sur $K^{\mathcal{U}}$ tel que M ne soit pas de la forme $X \underset{K}{\otimes} K^{\mathcal{U}}$, où X est A-module.*

Preuve : Par hypothèse il existe une suite X_p de A-modules indécomposables de dimension $\leq n$ deux à deux non isomorphes. Soit $\mathcal{U}$ un ultrafiltre non principal sur $\mathbb{N}$ et soit $M = \prod_{\mathcal{U}} X_p$: c'est un $K^{\mathcal{U}}$-espace vectoriel de dimension $\leq n$. Si M était de la forme $X \underset{K}{\otimes} K^{\mathcal{U}}$, alors $M \simeq \prod_{\mathcal{U}} X'_p$ avec $X'_p = X$ pour tout p. D'après la proposition 2.1 on a : $X_p \simeq X'_p = X$ pour $\mathcal{U}$-presque tous les X_p.

Modules sur l'algèbre $A_L = A \underset{K}{\otimes} L$

Soit K un corps commutatif, algébriquement clos, L une extension de K et A un K-algèbre (de dimension finie sur K).

Proposition 2.9 - *Soient A une algèbre sur le corps K algébriquement clos, M et N deux A-modules. On considère une extension L de K telle que $M \underset{K}{\otimes} L$ et $N \underset{K}{\otimes} L$ sont des $A_L (= A \underset{K}{\otimes} L)$-modules isomorphes. Alors M et N sont isomorphes.*

Preuve : (i) Supposons d'abord que $L = K^{\mathcal{U}}$. Alors, d'après la proposition 2.7, $M \otimes K^{\mathcal{U}} \simeq M^{\mathcal{U}}$, $N \otimes K^{\mathcal{U}} \simeq N^{\mathcal{U}}$, d'où un $A^{\mathcal{U}} (\simeq A \underset{K}{\otimes} K^{\mathcal{U}})$-isomorphisme de $M^{\mathcal{U}}$ sur $N^{\mathcal{U}}$. Il en résulte que (corollaire 2.2), M est isomorphe à N comme A-modules.

(ii) Considérons le cas général. Comme K est algébriquement clos et L une extension de K, il existe un K-isomorphisme de L dans une ultrapuissance $K^{\mathcal{U}}$ (proposition 1.10). De l'isomorphisme $M \underset{K}{\otimes} L \simeq N \underset{K}{\otimes} L$, on déduirait par l'extension de scalaires un isomorphisme: $(M \underset{K}{\otimes} L) \underset{L}{\otimes} K^{\mathcal{U}} \simeq (N \underset{K}{\otimes} L) \underset{L}{\otimes} K^{\mathcal{U}}$. D'après l'associativité des produits tensoriels, on a un isomorphisme :

$$M \underset{K}{\otimes} (L \underset{L}{\otimes} K^{\mathcal{U}}) \simeq N \underset{K}{\otimes} (L \underset{L}{\otimes} K^{\mathcal{U}}),$$

'est-à-dire que $M \underset{K}{\otimes} K^{\mathcal{U}} \simeq N \underset{K}{\otimes} K^{\mathcal{U}}$, et on a donc $M \simeq N$.

Proposition 2.10 - *Soient L une extension d'un corps algébriquement clos K, M un A-module indécomposable (de dimension finie sur K). Le $A_L (= A \underset{K}{\otimes} L)$-module $M \underset{K}{\otimes} L$ est indécomposable.*

Preuve : (i) Supposons que $L = K^{\mathcal{U}}$: on a démontré (corollaire 2.2) que M est indécomposable si et seulement si $M^{\mathcal{U}} \cong M \otimes_K K^{\mathcal{U}}$ est $A^{\mathcal{U}}$-module indécomposable.

(ii)Considérons le cas général. Comme K est algébriquement clos, il existe un K-isomorphisme de L dans une ultrapuissance $K^{\mathcal{U}}$ de K. Supposons que le A_L-module $M \otimes_K L$ admette une décomposition, $M \otimes L = X \oplus Y$, où X et Y sont A_L-modules. Par extension de scalaires, $(M \otimes_K L) \otimes_L K^{\mathcal{U}} = (X \oplus Y) \otimes_L K^{\mathcal{U}}$. Par associativité des produits tensoriels et distributivité $\otimes$ par rapport à $\oplus$, on a

$$M \otimes_K (L \otimes_L K^{\mathcal{U}}) \cong (X \otimes_L L^{\mathcal{U}}) \oplus (Y \otimes_L K^{\mathcal{U}}).$$

Mais on a $M \otimes_K K^{\mathcal{U}} \simeq M^{\mathcal{U}}$, qui est indécomposable comme $A^{\mathcal{U}}$-module, puisque M l'est; l'un de ces deux modules, $X \otimes_L K^{\mathcal{U}}$ ou $Y \otimes K^{\mathcal{U}}$, est nul, par exemple $Y \otimes_L K^{\mathcal{U}}$, donc $Y = 0$.

Algèbre de Représentation Finie.

Soit A une algèbre de dimension finie sur un corps commutatif algébriquement clos K. L'algèbre A est dite de représentation finie si le nombre de types (= classes d'isomorphismes) d'indécomposables $M \in \text{mod } A$ est fini.

Proposition 2.11 - L'algèbre A est de représentation finie si et seulement si l'algèbre $A^{\mathcal{U}} \cong A \otimes_K K^{\mathcal{U}}$ est de représentation finie.

Preuve : (i) Supposons que A soit de représentation finie et désignons par $M_1, \dots, M_n$ des modules représentant les types d'indécomposables de A. D'après le corollaire 2.2, les $M_1^{\mathcal{U}}, \dots M_n^{\mathcal{U}}$ sont des $A^{\mathcal{U}}$-modules indécomposables. Soit P un $A^{\mathcal{U}}$-module indécomposable, il existe, (Proposition 2.4), une famille $(X_i)_{i \in I}$ de A-modules indécomposables tels que $P \simeq \prod_{\mathcal{U}} X_i$. Chaque X_i est isomorphe à un des M_j, $1 \leq j \leq n$. Posons :

$$I_1 = \{i \in I \mid X_i \simeq M_1\}, \dots, I_n = \{i \in I \mid X_i \simeq M_n\}.$$

On a une partition finie $I = I_1 \cup \dots \cup I_n$. Comme $\mathcal{U}$ est un ultrafiltre, pour un k unique, $1 \leq k \leq n$, on a $I_k \in \mathcal{U}$, d'où $P \simeq M^{\mathcal{U}}$. C'est dire que l'ensemble $\{M_i^{\mathcal{U}} \mid 1 \leq i \leq n\}$ est un ensemble de représentants des types de $A^{\mathcal{U}}$-modules indécomposables.

(ii) Si A est de représentation infinie, il existe une famille infinie de A-modules indécomposables, deux à deux, non-isomorphes $(M_i)_{i \in I}$. D'après les propositions 2.1 et 2.4, les $M_i^{\mathcal{U}}$ sont des $A^{\mathcal{U}}$-modules indécomposables, deux à deux non-isomorphes et l'algèbre $A^{\mathcal{U}}$ est de représentation infinie.∎

<u>Proposition</u> 2.12 - <u>Soient</u> L <u>une extension algébriquement close d'un corps algébriquement clos</u> K, A <u>une</u> K-<u>algèbre et</u> P <u>un</u> $A_L = A \underset{K}{\otimes} L$-<u>module indécomposable</u>. <u>Pour tout</u> K-<u>isomorphisme de</u> L <u>dans une ultrapuissance</u> $K^{\mathfrak{U}}$, <u>le</u> $A_L \otimes K^{\mathfrak{U}} \simeq A^{\mathfrak{U}}$-<u>module</u> $P \underset{L}{\otimes} K^{\mathfrak{U}}$ <u>est indécomposable</u>.

<u>Preuve</u> : Notons d'abord que, par associativité des produits tensoriels, $A_L \underset{L,j}{\otimes} K^{\mathfrak{U}}$ est isomorphe à $A \underset{K,\delta_K}{\otimes} K^{\mathfrak{U}}$, où j le K-isomorphisme de L dans K et δ_K l'injection canonique de K dans $K^{\mathfrak{U}}$.

Considérons une décomposition en somme directe

$$P \underset{L,j}{\otimes} K^{\mathfrak{U}} = X \oplus Y.$$

Introduisons un isomorphisme ϕ de $K^{\mathfrak{U}}$ dans une ultrapuissance $L^{\mathfrak{V}}$ de L tel que le diagramme suivant :

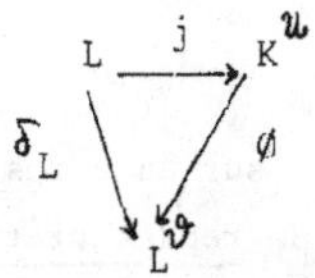

soit commutatif à l'aide de ϕ on a :

$$(P \underset{L,j}{\otimes} K^{\mathfrak{U}}) \underset{K^{\mathfrak{U}},\phi}{\otimes} L^{\mathfrak{V}} \cong (X \underset{K^{\mathfrak{U}},\phi}{\otimes} L^{\mathfrak{V}}) \oplus (Y \underset{K^{\mathfrak{U}},\phi}{\otimes} L^{\mathfrak{V}}).$$

Mais $(P \underset{L,j}{\otimes} K^{\mathfrak{U}}) \underset{K^{\mathfrak{U}},\phi}{\otimes} L^{\mathfrak{V}}$ est isomorphe à $P \underset{L,\delta_L}{\otimes} L^{\mathfrak{V}}$ et $P \underset{L,\delta_L}{\otimes} L^{\mathfrak{V}} \simeq P^{\mathfrak{V}}$ est indécomposable (cf. proposition 2.4). Alors on a soit $X \underset{K^{\mathfrak{U}},\phi}{\otimes} L^{\mathfrak{V}} = 0$ soit $Y \underset{K^{\mathfrak{U}},\phi}{\otimes} L^{\mathfrak{V}} = 0$, c'est dire que $X = 0$ ou $Y = 0$.

<u>Proposition</u> 2.13 - <u>Soit</u> A <u>une</u> K-<u>algèbre de représentation finie</u>. <u>Pour toute extension algébriquement close</u> L <u>de</u> K <u>et pour tout</u> $A_L = (A \underset{K}{\otimes} L)$-<u>module</u> P <u>il existe un</u> A-<u>module</u> M <u>tel que</u> P <u>soit isomorphe à</u> $M \underset{K}{\otimes} L$.

<u>Preuve</u> : Par le théorème de Krull-Schmidt, on peut supposer que le A_L-module P est indécomposable. Si j est un K-isomorphisme de L dans une ultrapuissance $K^{\mathfrak{U}}$ de K, $P \underset{L,j}{\otimes} K^{\mathfrak{U}}$ est un $A_L \underset{L,j}{\otimes} K^{\mathfrak{U}}$-module indécomposable (cf. proposition 2.12) ; remarquons que $A_L \underset{L,j}{\otimes} K^{\mathfrak{U}} \simeq A \underset{K,\delta_K}{\otimes} K^{\mathfrak{U}} \cong A^{\mathfrak{U}}$. Alors d'après la proposition 2.11, il existe un A-module indécomposable M tel que $P \underset{L,j}{\otimes} K^{\mathfrak{U}}$ soit

isomorphe à $M \underset{K,\delta_K}{\otimes} K^{\mathcal{U}} \cong M^{\mathcal{U}}$. Soit ϕ un isomorphisme de $K^{\mathcal{U}}$ dans une ultrapuissance $L^{\mathcal{V}}$ de L tel que le diagramme suivant :

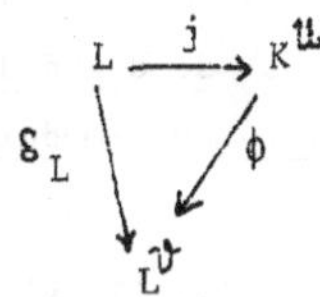

soit commutatif. Alors $(P \underset{L,j}{\otimes} K^{\mathcal{U}}) \underset{K^{\mathcal{U}},\phi}{\otimes} L^{\mathcal{V}} \simeq (M \underset{K,\delta_K}{\otimes} K^{\mathcal{U}}) \underset{K^{\mathcal{U}},\phi}{\otimes} K^{\mathcal{V}}$.

D'après l'associativité des produits tensoriels, $(P \underset{L,j}{\otimes} K^{\mathcal{U}}) \underset{K^{\mathcal{U}},\phi}{\otimes} L^{\mathcal{V}}$ est isomorphe à $P \underset{L,\delta_L}{\otimes} L^{\mathcal{V}} \simeq P^{\mathcal{V}}$; de même on a :

$(M \underset{K,\delta_K}{\otimes} P^{\mathcal{U}}) \underset{K^{\mathcal{U}},\phi}{\otimes} L^{\mathcal{V}} \simeq M \underset{K,\phi\circ\delta_K}{\otimes} L^{\mathcal{V}} \simeq (M \underset{K,i}{\otimes} L) \underset{L,\delta_L}{\otimes} L^{\mathcal{V}}$, où i est l'injection canonique de K dans L. Alors P est isomorphe à $(M \underset{K,i}{\otimes} L) \underset{L,\delta_L}{\otimes} L^{\mathcal{V}}$ comme $(A \underset{K,i}{\otimes} L) \underset{L,\delta_L}{\otimes} L^{\mathcal{V}}$-modules ; c'est-à-dire que $P^{\mathcal{V}} \simeq (M \underset{K,i}{\otimes} L)^{\mathcal{V}}$ et, d'après la proposition 1.11, on a $P \simeq M \underset{K,i}{\otimes} L$.

Algèbres Closes

Soit A une algèbre de dimension finie sur un corps algébriquement clos K. Il est bien connu que si L est une extension de K et si P est un $A_L (= A \underset{K}{\otimes} L)$-module semi-simple de dimension finie sur L, il existe un A-module (semi-simple) M tel que $P \cong M \underset{K}{\otimes} L$.

<u>Définition</u> - On dira que l'algèbre est <u>close</u> si elle satisfait la condition suivante :

"<u>Pour toute extension algébriquement close</u> L <u>de</u> K <u>et tout</u> $A_L (= A \underset{K}{\otimes} L)$-<u>module</u> P <u>de dimension finie sur</u> L, <u>il existe un A-module</u> M <u>de dimension finie sur</u> K <u>tel que</u> $P \cong M \underset{K}{\otimes} L$".

Toute algèbre de représentation finie est une algèbre close, d'après la proposition 2.13 .

<u>Proposition</u> 2.14 - <u>Soit</u> A <u>une algèbre sur un corps algébriquement clos</u> K. <u>Il y a équivalence entre</u> :

(a) A <u>est close</u>.

(b) Pour toute extension algébriquement close L de K et tout $A_L(= A \otimes_K L)$-module indécomposable P, il existe un A-module M tel que $P \cong M \otimes_K L$.

(c) Pour toute extension algébriquement close L de degré de transcendance fini sur K et tout A_L-module P, il existe un A-module M tel que $P \cong M \otimes_K L$.

(d) Pour tout ultrafiltre $\mathcal{U}$, si Q est un $A^{\mathcal{U}}$-module, il existe un A-module M tel que $Q \simeq M^{\mathcal{U}}$.

Preuve : (a) $\Longrightarrow$ (b) est évident.

(b) $\Longrightarrow$ (c) Soient L une extension algébriquement close de K de degré de transcendance finie sur K, et P un A_L-module ; alors, par le théorème de Krull-Schmidt, on a :

$$P = P_1 \oplus \ldots \oplus P_n.$$

Comme tous les P_r, $1 \leq r \leq n$, sont des A_L-modules indécomposables, par (b) il existe $M_1, \ldots, M_n$ tels que $P_r \cong M_r \otimes_K L$, $1 \leq r \leq n$; d'où $P \cong (M_1 \otimes_K L) \oplus \ldots \oplus (M_n \otimes_K L)$ et, par la distributivité des produits tensoriels par rapport à $\oplus$, on a :

$$P \cong (M_1 \oplus \ldots \oplus M_n) \otimes_K L \cong M \otimes_K L \text{ , où } M = M_1 \oplus \ldots \oplus M_n \quad .$$

(c) $\Longrightarrow$ (d) Soient $\mathcal{U}$ un ultrafiltre et Q un $A^{\mathcal{U}}$-module indécomposable sur $K^{\mathcal{U}}$. Fixons une base $e_1, \ldots, e_q$ de Q, une base $a_1, \ldots, a_n$ de A sur K qui est aussi une base de $A^{\mathcal{U}}$ sur $K^{\mathcal{U}}$, et considérons les constantes de structure γ_{ij}^k de Q dans ces bases.

$$a_i e_j = \sum \gamma_{ij}^k \, e_k \quad .$$

Si L est la fermeture algébrique dans $K^{\mathcal{U}}$ du corps obtenu par adjonction les (γ_{ij}^k) à l'image de K par l'injection diagonale de K dans $K^{\mathcal{U}}$, L est de degré de transcendance finie sur K. Soit Q' l'ensemble des combinaisons linéaires à coefficients dans L des (e_i) ; comme les $\gamma_{ij}^k \in L$, Q' est un A_L-module et $Q' \otimes K^{\mathcal{U}} \cong Q$. D'après (c), il existe un A-module M (sur K) tel que $Q' \cong M \otimes_K L$. Alors $Q \cong (M \otimes_K L) \otimes_L K^{\mathcal{U}} \cong M \otimes_K K^{\mathcal{U}}$; l'homomorphisme composé de $K \longrightarrow L$ et de l'injection de L dans $K^{\mathcal{U}}$ étant l'isomorphisme diagonal.

(d) $\Longrightarrow$ (a) Soit L une extension algébriquement close de K, i étant l'injection de K dans L. Il existe un K-isomorphisme j de L dans une ultrapuissance $K^{\mathcal{U}}$ de K et un L-isomorphisme ϕ de $K^{\mathcal{U}}$ dans une ultrapuissance $L^{\mathcal{V}}$ de L tel que le diagramme suivant :

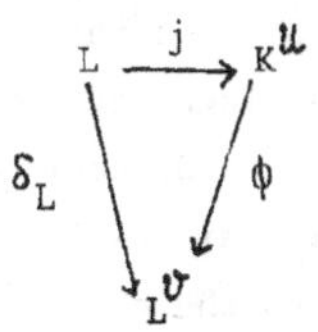

soit commutatif. Soit P un A_L-module, remarquons que l'algèbre $(A \underset{K,i}{\otimes} L) \underset{L,j}{\otimes} K^{\mathcal{U}}$ est isomorphe à $A \underset{K,\delta_K}{\otimes} K^{\mathcal{U}} \cong A^{\mathcal{U}}$. Donc le module $P \underset{L,j}{\otimes} K^{\mathcal{U}}$ est un $A^{\mathcal{U}}$-module et d'après (d), il existe un A-module M et un $A^{\mathcal{U}}$-isomorphisme entre $P \underset{L,j}{\otimes} K^{\mathcal{U}}$ et $M \underset{K,\delta_K}{\otimes} K^{\mathcal{U}}$. A l'aide de ϕ on a un $(A \underset{K,\delta_K}{\otimes} K^{\mathcal{U}}) \underset{K^{\mathcal{U}},\phi}{\otimes} L^{\vartheta}$-isomorphisme entre $(P \underset{L,j}{\otimes} K^{\mathcal{U}}) \underset{K^{\mathcal{U}},\phi}{\otimes} L^{\vartheta}$ et $(M \underset{K,\delta_K}{\otimes} K^{\mathcal{U}}) \underset{K^{\mathcal{U}},\phi}{\otimes} L^{\vartheta}$. Par associativité des produits tensoriels, on a : $(P \underset{L,j}{\otimes} K^{\mathcal{U}}) \underset{K^{\mathcal{U}},\phi}{\otimes} L^{\vartheta} \simeq P \underset{L,\delta_L}{\otimes} L^{\vartheta} \simeq P^{\vartheta}$. De même :

$$(M \underset{K,\delta_K}{\otimes} K^{\mathcal{U}}) \underset{K^{\mathcal{U}},\phi}{\otimes} L^{\vartheta} = ((M \underset{K,i}{\otimes} L) \underset{L,j}{\otimes} K^{\mathcal{U}}) \underset{K^{\mathcal{U}},\phi}{\otimes} L^{\vartheta} \simeq (M \underset{K,i}{\otimes} L) \underset{L,\delta_L}{\otimes} L^{\vartheta} \cong (M \underset{K,i}{\otimes} L)^{\vartheta}.$$

Alors $P^{\vartheta} \simeq (M \underset{K,i}{\otimes} L)^{\vartheta}$ et d'après la proposition 1.11, on a $P \simeq M \underset{K,i}{\otimes} L$.

Remarque : Pour (d) $\Longrightarrow$ (a) de la proposition précédente, on peut supposer que (d) est valable pour les ultrapuissances $K^{\mathcal{U}}$ de K telles que $|K^{\mathcal{U}}| > |K|$.

Proposition 2.15 - Soit A une algèbre sur un corps algébriquement clos K. Les assertions suivantes sont équivalentes :

(a) L'algèbre A est close.

(b) Soit Ω une extension algébriquement close de K de degré de transcendance infinie dénombrable. Si P est un A-module, il existe un A-module M tel que $P = M \underset{K}{\otimes} \Omega$.

(a) $\Longrightarrow$ (b) est évident.

(b) $\Longrightarrow$ (a) On va montrer que (b) implique la condition (d) de la proposition (2.14), pour les ultrapuissances $K^{\mathcal{U}}$ telle que $|K^{\mathcal{U}}|>|K|$. Alors $K^{\mathcal{U}}$ est de degré de transcendance infinie $> |K|$ sur K. Soit B une base de transcendance de $K^{\mathcal{U}}$ sur K. Introduisons une base $e_1,\dots,e_q$ du $(K^{\mathcal{U}})$-espace vectoriel Q et une base $a_1,\dots,a_m$ de A sur K (qui est aussi une base de $A^{\mathcal{U}}$ sur $K^{\mathcal{U}}$). Si les γ^k_{ij} sont les constantes de structure de Q associées à ces bases, si L est la fermeture algébrique dans $K^{\mathcal{U}}$ de corps obtenu par adjonction des γ^k_{ij} à $\delta_K(K)$, L est de degré de transcendance finie sur K. Soit $\eta_1,\dots,\eta_n$ une base de transcendance de L sur K; d'après le théorème d'échange, il existe un sous-ensemble B' de B tel que $\{\eta_1,\dots,\eta_n\} \cup B'$ soit une base de transcendance de $K^{\mathcal{U}}$ sur K. L'ensemble B' est infini, il contient donc un sous-ensemble dénombrable D. La fermeture algébrique Ω dans $K^{\mathcal{U}}$ du corps engendré sur $\delta(K)$ par $\{\eta_1,\dots,\eta_r\} \cup D$ est de degré de transcendance infinie dénombrable. Si Q' est l'ensemble des combinaisons linéaires des e_j à coefficients dans Ω, $Q \cong Q' \underset{\Omega}{\otimes} K^{\mathcal{U}}$, par hypothèse, il existe un A-module M tel que $Q' \cong M \underset{K}{\otimes} \Omega$. D'où $Q \cong (M \underset{K}{\otimes} \Omega) \underset{K}{\otimes} K^{\mathcal{U}}$, c'est-à-dire que $Q \simeq M \underset{K}{\otimes} K^{\mathcal{U}} \cong M^{\mathcal{U}}$.

<u>Proposition</u> 2.16 - <u>Soit</u> A <u>une algèbre sur un corps algébriquement clos</u> K. A <u>est close si et seulement si pour tout</u> $n \in \mathbb{N}$, <u>l'ensemble de types de A-modules (indécomposables) de dimension finie</u> $\leq n$ <u>est fini</u>.

<u>Preuve</u> : Supposons qu'il existe un $n \in \mathbb{N}$ et une suite infinie $M_1, M_2, \dots, M_i, \dots$ de A-modules (indécomposables) de dimension $\leq n$, deux à deux non isomorphes. L'ultraproduit $\prod_{\mathcal{U}} M_i$ est $A^{\mathcal{U}}$-module (indécomposable) de dimension $\leq n$ sur K ; $\prod_{\mathcal{U}} M_i$ n'est pas de la forme $X^{\mathcal{U}}$, (sinon les M_i seraient presque tous isomorphes à X), et l'algèbre A ne serait pas close. Réciproquement supposons qu'il n'y ait, pour chaque $n \in \mathbb{N}$, qu'un nombre fini de types de modules indécomposables de dimension $\leq n$. Soit P un $A^{\mathcal{U}}$-module indécomposable de dimension p, il existe une famille $(X_i)_{i \in I}$ de A-modules, indécomposables de dimension $\leq p$, sur A, tels que $P \simeq \prod_{\mathcal{U}} X_i$. Par hypothèse, il existe un A-module indécomposable X de dimension $\leq p$ tel que presque tous les X_i sont isomorphes à X. D'où $P \simeq X^{\mathcal{U}}$ et A est close (condition (d) de la proposition 2.14).

<u>Caractérisation des Algèbres de Représentation Finie</u>.

Le résultat suivant est établi par L.A. Nazarova, et A. Roiter [7].

<u>2ème Conjecture de Brauer-Thrall</u> :

<u>Soit</u> A <u>une algèbre sur un corps algébriquement clos</u> K. <u>Si</u> A <u>est de représentation infinie, il existe une infinité d'entiers</u> n_k, <u>telle que, pour tout</u> n_k, <u>il existe une famille infinie de A-modules indécomposables, deux à deux non isomorphes, de dimension</u> n_k <u>sur</u> K.

On a montré que toute algèbre de représentation finie est une algèbre close. En utilisant le théorème de Nazarova-Roiter précédent, on va établir la réciproque :

Proposition 2.17 - Toute algèbre close sur un corps algébriquement close est de représentation finie.

Preuve : Si A n'est pas de représentation finie, il existe pour un entier n une infinité de types d'indécomposables de dimension $\leq n$, alors, d'après la proposition 2.15, l'algèbre A n'est pas close.

On peut résumer les résultats précédents par le théorème suivant :

Théorème 2.18 - Soit A une algèbre sur un corps algébriquement clos K. Les assertions suivantes sont équivalentes :

(i) A est de représentation finie.

(ii) A est close, c'est-à-dire, pour toute extension algébriquement close L de K et tout A_L-module P, il existe un A-module M tel que $P \cong M \otimes_K L$.

(iii) Même énoncé avec P indécomposable (comme A_L-module).

(iv) Même énoncé que (ii) mais L est supposé de degré de transcendance finie sur K.

(v) Même énoncé que (ii) mais L supposé de degré de transcendance infinie dénombrable sur K.

(vi) Pour tout ultrafiltre $\mathcal{U}$, si Q est un $A^{\mathcal{U}}$-module de dimension finie sur $K^{\mathcal{U}}$, il existe un A-module M tel que Q soit isomorphe à $M^{\mathcal{U}}$.

(vii) Pour tout $n \in \mathbb{N}$ il existe un nombre fini de types de A-modules indécomposables de dimension $\leq n$ sur K.

§.3 ALGEBRES DE REPRESENTATION INFINIE.

Soient $(K_i)_{i \in \mathbb{N}}$ une famille dénombrable de corps algébriquement clos dénombrables et de caractéristique 0, $\mathcal{U}$ un ultrafiltre non-principal sur $\mathbb{N}$.

Nous avons démontré (proposition 1.9) que le corps ultraproduit $\prod_{\mathcal{U}} K_i$ est isomorphe à $\mathbb{C}$. En particulier si $\bar{\mathbb{Q}}$ est le corps de nombres algébriques, l'ultrapuissance $\bar{\mathbb{Q}}^{\mathcal{U}}$ est isomorphe à $\mathbb{C}$.

Proposition 3.1 - Soit $(A_i)_{i \in \mathbb{N}}$ une famille de $\bar{\mathbb{Q}}$-algèbres de dimension finie $\leq m$. On suppose que, pour tout $i \in \mathbb{N}$, A_i est de représentation finie avec $n(i)$ types de modules indécomposables. Si $\text{Sup}\, n(i) < \aleph_0$, l'algèbre ultraproduit $\prod_{\mathcal{U}} A_i$ est une $\bar{\mathbb{Q}}^{\mathcal{U}}$-algèbre de représentation finie.

Preuve : Pour tout A_i, on désigne par $M_{i1},\dots,M_{in(i)}$ une famille de représentants des types l'indécomposables. Notons que l'algèbre $\prod_{\mathcal{U}} A_i$ est de dimension finie $\leq m$ sur $\bar{Q}^{\mathcal{U}}$.

Soit P un $\prod_{\mathcal{U}} A_i$-module indécomposable. Il existe une famille $(X_i)_{i\in\mathbb{N}}$ de A_i-modules indécomposables telle que $P \simeq \prod_{\mathcal{U}} X_i$, X_i étant un A_i-module indécomposable; et il existe un $k(i)$, $1 \leq k(i) \leq n$ tel que $X_i \simeq M_{ik(i)}$. On pose $I_1 = \{i\in\mathbb{N} \mid k(i) = 1\},\dots,I_n = \{i\in\mathbb{N} \mid k(i) = n\}$; alors $I_1 \cup \dots \cup I_n = \mathbb{N} \in \mathcal{U}$ et, comme $\mathcal{U}$ est un ultrafiltre, il existe un k, $1 \leq k \leq n$, tel que $I_k \in \mathcal{U}$. Il en résulte que $P \simeq \prod_{\mathcal{U}} M_{ik}$. Tout $A^{\mathcal{U}}$-module est donc isomorphe à un module du type $\prod_{\mathcal{U}} M_{ik}$, $1 \leq k \leq n$. Ces modules étant en nombre fini (au plus égal à n), $\prod_{\mathcal{U}} A_i$ est de type de représentation finie.

Le résultat précédent entraîne que, si l'algèbre A est de représentation infinie, on a ou bien :

(i) les algèbres $(A_i)_{i\in\mathbb{N}}$ sont $\mathcal{U}$-presque toute de représentation infinie, ou bien :

(ii) les algèbres $(A_i)_{i\in\mathbb{N}}$ sont $\mathcal{U}$-presque toute de représentation finie et $\operatorname{Sup} n(i) = \aleph_0$.

<u>Proposition</u> 3.2 - <u>Soit</u> A <u>une algèbre sur un corps infini</u> K. <u>L'ensemble des types de A-modules est de cardinal</u> $\leq$ card K.

<u>Preuve</u> : Soient $a_1,\dots,a_m$ une base de A sur K. Tout A-module M est isomorphe à un module dont l'espace vectoriel sous jacent est un K^p. Une structure de A-module sur K est entièrement caractérisée par la donnée des m applications linéaires de K^p dans lui-même définis par les a_i, applications qu'on peut identifier à des matrices $M_p(K)$. L'ensemble des types de A-modules est donc de cardinalité inférieure à :

$$\operatorname{Card} \bigcup_p |M_p(K)^m \times K^p| = \operatorname{Card} \bigcup_p |K^{mp^2+p}|,$$

ensemble qui a la même cardinalité que K.

<u>Théorème</u> 3.3 - <u>Soit</u> A <u>une algèbre sur un corps algébriquement clos ayant la puissance du continu. S'il existe une infinité de</u> A-<u>modules de dimension finie</u> n <u>deux à deux non isomorphes, il existe une infinité, ayant la puissance du continu, de</u> A-<u>modules de dimension</u> n <u>deux à deux non isomorphes</u>.

<u>Preuve</u> : Soient $M_1,\dots,M_p,\dots$ une suite de A-modules deux à deux non isomorphes de dimension n et A une algèbre sur $\bar{Q}^{\mathcal{U}}$. L'algèbre A s'identifie à un ultraproduit $\prod_{\mathcal{U}} A_i$, A_i étant $\bar{Q}$-algèbre, et pour tout A-module M_p il existe

une famille $(M_{i,p})_{i\in\mathbb{N}}$ de A_i-modules tel que $M_p \simeq \prod_{\mathcal{U}} M_{i,p}$. Soit $\sigma_{i,n}$ l'ensemble des types de A_i-modules indécomposables de dimension n.

(i) On ne peut avoir $\sup_i |\sigma_{i,n}| < \aleph_0$; en effet, si $\sup |\sigma_{i,n}| = r$, on définit une partition finie de $\mathbb{N}$ en posant :

$$I_k = \{ i \in \mathbb{N} \mid M_{i,p} \text{ a pour type le } k\text{-ième élément de } \sigma_{i,n} \},\ 1 \leq k \leq r.$$

Comme $\mathcal{U}$ est un ultrafiltre sur $\mathbb{N}$, il existe un $I_k \in \mathcal{U}$; c'est-à-dire que les $M_{i,p}$ sont presque tous isomorphes à un M_s (M_s k-ième élément de $\sigma_{i,n}$). Alors $M_p \simeq \prod_{\mathcal{U}} M_{i,p} \simeq M_s^{\mathcal{U}}$, et d'après la proposition 3.1, presque tous les M_p sont isomorphes. Par suite, ou bien presque tous les $\sigma_{i,n}$ sont infinis, ou bien $\sup_{\mathcal{U}} |\sigma_{i,n}| = \aleph_0$.

Appliquons le lemme 1.4, l'ultraproduit des $\sigma_{i,n}$, $\prod_{\mathcal{U}} \sigma_{i,n}$, a la puissance du continu (on a remplacé les ensembles $\sigma_{i,n}$ qui seraient vides par $\{\emptyset\}$). On définit une application de $\prod_{\mathcal{U}} \sigma_{i,n}$ dans V, ensemble des types de A-modules indécomposables, en associant à $\widetilde{(\tau_i)}$ le type du A-module ultraproduit $\prod_{\mathcal{U}} X_i$ où X_i est de type τ_i (notons que $\prod_{\mathcal{U}} X_i$ est de dimension $\leq$ n, puisque $\tau_i \in \sigma_{i,n}$ pour tout i). Cette application est injective : soient $\prod_{\mathcal{U}} X_i$ et $\prod_{\mathcal{U}} X'_i$ dans V, associés par l'application précédente à $\widetilde{(\tau_i)}$ et $\widetilde{(\tau'_i)}$, tels que $\prod_{\mathcal{U}} X_i \simeq \prod_{\mathcal{U}} X'_i$. D'après la proposition 2.1, pour $\mathcal{U}$-presque tous les i, $X_i \simeq X'_i$; c'est dire que $\widetilde{(\tau_i)} = \widetilde{(\tau'_i)}$. Alors l'ensemble des types de A-modules indécomposables de dimension $\leq$ n a au moins la puissance du continu. D'après la proposition 3.2, l'ensemble des types de A-modules a au plus la puissance du continu. Alors la puissance de l'ensemble des types de A-modules est exactement $2^{\aleph_0}$.

Exemples (1) Soit A une algèbre complexe de représentation infinie. On peut trouver un corps K_A de degré de transcendance fini sur $\bar{\mathbb{Q}}$: $K_A = \bar{\mathbb{Q}}(\alpha_{ij}^k)$ où les (α_{ij}^k) sont les constantes de structure de l'algèbre A, tel que $A = \tilde{A} \otimes_{K_A} \mathbb{C}$. Le corps K_A est dénombrable il en résulte que l'ensemble des types de $\tilde{A}$-modules indécomposables est dénombrable. L'ensemble des types de A-modules est de la forme $M \otimes_{K_A} \mathbb{C}$, où M est un $\tilde{A}$-module indécomposable, est aussi dénombrable.

Comme l'ensemble des types de A-modules indécomposables a la puissance du continu, il en résulte qu'il existe des A-modules qui ne sont pas de la forme $M \otimes_{K_A} \mathbb{C}$.

(2) Soient A une K-algèbre de dimension finie et $\mathfrak{U}$ un ultrafiltre non-principal sur $\mathbb{N}$. L'algèbre A est de représentation infinie si et seulement si $A^{\mathfrak{U}}$ est de représentation infinie (Prop.2.11). Si l'ensemble des types de A-modules indécomposables est dénombrable, si $(M_i)_{i\in\mathbb{N}}$ est une famille de représentants, des types de A-modules indécomposables, alors les $(M_i^{\mathfrak{U}})_{i\in\mathbb{N}}$ sont de $A^{\mathfrak{U}}$-modules indécomposables deux à deux non isomorphes. Comme l'ensemble des types de $A^{\mathfrak{U}}$-modules a la puissance du continu, il en résulte qu'il existe des $A^{\mathfrak{U}}$-modules indécomposables qui ne sont pas de la forme $M \otimes_K K^{\mathfrak{U}} = M^{\mathfrak{U}}$; il est entendu que le corps K est algébriquement clos dénombrable.

Ainsi, soit $A = \bar{\mathbb{Q}}[X,Y]/(X^2,XY,Y^2)$. L'algèbre A est de représentation infinie, et on obtient une infinité de types d'indécomposables en faisant opérer X et Y sur $\bar{\mathbb{Q}}^{2m}$ au moyen des matrices $\begin{bmatrix} 0 & 0 \\ 1_m & 0 \end{bmatrix}$ et $\begin{bmatrix} 0 & 0 \\ J(m,\lambda) & 0 \end{bmatrix}$, où $J(m,\lambda)$ désigne la matrice de Jordan de valeur propre λ et de taille $m\times m$:

$$J(m,\lambda) = \begin{bmatrix} \lambda & 0 & 0 & \cdots \\ 1 & \lambda & 0 & \cdots \\ 0 & 1 & \lambda & \cdots \\ \cdots & \cdots & \cdots & \cdots \end{bmatrix} \in \bar{\mathbb{Q}}^{m\times m} .$$

Pour tout $\lambda \in \bar{\mathbb{Q}}$ on peut trouver un A-module indécomposable M_λ de dimension finie sur $\bar{\mathbb{Q}}$. Alors $(M_\lambda^{\mathfrak{U}})_{\lambda\in\bar{\mathbb{Q}}}$ sont des $A^{\mathfrak{U}}$-modules indécomposables deux à deux non-isomorphes. Comme l'ensemble des types de $A^{\mathfrak{U}}$-modules a la puissance du continu, alors , pour tout $\mu \in (\bar{\mathbb{Q}}^{\mathfrak{U}} - \delta(\bar{\mathbb{Q}}))$, les $A^{\mathfrak{U}}$-modules X_μ ne sont pas de la forme $M_\lambda^{\mathfrak{U}}$ pour tout $\lambda \in \bar{\mathbb{Q}}$.

Bibliographie

[1] M. AUSLANDER : Large Modules over Artin Algebras, A collection of Papers in Honour of S. Eilenberg, Academic Press, 1975.

[2] J.L. BELL et A.B. SLOMSON : Models and ultraproducts, An introduction, North - Holland, Amsterdam , 1969.

[3] N. BOURBAKI : Algèbre, Herman, Paris, 1967.

[4] C.W. CURTIS et J.P. JANS : On algebras with a finite number of indecomposable modules. Trans. Amer. Math. Soc. 114 (1965), 122-132.

[5] P. GABRIEL : Indecomposable representation II, Symp. Math. Ist. Naz. Alta Math. (1973), 81-104.

[6] N. JACOBSON : Lectures in Abstract Algebra, vol. III, Van Nostrand Comp. Inc, 1964.

[7] L.A. NAZAROVA A. ROITER : Categorial matricial Problems and the Conjecture of Brauer - Thrall, Preprint Institute of Maths. of the Academy of Sciences of Ukrania, Kiev, 1974.

[8] F. TAHA : Algèbres simples centrales sur les corps ultraproduits de corps p-adiques (ce fascicule).

[9] K. YAMAGATA : On Artinian Rings of Finite Representation Type, J. of Algebra 50, 276-283, (1978).

IDEAUX PRIMITIFS DANS DES ALGEBRES UNIVERSELLES

Sleiman Yammine
Université Pierre et Marie Curie
- Paris 6 -

INTRODUCTION :

Soit $A = U(\mathfrak{g})$ l'algèbre enveloppante d'une algèbre de Lie $\mathfrak{g}$ de dimension finie sur un corps commutatif k de caractéristique 0 . Nous trouvons dans ([1], 4.5.7), lorsque $\mathfrak{g}$ est résoluble, différentes caractérisations d'un idéal primitif de A . Nous retrouvons, lorsque k est non dénombrable et algébriquement clos et en combinant ([2], Théorème C),([1],3.1.15,4.1.7) et ([3], 4.6(i)), les mêmes caractérisations. On étend, dans ce travail, le résultat qui décrit la quasi-totalité de ces caractérisations (resp. toutes ces caractérisations et le théorème B de [2]) au cas ou k est quelconque (resp. au cas où k est seulement non dénombrable).

On suppose dans la suite que les anneaux sont unitaires, les corps sont commutatifs et les algèbres sont associatives. La notion de primitivité est toujours traitée à gauche.

On considère les conditions suivantes concernant un idéal premier p d'un anneau A :

$1(A,p)$: L'idéal p est primitif.

$2(A,p)$: L'intersection des idéaux premiers de A contenant strictement p est distincte de p.

$3(1,p)$: L'intersection des idéaux primitifs de A contenant strictement p est distincte de p.

$4(A,p)$: il existe une famille dénombrable $\mathcal{E}$ d'idéaux bilatères de A contenant strictement p, telle que tout idéal bilatère de A contenant strictement p contienne au moins un $I \in \mathcal{E}$.

Si de plus A est une algèbre noethérienne à gauche sur un corps k, on ajoute les conditions suivantes :

$5(A,k,p)$: Le centre de Fract (A/p) est algébrique sur k.

$6(A,k,p)$: Le centre de Fract (A/p) est de degré fini sur k.

<u>PROPOSITION 1. - Soient</u> k <u>un corps,</u> k' <u>une extension algébrique de</u> k <u>et</u> A <u>une k-algèbre. On note</u> $B = k' \otimes_k A$ <u>et</u> $f : A \to B$ <u>le monomorphisme canonique de k-algèbres. Soient</u> p'' <u>un idéal premier de</u> B <u>et</u> $p = f^{-1}(p'')$.

1) <u>On suppose que l'anneau</u> A <u>est noethérien à gauche. Si</u> $2(A,p)$ <u>(resp.</u> $4(A,p)$<u>) est réalisée, alors</u> $2(B,p'')$ <u>(resp.</u> $4(B,p'')$<u>) est réalisée.</u>

2) <u>On suppose que</u> k' <u>est une extension séparable de</u> k <u>et</u> B <u>un anneau noethérien à gauche.</u> $2(A,p)$ (<u>resp.</u> $4(A,p)$; $5(A,k,p)$) <u>est alors réalisée si et seulement si</u> $2(B,p'')$ (<u>resp.</u> $4(B,p'')$; $5(B,k',p'')$),<u>est réalisée</u>. Si $6(A,k,p)$ <u>est réalisée, alors</u> $6(B,k',p'')$ <u>est réalisée. Si de plus</u> k' <u>est de degré fini sur</u> k <u>et</u> $6(B,k',p'')$ <u>est réalisée, alors</u> $6(A,k,p)$ <u>est réalisée</u>.

3) <u>On suppose que</u> k <u>est de caractéristique</u> 0 <u>et l'anneau</u> B <u>est noethérien à gauche. Alors</u> $1(A,p)$ (<u>resp.</u> $3(A,p)$) <u>est réalisée si et seulement si</u> $1(B,p'')$ (<u>resp.</u> $3(B,p'')$) <u>est réalisée</u>.

<u>Preuve.</u> 1)- Supposons que $2(A,p)$ est réalisée et désignons par $\mathcal{L}$ (resp. $\mathcal{L}''$) l'ensemble des idéaux premiers de A(resp. B) contenant strictement p (resp. p''). Si $\mathcal{L}'' = \phi$, alors $\bigcap_{g'' \in \mathcal{L}''} g'' = B \supsetneq p''$. Si $\mathcal{L}'' \neq \phi$, alors ([6], corollaire 1.3), pour tout $g'' \in \mathcal{L}''$, $f^{-1}(g'') \in \mathcal{L}$ et $f^{-1}(\bigcap_{g'' \in \mathcal{L}''} g'') \supseteq \bigcap_{g \in \mathcal{L}} g \supsetneq p$. Donc $p'' \subsetneq \bigcap_{g'' \in \mathcal{L}''} g''$ et $2(B,p'')$ est réalisée.

Supposons que $4(A,p)$ est réalisée c'est-à-dire qu'il existe une famille dénombrable $\mathcal{E}$ d'idéaux bilatères de A contenant strictement p , telle que tout idéal bilatère de A contenant strictement p contienne au moins un $I \in \mathcal{E}$. Il est alors aisé de voir, en considérant la famille $\mathcal{E}'' = ((k' \otimes_k I)+p'')_{I \in \mathcal{E}}$ d'idéaux bilatères de B et en appliquant ([6], corollaire 1.3), que $4(B,p'')$ est réalisée.

2) Supposons que $2(B,p'')$ est réalisée et conservons les notations de 1). Si $\mathcal{L} = \phi$, alors $\bigcap_{g \in \mathcal{L}} g = A \supsetneq p$. Si $\mathcal{L} \neq \phi$, alors ([4], corollaire 3.13), pour tout $g \in \mathcal{L}$ il existe $g'' \in \mathcal{L}''$

tel que $\mathcal{g} = f^{-1}(\mathcal{g}'')$. Il s'en suit, d'après l'hypothèse et ([6], corollaire 1.3), que $p \subsetneq \bigcap_{\mathcal{g} \in \mathcal{L}} \mathcal{g}$ et $2(A,p)$ est réalisée.

Supposons que $4(B,p'')$ est réalisée c'est-à-dire qu'il existe une suite $(I''_n)_{n \in \mathbb{N}}$ d'idéaux bilatères de B contenant strictement p'' telle que tout idéal bilatère de B contenant strictement p'' contienne au moins l'un des I''_n. Nous remarquons alors, en considérant la famille $\mathcal{E} = (I''_{n,m})_{(n,m) \in \mathbb{N} \times \mathbb{N}^*}$ où $I''_{n,m} = (f^{-1}(I''_n))^m + p$ pour tout $(n,m) \in \mathbb{N} \times \mathbb{N}^*$ et en revenant à la démonstration de la proposition 4.7 de [5], que $4(A,p)$ est réalisée.

Réservons dans la suite la notation $Z(R)$ au centre d'un anneau quelconque R et la notation $C_R(\theta)$ ou tout simplement $C(\theta)$ à l'ensemble des éléments de R non diviseurs de zéro modulo un idéal bilatère θ de R. Posons $p' = k' \otimes_k p$, $\bar{A} = A/p$, $\bar{B} = B/p'$, $\bar{\bar{B}} = B/p''$ et considérons le diagramme commutatif suivant :

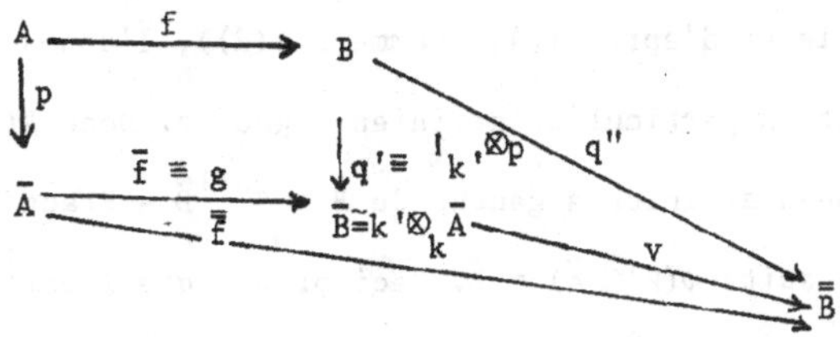

où p(resp. q',q'' et v) désigne l'épimorphisme canonique de k-algèbres (resp. désignent les épimorphismes canoniques de k'-algèbres), $g : \bar{A} \to k' \otimes_k \bar{A}$ le monomorphisme canonique de k-algèbres, $\bar{f}$ et $\bar{\bar{f}}$ les monomorphismes de k-algèbres déduits de f par passage au quotient. En posant $S = C_{\bar{A}}(\bar{0})$ et $S'' = C_{\bar{\bar{B}}}(\bar{0})$, nous obtenons, d'après ([5], Proposition 1.1(3)) et vu les relations

$C(p) = p^{-1}(C_{\bar{A}}(\bar{0}))$ et $C(p'') = q''^{-1}(C_{\bar{\bar{B}}}(\bar{0}))$, $S = \bar{\bar{f}}^{-1}(S'')$ et par suite $\bar{\bar{f}}(S) \subseteq S''$. Ceci permet de dresser le diagramme commutatif suivant :

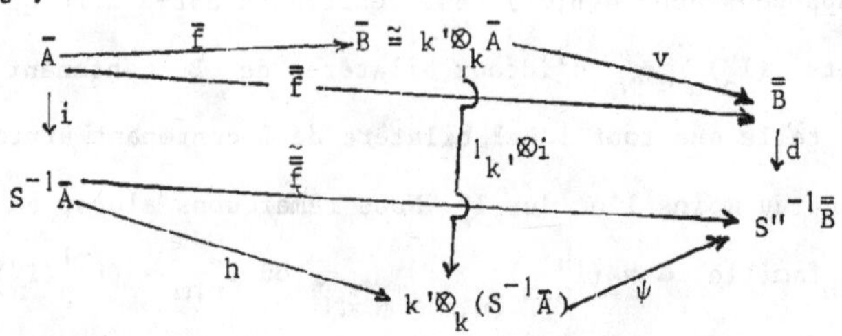

où i et d désignent les injections canoniques, $\tilde{\bar{\bar{f}}}$ l'unique homomorphisme d'anneaux prolongeant $\bar{\bar{f}}$, h le monomorphisme canonique de k-algèbres et ψ , vu que $\bar{f}$ est en particulier un monomorphisme de k-algèbres, l'unique homomorphisme de k'-algèbres vérifiant $\tilde{\bar{\bar{f}}} = \psi \circ h$. D'après ([4], Lemme 3.5), $k'\otimes_k(S^{-1}\bar{A}) \cong S'^{-1}\bar{B}$ où $S' = \bar{f}(S)$ et $k'\otimes_k(S^{-1}\bar{A})$ est alors un anneau noethérien à gauche. Par conséquent, du fait que la k-algèbre $\tilde{A} = S^{-1}\bar{A} = \mathrm{Fract}(\bar{A})$ est simple et d'après ([4], Lemme 3.4(2)), l'anneau $k'\otimes_k\tilde{A}$ est semi-simple et en particulier artinien à gauche. Donc $\psi(k'\otimes_k\tilde{A})$ est un sous-anneau artinien à gauche de $\tilde{\bar{\bar{B}}} = S''^{-1}\bar{\bar{B}} = \mathrm{Fract}(\bar{\bar{B}})$ qui contient $\bar{\bar{B}}$ et par suite $\psi(k'\otimes_k\tilde{A}) = \tilde{\bar{\bar{B}}}$. Ceci prouve que ψ est surjectif, donc que la k'-algèbre $\tilde{\bar{\bar{B}}}$ est isomorphe à la k'-algèbre $(k'\otimes_k D)\otimes_D\tilde{A})/_Q$ où $D = Z(\tilde{A})$ et Q un idéal bilatère de la k'-algèbre $(k'\otimes_k D)\otimes_D\tilde{A}$. Il s'en suit ([1], 4.5.1) que $\tilde{\bar{\bar{B}}}$ est isomorphe à la k'-algèbre $(k'\otimes_k D/_P)\otimes_D\tilde{A}$ où P est un idéal de $k'\otimes_k D$ et $Z(\tilde{\bar{\bar{B}}}) \cong (k'\otimes_k D)/_P$. D'autre part il est aisé de voir que $\tilde{\bar{\bar{f}}}$ est central (i.e. $\tilde{\bar{\bar{f}}}(Z(\tilde{A})) \subseteq Z(\tilde{\bar{\bar{B}}})$) . Si 5(A,k,$p$) est réalisée,alors $k'\otimes_k D$ est entier sur k' et par conséquent $Z(\tilde{\bar{\bar{B}}}) \cong (k'\otimes_k D)/_P$ est algébrique sur k' c'est-à-dire $5(B,k',p'')$ est réalisée. Inversement, si $5(B,k',p'')$ est réalisée, alors $Z(\tilde{\tilde{B}})$ est algébrique sur k et par conséquent Z(A) est algébrique sur k c'est-à-dire $5(A,k,p)$ est réalisée. Si $6(A,k,p)$ est réalisée, alors $Z(\tilde{\tilde{B}}) = (k' \otimes_k D)/_P$ est de degré fini sur k' et $6(B,k',p'')$ est réalisée. Supposons de plus que k' est de degré fini sur k et que $6(B,k',p'')$ est réalisée, alors $Z(\tilde{\tilde{B}})$ est de degré fini sur k et par conséquent $Z(\tilde{A})$ est de degré fini sur k et $6(A,k,p)$ est réalisée.

3) D'après ([5], Proposition 4.6), $1(A,p)$ est réalisée si et seulement si $1(B,p'')$ est réalisée. En raisonnant enfin comme dans 1) et 2) pour $2(A,p)$ et $2(B,p'')$ et en utilisant ([5], Proposition 4.6), nous démontrons que $3(A,p)$ est réalisée si et seulement si $3(B,p'')$ est réalisée. ||

PROPOSITION 2. - Soient k un corps algébriquement clos, k' une extension de k , A une k-algèbre et $B = k' \otimes_k A$. Pour un idéal p de A , les conditions suivantes sont équivalentes :

a) p est un idéal premier de A ; b) $p' = k' \otimes_k p$ est un idéal premier de B. De plus, dans ces conditions, si $2(B,p')$ est réalisée, alors $2(A,p)$ est réalisée.

Preuve. - b) ⇒ a) est évidente et a) ⇒ b) découle de ([7], p. 70, Proposition 17.2). Supposons que a) et $2(B,p')$ sont réalisées et notons $\mathcal{L}$ (resp. $\mathcal{L}''$) l'ensemble des idéaux premiers de A(resp. B) contenant strictement p(resp. p'). Il est aisé de voir que, pour tout $g \in \mathcal{L}$, $g' = k' \otimes_k g \in \mathcal{L}''$. Donc $p' \subsetneq \bigcap_{g \in \mathcal{L}} g' = k' \otimes_k (\bigcap_{g \in \mathcal{L}} g)$ et par suite $p \subsetneq \bigcap_{g \in \mathcal{L}} g$ et $2(A,p)$ est réalisée. ||

On désigne par $W_A(M)$ l'ensemble des idéaux primitifs d'un anneau A contenant une partie M de A .

PROPOSITION 3. - Soient k un corps de caractéristique 0 , $\mathfrak{g}$ une k-algèbre de Lie de dimension finie sur k et $A = U(\mathfrak{g})$ l'algèbre enveloppante de $\mathfrak{g}$. Soit p un idéal premier de A .

1) Si k est non dénombrable, alors pour toute suite $(X_n)_{n \in \mathbb{N}}$ de parties de $W_A(p)$ recouvrant $W_A(p)$, il existe $r \in \mathbb{N}$ tel que $p = \bigcap_{g \in X_r} g$.

2) Les conditions 1(A,p) , 2(A,p) , 3(A,p) , 5(A,k,p) et 6(A,k,p) sont équivalentes et, lorsque k est non dénombrable, elles sont équivalentes à 4(A,p) .

Preuve. - Soit k' une clôture algébrique de k , posons $B = k' \otimes_k A \cong U(k' \otimes_k \mathfrak{g})$ et désignons par $f : A \to B$ le monomorphisme canonique de k-algèbres. Choisissons arbitrairement ([4], Proposition 1.2 (2)) un idéal premier p'' de B tel que $p = f^{-1}(p'')$.

1) Soit $(X_n)_{n \in \mathbb{N}}$ une suite de parties de $W_A(p)$ recouvrant $W_A(p)$. Pour tout $n \in \mathbb{N}$ considérons l'ensemble $Y_n = \{g'' \in W_B(p'')$ tel que $f^{-1}(g'') \in X_n\}$. Nous avons tout d'abord $X_n = \{g = f^{-1}(g'')$ où $g'' \in Y_n\}$ $(n \in \mathbb{N})$. L'inclusion $\{g = f^{-1}(g'')$ où $g'' \in Y_n\} \subseteq X_n$ est évidente ; si $g \in X_n$, il existe ([5], 5.2) $g'' \in W_B(p'')$ tel que $g = f^{-1}(g'')$ et l'inclusion $X_n \subseteq \{g = f^{-1}(g'')$ où $g'' \in Y_n\}$ est établie. D'autre part, si $g'' \in W_B(p'')$, alors ([5], Proposition 4.6) $g = f^{-1}(g'') \in W_A(p)$ et par suite il existe $n \in \mathbb{N}$ tel que $g \in X_n$ c'est-à-dire tel que $g'' \in Y_n$. Donc $(Y_n)_{n \in \mathbb{N}}$ est une suite de parties de $W_B(p'')$ recourvrant $W_B(p'')$ et, d'après ([2], Théorème B), il existe $r \in \mathbb{N}$ tel que $p'' = \bigcap_{g'' \in Y_r} g''$. D'où

$$p = \bigcap_{\mathfrak{q}'' \in Y_r} f^{-1}(\mathfrak{q}'') = \bigcap_{\mathfrak{q} \in X_r} \mathfrak{q} .$$

2) Nous avons : $1(A,p) \Rightarrow 6(A,k,p)$ ([1], 4.1.7) ; $6(A,k,p) \Rightarrow 5(A,k,p)$ et $2(A,p) \Rightarrow 3(A,p)$ sont évidentes ; $3(A,p) \Rightarrow 1(A,p)$ ([1], 3.1.15). Remarquons, pour établir $5(A,k,p) \Rightarrow 2(A,p)$, que nous pouvons toujours choisir $k' \subseteq k''$ où k'' est une extension algébriquement close non dénombrable de k . Posons $C = k'' \otimes_k A \cong k'' \otimes_{k'} B$ et supposons que $5(A,k,p)$ est réalisée. D'après la Proposition 1(2), $5(B,k',p'')$ est réalisée et la k'-algèbre Fract (B/p'') a alors pour centre k' . Donc Fract $(C/k'' \otimes_{k'} p'')$ a pour centre k'' et, vu que $k'' \otimes_{k'} p''$ est (Proposition 2) un idéal premier de C , $5(C,k'',k'' \otimes_{k'} p'')$ est réalisée. Par conséquent, d'après ([2], Théorème C) et ([3], Théorème 4.6 (i)), $2(C, k'' \otimes_{k'} p'')$ est réalisée. Donc (Proposition 2) $2(B,p'')$ est réalisée et (Proposition 1 (2)) $2(A,p)$ est réalisée. Lorsque, au départ, k est non dénombrable, alors $1(B,p'')$, $2(B,p'')$, $3(B,p'')$, $4(B,p'')$, $5(B,k',p'')$ et $6(B,k',p'')$ sont équivalentes et nous obtenons la dernière assertion en utilisant la proposition 1.

REFERENCES

[1] J. Dixmier, Algèbres enveloppantes, Gauthier-Villars, 1974.

[2] J. Dixmier, Idéaux primitifs dans les algèbres enveloppantes, J. of Algebra, 48, 96-112, 1977.

[3] C. Moeglin, Thèse de Doctorat d'état, Université Pierre et Marie Curie, Paris VI.

[4] S. Yammine, Les Théorèmes de Cohen-Seidenberg en algèbre non commutative, Lecture notes in Mathematics, 740, 120-169, Springer-Verlag.

[5] S. Yammine, Localisation des idéaux semi-premiers et extension des scalaires dans les algèbres noethériennes sur un corps, Lecture notes in mathematics, 795, 251-290, Springer-Verlag.

[6] S. Yammine, Théorèmes d'incomparabilité et de descente en algèbre non commutative, à paraître dans les comptes rendus du 105ème congrès national des sociétés savantes (Caen).

[7] G.M. Bergman, Zero-divisors in tensor products, Lecture notes in mathematics, 545, 32-82, Springer-Verlag.

=-=-=-=-=-=-=-=-=-=

ULTRA-PRODUITS D'ALGEBRES DE LIE

par Marie Paule MALLIAVIN

Depuis 1965, date à laquelle J. Ax et S. Kochen obtinrent, au moyen des ultra-produits, un résultat de théorie algébrique des nombres, (cf. F. Taha [10]), il est à noter le théorème d'Amitsur (cf [1]) généralisant celui de Posner en théorie des algèbres à identités polynomiales. Deux des conférences précédentes de ce fascicule sont d'autres applications de la théorie des ultra-produits à certaines parties de l'algèbre. L'objet de cet article est d'en donner de nouvelles applications, spécialement dans le cadre des représentations d'algèbre de Lie. De nombreux résultats récents [7] [8] ont lieu pour des algèbres de Lie, sur un corps non dénombrable. Nous verrons qu'au moyen des ultra-produits certains de ces résultats s'étendent au cas d'un corps de base dénombrable.

§.1 - <u>Rappels sur les ultra-produits</u>.

Bien que les notations et définitions soient les mêmes que celles de [10] , dont nous utiliserons les résultats ainsi que ceux de [13], rappelons pour la commodité du lecteur que par <u>ultrafiltre</u> $\mathfrak{U}$ sur $\mathbb{N}$ (ceux sont à peu près les seuls que nous envisagerons) nous entendrons un élément maximal de l'ensemble ordonné de tous les <u>filtres</u> sur $\mathbb{N}$ et un <u>filtre</u> $\mathfrak{U}$ sur $\mathbb{N}$ possède les trois propriétés suivantes :

(1) Si $E, F \in \mathfrak{U}$ alors $E \cap F \in \mathfrak{U}$;

(2) $\emptyset \notin \mathfrak{U}$;

(3) Si $E \in \mathfrak{U}$ et $\mathbb{N} \supseteq F \supset E$ alors $F \in \mathfrak{U}$.

On supposera toujours que l'ultrafiltre en question n'est pas <u>principal</u> (ou <u>trivial</u> [cf. Bourbaki Topologie]), un filtre $\mathfrak{U}$ étant principal s'il existe $i \in \mathbb{N}$ tel que $\mathfrak{U} = \{E \ni i \mid E \subseteq \mathbb{N}\}$.

Soit $\{A_i, i \in \mathbb{N}\}$ une famille d'ensembles ou de groupes, ou d'anneaux et soit $\mathfrak{U}$ un ultrafiltre sur $\mathbb{N}$ alors $\prod_{\mathfrak{U}} A_i$, l'ultraproduit des A_i, est définit comme le quotient du produit direct (d'ensembles, de groupes ou d'anneaux) $\prod_{i\in\mathbb{N}} A_i$, par la relation d'équivalence $\mathcal{R}$

$$(x_i)_{i\in\mathbb{N}} \equiv (y_i)_{i\in\mathbb{N}} \pmod{\mathcal{R}} \text{ si et seulement si}$$

$$\{i\in\mathbb{N} \quad x_i = y_i\} \text{ est un ensemble de } \mathcal{U} .$$

Evidemment une telle définition est possible pour un filtre ordinaire mais si $\mathcal{U}$ est un ultrafiltre on a la propriété plus forte suivante :

$$(x_i)_{i\in\mathbb{N}} \not\equiv (y_i)_{i\in\mathbb{N}} \pmod{\mathcal{R}} \text{ si et seulement si}$$

$$\{i \mid x_i \neq y_i\} \text{ est un ensemble de } \mathcal{U} .$$

On obtient ainsi un ensemble ou un groupe ou un anneau qui est noté $\prod_{\mathcal{U}} A_i$. Si A_i est un même objet A pour tout $i\in\mathbb{N}$, on note $\prod_{\mathcal{U}} A_i$ par $A^{\mathcal{U}}$ et on appelle $A^{\mathcal{U}}$ <u>l'ultrapuissance</u> de A. Il existe un application injective (l'application diagonale) : $A \longrightarrow A^{\mathcal{U}}$ qui est un homomorphisme de groupe, ou d'anneau, si A est un groupe ou un anneau. Le passage de l'anneau A à l'anneau $A^{\mathcal{U}}$ conserve beaucoup des propriétés de A ; par exemple si A est commutatif, il en est de même de $A^{\mathcal{U}}$ et plus précisément le centre de $A^{\mathcal{U}}$ coïncide avec $Z(A)^{\mathcal{U}}$ où $Z(A)$ désigne le centre de A. Si A est un corps (gauche) il en est de même de $A^{\mathcal{U}}$. T. Amitsur [1] a prouvé que si A est un anneau primitif alors $A^{\mathcal{U}}$ est primitif et plus précisément si $A = M_n(D)$ est un anneau de matrices $n\times n$ à coefficients dans un corps gauche D alors $A^{\mathcal{U}} = M_n(D^{\mathcal{U}})$ et si e_i $1\leq i\leq n$ est un système complet d'idempotents mutuellement orthogonaux de A alors, les $\{e_i \quad 1\leq i\leq n\}$ forment aussi un système complet d'idempotents mutuellement orthogonaux de $A^{\mathcal{U}}$. (cf [1]).

Si k est un corps commutatif, A un k-espace vectoriel de dimension finie sur k alors $A^{\mathcal{U}} \simeq k^{\mathcal{U}} \otimes_k A$ (cf. [10]) et si A est une algèbre associative (resp. de Lie) il en est de même de $A^{\mathcal{U}}$ et l'isomorphisme $A^{\mathcal{U}} \simeq k^{\mathcal{U}} \otimes_k A$ est un isomorphisme d'algèbres : ceci a été démontré en [10] pour les algèbres associatives et se démontre de la même façon pour les algèbres de Lie.

<u>Proposition</u> 1.10 - <u>Si V est un k-espace vectoriel de dimension finie on a</u> :
$\mathrm{Hom}_{k^{\mathcal{U}}}(V^{\mathcal{U}}, k^{\mathcal{U}}) \simeq \mathrm{Hom}_k(V,k)^{\mathcal{U}} \simeq \mathrm{Hom}_k(V,k) \otimes_k k^{\mathcal{U}}$.

<u>Preuve</u> : Comme $\mathrm{Hom}_k(V,k)$ est un k-espace vectoriel de dimension finie, le second isomorphisme en résulte. D'autre part on a $V^{\mathcal{U}} = k^{\mathcal{U}} \otimes_k V$; donc $\mathrm{Hom}_{k^{\mathcal{U}}}(V^{\mathcal{U}}, k^{\mathcal{U}}) \simeq k^{\mathcal{U}} \otimes \mathrm{Hom}_k(V,k)$.

§.2 - Ultra produits et idéaux primitifs.

Le résultat suivant a été démontré par C. Moeglin [8]) dans le cas où le corps de base k est algébriquement clos non dénombrable.

Proposition 2.1 - Soit k un corps de caractéristique 0, $\mathfrak{g}$ un k-algèbre de Lie de dimension finie sur k. Soit P un idéal premier de l'algèbre enveloppante $U(\mathfrak{g})$ de $\mathfrak{g}$. Alors les conditions suivantes sont équivalentes :

(1) P est un idéal primitif.

(2) Le centre de l'anneau total des fractions de $\frac{U(\mathfrak{g})}{P}$ est une extension de degré fini de k.

(3) Le centre de l'anneau total des fractions de $U(\mathfrak{g})/p$ est une extension algébrique de k.

(4) P est localement fermé dans Spec $U(\mathfrak{g})$.

Ce résultat vient d'être aussi démontré par R. Irving-L. Small [6] . Nous donnerons ici une démonstration utilisant les ultra filtres. Un idéal premier P est localement fermé dans Spec $U(\mathfrak{g})$ si l'intersection des idéaux premiers de $U(\mathfrak{g})$ qui contiennent strictement P contient aussi strictement P, c'est-à-dire que $\{P\}$ est localement fermé, pour la topologie de Zariski, dans Spec $U(\mathfrak{g})$.

Preuve du 2.1

(4) $\Longrightarrow$ (1) Résulte de [4] du fait que tout idéal semi-premier de $U(\mathfrak{g})$ est intersection des idéaux primitifs qui le contiennent, ceci quelque soit le corps k et l'algèbre de Lie $\mathfrak{g}$.

(1) $\Longrightarrow$ (2) est démontré en 4.17 [2] et résulte du lemme de Quillen.

(2) $\Longrightarrow$ (3) est évident.

(3) $\Longrightarrow$ (4) On suppose d'abord que k est dénombrable et algébriquement clos. Alors $k^{\mathcal{U}}$ est isomorphe à $\mathbb{C}$, (cf. [13]) donc $k^{\mathcal{U}}$ est non dénombrable et algébriquement clos Soit P un idéal premier de $U(\mathfrak{g})$ tel que le centre de l'anneau total des fractions de U/p coïncide avec k. D'après [2] $k^{\mathcal{U}} \otimes P$ est un idéal semi-premier de $k^{\mathcal{U}} \otimes_k U(\mathfrak{g})$ (ceci parce que l'on ait en caractéristique 0) et l'on a : $P = (k^{\mathcal{U}} \otimes_k P) \cap U(\mathfrak{g})$. On en déduit l'inclusion $\frac{U(\mathfrak{g})}{P} \hookrightarrow k^{\mathcal{U}} \otimes_k \frac{U}{P} = \frac{k^{\mathcal{U}} \otimes U(\mathfrak{g})}{k^{\mathcal{U}} \otimes P}$. D'autre part $U(\mathfrak{g})/P$ est une sous k-algèbre de son anneau total de fractions, noté Fr(U/P), lequel est un anneau de matrices $n \times n$ à coefficients dans un corps gauche D, soit $\mathbb{M}_n(D)$, et on a $Fr(\frac{U}{P}) = \mathbb{M}_n(D) \subset [Fr(\frac{U}{P})]^{\mathcal{U}} = \mathbb{M}_n(D^{\mathcal{U}})$, où $\mathbb{M}_n(-)$ correspond au système complet d'idempotents orthogonaux $e_1, e_2, \ldots, e_n$. Comme $[Fr(\frac{U}{P})]^{\mathcal{U}}$ est une $k^{\mathcal{U}}$-algèbre, on obtient un homomorphisme de $k^{\mathcal{U}}$-algèbre par extension des

scalaires :

$$\varphi : k^{\mathcal{U}} \otimes_k \frac{U}{P} \longrightarrow [\mathrm{Fr}(\frac{U}{P})]^{\mathcal{U}}$$

$$(\lambda_i) \otimes x \longmapsto (\lambda_i x)$$

En effet si $\widetilde{(\lambda_i)} = \widetilde{(\mu_i)}$ dans $k^{\mathcal{U}}$, alors $\{i | \lambda_i = \mu_i\} \in \mathcal{U}$. Donc si $x \in U/P$ on a $\{i \; \lambda_i\, x = \mu_i\, x\} \supset \{i \,|\, \lambda_i = \mu_i\}$ et donc $\{i \,|\, \lambda_i\, x = \mu_i\, x\}$ appartient à $\mathcal{U}$. Donc $\widetilde{(\lambda_i x)} \in \mathrm{Fr}(\frac{U}{P})^{\mathcal{U}}$ est bien défini. Il est facile de vérifier que φ est injective ; en effet si $\sum_{j=1}^{n} \widetilde{(\lambda_i^{(j)})} \otimes x_j \in k^{\mathcal{U}} \otimes_k \frac{U}{P}$ a pour image 0, on peut supposer les x_j linéairement indépendants sur k. On a alors $\sum_{j=o}^{n} \widetilde{(\lambda_i^{(j)}\, x_j)} = 0$ dans $[\mathrm{Fr}(\frac{U}{P})]^{\mathcal{U}}$. Donc $\{i \in \mathbb{N} | \sum_{j=o}^{n} \lambda_i^{(j)}\, x_j = 0\}$ est un ensemble de $\mathcal{U}$. Comme $\lambda_i^{(j)} \in k$ pour tout i et tout j et que les x_j sont linéairement indépendants sur k, l'ensemble $\{i \in \mathbb{N} | \sum_{j=o}^{n} \lambda_i^{(j)}\, x_j = 0\}$ coïncide avec l'ensemble $E = \{i \in \mathbb{N} \mid \lambda_i^{(j)} = 0 \quad j = 1,\ldots,n\}$. Donc pour chaque $j \in \{1,\ldots,n\}$ l'ensemble $\{i \in \mathbb{N} | \lambda_i^{(j)} = 0\}$ contient E et appartient donc à $\mathcal{U}$. Par suite $\widetilde{(\lambda_i^{(j)})} = 0$ dans $k^{\mathcal{U}}$ pour chaque j. Donc $\ker \varphi = 0$ et $k^{\mathcal{U}} \otimes_k \frac{U}{P}$ est une sous-$k^{\mathcal{U}}$-algèbre de $[\mathrm{Fr}(\frac{U}{P})]^{\mathcal{U}}$. On a donc les inclusions :

$$(*) \quad \frac{U}{P} \subset k^{\mathcal{U}} \otimes_k \frac{U}{P} \subset [\mathrm{Fr}(\frac{U}{P})]^{\mathcal{U}} \quad .$$

<u>Lemme intermédiaire</u>. <u>Soit</u> S (resp. T) <u>l'ensemble des éléments réguliers de</u> U/P (resp. $k^{\mathcal{U}} \otimes_k \frac{U}{P}$). <u>Alors</u> $S \subset T$ <u>et chaque élément</u> t <u>de</u> T <u>est non diviseur de zéro dans</u> $[\mathrm{Fr}(\frac{U}{P})]^{\mathcal{U}}$.

<u>Preuve du lemme</u>. Il est évident que $S \subset T$. Soit $t \in T$; supposons que $\{y \in [\mathrm{Fr}(\frac{U}{P})]^{\mathcal{U}} \mid yt = 0\} = I \neq 0$; alors I est un idéal à gauche non nul de $M_n(D^{\mathcal{U}})$. Il est donc de la forme $M_n(D^{\mathcal{U}})e_{i_1} \oplus \ldots \oplus M_n(D^{\mathcal{U}})e_{i_s}$ où $e_{i_j} \in \mathrm{Fr}(\frac{U}{P})$ sont des idempotents deux à deux orthogonaux. Donc $(M_n(D)e_{i_1} \oplus \ldots \oplus M_n(D)e_{i_s})t = 0$, ce qui est impossible car $t \in T$. Donc $I = 0$ et t est non diviseur de zéro dans $[\mathrm{Fr}(\frac{U}{P})]^{\mathcal{U}}$.

<u>Fin de la démonstration de la proposition</u> 2.1 - D'après le lemme précédent on a les inclusions :

$$\mathrm{Fr}(\frac{U}{P}) \subset T^{-1}(k^{\mathcal{U}} \otimes_k \frac{U}{P}) \subset [\mathrm{Fr}(\frac{U}{P})]^{\mathcal{U}}$$

car dans $(\mathrm{Fr}(\frac{U}{P}))^{\mathcal{U}}$ les éléments non diviseurs de zéro sont inversibles. Par définition de T, $T^{-1}(k^{\mathcal{U}} \otimes_k \frac{U}{P})$ est l'anneau total des fractions de $k^{\mathcal{U}} \otimes_k \frac{U}{P}$. Par hypothèse k est le centre de $\mathrm{Fr}(\frac{U}{P})$. Donc $k^{\mathcal{U}}$ est le centre de

$[Fr(\frac{U}{P})]^{\mathcal{U}}$. Soit x un élément du centre de $T^{-1}(k^{\mathcal{U}} \otimes_k \frac{U}{P})$, alors x est un élément de $[Fr(\frac{U}{P})]^{\mathcal{U}}$ qui commute avec chaque élément de $Fr(\frac{U}{P})$. Donc x appartient au centre de $[Fr(\frac{U}{P})]^{\mathcal{U}}$ c'est-à-dire à $k^{\mathcal{U}}$. Par conséquent le centre de $T^{-1}(k^{\mathcal{U}} \otimes \frac{U}{P}) = Fr(k^{\mathcal{U}} \otimes \frac{U}{P})$ est égal à $k^{\mathcal{U}}$. Ceci prouve que $T^{-1}(k^{\mathcal{U}} \otimes \frac{U}{P})$ est un anneau simple, donc que $k^{\mathcal{U}} \otimes P$ est un idéal premier de $k^{\mathcal{U}} \otimes U$ et que le centre de $Fr(\frac{k^{\mathcal{U}} \otimes U}{k^{\mathcal{U}} \otimes P})$ est $k^{\mathcal{U}}$. Puisque $k^{\mathcal{U}}$ n'est pas dénombrable, $k^{\mathcal{U}} \otimes P$ est localement fermé dans $Spec(k^{\mathcal{U}} \otimes_k U(\mathfrak{g}))$. Il est alors facile de conclure comme dans [6] en choisissant un élément $a \in k^{\mathcal{U}} \otimes U(\mathfrak{g})$, $a \notin k^{\mathcal{U}} \otimes P$ et $a \in Q$ pour chaque idéal premier Q de $k^{\mathcal{U}} \otimes U(\mathfrak{g})$ qui contient strictement $k^{\mathcal{U}} \otimes P$. Notons K le surcorps de k engendré par les scalaires qui apparaissent dans a :

$$a = \sum_{\text{finie}} \lambda_i \otimes a_i \qquad \lambda_i \in k^{\mathcal{U}} \qquad a_i \in U(\mathfrak{g}) .$$

Il est facile de vérifier que :

$$(k^{\mathcal{U}} \otimes_k P) \cap (K \otimes U(\mathfrak{g})) = K \otimes_k P$$

est un idéal premier, tel que $a \in K \otimes_k U(\mathfrak{g})$ et $a \notin K \otimes P$. Soit Q' un idéal premier de $K \otimes_k U(\mathfrak{g})$ tel que $K \otimes_k P \subsetneq Q'$; alors $k^{\mathcal{U}} \otimes_K Q'$ est un idéal semi premier de $k^{\mathcal{U}} \otimes_K U(\mathfrak{g})$ tel que $k^{\mathcal{U}} \otimes_k P \subsetneq k^{\mathcal{U}} \otimes_K Q'$ et par suite $a \in k^{\mathcal{U}} \otimes_K Q'$; donc $a \in Q'$. On a donc une extension de type fini K de k telle que l'idéal premier $K \otimes_k P$ de $K \otimes_k U(\mathfrak{g})$ est localement fermé. On peut supposer que K est une extension transcendante finie de k, $K = k(X_1,\dots,X_n)$ car K est une extension de type fini de k et que le comportement des idéaux premiers par extension de degré fini du corps de base est bien connu [11]. Ensuite de proche en proche il est possible de prouver (lemme 3.1 [6]) le résultat. Passons au cas où k est non algébriquement clos et soit k'' une clôture algébrique de k. Alors d'après cor. 3.13 de [11] il existe un idéal premier P'' de $k'' \otimes_k U(\mathfrak{g}) = U(k'' \otimes_k \mathfrak{g})$ tel que $P'' \cap U(\mathfrak{g}) = P$. D'après la proposition 1.2) de [12] le centre de l'anneau total des fractions de $\frac{k'' \otimes U(\mathfrak{g})}{P''}$ est égal à k''. Donc par ce qui précède P'' est primitif et d'après la proposition 1 3°) de [12] P est lui même primitif.

<u>Corollaire</u> 2.2 - <u>Si</u> k <u>est un corps de caractéristique</u> 0, $\mathfrak{g}$ <u>une</u> k-<u>algèbre de Lie de dimension finie sur</u> k <u>et</u> P <u>un idéal premier de</u> $U(\mathfrak{g})$ <u>alors</u> $k^{\mathcal{U}} \otimes_k P$ <u>est un idéal premier de</u> $k^{\mathcal{U}} \otimes U(\mathfrak{g})$.

<u>Preuve</u> : La démonstration comme dans la proposition 3.1 consiste à prouver que le centre de l'anneau total des fractions de $k^{\mathcal{U}} \otimes \frac{U}{P}$ est un corps ; donc cet anneau est simple. Par suite $k^{\mathcal{U}} \otimes P$ est premier.

<u>Corollaire</u> 2.3 - <u>Soit</u> k <u>un corps algébriquement clos de caractéristique</u> 0 <u>alors si</u> L <u>est une extension quelconque de</u> k, <u>et si</u> P <u>est un idéal premier de</u> $U(\mathfrak{g})$, <u>l'idéal</u> $L \otimes P$ <u>de</u> $U(L \otimes \mathfrak{g})$ <u>est premier</u>.

<u>Preuve</u> : D'après [13] il existe un k-isomorphisme de L dans une ultra puissance, de k. Donc d'après le corollaire précédent $k^{\mathcal{U}} \otimes P$ est un idéal premier de $U(k^{\mathcal{U}} \otimes \mathfrak{g})$ et comme $(k^{\mathcal{U}} \otimes P) \cap (L \otimes U(\mathfrak{g})) = L \otimes P$, ce dernier est premier ([2] 3.4.1).

Le corollaire 2.3 pour un idéal primitif est trivial vue la caractérisation 2.1. Le même corollaire 2.3 est vrai pour une k-algèbre A noethérienne à gauche, un idéal premier P de A, pourvu que $L \otimes_k A$ reste noethérienne pour toute extension L du corps de base k.

On utilisera une partie du lemme suivant au paragraphe 3.

<u>Lemme</u> 2.4 - <u>Soit</u> I <u>un idéal bilatère de</u> $U(\mathfrak{g})$, <u>alors</u> $k^{\mathcal{U}} \otimes_k I$ <u>est contenu dans l'idéal bilatère</u> $I^{\mathcal{U}}$ <u>de</u> $U(\mathfrak{g})^{\mathcal{U}}$ <u>et l'on a</u> : $I^{\mathcal{U}} \cap (k^{\mathcal{U}} \otimes U(\mathfrak{g})) = k^{\mathcal{U}} \otimes I$.

<u>Preuve</u> : Comme $I \subset I^{\mathcal{U}}$ et que $I^{\mathcal{U}}$ est un $k^{\mathcal{U}}$-espace vectoriel, on a un homomorphisme

$$\varphi : k^{\mathcal{U}} \otimes I \longrightarrow I^{\mathcal{U}}$$

tel que $\widetilde{(\lambda_i)} \otimes x \longmapsto \widetilde{(\lambda_i x)}$ dont on démontre facilement qu'il est injectif. En identifiant $k^{\mathcal{U}} \otimes I$ à son image par φ, il est évident que l'on a : $k^{\mathcal{U}} \otimes I \subseteq I^{\mathcal{U}} \cap (k^{\mathcal{U}} \otimes U(\mathfrak{g}))$. D'autre part on a :

$$I^{\mathcal{U}} \cap (k^{\mathcal{U}} \otimes U(\mathfrak{g})) = I^{\mathcal{U}} \cap U_{k^{\mathcal{U}}}(\mathfrak{g}^{\mathcal{U}}) = I^{\mathcal{U}} \cap (\bigcup_{m \geq 0} F^m U(\mathfrak{g}^{\mathcal{U}})$$

où $F^m(-)$ désigne les espaces vectoriels de la filtration naturelle de $U_{k^{\mathcal{U}}}(\mathfrak{g}^{\mathcal{U}})$. D'où $I^{\mathcal{U}} \cap (k^{\mathcal{U}} \otimes U(\mathfrak{g})) = \bigcup_{m \geq 0} [I^{\mathcal{U}} \cap F^m U(\mathfrak{g}^{\mathcal{U}})]$. Puisque $F^m U_k(\mathfrak{g})$ est de dimension finie sur k on a :

$$F^m U_{k^{\mathcal{U}}}(\mathfrak{g}^{\mathcal{U}}) = k^{\mathcal{U}} \otimes F^m U_k(\mathfrak{g}) = (F^m U_k(\mathfrak{g}))^{\mathcal{U}}$$

Donc

$$I^{\mathcal{U}} \cap F^m U_{k^{\mathcal{U}}}(\mathfrak{g}^{\mathcal{U}}) = k^{\mathcal{U}} \otimes (I^{\mathcal{U}} \cap F^m U(\mathfrak{g}))$$

Posons $V = U_k(\mathfrak{g})$: c'est un k-espace vectoriel de dimension finie. On a :

$$I^{\mathcal{U}} \cap V^{\mathcal{U}} = I^{\mathcal{U}} \cap (k^{\mathcal{U}} \otimes V) \supset (k^{\mathcal{U}} \otimes I) \cap (k^{\mathcal{U}} \otimes V)$$

$= k^{\mathcal{U}} \otimes_k (I \cap V)$ par platitude

$= (I \cap V)^{\mathcal{U}}$ car $I \cap V$ est de dimension finie sur k.

Soit $x = \widetilde{(x_i)} \in I^{\mathcal{U}} \cap V^{\mathcal{U}}$. Alors

$\{i \mid x_i \in I\} \in \mathcal{U}$, $\{x \mid x_i \in V\} \in \mathcal{U}$. Donc $\{i \mid x_i \in I \cap V\} \in \mathcal{U}$. Donc $(\widetilde{x_i}) \in (I \cap V)^{\mathcal{U}}$. Donc $I^{\mathcal{U}} \cap (k^{\mathcal{U}} \otimes V) = (I \cap V)^{\mathcal{U}}$ et $I^{\mathcal{U}} \cap (F^m\, U(\mathfrak{g}^{\mathcal{U}}) = (I \cap F^m\, U(\mathfrak{g}))^{\mathcal{U}}$. Donc $\bigcup_m I^{\mathcal{U}} \cap F^m\, U(\mathfrak{g}^{\mathcal{U}}) = \bigcup_m (I \cap F^m\, U(\mathfrak{g}))^{\mathcal{U}}$. Donc

$$I^{\mathcal{U}} \cap (k^{\mathcal{U}} \otimes U(\mathfrak{g})) = \bigcup_m k^{\mathcal{U}} \otimes (I \cap F^m\, U(\mathfrak{g})) \subsetneq k^{\mathcal{U}} \otimes I.$$ D'où l'égalité cherchée.

§.3 - Le semi-centre.

Le résultat suivant a été obtenu directement par V. Ginsburg [5]. Ici nous partirons du cas non dénombrable considéré par C. Moeglin [7].

<u>Proposition</u> 3.1 - <u>Soit</u> k <u>un corps de caractéristique 0,</u> $\mathfrak{g}$ <u>une algèbre de Lie de dimension finie sur</u> k, $U(\mathfrak{g})$ <u>son algèbre enveloppante et</u> $E(\mathfrak{g})$ <u>l'anneau des semi-invariants de</u> $U(\mathfrak{g})$. <u>Alors tout idéal bilatère non nul de</u> $U(\mathfrak{g})$ <u>a une intersection non nul avec</u> $E(\mathfrak{g})$.

<u>Preuve</u> : On commencera par supposer k algébriquement clos. On peut supposer k dénombrable, sinon le résultat est démontré dans [7] . Si $\mathcal{U}$ est un ultra-filtre sur $\mathbb{N}$ on a $k^{\mathcal{U}} \simeq \mathbb{C}$ ([cf. [13]) et $U_{k^{\mathcal{U}}}(\mathfrak{g}^{\mathcal{U}}) = U_{k^{\mathcal{U}}}(k^{\mathcal{U}} \otimes_k \mathfrak{g}) = k^{\mathcal{U}} \otimes_k U(\mathfrak{g})$. Soit I un idéal bilatère non nul de $k^{\mathcal{U}} \otimes U(\mathfrak{g})$. Alors $k^{\mathcal{U}} \otimes I$ est un idéal bilatère non nul de $k^{\mathcal{U}} \otimes U(\mathfrak{g})$. Donc par [7] on a $E(\mathfrak{g}^{\mathcal{U}}) \cap (k^{\mathcal{U}} \otimes I) \neq 0$, où $E(\mathfrak{g}^{\mathcal{U}})$ est l'anneau des semi-invariants de $U(\mathfrak{g}^{\mathcal{U}})$. On a $k^{\mathcal{U}} \otimes I \subset I^{\mathcal{U}}$, d'après le lemme 2.4, , donc $I^{\mathcal{U}} \cap E(\mathfrak{g}^{\mathcal{U}}) \neq 0$. Soit $x = (\widetilde{x_i}) \in E(\mathfrak{g}^{\mathcal{U}}) \cap I^{\mathcal{U}}$, $x \neq 0$, alors $\{i \mid x_i \in I\ x_i \neq 0\}$ est un élément de $\mathcal{U}$. D'autre part il existe $(\widetilde{\lambda_i}) = \lambda \in (\mathfrak{g}^{\mathcal{U}})^* = (\mathfrak{g}^*)^{\mathcal{U}}$ tel que $[g,\lambda] = \lambda(g)x$, pour tout $g \in \mathfrak{g}^{\mathcal{U}}$; en particulier pour tout $g \in \mathfrak{g}$ on a : $[g,x] = \lambda(g)x$; d'où $\{i \mid [g,x_i] = \lambda_i(g)x_i\}$ est un élément de $\mathcal{U}$. Donc si $g \in \mathfrak{g}$, $\{i,\ x_i \neq 0\ \ x_i \in I$ et $[g,x_i] = \lambda_i(g)x_i\} \in \mathcal{U}$. Soit $e_1, e_2, \ldots, e_n$ une base de $\mathfrak{g}$ sur k. Alors pour chaque e_j, $1 \leq j \leq n$, $X_j = \{i \quad x_i \in I\setminus(0)\ [e_j,x_i] = \lambda_i(e_j)x_i\}$ est un élément de $\mathcal{U}$. Donc il existe $y \in X_1 \cap \ldots \cap X_n$ c'est-à-dire $y \in I\setminus(0)$ et $\mu \in \mathfrak{g}^*$ et tel que $[e_j,y] = \mu(e_j)y$, $j = 1,\ldots,n$. Donc si $g = \sum_{j=1}^{n} \nu_j e_j$ est un élément de $\mathfrak{g}$ on a :

$$[g,y] = \Big[\sum_{j=1}^{n} \nu_j e_j,\ y\Big] = \sum_{j=1}^{n} \nu_j [e_j,y]$$
$$= \sum_{j=1}^{n} \nu_j \mu(e_j) y = \mu\Big(\sum_{j=1}^{n} \nu_j e_j\Big) y = \mu(g) y$$

et par suite $y \in E(\mathfrak{g}) \cap I$ $y \neq 0$. Si k n'est pas algébriquement clos, soit $\tilde{k}$ une clôture algébrique de k et $E(\tilde{\mathfrak{g}})$ le semi-centre de $U(\tilde{\mathfrak{g}})$. On a $\tilde{k} \otimes_k E(\mathfrak{g}) \subset E(\tilde{\mathfrak{g}})$ et $\tilde{k} \otimes I$ est un idéal bilatère non nul de $U(\tilde{\mathfrak{g}})$. Donc il existe $x \in (\tilde{k} \otimes I) \cap E(\tilde{\mathfrak{g}})_\lambda$, $x \neq 0$, d'après la démonstration précédente, où :

$$\lambda : \tilde{\mathfrak{g}} \longrightarrow \tilde{k} \qquad \lambda \in \tilde{k} \otimes_k \mathrm{Hom}(\mathfrak{g},k),$$

$\lambda = \sum_i \beta_i \lambda_i$, $\beta_i \in \tilde{k}$ $\lambda_i \in \mathrm{Hom}(\mathfrak{g},k)$. On a $[g,x] = \lambda(g)x$ pour tout $g \in \tilde{\mathfrak{g}}$.

Ecrivons :

$$x = \sum \alpha_i \otimes x_i \qquad \alpha_i \in \tilde{k} \qquad x_i \in \mathfrak{g} \quad .$$

On se ramène au cas où $\tilde{k}$ est le surcorps de k engendré par les β_i et les α_i; quitte à agrandir ce surcorps on peut supposer que $k \subset k(\xi)$ où $k(\xi)$ est une extension galoisienne de k de groupe de Galois G. Si N est la norme sur $k(\xi)$ étendu à $k(\xi) \otimes \mathfrak{g}$ et $k(\xi) \otimes \mathrm{Hom}_k(\mathfrak{g},k)$ on a :

$$[g,N(x)] = N(\lambda)(g)\,N(x) .$$

En effet

$$gx - xg = \sum \alpha_i (gx_i - x_i g) = \sum \beta_i \lambda_i (g)$$
$$= \sum_{i,j} \alpha_i \beta_j \lambda_j (g)\, x_i .$$

Donc

$$gN(x) - N(x)g = \sum_{i,j} N(\alpha_i) N(\beta_j) \lambda_j (g)\, x_i$$
$$= N(\lambda)(g) N(x)$$

§.4 - Idéaux primitifs induits.

Le résultat suivant (cf. [8] th. 1.2) est encore vrai pour un corps de base algébriquement clos, dénombrable, de caractéristique 0. Dans tout le paragraphe nous supposerons le corps de base algébriquement clos de caractéristique 0.

Proposition 4.1 - Tout idéal primitif minimal de $U(\mathfrak{g})$ est un idéal induit de Duflo (donc est complètement premier).

Preuve : La démonstration des th. 1.2 de [8] s'applique mot pour mot, compte tenu des remarques suivantes :

1°) Si un idéal I de $U(\mathfrak{g})$ est primitif, il satisfait au critère b' du théorème C de [3] d'après la remarque 1.7 de cet article.

2°) Si z appartient au centre d'une algèbre de Lie $\mathfrak{g}$, tout idéal primitif I de $U(\mathfrak{g})$ inclus dans un idéal primitif contenant $z-1$, contient aussi $z-1$, car un idéal primitif est rationnel lorsque le corps de base est algébriquement clos.

3°) Dans la démonstration du cas (3) du théorème 1.2 de [8] il suffit, au lieu de renvoyer au théorème A de [3] d'appliquer [9].

<u>Proposition</u> 4.2 - <u>Soit</u> $I = I_{\mathfrak{g}}(\mu)$ <u>un idéal induit de Duflo. On note</u> ρ <u>l'antiautomorphisme principal de</u> $U(\mathfrak{g})$. <u>Alors</u> $\rho I = I_{\mathfrak{g}}(-\mu)$.

<u>Preuve</u> : Soit $k^{\mathcal{U}}$ où $\mathcal{U}$ est un ultrafiltre sur $\mathbb{N}$. Alors μ définit $(\mu) \in (\mathfrak{g}^{\mathcal{U}})^* = (\mathfrak{g}^*)^{\mathcal{U}}$. Donc $k^{\mathcal{U}} \otimes I = I_{\mathfrak{g}^{\mathcal{U}}}((\mu))$. Si ρ est l'antiautomorphisme principal de $U(\mathfrak{g})$ on peut le considérer comme la restriction à $U(\mathfrak{g})$ de l'anti-automorphisme principal de $U(\mathfrak{g}^{\mathcal{U}})$. On a $\rho(k^{\mathcal{U}} \otimes I) = I_{\mathfrak{g}^{\mathcal{U}}}(-(\mu))$. Donc $k^{\mathcal{U}} \otimes \rho(I) = I_{\mathfrak{g}^{\mathcal{U}}}(-(\mu)) = k^{\mathcal{U}} \otimes I_{\mathfrak{g}}(-\mu)$. D'où $\rho(I) = I_{\mathfrak{g}}(-\mu)$.

<u>Proposition</u> 4.3 - <u>Si</u> $\mathfrak{g}$ <u>est algébrique et si</u> I <u>est un idéal primitif de</u> $U(\mathfrak{g})$ <u>alors</u> $d(U(\mathfrak{g})/I)$ <u>est pair ; si</u> $I = I_{\mathfrak{g}}(\nu)$ <u>est un idéal induit de Duflo, on a</u> : $d(U(\mathfrak{g})/I) = \dim \mathfrak{g}.\nu$.

<u>Preuve</u> : On sait que $k^{\mathcal{U}} \otimes I$ est primitif et que $d_k(U(\mathfrak{g})/I) = d_{k^{\mathcal{U}}}(k^{\mathcal{U}} \otimes U(\mathfrak{g})/I)$ est pair. Si $I = I_{\mathfrak{g}}(\nu)$, alors $k^{\mathcal{U}} \otimes I = I_{\mathfrak{g}^{\mathcal{U}}}((\nu))$. Donc ([8] Th. 4.2) on a $\dim \mathfrak{g}^{\mathcal{U}}.(\nu) = d(U(\mathfrak{g})/I)$. Il suffit de prouver que $\dim_{k^{\mathcal{U}}} \mathfrak{g}^{\mathcal{U}}.(\nu) = \dim_k \mathfrak{g}.\nu$. On note B_ν (resp. $B_{(\nu)}$) la bilinéaire alternée : $(X,Y) \longrightarrow \nu([X,Y])$ sur $\mathfrak{g}$, (resp. $(X,Y) \longrightarrow (\nu)([X,Y])$ sur $\mathfrak{g}^{\mathcal{U}}$). Alors $B_{(\nu)}$ est l'extension de B_ν à $k^{\mathcal{U}} \otimes \mathfrak{g}$; donc $\ker B_{(\nu)} = k^{\mathcal{U}} \otimes \ker B_\nu$ et $\dim_k \mathfrak{g}.\nu = \dim_k \mathfrak{g} - \dim_k \ker B_\nu$ $= \dim_{k^{\mathcal{U}}} \mathfrak{g}^{\mathcal{U}}.(\nu)$.

Bibliographie

[1] T. AMITSUR : Prime ideals having P.I. with arbitrary coefficient Proc. London, Math. Soc (3) 17 (1962) 470-486.

[2] J. DIXMIER : Algèbres enveloppantes. Gauthier Villars, 1974.

[3] J. DIXMIER : Idéaux primitifs dans les algèbres enveloppantes. J. of Algebra 48, p.96-112, 1977.

[4] M. DUFLO : Certaines algèbres de type fini sont des algèbres de Jacobson. J. of Algebra, 27, 1973, p.358-365.

[5] V. GINSBURG : On the ideals of $U(\mathfrak{g})$ (à paraître).

[6] R. IRVING-L. SMALL : On the characteriszation of primitive ideals in enveloping algebras, Math. Z. 173, p.217-221, (1980).

[7] C. MOEGLIN : Idéaux bilatères des algèbres enveloppantes. Bull. Soc. Math. France 108, 1980, p.143-186.

[8] C. MOEGLIN : Idéaux primitifs des algèbres enveloppantes J. de Math. Pures et Appliquées 59, 1980, p.265-336.

[9] C. MOEGLIN-R. RENTSCHLER : Orbites d'un groupe algébrique dans l'espace des idéaux rationnels d'une algèbre enveloppante (à paraître).

[10] F. TAHA : Algèbres simples centrales sur les corps ultraproduits de corps p-adiques (ce fascicule).

[11] S. YAMMINE : Les théorèmes de Cohen-Seidenberg en algèbre non commutative séminaire d'algèbre P. Dubreil 77-78, Lectures Notes in Math. 740.

[12] S. YAMMINE : Idéaux Primitifs dans des algèbres universelles (ce fascicule).

[13] M. ZAYED : Caractérisation des algèbres de représentation finie sur des corps algébriquement clos (ce fascicule).

GRADE ET THEOREME D'INTERSECTION EN

ALGEBRE NON COMMUTATIVE

Erratum

Marie-Paule MALLIAVIN

Soit k un corps algébriquement clos de caractéristique 0, $\mathfrak{g}$ une k-algèbre de Lie résoluble de dimension finie n sur k et soit $x_1,\ldots,x_n$ une base de $\mathfrak{g}$ adaptée à un drapeau d'idéaux de $\mathfrak{g}$. Soit U l'algèbre enveloppante de $\mathfrak{g}$, P et Q des idéaux premiers de $U(\mathfrak{g})$ tels que $P + Q \neq U(\mathfrak{g})$. On note $(U)^{opp}$ l'algèbre U muni de la multiplication opposée $\star$, $\otimes = \otimes_k$. Soit I l'idéal à gauche de $U \otimes U^{opp}$ engendré par $x_1 \otimes 1 - 1 \otimes x_1 \ldots x_n \otimes 1 - 1 \otimes x_n$; c'est aussi le noyau de l'application multiplication φ :

$$0 \to I \to U \otimes U^{opp} \xrightarrow{\varphi} U \to 0,$$

la suite précédente étant une suite exacte de $U \otimes U^{opp}$-modules à gauche.

Posons $A = \dfrac{B}{P\otimes U^{opp}+U\otimes Q^{opp}}$ où $B = U \otimes U^{opp}$;

$$\Gamma = P \otimes U^{opp} + U \otimes Q^{opp} \quad \text{et} \quad \bar{I} = \frac{I+\Gamma}{\Gamma}$$

On a $A/\bar{I} = \dfrac{U}{P+Q}$ car il est facile de vérifier que le diagramme suivant est à lignes et colonnes exactes :

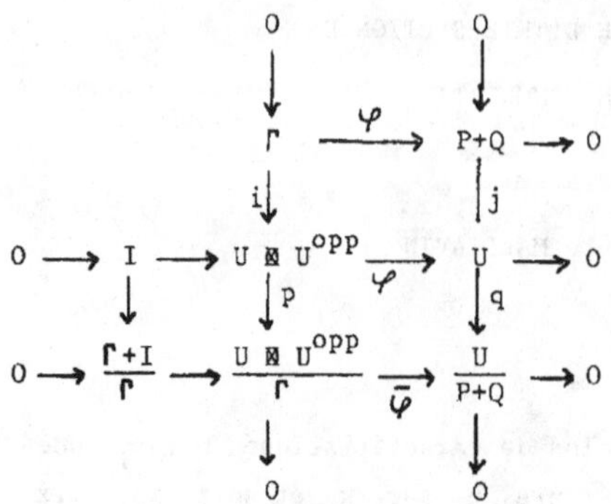

La démonstration de la proposition 2.4 de [2] est incomplète ainsi que me l'a fait remarquer A. Joseph, que je remercie. Il s'agissait de démontrer l'inégalité suivante $ht(P + Q) \leq ht P + ht Q$. En utilisant [4] il suffit de supposer le corps k algébriquement clos.

On note GK-dim la dimension de Gelfand-Kirillov sur k d'une algèbre ou d'un module. Le résultat affirmé et non démontré dans la proposition citée est l'inégalité :

$$(*) \qquad \text{GK-dim } A \leq n + \text{GK-dim } A/\bar{I}.$$

<u>Lemme</u> 1 : <u>On a, en posant</u> $y_i = x_i \otimes 1 - 1 \otimes x_i$, $i = 1,2,\dots,n$ <u>et</u> $B = U \otimes U^{opp}$,

$$(By_1+\dots+By_{i-1})y_i \subseteq By_1+\dots+By_{i-1}$$

<u>pour</u> $i=2,\dots,n$.

<u>Preuve</u> : Il suffit de prouver que si $j < i$ on a

$$y_j y_i \in By_1+\dots+By_j \qquad 1 \leq j < i \leq n$$

$$y_j y_i - y_i y_j = (x_j \otimes 1 - 1 \otimes x_j)(x_i \otimes 1 - 1 \otimes x_i) - (x_i \otimes 1 - 1 \otimes x_i)(x_j \otimes 1 - 1 \otimes x_j)$$

$$= (x_j x_i - x_i x_j) \otimes 1 + 1 \otimes (x_j * x_i - x_i * x_j) - x_j \otimes x_i - x_i \otimes x_j$$

$$+ x_i \otimes x_j + x_j \otimes x_i = \sum_{l=1}^{j} a_l x_l \otimes 1 - 1 \otimes \sum_{l=1}^{j} a_l x_l$$ car $*$ est la multiplication opposée

$$= \sum_{l=1}^{j} a_l (x_l \otimes 1 - 1 \otimes x_l) = \sum_{l=1}^{j} a_l y_l \quad \text{où} \quad a_l \in k .$$

Pour démontré $(*)$ on pose

$$M_{i-1} = A/A\bar{y}_1+\ldots+A\bar{y}_{i-1} = \frac{B}{\Gamma+By_1+\ldots+By_{i-1}}$$

D'après le lemme 1, la multiplication à droite par y_i induit un endomorphisme que l'on notera $\tilde{y}_i$ sur M_{i-1} ; il suffit de démontrer l'inégalité :

$$(**) \qquad \text{GK-dim } M_{i-1} \leq 1 + \text{GK-dim } \frac{M_{i-1}}{\tilde{y}_i(M_{i-1})}$$

pour $i=2,\ldots,n$.

<u>Lemme</u> 2 <u>soient</u> N <u>un</u> <u>B-module à gauche de type fini et</u> ψ <u>un</u> <u>B-endomorphisme de</u> N. <u>S'il existe un entier</u> s <u>tel que</u> $\psi^s(N) = 0$ <u>alors</u> GK-dim N = GK-dim $N/\psi(N)$.

<u>Preuve</u> : On raisonne par récurrence sur s partant du cas s=1 où il n'y a rien à démontrer. Supposons $s > 1$ et l'égalité vraie jusqu'à l'ordre s-1. On a donc GK-dim $N/\psi(N)$ = GK-dim $N/\psi^{s-1}(N)$, car $\psi^{s-1}(N/\psi^{s-1}(N)) = 0$ et donc GK-dim $N/\psi^{s-1}(N)$ = GK-dim $\frac{N/\psi^{s-1}(N)}{\psi(N)/\psi^{s-1}(N)}$ = GK-dim $N/\psi(N)$.

D'autre part, d'après le lemme 2.4 de [3], on a :

$$\text{GK-dim } N = \text{Sup}(\text{GK-dim } \frac{N}{\psi(N)}, \text{ GK-dim } \psi(N)) .$$

Si l'on avait :

$$\text{GK-dim } \frac{N}{\psi(N)} \lneq \text{GK-dim } \psi(N) = \text{GK-dim } N$$

alors on aurait :

$$\text{GK-dim } N/\psi(N) \lneq \text{GK-dim } \frac{\psi(N)}{\psi^2(N)} = \text{GK-dim } \psi(N), \text{ car}$$

$\psi^{s-1}(\psi(N)) = 0$. D'autre part, il existe une suite exacte de B-modules à gauche :

$$0 \longrightarrow \ker\theta \longrightarrow N/\psi(N) \xrightarrow{\theta} \psi(N)/\psi^2(N) \longrightarrow 0$$

où θ fait correspondre à la classe modulo $\psi(N)$ de $x \in N$, la classe modulo $\psi^2(M)$ de $\psi(x)$; il résulte de cette suite l'inégalité :

$$\text{GK-dim } \frac{\psi(N)}{\psi^2(N)} \leq \text{GK-dim } \frac{N}{\psi(N)}$$

que contre dit GK-dim $N/\psi(N) \lneq$ GK-dim $\frac{\psi(N)}{\psi^2(N)}$.

<u>Corollaire</u> : <u>Si</u> N <u>est un B-module de type fini et</u> ψ <u>un endomorphisme de</u> N <u>alors</u> :

1°) GK-dim $N/\psi^s N$ = GK-dim $N/\psi(N)$ <u>pour tout</u> $s \geq 1$.

2°) Si $\psi^s(N) = 0$ on a GK-dim $N \leq 1$ + GK-dim $N/\psi(N)$.

Preuve : 1°) Il résulte du lemme 2 appliqué à $N/\psi^s(N)$ et 2°) est immédiat à partir du lemme 2.

Lemme 3 : Soit N un B-module de type fini et ψ un endomorphisme de N alors
1°) Si ψ n'est pas nilpotent, il existe s tel que ψ est un endomorphisme injectif de $\psi^s(N)$.
2°) Si GK-dim $\psi^s(N) \leq 1$ + GK-dim $\psi^s(N)/\psi^{s+1}(N)$ alors on a :
GK-dim $N \leq 1$ + GK-dim $N/\psi(N)$.

Preuve : On a $\ker\psi \subseteq \ker\psi^2 \subseteq \ldots$ et puisque N est noethérien, il existe s tel que $\ker\psi^s = \ker\psi^{s+1}$. Donc si $\psi(\psi^s(x)) = 0$ alors $\psi^s(x) = 0$ d'où 1°).
2°) Si GK-dim $\psi^s(N) <$ GK-dim N, alors d'après [3] (loc. cit) on a
GK-dim N = GK-dim $N/\psi^s(N) \leq 1$ + GK-dim $N/\psi(N)$ d'après le corollaire du lemme 2. Si GK-dim N = GK-dim $\psi^s(N)$, et si l'on suppose que
GK-dim $\psi^s(N) \leq 1$ + GK-dim $\psi^s(N)/\psi^{s+1}(N)$, alors on a
GK-dim $N \leq 1$ + GK-dim $\psi^s(N)/\psi^{s+1}(N)$. On considère la suite exacte :

$$0 \longrightarrow \ker\psi^s \longrightarrow \frac{N}{\psi(N)} \xrightarrow{\tilde{\psi}^s} \frac{\psi^s(N)}{\psi^{s+1}(N)} \longrightarrow 0$$

où $\tilde{\psi}^s$ est induit par ψ^s ; il en résulte que

$$\text{GK-dim}\ \frac{\psi^s(N)}{\psi^{s+1}(N)} \leq \text{GK-dim}\ \frac{N}{\psi(N)} \quad .$$

D'où l'inégalité : GK-dim $N \leq 1$ + GK-dim $N/\psi(N)$.

Pour démontrer (**) (p.3) il suffit d'après le lemme 3.2°) de démontrer que si $\tilde{y}_i$ est injectif dans $\tilde{y}_i^s(M_{i-1})$ alors

$$\text{GK-dim}\ M_{i-1}\, y_i^s \leq 1 + \text{GK-dim}\ \frac{M_{i-1}\, y_i^s}{M_{i-1}\, y_i^{s+1}} \quad .$$

Pour celà il suffit de démontrer:

Proposition 1 : Si N est un B-module à gauche de type fini, si $y \in B$ est de degré de filtration 1 et si la multiplication à droite par y dans N détermine un B-homomorphisme injectif ψ, alors GK-dim $N \leq 1$ + GK-dim N/Ny.

Preuve : L'inégalité à démontrer est équivalente à $gr_B(N/\psi(N)) \leq 1 + gr_B(N)$ où $gr_B(-)$ est le grade sur B du B-module à gauche (-).

Posons $t = gr_B(N)$; alors par définition du grade on a :

$$Ext^i_B(N,B) = 0 \text{ si } t < t \text{ et } Ext^t_B(N,B) \neq 0 .$$

Supposons que $1+gr_B(N) < gr_B(N/\psi(N))$; comme le grade ne varie pas par passage aux gradués on a donc $1+gr_{GrB}(Gr\ N) < gr_{GrB}(Gr\ \frac{N}{\psi(N)})$. De plus ψ induit par passage au gradué la multiplication par Cry et $Gr(N/\psi(N)) = \frac{GrN}{Gry.GrN}$ car les filtrations sont déduites de la filtration de B.

On a donc la suite exacte :

$$0 \longrightarrow \ker Gry \longrightarrow GrN \xrightarrow{Gry} GrN \longrightarrow Gr(N/Gr(y)\ GrN) \longrightarrow 0.$$

On obtient de l'inégalité stricte

$$1+gr_{GrB}\ GrN < gr_{GrB}\ \frac{GrN}{Gry.GrN}$$

les suites exactes :

$$0 \longrightarrow Ext^t(GrM,GrB) \xrightarrow{\text{mult } Gry} Ext^t(\frac{GrM}{\ker Gry}, GrB) \longrightarrow 0$$

et

$$0 \longrightarrow Ext^t(\frac{GrM}{\ker Gry}, GrB) \xrightarrow[\varphi]{} Ext^t(GrM,GrB) \longrightarrow 0$$

où $Ext^t(GrM,GrB)$ est $\neq 0$ donc aussi $Ext^t(\frac{GrM}{\ker Gry}, GrB)$.

En composant $\varphi \circ$ mult Gry on obtient un isomorphisme

$$Ext^t(GrN,GrB) \longrightarrow Ext^t\ GrN,GrB).$$

Puisque $Ext^t(GrM,GrB)$ est $\neq 0$ son annulateur est un idéal gradué de GrB, distinct de GrB, donc contenu dans l'idéal d'augmentation $\mathfrak{m}$ de GrB. Si P est un idéal premier de GrB ne contenant pas $\mathfrak{m}$ on a alors

$$Ext^t(GrN,CrB)_{\mathfrak{m}} = 0$$

De plus $Ext^t(GrN,GrB)_{\mathfrak{m}} \xrightarrow{\varphi_{\mathfrak{m}} \circ \text{ mult.Gry}} Ext^t(GrN,GrB)$ où $\varphi_{\mathfrak{m}}$ est le prolongement de φ à $Ext^t(\frac{GrM}{\ker Gry}, GrB)_{\mathfrak{m}}$. Posons $Q = Ext^t(\frac{GrM}{\ker Gry}, GrB)_{\mathfrak{m}}$

On a $\varphi_{\mathfrak{m}}(\mathfrak{m}Q) \subsetneq \mathfrak{m}\ \varphi_{\mathfrak{m}}(Q)$. Donc $\varphi_{\mathfrak{m}} \circ$ mult.Gry envoie Q dans $\mathfrak{m}Q$. D'où $Q = \mathfrak{m}Q$ ce qui n'est possible que si $Q = 0$ car Q est de type fini.

Je remercie P. Tauvel dont les remarques m'ont permis de simplifier une première version

Bibliographie

[1] J.E. Björk - Rings of differential operators. North. Holland, 1979.

[2] M.P. Malliavin - Grade et théorème d'intersection en algèbre non commutative p.76-87, Ring Theory Antwerp 1980, Lecture Notes 825, Springer-Verlag.

[3] D.A. Vogan - Gelfand - Kirillov - dimension for Harish-Chandra modules, Inventiones, Math. 48, 1978, p.75-98.

[4] S. Yammine - Théorèmes de Cohen-Seidenberg en algèbre non commutative, Séminaire d'Algèbre Paul Dubreil 1977-78, Lecture Notes 740, Springer-Verlag.

Sur la dimension de Krull de l'algèbre enveloppante d'une algèbre de Lie semi-simple

par Thierry LEVASSEUR

1) Introduction et notations

1.1 - Soit k un corps algébriquement clos de caractéristique 0 et $\underline{g}$ une algèbre de Lie semi-simple sur k . Nous noterons $U(\underline{g})$ l'algèbre enveloppante de $\underline{g}$ et si M est un $U(\underline{g})$-module de type fini (à gauche) nous désignerons par K dim M (resp. d(M)) sa dimension de Krull-Gabriel-Rentschler (resp. sa dimension de Gelfand-Kirillov), cf. [J.2] pour les définitions.
Rappelons que les dimensions de Gelfand-Kirillov de $U(\underline{g})$ et de ses quotients primitifs sont connues mais qu'il n'en est pas de même pour la dimension de Krull, en particulier il est conjecturé que : K dim $U(\underline{g})$ = dimension d'une sous-algèbre de Borel de $\underline{g}$. Seules des réponses partielles ont été données sur cette question, ([J.2] , [L.] , [Sm]) ; en particulier S.P. Smith ([Sm]) a montré que l'on a l'inégalité K dim $U(\underline{g}) \leqslant \dim_k \underline{g} - r$ où 2r désigne l'inf. des d(U(g)/I) avec I idéal primitif de codimention infinie.
Nous voulons ici donner une démonstration sensiblement différente de ce résultat et montrer comment la connaissance des opérateurs différentiels sur "l'espace affine basique" (cf. [G.K]) permet d'obtenir le cas échéant des informations sur K dim $U(\underline{g})$. Comme application on obtient la dimension de Krull de l'algèbre enveloppante d'une somme directe de copies de sl(2;k), problème posé dans [Sm] § 3.3.

1.2 - Notations et rappels

Donnons les notations que nous garderons dans tout ce qui suit, (en plus de celles de 1.1) :

- $\underline{g}$ algèbre de Lie semi-simple, de dimension $2n + \ell$, que l'on prendra écrite sous la forme : $\underline{g} = \underline{g}_{\mathbb{Q}} \otimes_{\mathbb{Q}} k$ où $\underline{g}_{\mathbb{Q}}$ est une algèbre de Lie semi-simple scindée sur $\mathbb{Q}$, (à l'aide d'une base de Cartan)
- $\underline{h}_{\mathbb{Q}}$ une sous-algèbre de Cartan de $\underline{g}_{\mathbb{Q}}$ et $\underline{h} = \underline{h}_{\mathbb{Q}} \otimes_{\mathbb{Q}} k$
- $\ell = \dim_{\mathbb{Q}} \underline{h}_{\mathbb{Q}} = \dim_k \underline{h}$, h^* le dual de $\underline{h}$
- Δ(resp. Δ^+) l'ensemble des racines (resp. des racines positives)

de $\underline{g}$ associées à la paire $(g_Q, \underline{h}_Q)$ et ρ la demi-somme des racines positives.

- $\underline{g} = \mathfrak{n}^+ \oplus \underline{h} \oplus \mathfrak{n}^-$ une décomposition triangulaire correspondante.

- $\{Q_1, \ldots, Q_\ell\}$ ℓ éléments algébriquement indépendants engendrant le centre de l'algèbre enveloppante de $\underline{g}_Q$ sur Q. Ainsi le centre $Z(\underline{g})$ de $U(\underline{g})$ est $k[Q_1, \ldots, Q_\ell]$.

- Si $\lambda \in \underline{h}^*$ on note $M(\lambda)$ le module de Verma de plus haut poids $\lambda - \rho$, χ_λ son caractère central (qui est aussi celui de son unique quotient simple $L(\lambda)$) ; $J(\lambda) = U(g)\ \mathrm{Ker}\ \chi_\lambda$ l'annulateur de $M(\lambda)$ i.e. $J(\lambda) = U(\underline{g})\ (Q_1 - \chi_\lambda(Q_1), \ldots, Q_\ell - \chi_\lambda(Q_\ell))$

- $B_\lambda = U(\underline{g})/J(\lambda)$.

- G le groupe adjoint de $\underline{g}$. Si g^* est le dual de $\underline{g}$, on l'identifie à $\underline{g}$ par la forme de Killing et si $X \in \underline{g}$, on note $\dim G.X$ la dimension de l'orbite $G.X$; de sorte que $2r = \inf\{\dim G.X,\ X \in \underline{g} \setminus \{0\}\}$ (cf. 1.1).

- Rappelons pour finir que $U(\underline{g})$ est muni d'une filtration canonique par des $U_n(\underline{g})$ ([D] 2.3.1) et que le gradué associé est $S(\underline{g})$ l'algèbre symétrique de $\underline{g}$. De plus si M est un $U(g)$-module de type fini (à gauche par exemple) il peut être muni d'une filtration compatible avec celle de $U(\underline{g})$ qui fait du gradué de M, $\mathrm{gr}\, M$, un $S(g)$-module de type fini et alors

$$d(M) = \dim_{S(\underline{g})} \mathrm{gr}\, M = \dim \frac{S(g)}{\mathrm{Ann}_{S(g)}\ \mathrm{gr}\, M} = \dim V(\mathrm{ann}_{S(g)}\ \mathrm{gr}\, M)$$

en notant $V(\mathrm{Ann}_{S(g)}\ \mathrm{gr}\, M)$ la variété des zéros dans $\underline{g}^*$ (donc dans $\underline{g}$) de l'annulateur de $\mathrm{gr}\, M$ dans $S(g)$.

Les modules seront pris à gauche sauf indication contraire ; si $U(\underline{g})/I$ est un quotient de $U(\underline{g})$ par un idéal bilatère $K \dim U(\underline{g})/I$ sera donc la dimension de Krull à gauche de $U(\underline{g})/I$, ce qui n'a pas d'importance car elle coïncide avec la dimension de Krull à droite. En effet on a $K \dim U(\underline{g})/I = K \dim U(\underline{g})/\sqrt{I}$ (à droite et à gauche) et par [Du] § II Corollaire 2, il existe un anti-automorphisme d'ordre 2 de $U(\underline{g})$ qui laisse stable les idéaux semi-premiers, donc les dimensions de Krull à droite et à gauche coïncident.

Pour un $U(\underline{g})$-module M on note $\mathrm{Ann}\, M$ l'annulateur de M dans $U(\underline{g})$.

2) Un théorème de S.P. Smith

Rappelons l'énoncé du théorème à démontrer ; ([Sm] 3.2.4 ; 3.2.9 ; 3.3.4) :

Théorème 2.1 - 1) Si M est un $U(\underline{g})$-module de type fini vérifiant $d(M) \leq r$ alors M est artinien

2) $K \dim U(\underline{g}) \leq \dim_k \underline{g} - r$.

Il est à noter que (2) est une conséquence facile de (1), (cf. [Sm] 2.2.8 ou [J.2]). Le point délicat se trouve dans la démonstration de (1).
Nous aurons à utiliser fréquemment le théorème suivant dû à A. Joseph et O. Gabber ([J.2] proposition 6.1.4) :

Théorème 2.2 - Si M est un $U(\underline{g})$-module de type fini : $2d(M) \geq d(U(\underline{g})/\mathrm{Ann}\, M)$.

Rappelons que si I est un idéal bilatère de $U(\underline{g})$ le gradué de I , $\mathrm{gr}\, I$, pour la filtration canonique de $U(\underline{g})$, est un idéal de $S(\underline{g})$ qui est G-stable, donc $\sqrt{(\mathrm{gr}\, I)}$ est une réunion d'orbites dans $\underline{g} \simeq \underline{g}^*$; on en déduit :

Lemme 2.3 - ([J.1] 10.1) - Si I est un idéal de $U(\underline{g})$ tel que $d(U(\underline{g})/I) < 2r$ on a $\dim_k U(\underline{g})/I < \infty$.

Preuve - Ecrivons $\sqrt{(\mathrm{gr}\, I)} = \bigcup_{X \in \mathcal{H}} G.X$ comme une réunion d'orbites avec $\mathcal{H} \subset \underline{g}$. Alors : $\forall X \in \mathcal{H}$, $\dim G.X \leq \dim \sqrt{(\mathrm{gr}\, I)} < 2r$, donc $G.X = \{0\}$ d'où $\sqrt{(\mathrm{gr}\, I)} = \{0\}$ et ainsi $d(U(\underline{g})/I) = 0$.

Nous aurons besoin du lemme suivant :

Lemme 2.4 - Soit M un $U(\underline{g})$-module de type fini tel que $\mathrm{Ann}\, M \cap Z(\underline{g})$ ne soit pas un idéal maximal de $Z(\underline{g})$; il existe alors $i \in \{1, \ldots, \ell\}$ tel que pour toute partie infinie $\mathcal{P}$ de k $M \neq (Q_i - \chi) M$, pour au moins un χ de $\mathcal{P}$.

Preuve - Puisque tout idéal maximal de $Z(\underline{g})$ est de la forme $(Q_1 - \chi_1, \ldots, Q_\ell - \chi_\ell)$ avec $(\chi_1, \ldots, \chi_\ell) \in k^\ell$, l'hypothèse faite sur l'annulateur de M peut s'écrire :

$$\text{(H)} \quad \exists\, i \in \{1, \ldots, \ell\} \text{ tel que } \forall \mu \in k \quad Q_i - \mu \notin \mathrm{Ann}\, M \cap Z(\underline{g}) .$$

La preuve du lemme est une simple adaptation du lemme de Quillen (cf. [D] 2.6.4) ; pour tout $Z \in Z(\underline{g})$ nous noterons h_Z l'homothétie de rapport Z de M , c'est un élément de $\mathrm{End}_{U(g)}(M)$. Notons que M étant un $U(g)$-module noethérien tout

U(g)-homomorphisme surjectif de M dans M est un isomorphisme, donc si l'on suppose $M = (Q_i - \chi) M$ pour tout $\chi \in \mathcal{P}$, $h^{-1}_{Q_i - \chi}$ existe pour tout $\chi \in \mathcal{P}$, montrons que cette hypothèse implique que h_{Q_i} est algébrique sur k. On pose $A = k[h_{Q_i}]$ (c'est une sous-algèbre de $\mathrm{End}_{U(g)}(M)$), supposons h_{Q_i} transcendant sur k ; comme dans le lemme de Quillen on montre qu'il existe $f \in A \setminus \{0\}$ tel que $M \otimes_A A_f$ soit libre sur A_f. Puisque $\mathcal{P}$ est infinie il existe un $\chi \in \mathcal{P}$ tel que $h_{Q_i} - \chi$ ne divise pas f ;
donc la multiplication par $Q_i - \chi$ n'est pas surjective dans $M \otimes_A A_f$. Mais si $a \in A_f$, $m \in M$ $(m \otimes a) h_{Q_i - \chi} = h_{Q_i - \chi} \cdot m \otimes a$ et $h_{Q_i - \chi}$ inversible dans $\mathrm{End}_{U(g)}(M)$ fournit une contradiction.

Ainsi h_{Q_i} est algébrique sur k ou encore, k étant algébriquement clos : il existe $\mu \in k$ tel que $h_{Q_i} - \mu = 0$ c'est-à-dire $(Q_i - \mu) M = (0)$ ce qui contredit (H). Donc l'hypothèse faite : $M = h_{Q_i - \chi}(M)$ pour tout $\chi \in \mathcal{P}$ n'est pas vraie ce qui termine la preuve du lemme 2.4.

Nous dirons que M est un U(g)-module d-*critique* si $d(\overline{M}) < d(M)$ pour tout quotient propre de M, nous avons alors :

Proposition 2.5 - *Soit* M *un* U(g)-*module de type fini* d-*critique vérifiant* $d(M) \leq r$. *On a* :

(i) *Tout quotient propre de* M *est de dimension finie sur* k.

(ii) Ann M *est un idéal primitif de* U(g).

Preuve - (i) Si $\overline{M}$ est un quotient de M, distinct de M, on a :

$$d(U(g)/\mathrm{Ann}\,\overline{M}) \leq 2d(\overline{M}) \leq 2(d(M) - 1) \leq 2r - 2 < 2r ,$$

en utilisant le théorème 2.2, et grâce au lemme 2.3 : $\dim_k(U(g)/\mathrm{ann}\,\overline{M}) < \infty$. On en déduit $\dim_k \overline{M} < \infty$ et si $\overline{M}$ n'est pas nul il contient un sous-module simple de dimension finie.

(ii) Puisque $J = \mathrm{Ann}\, M$ est premier ([J.2] 6.5.1) il suffit de montrer que $I = J \cap Z(g)$ est un idéal maximal du centre $Z(g)$.
Soit B une base de Δ, $(\bar{\omega}_\alpha)_{\alpha \in B}$ les poids fondamentaux relatifs à B et $(H_\alpha)_{\alpha \in B}$ la base de $\underline{h}$ vérifiant $\bar{\omega}_\alpha(H_\beta) = 0$ si $\alpha \neq \beta$, 1 si $\alpha = \beta$.
On note $P(\Delta)_+ = \{\sum_{\alpha \in B} n_\alpha \bar{\omega}_\alpha ,\ n_\alpha \in \mathbb{N}\ \forall \alpha \in B\}$ les poids entiers dominants de Δ.

Si $\lambda \in \underline{h}^*$ rappelons que $\chi_{\lambda+\rho}(Z) = \varphi(Z)(\lambda)$ pour tout Z dans $Z(\underline{g})$ où l'on désigne par φ l'homomorphisme d'Harish-Chandra de $Z(\underline{g})$ vers $U(\underline{h})$. Puisque Q_i est à coefficients dans $\mathbb{Q}$, $\varphi(Q_i)$ est un polynôme en les H_α à coefficients dans $\mathbb{Q}$, donc si $\lambda \in P(\Delta)_+$ on a $\varphi(Q_i)(\lambda) \in \mathbb{Q}$, $(i \in \{1,\ldots,\ell\})$. Posons $D_i = \{\chi_{\lambda+\rho}(Q_i)\ ;\ \lambda \in P(\Delta)_+\}$; cet ensemble est contenu dans $\mathbb{Q}$ et $\mathcal{P}_i = k \setminus D_i$ est une partie infinie de k. Rappelons que si $\lambda \in P(\Delta)_+$ le module $L(\lambda+\rho)$ est de dimension finie sur k et $Q_i - \chi_{\lambda+\rho}(Q_i)$ est dans l'annulateur de $L(\lambda+\rho)$.

Supposons que I ne soit pas un idéal maximal de $Z(\underline{g})$, le lemme 2.4 nous donne : $M \neq (Q_i - \chi)M$ pour un $i \in \{1,\ldots,\ell\}$ et un $\chi \in \mathcal{P}_i$.
Soit $\overline{M} = M/(Q_i - \chi)M \neq 0$, par (i) il existe $\overline{N}$, sous-module simple de dimension finie, contenu dans $\overline{M}$. Puisque $\operatorname{Ann}\overline{N} \supset \operatorname{Ann}\overline{M} \supset (Q_i - \chi)$ on déduit que $Q_i - \chi$ est dans l'annulateur d'un module simple de dimension finie c'est-à-dire χ de la forme $\chi_{\lambda+\rho}(Q_i)$ pour un λ de $P(\Delta)_+$ ce qui contredit $\chi \in \mathcal{P}_i$.

Avant de terminer la preuve du théorème 2.1 signalons le résultat suivant que nous aurons à utiliser ; (cf. [Du] ,[J.2] 6.5.4) :

<u>Proposition</u> 2.6 - <u>Soit $\lambda \in \underline{h}^*$ et $\bar{J} \in \operatorname{Spec} B_\lambda$, il existe un unique idéal minimal $\tilde{J}$ dans B_λ tel que $\tilde{J} \supsetneq \bar{J}$ et $\tilde{J}$ vérifie $\tilde{J}^2 + \bar{J} = \tilde{J}$.</u>

<u>Preuve</u> - L'existence de $\tilde{J}$ est assurée par le fait que B_λ est artinien pour les idéaux bilatères, son unicité par celui que $\bar{J}$ est premier et la dernière égalité vient des inclusions $\bar{J} \subset \bar{J} + \tilde{J}^2 \subset \tilde{J}$.

<u>Preuve du Théorème</u> 2.1 - (1) - Puisque $U(\underline{g})$ est noethérien il suffit de prouver que M contient un module simple ; on peut donc supposer que M est un module d-critique avec $d(M) \leq r$ (cf. [J.2] 3.1.2). La proposition 2.5 nous assure que $J = \operatorname{Ann} M$ est un idéal primitif de $U(\underline{g})$ donc $\bar{J} = J/J(\lambda) \in \operatorname{Spec} B_\lambda$ pour un $\lambda \in \underline{h}^*$. Nous allons nous inspirer de ([J.2] 6.5.5) pour obtenir le résultat cherché. Si $\bar{J}$ est de codimension finie, M est de dimension finie et si M n'est pas nul il contient un module simple ; nous supposerons donc que $\bar{J}$ n'est pas de codimension finie.
Posons $N = \tilde{J}M$ avec $\tilde{J}$ comme dans la proposition 2.6, alors N n'est pas nul sinon $\tilde{J}.M = (0)$ et $\bar{J} = \tilde{J}$ ce qui est impossible.
Remarquons que N est d-critique avec $d(N) \leq r$ car $N \subset M$. Montrons que N est simple : si $(0) \neq N' \subseteq N$ soit $J' = \operatorname{Ann} N/N'$ et $\bar{J}'$ son image dans B_λ. Par la proposition 2.5 (i) J' est de codimension finie d'où $\bar{J}' \supsetneq \bar{J}$ ce qui implique

$\tilde{J}' \supset \tilde{J}$. Ainsi $\tilde{J}$ annule N/N' ce qui signifie : $\tilde{J}N = \tilde{J}^2 M \subset N'$, mais $N = \tilde{J}M = (\tilde{J}^2 + \bar{J})M = \tilde{J}^2 M$, puisque $\bar{J}M = (0)$, donc $N \subset N'$ et N est simple.

Ceci achève la démonstration du théorème 2.1.

3) Anneau des opérateurs différentiels sur "l'espace affine basique" et application.

3.1 - Nous adoptons les notations de [G.K] à savoir :

G est un groupe algébrique simplement connexe correspondant à $\underline{g}$.

H, N^+, N^- sont les sous-groupes de G correspondant aux sous-algèbres $\underline{h}, \underline{\mathfrak{n}}^+, \underline{\mathfrak{n}}^-$.

$n = \dim \mathfrak{n}^+ = \dim N^+$.

$A = N^- \backslash G$ "l'espace affine basique" (classes à gauche selon N^-)

$k[G]$ (resp. $k[A]$) l'anneau des fonctions régulières sur G (resp. A)

$X(H)$ les caractères de H .

Λ les poids dominants entiers de G (correspondant à $P(\Delta)_+$ de 2)).

Nous savons que A est une variété quasi-affine lisse mais que $k[A]$ n'est pas un anneau régulier en général (cf. Remarque 3.4 ci-dessous) et la dimension de A est $n+\ell$, la dimension d'une sous-algèbre de Borel de $\underline{g}$.

Nous noterons $\mathcal{D}(A)$ l'anneau des opérateurs différentiels sur A (cf. [B.G.G.] définition 3.1) et si $m \in \mathbb{N}$, $\mathcal{D}(A)_m$, le $k[A]$-module des opérateurs différentiels d'ordre inférieur ou égal à m .

Rappelons que N^- opère sur $k[G]$ par :

$$L_g^G f(x) = f(g^{-1}x) \quad \text{si} \quad x \in G \ , \ g \in N^- \ , \ f \in k[G] \quad .$$

L'anneau $k[A]$ est égal à l'anneau des invariants, $k[G]^{N^-}$, sous cette action. D'autre part si $h \in H$ on note t_h la translation $x \longmapsto hx$ de G ;

Soit $N^-.g \in A$, puisque $hN^-.g = N^-h.g$, t_h s'étend en une translation sur A . Si $f \in k[A]$ on pose alors : $S(h).f = f \circ t_h$, ce qui définit une représentation de H dans $k[A]$. Cette représentation permet de montrer que $k[A] = \bigoplus_{\lambda \in \Lambda} k[A]^\lambda$ où l'on pose, pour $\lambda \in \Lambda$: $k[A]^\lambda = \{f \in k[A] / S(h).f = \lambda(h).f \ , \ \forall h \in H\}$ (cf. [G.K] §4). L'action de H induit une action sur $\mathcal{D}(A)$ en posant :

$$^hD = S(h) \circ D \circ S(h^{-1}) \quad \text{pour} \quad D \in \mathcal{D}(A) \ , \ h \in H \ .$$

Si $\lambda \in X(H)$ on note : $\mathcal{D}(A)^\lambda = \{D \in \mathcal{D}(A) / {}^hD = \lambda(h) D \ , \ \forall h \in H\}$, (c'est aussi $\{D \in \mathcal{D}(A) / D(k[A]^\mu) \subset k[A]^{\lambda+\mu}$ pour tout $\mu \in \Lambda\}$), $\mathcal{D}(A)_m^\lambda = \mathcal{D}(A)^\lambda \cap \mathcal{D}(A)_m$ si $m \in \mathbb{N}$; il est démontré dans [G.K] §6 que $\mathcal{D}(A)_m^\lambda$ est un k-espace vectoriel de dimension finie et que $\mathcal{D}(A) = \bigoplus_{\lambda \in X(H)} \mathcal{D}(A)^\lambda$.

En particulier $\mathcal{D}(A)^{o} = \{D \in \mathcal{D}(A) / {}^{h}D = D \quad \forall h \in H\}$ n'est autre que l'anneau des invariants de $\mathcal{D}(A)$ sous l'action de H.

D'autre part on peut définir $\tilde{U}(\underline{g}) = U(\underline{g}) \otimes_{Z(\underline{g})} U(\underline{h})$ en considérant $U(\underline{h})$ comme un $Z(\underline{g})$-module à l'aide de l'isomorphisme de Harish-Chandra de $Z(\underline{g})$ sur $U(\underline{h})^{W}$, en notant $U(\underline{h})^{W}$ les éléments invariants de $U(\underline{h})$ dans W, le groupe de Weyl de Δ, (cf. [D] 7.4.5). On a ainsi de manière évidente une opération de W sur $\tilde{U}(\underline{g})$ et $\tilde{U}(\underline{g})^{W} = U(\underline{g})$.

3.2 - Nous allons donner quelques résultats, bien connus, dont nous aurons besoin :

Proposition 3.1 - *Soit R une k-algèbre et H un groupe affine linéairement réductif de k-automorphismes de R. Supposons que H opère rationnellement dans R (i.e. R est réunion de sous H-modules de dimension finie sur k). Si $R^{H} = \{r \in R / h.r = r$ pour tout $h \in H\}$ on a $R = R^{H} \oplus R_{H}$ somme directe de R^{H}-modules. Si la dimension de Krull (Gabriel-Rentschler) de R existe (à gauche par exemple), celle de R^{H} aussi et l'on a $K \dim R^{H} \leq K \dim R$.*

Preuve - La première partie est bien connue (cf. [Fo]) et la seconde vient du fait que si N est un R^{H}-module l'application canonique $N \longrightarrow R \otimes_{R^{H}} N$ est injective.

3.3 - Reprenons la situation de 3.1.

Proposition 3.2 - *Si la dimension de Krull de $\mathcal{D}(A)$ (à gauche par exemple) existe alors :*

$$K \dim U(\underline{g}) \leq K \dim \mathcal{D}(A)$$

Preuve - Puisque $\tilde{U}(\underline{g})^{W} = U(\underline{g})$ on a (cf. [F.O]) $K \dim \tilde{U}(\underline{g}) = K \dim U(\underline{g})$. D'après [Sh], $\tilde{U}(\underline{g})$ n'est autre que $\mathcal{D}(A)^{H}$ (à un isomorphisme près). Si l'on prouve que l'action de H est rationnelle sur $\mathcal{D}(A)$ la proposition 3.1 nous donnera la conclusion voulue.

Soit $P \in \mathcal{D}(A)$ et montrons que $\dim_{k} < {}^{h}P \; ; \; h \in H > < \infty$.

Ecrivons $P = \sum_{\text{finie}} P^{\lambda_i}$ avec $P^{\lambda_i} \in \mathcal{D}(A)^{\lambda_i}$, $\lambda_i \in X(H)$.

Si l'ordre de P est $\leq m$, l'ordre de chaque P^{λ_i} aussi, donc P^{λ_i} est dans $\mathcal{D}(A)_m^{\lambda_i}$ pour tout i. Puisque $^hP = \sum_{\text{finie}} {}^h(P^{\lambda_i})$ et que $\mathcal{D}(A)_m^{\lambda_i}$ est de dimension finie sur k il suffit de montrer que $\mathcal{D}(A)_m^{\lambda_i}$ est un H-module pour avoir le résultat cherché. Soit donc $D \in \mathcal{D}(A)_m^{\lambda}$ avec $m \in \mathbb{N}$, $\lambda \in X(H)$; si $f \in K[A]$ et $h \in H$:

$$^h[D, {}^{h^{-1}}f] = {}^h\{D.{}^{h^{-1}}f - {}^{h^{-1}}f.D\} = {}^hD.f - f.{}^hD = [{}^hD, f] \quad .$$

Procédons par récurrence sur m : $[{}^hD, f] = {}^h[D, {}^{h^{-1}}f]$ est d'ordre $\leq m-1$, et ^hD est d'ordre $\leq m$ par définition de l'ordre ; ainsi $^h(\mathcal{D}(A)_m) \subset \mathcal{D}(A)_m$ pour tout m. Montrons $^h(\mathcal{D}(A)^{\lambda}) \subset \mathcal{D}(A)^{\lambda}$ pour tout $h \in H$; il s'agit de montrer que si $D \in \mathcal{D}(A)^{\lambda}$ et $h \in H$ alors $^hD \in \mathcal{D}(A)^{\lambda}$ ce qui est clair par le calcul suivant :

$$^{h'}({}^hD) = {}^{h'h}D = \lambda(h'h)\, D = \lambda(h')\,\lambda(h)\, D = \lambda(h')\, {}^hD \quad \text{pour tout } h' \in H .$$

Comme le montre la proposition 3.2 la connaissance de la dimension de Krull de $\mathcal{D}(A)$ donne une majoration de la dimension de Krull de $U(\underline{g})$; le problème est alors le calcul de cette dimension. Lorsque l'anneau $k[A]$ est régulier on a $K \dim \mathcal{D}(A) = \dim A = n+\ell$, mais en général $\operatorname{Spec} k[A]$ possède des singularités. Néanmoins la proposition 3.2 permet de calculer la dimension de Krull de $U(\underline{g})$ lorsque $\underline{g}$ est une somme directe de copies de $sl(2,k)$; on a :

<u>Proposition</u> 3.3 - <u>Soit</u> $\underline{g} = \bigoplus_{i=1}^{m} \underline{g}_i$, <u>avec</u> $\underline{g}_i = sl(2,k)$; <u>alors</u> :

$$K \dim U(\underline{g}) = 2m \quad .$$

<u>Preuve</u> - Avec les notations de 3.1 on peut prendre $G = G_1 \times \ldots \times G_m$ où $G_i = SL(2,k)$, c'est-à-dire que G est le sous-groupe de $SL(2m,k)$:

$$G = \begin{bmatrix} G_1 & 0 & \ldots & 0 \\ 0 & G_2 & \ldots & 0 \\ \vdots & & & \\ 0 & \ldots\ldots & & G_m \end{bmatrix}$$

Posons

$$G_i = \left\{ g = \begin{pmatrix} u_1^{(i)} & u_2^{(i)} \\ z_1^{(i)} & z_2^{(i)} \end{pmatrix} \quad ; \ \det g = 1 \right\}$$

On vérifie facilement que si $N_i^- = \left\{ g = \begin{pmatrix} 1 & 0 \\ x & 1 \end{pmatrix} \in G_i \right\}$ alors

$N^- = \prod_{i=1}^{m} N_i^-$ et $A = N^- \backslash G$ est l'ouvert de k^{2m} : $k^{2m} \backslash F$ où F est le fermé définit par :

$$F = \bigcup_{i=1}^{m} P_i \quad \text{avec} \quad P_i = \left\{ (u_1^{(1)}, u_2^{(1)}, \dots, u_1^{(m)}, u_2^{(m)}) \in k^{2m} / u_1^{(i)} = u_2^{(i)} = 0 \right\} .$$

Comme $\operatorname{codim}_{k^{2m}} F = 2$, on a $k[A] = k[u_i^{(j)} ; i = 1,2 ; j = 1, \dots, m]$ qui est un anneau de polynômes en $2m$ indéterminées. Donc $\mathcal{D}(A)$ n'est autre que l'algèbre de Weyl $A_{2m}(k) = k[u_i^{(j)}, \frac{\partial}{\partial u_i^{(j)}} ; i = 1,2 ; j = 1, \dots, m]$ dont on connait la dimension de Krull : $2m$.

La proposition 3.2 nous donne ainsi $K \dim U(\underline{g}) \leq 2m$; comme $2m$ est la dimension d'une sous-algèbre de Borel de $\underline{g}$ on a $K \dim U(\underline{g}) \geq 2m$ et donc $K \dim U(\underline{g}) = 2m$.

La proposition 3.3 donne en particulier une démonstration assez simple de $K \dim U(sl(2,k)) = 2$, (cf. [Sm] , [J.2]), et répond à la question posée dans [Sm] §3.3 . Elle montre également que la majoration $K \dim U(\underline{g}) \leq \dim \underline{g} - r$ du théorème 2.1 est loin d'être satisfaisante dans le cas général : si $\underline{g}$ est comme dans la proposition 3.3 on a $r = 1$, $\dim \underline{g} = 3m$ mais $K \dim U(\underline{g}) = 2m$.

<u>Remarque</u> 3.4 - L'anneau $k[A]$ n'est en général pas régulier et la présence de singularités rend difficile le calcul des opérateurs différentiels sur $\mathcal{D}(A)$: On ne peut en effet les exprimer à l'aide des dérivations de $k[A]$ (cf. [13 G.G] page 33). Dans le cas où $\underline{g} = sl(3,k)$ la situation est la suivante : soit

$$G = SL(3,k) = \left\{ g = \begin{pmatrix} u_1 & u_2 & u_3 \\ t_1 & t_2 & t_3 \\ z_1 & z_2 & z_3 \end{pmatrix} ; \det g = 1 \right\} \quad ; \text{on a :}$$

$k[A] = k[u_1, u_2, u_3 , v_1 = u_2 t_3 - u_3 t_2 , v_2 = u_3 t_1 - u_1 t_3 , v_3 = u_1 t_2 - u_2 t_1]$ avec la relation $u_1 v_1 + u_2 v_2 + u_3 v_3 = 0$. C'est-à-dire, si X_i, Y_i $i = 1,2,3$, sont des variables

$$k[A] = \frac{k[X_1, Y_1, X_2, Y_2, X_3, Y_3]}{(X_1 Y_1 + X_2 Y_2 + X_3 Y_3)} \quad ;$$

ainsi Spec $k[A]$ possède une singularité en 0 . Remarquons pour finir que le théorème 2.1 donne ici :

$$5 \leq K \dim U(sl(3;k)) \leq 6 \quad .$$

Références

[B.G.G] I.N. Bernstein, I.M. Gelfand and S.I. Gelfand - Differential operators on the base affine space and a study of g-modules. In "Lie Groups and their representations" Proceedings, Bolyai János Math. Soc. Budapest (1971).

[D] J. Dixmier - Algèbres Enveloppantes Gauthier Villars (1974).

[Du] M. Duflo - Sur la classification des idéaux primitifs dans l'algèbre enveloppante d'une algèbre de Lie semi-simple. Ann. of Math. 105 (1977) 107-120.

[Fo] J. Fogarty - Invariant theory. Benjamin. New York (1969).

[F-O] J.W. Fisher, J. Osterburg - Some Results on rings with finite group actions. L.N. in Math. n° 25 (1976). Springer Verlag.

[G.K] I.M. Gelfand and A.A Kirillov - The structure of the Lie field connected with a split semi-simple lie algebra. Funk. Analiz i Ego. Prilozhen 3 n° 1 7-26 (1969).

[J.1] A. Joseph - The minimal orbit in a simple Lie algebra and its associated maximal ideal. Ann. Sci. Ec. Normale Sup. 9 (1976) 1-30.

[J.2] " - Cours de 3ème cycle. Université Paris VI (1981).

[L] T. Levasseur - Dimension injective des quotients primitifs minimaux de l'algèbre enveloppante d'une algèbre de Lie semi-simple. C.R. Acad. Sci. t. 292 - Série I (1981) 385-387.

[Sh] N.N Shapovalov - On a conjecture of Gelfand-Kirillov. Funk. Analiz i Ego Pril. vol 7 n° 2 (1973) 93-94.

.../

[Sm] S.P. Smith - Krull dimension and Gelfand-Kirillov dimension of modules over enveloping algebras.
Ph. Doc. Leeds (1981).

Université Pierre et Marie Curie
Institut de Mathématiques Pures et Appliquées
U.E.R. 47 - Tour 45-46
4, Place Jussieu

75230 PARIS CEDEX 05

UN THEOREME DE L'IDEAL A GAUCHE PRINCIPAL DANS CERTAINS ANNEAUX

Guy Maury

INTRODUCTION.

Le théorème de l'idéal principal dans un domaine commutatif noethérien est susceptible de plusieurs généralisations non équivalentes à un anneau premier noethérien à gauche R [10], [8], [16], [7]. Soit a un élément de R non diviseur de zéro et non inversible. Soit $Ra = \bigcap_{j=1}^{n} X_j$, X_j P_j-tertiaire, $j = 1,\ldots,n$ une décomposition tertiaire réduite de Ra, [13]. Les idéaux premiers $P_1,\ldots,P_n$ sont dits associés à Ra. Dans cet article nous adoptons la définition suivante d'un anneau vérifiant le théorème de l'idéal à gauche principal.

DEFINITION 0.1 (Théorème de l'idéal à gauche principal en abrégé T.I.G.P.) - *L'anneau* R *premier, noethérien à gauche, est dit vérifier le théorème de l'idéal à gauche principal (T.I.G.P.) si pour tout* a *non diviseur de zéro et non inversible de* R, $P_1,\ldots,P_n$ *désignant les idéaux premiers associés à* Ra, Z *désignant le centre de* R, *tout idéal premier minimal dans l'ensemble* $(P_j \cap Z)_{j=1,\ldots,n}$

d'idéaux premier de Z *est un idéal de hauteur* 1 *de* Z *et si l'idéal premier* P_j *correspondant est aussi de hauteur* 1.

Nous avons démontré en [16] qu'un Z-ordre classique dans une algèbre centrale simple Σ de dimension n^2 sur le corps K des fractions du domaine d'intégrité noethérien intégralement clos Z vérifiait le T.I. G.P. et plus généralement en [7] qu'un anneau à identité polynômiale (en abrégé PI-anneau) premier et noethérien dont le centre Z est un domaine de Krull le vérifiait aussi.

Dans un premier paragraphe nous donnons d'autres exemples d'anneaux premiers noethériens à gauche vérifiant le T.I. G.P. : en particulier un PI-anneau R premier, noethérien, entier sur son centre, dont la clôture intégrale du centre est caténaire (par exemple, cela a lieu si R est affine), le vérifie. La méthode est celle de [16] et [7] : elle consiste à plonger ces anneaux dans un ordre maximal convenable et à utiliser un théorème de l'idéal à gauche principal dans les ordres maximaux réguliers noethériens à gauche ([15] ou pour une démonstration plus détaillée [17] ch. 6 th. 2.4). Dans un second paragraphe nous étudions un anneau premier noethérien à gauche admettant un anneau de fractions artinien simple Q (donc de Goldie à droite) qui se plonge dans un ordre maximal régulier de Q qui lui est équivalent.

Nous renvoyons à [17] pour les définitions concernant la théorie des ordres maximaux.

I.

DEFINITION I.1 - *Un anneau* R *sera dit localement fini sur un sous-anneau* A *de son centre si pour toute famille* $\{x_1,\dots,x_k\}$ *d'éléments de* R *l'anneau* $A\{x_1,\dots,x_k\}$ *engendré par* A *et* $x_1,\dots,x_k$ *est un* A*-module de type fini.*

REMARQUE I.2 - Il est clair qu'un anneau localement fini sur A, sous-anneau de son centre est entier sur A. Par exemple un PI-anneau premier entier sur son centre Z est localement fini sur

Z ([18] et [17] ch. 8, th. 2.3). Mais on peut trouver des anneaux localement finis sur leur centre qui ne sont pas des PI-anneaux : soit k' un corps de dimension infinie sur leur centre K tel que toute famille finie d'éléments de k' appartienne à un sous-corps de k' de même centre k et qui soit un k-espace vectoriel de dimension finie (Köthe [11] a construit un exemple de tel corps). Soient $X_1,\ldots,X_r$ des indéterminées commutant entre elles et avec tout élément de k'. L'anneau $k'[X_1,\ldots,X_r]$ n'est pas un PI-anneau, c'est un anneau premier noethérien sans diviseurs de zéro localement fini sur son centre $k[X_1,\ldots,X_r]$.

DEFINITION I.3 - *Etant donné un anneau* R *on note* dim R *la borne supérieure des nombres (finis ou non) d'éléments des suites croissantes strictes d'idéaux premiers de* R *et, si* P *est un idéal de* R*,* ht P *la borne supérieure des nombres (finis ou non) d'éléments des suites croissantes strictes d'idéaux premiers strictement contenus dans* P. *Nous dirons que* R *est caténaire si pour tout idéal premier* P *de* R *on a* $\dim R = \dim R/_P + \mathrm{ht}\, P$, *avec* dim R *finie.*

Rappelons les lemmes suivants :

LEMME I.4 - *Si* R *est un anneau premier de centre* Z *et si* R' *est un sur-anneau de* R *entier sur* Z *(unité de* R = *unité de* Z*) tel que* Z *soit contenu dans le centre de* R'. *Alors* R'a = R' *pour* $a \in R$ *implique* Ra = R.

DEMONSTRATION. - Si R'a = R' il existe $a^{-1} \in R'$ avec $a^{-1}a = 1$. Comme R' est entier sur Z il existe des éléments de Z $r_{n-1},\ldots,r_o$ tels que : $(a^{-1})^n + r_{n-1}(a^{-1})^{n-1} + \ldots + r_o = 0$. Multiplions par a^{n-1} à droite il vient :

$$a^{-1} + r_{n-1} + \ldots + r_o\, a^{n-1} = 0$$

et ceci prouve $a^{-1} \in R$ donc Ra = R.

LEMME I.5 - *Si* R *est un anneau noethérien à gauche de centre* Z *et si* T *est un idéal à gauche* P*-tertiaire de* R, *alors pour tout* $x \in P \cap Z$, *il existe* $n \in \mathbb{N}$ *avec* $x^n \in T$.

DEMONSTRATION. - Voir [17] ch. 9 lemme 1.2 ou [16].

LEMME I.6 - *Si* R *est un anneau premier noethérien à gauche de centre* Z *et si* T *est un idéal à gauche* P*-tertiaire. Posons* $p = P \cap R$ *alors* $T_p \cap R = T$ *et* $T_p \neq R_p$.

DEMONSTRATION. - Voir [17] ch. 9 lemme 1.4 ou [16].

LEMME I.7 (légère généralisation du th. 3.3 de [1]) - *Soit* R *un anneau premier entier sur un sous-anneau* A *de son centre tel que* $R/_P$ *soit de Goldie pour tout idéal premier* P *de* R :

1). *Si* $p_1 \subseteq \ldots \subseteq p_k$ *est une suite* (s) *ascendante d'idéaux premiers de* A *et soit* P_1 *idéal premier de* R *avec* $P_1 \cap A = p_1$. *Il existe une suite ascendante stricte d'idéaux premiers de* R, $(\mathcal{S})$, $P_1 \subseteq \ldots \subseteq P_k$ *avec* $P_i \cap A = p_i$. *Réciproquement à une suite* $(\mathcal{S})$ *correspond par intersection avec* A *une suite* (s). *On a* dim A = dim R.

2). *Pour tout* P *idéal premier de* R *tel que* $P \cap A = p$, dim $R/_P$ = dim $A/_p$ *et* ht p $\geqslant$ ht P. *Si* ht p *ou* dim R *est fini alors* ht p = ht P *pour un certain idéal premier* P *de* R *avec* $P \cap R = p$.

3). *Si* R *est caténaire alors* A *l'est et* ht P = ht p *pour tout couple d'idéaux premiers tels que* $P \cap A = p$.

4). *Si* A *est intégralement clos et* R *localement fini sur* A *alors* ht p = ht P *pour tout couple* (P,p) *d'idéaux premiers tels que* $P \cap A = p$ *et* R *est caténaire si* A *l'est.*

DEMONSTRATION. - La démonstration est laissée au lecteur. Remarquer que le going-up theorem et le théorème d'incomparabilité sont valables pour le couple (R,A) ([2] et [17] ch. 3 lemme 4.2 et 4.3) et que sous les hypothèses de 4) le going-down theorem est valable pour le couple (R,A) ([5] ou [17] ch. 8 prop. 1.1). On pourra se reporter à la démonstration du th. 3.3 de [1].

THEOREME I.8 - *Soit* R' *un anneau premier noethérien à gauche localement fini sur son centre* Z', *ordre maximal (régulier) de son anneau des fractions* Q. *Soit* R *un ordre de* Q, *contenu dans* R', *noethérien à gauche et de centre* Z. *Alors* R *vérifie le T.I G.P. (définition* 0.1) *si l'une ou l'autre des deux conditions suivantes est vérifiée :*

1). Z = Z'; ou 2). Z' *est caténaire, entier sur* Z.

DEMONSTRATION. - Faisons d'abord un certain nombre de remarques préliminaires : R' est entier sur Z'. D'après [6] prop. 1.3 et 1.16 il admet un anneau de fractions des deux côtés artinien simple Q et si K désigne le corps des fractions de Z', on a KR' = R'K = Q et K est le centre de Q. Comme R est un ordre de Q on a $Z = R \cap K \subseteq Z'$. Comme R' est un ordre maximal noethérien à gauche on sait que Z' est un domaine de Krull ([5] ou [17] ch. 8 lemme 3.1). Pour tout idéal premier P' de R', $R'/_{P'}$ est entier sur $Z'/_{Z' \cap P'}$, sous-anneau de son centre, est noethérien à gauche donc admet un anneau de fractions artinien simple d'après [6] prop. 1.3 et donc est un anneau de Goldie. On sait alors ([5] ou [17] ch. 8 prop. 1.1) que le couple (R',Z') vérifie le going-down theorem. Si l'on suppose Z' entier sur Z, Z' est la clôture intégrale de Z. Remarquons qu'alors R' est entier sur Z et R est entier sur Z. Soit a un élément non diviseur de zéro et non inversible de R et $Ra = \bigcap_{j=1}^{n} X_j$, X_j P_j-tertiaire $j = 1,\dots,n$ une décomposition tertiaire réduite de Ra. Soit p un élément minimal de l'ensemble $p_1 = P_1 \cap Z,\dots,p_n = P_n \cap \mathbb{Z}$. Supposons que l'on ait $p = p_j$ pour $j = 1,\dots,k$ et $p \neq p_j$ donc $p_j \not\subseteq p$ pour $j = k+1,\dots,n$. On a $(X_j)_p = R_p$ pour $j = k+1,\dots,n$ d'après le lemme I.5. Soit $x \in p_p$ il existe $\ell \in \mathbb{N}$ avec $x^\ell \in R_p a = \bigcap_{j=1}^{k} (X_j)_p$ toujours d'après le lemme I.5 ; R_p est un sous-anneau de R'_p. D'après [6] prop. 1.4

R' est un ordre régulier de Q et même totalement borné. D'après [17] ch. 4, th. 3.10, R'_p est un ordre maximal régulier noethérien à gauche de Q de centre Z'_p. Il est clair que R'_p est localement fini sur Z'_p et que l'on a le going-down theorem pour le couple (R'_p, Z'_p). On a $R'_p a \neq R'_p$ d'après le lemme 1.4 car $R_p a \neq R_p$ d'après le lemme I.6. De plus a est évidemment non diviseur de zéro dans R'_p. On sait alors que $R'_p a$ a tous ses idéaux premiers associés de hauteur 1 ([17] ch. 6, th. 2.4 et remarque 2.5). Soit P' l'un d'eux, $P' \cap Z'_p = p'$ est aussi de hauteur 1 dans Z'_p (lemme I.7.4)). De $x^{\ell} \in R_p a \subseteq R'_p a$, on déduit ($x$ central) $x^{\ell} \in p'$ et $x \in p'$ donc $x \in p' \cap Z_p$ et $p' \cap Z_p = p_p$ puisque p_p est l'idéal maximum de Z_p.

a). Supposons d'abord $Z' = Z$ donc $Z'_p = Z_p$ et $p' = p_p$; p' étant de hauteur 1, il est est ainsi de p. On déduit du lemme I.7 2) que $P_1, \ldots, P_k$ sont de hauteur 1.

b). Supposons toujours Z' entier sur Z et de plus Z' caténaire. Alors d'après le lemme I.7 4) R' est caténaire. Soit $\bar{P}' = P' \cap R'$, $\bar{P}'$ est un idéal premier de R', il est clair que l'on a

$\text{ht } \bar{P}' = \text{ht } P' = 1$. Comme $\bar{P}' \cap Z = P' \cap Z = p$ et que R' est caténaire, entier sur Z, le lemme I.7 3) donne $\text{ht } p = \text{ht } \bar{P}' = 1$. Comme en a), on démontre alors que $P_1, \ldots, P_k$ sont de hauteur 1.

THEOREME I.9 - *Soit* R *un* PI-*anneau premier noethérien entier sur son centre* Z. *Soit* Z' *la clôture intégrale de* Z *dans son corps des fractions. Si l'une des deux conditions suivantes est vérifiée :*

1). $Z = Z'$; *ou* 2). Z' *est caténaire,* R *vérifie le T.I.G.P. (définition* 0.1*).*

DEMONSTRATION. - On sait d'après [3] th. 2.1 que R étant entier sur Z, Z' est un domaine de Krull. Si $Z' = Z$ c'est un théorème de [7] et sa démonstration (voir aussi [17] ch. 9, th. 1.5 et sa démonstration). Démontrons le théorème sous l'hypothèse "Z' caténaire". On sait que RZ' est un anneau dont l'anneau des fractions est l'anneau des fractions Q de R. Le centre de RZ' est Z' : Il est clair que le centre de RZ' est RZ' $\cap$ K où K est le corps des fractions de Z ; d'autre part tout élément $t = \sum_{i=1}^{r} z_i r_i$, $z_i \in Z'$, $r_i \in R$ est entier sur Z' car $Z'\{t\} \subseteq Z'.Z\{r_1,\ldots,r_r\}$, mais $Z\{r_1,\ldots,r_r\}$ est un Z-module de type fini car R est localement fini sur Z ([18] ou [17] ch. 8 th. 2.3). Ainsi Z'{t} est contenu dans une Z'-algèbre qui est un Z'-module de type fini $\sum_{j=1}^{\ell} Z'u_j$. On a donc $tu_i = \sum_{j=1}^{\ell} c_{ij} u_j$, $\sum_{j=1}^{\ell} (c_{ij} - \delta_{ij} t)u_j = 0$ et le raisonnement habituel du déterminant montre que t est entier sur Z'. Ainsi tout élément du centre de RZ' est entier sur Z' et est un élément de K, il est donc dans Z'. Ainsi RZ' se plonge dans un ordre maximal de Q qui est un Z'-ordre maximal R' de Q ([9] ; [17] ch. 8 prop. 3.3 et th. 3.2) ; soit p un idéal premier de Z, R'_p est un Z'_p-ordre maximal de Q donc aussi un ordre maximal régulier de Q et aussi un anneau de Krull régulier au sens de Marubayashi, premier ([14] ou [17] ch. 11 exemple 2 avant le théorème 1.1). D'autre part si a est un élément non diviseur de zéro dans R donc dans R'_p et non inversible de R tel que $R_p a \neq R_p$, on a aussi $R'_p a \neq R'_p$ (lemme I.4). Comme R'_p est un anneau de Krull premier régulier au sens de Marubayashi, $R'_p a$ est intersection d'un nombre fini r d'idéaux à gauche P'_i-primaires avec ht $P'_i = 1$ pour $i = 1,\ldots,r$ (d'après [17] ch. 11 remarque suivant la proposition 1.5). On raisonne alors comme au théorème I.1.

COROLLAIRE I.3 - *Un* PI-*anneau* R *premier noethérien, affine, entier sur son centre* Z *vérifie le T.I.G.P.*

DEMONSTRATION. - Dire que R est affine signifie que $R = k\{x_1,\ldots,x_t\}$ où k est un sous-corps de son centre. Il suffira pour utiliser le théorème I.2 de démontrer que la clôture intégrale Z' de Z dans le corps des fractions K de Z est caténaire. Soit T le sous-anneau de K engendré par Z et les coefficients des polynômes caractéristiques réduits des éléments de R, [1]. Comme R est entier sur Z on a $Z \subseteq T \subseteq Z'$. D'après la démonstration du th. 7.1 de [1] T est une extension affine de k (c'est-à-dire une k-algèbre finiment engendrée). D'après le théorème de normalisation ([19] vol. I page 266) il existe un anneau de polynômes $A = k[y_1,\dots,y_s]$ sous-anneau de T, tel que T soit entier sur A. D'après le lemme I.7 4) Z' étant entier sur T donc sur A est caténaire.

II.

Dans tout ce paragraphe R est un ordre dans un anneau artinien simple Q, R noethérien à gauche, qui se plonge dans un ordre maximal régulier équivalent R' de Q. Il existe alors un plus grand idéal bilatère non nul V commun à R et R' et R' est aussi noethérien à gauche. Si P est un idéal premier (soit de R soit de R'), P est dit régulier si $V \not\subseteq P$.

PROPOSITION II.1 - *Soient* R *et* R' *comme ci-dessus,* a *un élément non diviseur de zéro et non inversible de* R , $\beta(Ra)$ *le plus grand idéal bilatère (non nul) contenu dans* Ra *et* P *un idéal premier minimal au-dessus de* Ra. *Si* P *est régulier alors* P *est de hauteur* 1.

DEMONSTRATION. - Remarquons que R'a = R' est impossible car on aurait alors $VR'a = VR' = V \subseteq Ra$ et $V \subseteq \beta(Ra) \subseteq P$ impossible. Soient P'_j, $j = 1,\dots,r$, ht $P'_j = 1$ les idéaux premiers associés à $R'a = \bigcap_{j=1}^{r} Y'_j$, Y'_j P'_j-primaire, $j = 1,\dots,r$ ([17] ch. 6 th. 2.4 et 2.5). On a $\prod_{i=1}^{r} P'^{\rho_i}_i \subseteq R'a$ avec $\rho_i \in \mathbb{N}$, et $V\prod_{i=1}^{r} P'^{\rho_i}_i \subseteq VR'a \subseteq Ra$

$V\prod_{i=1}^{r} P'^{\rho_i}_i V \subseteq V\prod_{i=1}^{r} P'^{\rho_i}_i \subseteq Ra$. On a donc $V\prod_{i=1}^{r} P'^{\rho_i}_i V \subseteq P$ et il existe un certain i avec $P'_i \cap R \subseteq P$; P'_i est régulier comme P donc $P'_i \cap R$ est premier ([1] th. 2.6 1)). Soit $N = \beta(Ra)$, $VNV \subseteq N \subseteq Ra \subseteq R'a$ donc $VNV \subseteq \beta(R'a) \subseteq P'_i$ et $N \subseteq P'_i \cap R$. L'hypothèse sur P montre que $P = P'_i \cap R$ et que alors P est de hauteur 1 ([1] th. 2.5).

REMARQUE II.2 - Il existe un c-idéal V' de R' [17] tel que $V' \subseteq V$ et $V'R' = R'V' \subseteq R$, car R' est un ordre régulier de Q. Remarquons aussi que la proposition II.1 et sa démonstration restent valables si on remplace V par V', tel que $V' \subseteq V$ et $R'V' = V'R' = V' \cap R$, et "P régulier" par "$V' \not\subseteq P$".

PROPOSITION II.3 - *Soient* T *un ordre dans un anneau artinien simple* Q*;* S *un ordre de* Q *contenant* T *tel qu'il existe un plus grand idéal bilatère non nul* V *commun à* T *et* S $(V = VS = SV \subseteq T)$:

1). *Si* P *est un idéal premier régulier de* T *posons* $T_P = \{x \in Q | \exists$ N *idéal bilatère de* R*,* $N \not\subseteq P$, $Nx \subseteq T\}$. *Si* P' *est un idéal premier régulier de* S *tel que* $P' \cap T = P$*, alors on a* $T_P = S_{P'}$.

2). *Si* $\mathscr{C}(P) = \{c \in T | x \in T, cx \in P \Rightarrow x \in P\}$ *on a* $\mathscr{C}(P) = \mathscr{C}(P') \cap T$.

DEMONSTRATION.

1). Soit $x \in T_P$, il existe N idéal bilatère de T, $N \not\subseteq P$ avec $Nx \subseteq T$ donc $VNVx \subseteq Nx \subseteq T \subseteq S$ et $VNV \not\subseteq P'$ donc $x \in S_{P'}$. Réciproquement de $x \in S_{P'}$ on déduit $N'x \subseteq S$ pour N' idéal bilatère de S, $N' \not\subseteq P'$, donc $VN'x \subseteq T$ et VN' est un idéal bilatère de R non contenu dans $P' \cap T = P$ et $x \in T_P$.

2). Soit $c \in \mathscr{C}(P)$ et $cx \in P'$ avec $x \in S$, alors on a $cxV \subseteq P'V \subseteq P = P' \cap T$, donc $xV \subseteq P \subseteq P'$ et $x \in P'$ donc $\mathscr{C}(P) \subseteq \mathscr{C}(P') \cap T$. Réciproquement soit $x \in \mathscr{C}(P') \cap T$, considérons $x \in T$ et $cx \in P$, on déduit $cx \in P'$ donc $x \in P'$ et $x \in P' \cap T = P$ et ainsi $\mathscr{C}(P') \cap T \subseteq \mathscr{C}(P)$ et l'égalité.

PROPOSITION II.4 - *Soient* R *et* R' *comme ci-dessus avec de plus* R *noethérien des deux côtés. Pour tout* P *premier régulier de* R *de hauteur* 1*,* R *vérifie la condition de Ore par rapport à* $\mathscr{C}(P)$ *et le localisé correspondant* $R_{\mathscr{C}(P)} = R_P = R'_{\mathscr{C}(P')} = R'_{P'}$*, où* P' *est un idéal premier* P' *de* R' *avec* $P' \cap R = P$*. Presque tous les idéaux premiers de hauteur* 1 *de* R *sont réguliers.* (aussi M. CHAMARIE non publié).

DEMONSTRATION. - Soit P' un idéal premier régulier de R' tel que $R \cap P' = P$ (il en existe [1] th. 2.5) et on a ht P' = ht P = 1 (même référence). On sait que sous les hypothèses R' vérifie la condition de Ore des deux côtés par rapport à $\mathscr{C}(P')$ que les familles $\mathscr{F}_{P'}$ = {I' idéal à gauche de R', contenant un idéal bilatère non contenu dans P'} et $\mathscr{F}^{P'}$ = {I' idéal à gauche de R' coupant $\mathscr{C}(P')$} sont égales et que $R'_{P'} = R'_{\mathscr{C}(P')}$ ([4] et [17] ch. 4 2.20 et 2.15). D'après la proposition II.3 on a $R_P = R'_{P'} = R'_{\mathscr{C}(P')}$. Montrons que $R_P = R_{\mathscr{C}(P)}$: tout élément $c \in \mathscr{C}(P)$ étant un élément de $\mathscr{C}(P')$ d'après II.3 a un inverse dans $R'_{\mathscr{C}(P')} = R_P$. De plus soit $x \in R_P$, il existe un idéal bilatère N de R non contenu dans P tel que $Nx \subseteq R$, comme N contient un élément $c' \in \mathscr{C}(P)$ on a $c'x \subseteq R$ et donc R_P est un anneau de fractions à gauche de R selon $\mathscr{C}(P)$ (et de même à droite). Ainsi $R_P = R_{\mathscr{C}(P)}$. On sait qu'il existe un nombre fini d'idéaux premiers P_i de R tels que $\prod_{i=1}^{r} P_i \subseteq V \subseteq P_i$ (par exemple [12] prop. 2.4) $i = 1,\dots,r$. Ainsi un idéal premier P, ht P = 1, contenant V est un de ces P_i et donc presque tous les idéaux premiers de hauteur R sont réguliers.

LEMME II.5 - *Soit* X *un idéal à gauche* P*-tertiaire de* R *(*R *et* R' *comme en* II.4*)*, P *idéal premier régulier de* R *de hauteur* 1, *alors* X *est* P*-primaire.* (Aussi M. CHAMARIE non publié).

DEMONSTRATION. - D'après [1] th. 2.7 et 2.5 il existe P' idéal premier régulier de R' avec ht P' = ht P = 1 et $P' \cap R = P$. Posons $X' = R_P X = R'_{P'} X$ (II.3). On sait que X' est un idéal à gauche $P'_{P'}$-primaire de $R'_{P'}$ (car $R'_{P'}$ est un ordre maximal régulier de Q, principal ([4] ; [17] ch. 4, 2.20 et 2.15) et on lui applique le théorème de l'idéal à gauche principal ([15] ; [17] ch. 6 th. 2.4) en se souvenant que $P'_{P'}$ est le seul idéal premier non nul de $R'_{P'}$ ([17] ch. 4 th. 2.15 et sa démonstration). On a $X' \cap R = X$ (car soit $x \in X' \cap R = R_P X \cap R$; il existe un idéal bilatère N de R avec $N \not\subseteq P$ tel que $Nx \subseteq X$ et, X étant P-tertiaire, $x \in X$). On a

$P'_{P'} \cap R = P' \cap R = P$ et X est P-primaire (X étant P-tertiaire il suffit de démontrer qu'il contient une puissance P^n de P : comme X' est $P_{P'}$-primaire il existe $n \in \mathbb{N}$ avec $P'^{n}_{P'} \subseteq X'$ et donc $(P'_P \cap R)^n = P^n \subseteq X' \cap R = X$).

PROPOSITION II.6 - *Soient* R *et* R' *comme ci-dessus en* II.4 *et* V' *un* c-*idéal de* R' *tel que* $V' = V'R' = R'V' \subseteq R$. *Soit* $Ra = \bigcap_{j=1}^{m} X_j$, X_j P_j - *tertiaire*, $j = 1,\dots,m$ *une décomposition tertiaire réduite de* Ra *pour un élément* a *non diviseur de zéro et non inversible de* R. *Si* $V' \not\subseteq P_j$ *pour* $j = 1,\dots,m$ *alors* $\mathrm{ht}\, P_j = 1$ *et* X_j *est* P_j-*primaire pour* $j = 1,\dots,m$.

DEMONSTRATION. - On a $V'(R'a \cap R) \subseteq Ra \subseteq X_j$ et $V' \not\subseteq P_j$ implique $R'a \cap R \subseteq X_j$ et $R'a \cap R = Ra$ et donc $R'a \neq R'$. De plus a est un élément inversible dans Q donc non diviseurs de zéro dans R' et $R'a = \bigcap_{i=1}^{n} Y'_i$, Y'_i P'_i - primaire, $\mathrm{ht}\, P'_i = 1$, $i = 1,\dots,n$. A priori certains éléments de $\{P'_1,\dots,P'_n\}$ pourraient être associés à V' par exemple $\{P'_{k+1},\dots,P'_n\}$. Comme il existe $n_{k+1},\dots,n_n$ tels que $P'^{n_{k+1}}_{k+1} \dots P'^{n_n}_{n} \subseteq \bigcap_{i=k+1}^{n} Y'_i$, comme V' est un c-idéal et $\bigcap_{i=k+1}^{n} Y'_i$ est un c-idéal à gauche il existe $p \in \mathbb{N}$ avec $V'^p \subseteq \bigcap_{i=k+1}^{n} Y'_i$. Comme P'_i n'est pas pour $i = 1,\dots,k$ un idéal premier associé au c-idéal V' on a $V' \not\subseteq P'_i$ pour $i = 1,\dots,k$ ce qui implique que $Y'_i \cap R$ est un idéal à gauche de R $P'_i \cap R$-primaire (même démonstration qu'au lemme 1.8 ch. 9 de [17]). Posons $Y = \bigcap_{i=1}^{k} Y'_i \cap R$ et $Z = \bigcap_{i=k+1}^{n} (Y'_i \cap R)$. On a $Ra = Y \cap Z$. Aucun idéal premier associé à Z n'appartient à $\{P_1,\dots,P_m\}$ car sinon on déduirait de $V'^p \subseteq Z$, $V'^p \subseteq P_i$ et $V' \subseteq P_i$ ce qui n'est pas. On a donc $\mathrm{Ass}(R/_{Ra}) \cap \mathrm{Ass}(R/_Z) = \phi$. Mais $Y/_{Y \cap Z}$ est un sous-module de $R/_{Ra}$ et comme $Y/_{Y \cap Z}$ est isomorphe à un sous-module de $R/_Z$, on a $\mathrm{Ass}\, Y/_{Y \cap Z} \subseteq \mathrm{Ass}\, R/_{Ra} \cap \mathrm{Ass}\, R/_Z$ ce qui implique $\mathrm{Ass}\, Y/_{Y \cap Z} = \phi$ et $Y = Y \cap Z$ puisque R est noethérien. Ainsi $Ra = \bigcap_{i=1}^{k} (Y'_i \cap R)$ où $Y'_i \cap R$ est $P'_i \cap R$-primaire

$i = 1,\dots,k$. Comme P'_i est un idéal premier de R'ne contenant pas V' donc régulier $P'_i \cap R$ est un idéal premier régulier de R ([1] th. 2.6) et $ht(P'_i \cap R) = ht\ P'_i = 1$ ([1] th. 2.7). On a donc $k = m$, $P_i = P'_i \cap R$, $i = 1,\dots,m$ et $ht\ P_i = 1$. Il reste à prouver que X_i est P_i-primaire pour $i = 1,\dots,m$, or ceci résulte immédiatement de II.5.

La proposition II.4 généralise un résultat de [7] (voir aussi [17] ch. 9 , th. 1.7) et la proposition II.6 un résultat de [7], de [16] (voir aussi [17] ch. 9 th. 1.9).

BIBLIOGRAPHIE

[1] AMITSUR S.A. et SMALL L.W., Prime ideals in PI-rings J. of algebra, 62, 358-383, 1980.

[2] BLAIR W.D., Rigt noetherian rings integral over their centers : J. of Algebra, 27, 1973, 187-198.

[3] BRAUN A., A characterization of prime noetherian PI-rings and a theorem of Mori-Nagata Proc. of the amer. math. soc. 74 (1), 1979, 9-15.

[4] CHAMARIE M., Localisation dans les ordres maximaux, Comm. in Algebra, 2, 1974, 279-293.

[5] CHAMARIE M., Ordres maximaux et R-ordres maximaux, J. of algebra 58, 1979, 148-156.

[6] CHAMARIE M. et HUDRY A., Anneaux noethériens à droite entiers sur un sous-anneau de leur centre, Com. in algebra, 6, 1978, 203-222.

[7] CHAMARIE M. et MAURY G., Un théorème de l'idéal à gauche principal ..., C.R. Acad. Sci. Paris, série A, 286, (1978), 609-611.

[8] CHATTERS A.W., HAJARNAVIS C.R., LENAGHAN T.H., Reduced rank in noetherian rings, J. of algebra, 61, 2, 582-589.

[9] FOSSUM R., Maximal orders over a Krull domain, J. of Algebra, 10, 1968, 321-332.

[10] JATEGAONKAR A.V., Relative Krull dimension and prime ideals in right noetherian rings, Com. in algebra 2, (1974), 429-458.

[11] KOTHE G., Schiefkörper unendlichen Ranges über dem Zentrum, Math. Ann., 105, 1931, 15-39.

[12] LESIEUR L. et CROISOT R., Théorie noethérienne des anneaux, des demi-groupes, des modules dans le cas non commutatif I, Colloque d'algèbre supérieure de Bruxelles, 1956, 79-121.

[13] LESIEUR L. et CROISOT R., Théorie noethérienne des anneaux, des demi-groupes, des modules dans le cas non commutatif II, Math. Ann. 134, 1958, 458-476.

[14] MARUBAYASHI H., Non kommutative Krull rings, Osaka J. Math., 12, 1975, 703-714.

[15] MAURY G., La condition intégralement clo dans quelques structures algébriques, Ann. Sc. Ec. Norm. Sup., 3ème série, 78, 1961, 31-100.

[16] MAURY G., Un théorème de l'idéal à gauche principal dans les R-ordres, Com. in algebra, 7, 1979, 677-687.

[17] MAURY G. et RAYNAUD J., Ordres maximaux au sens d'Asano. Lecture Notes in mathematics n°808, 1980, 192 pages, Springer Verlag.

[18] SHELTER W., Integral extensions of rings satisfying a polynomial identity, J. of algebra, 40, 1976, 245-257.

[19] ZARISKI Q, SAMUEL P., Commutative algebra, Van Nostrand, Princeton, 1958 et 1960, Vol. I.

Département de Mathématiques
Université Claude-Bernard - Lyon I
43, boulevard du 11 novembre 1918
69622 Villeurbanne, France

The Artin-Rees Property

by P.F. Smith

All rings are assumed to be associative but not necessarily commutative and to have an identity element 1. All modules are right modules and are unital. In this paper we review some recent work concerning the weak Artin Rees property, called the AR property for short.

1. The Artin-Rees Theorem

1.1 Theorem (Artin, Rees). Let R be a commutative Noetherian ring, I an ideal of R, M a finitely generated R-module and N a submodule of M. Then there exists a positive integer k such that for all $n \geq k$

$$N \cap MI^n = (N \cap MI^k)I^{n-k}.$$

Proof (See (20)). Suppose first that M = R. There exists a finite collection of elements $a_1, \ldots, a_s$ in I such that $I = Ra_1 + \ldots + Ra_s$. Let x be an indeterminate and S the subring of the polynomial ring $R[x]$ generated by R and the elements $a_i x (1 \leq i \leq s)$. Then S consists of all polynomials of the form

$$c_0 + c_1 x + \ldots + c_n x^n$$

where $n \geq 0$, $c_0 \in R$ and $c_j \in I^j$ $(1 \leq j \leq n)$. If N is an ideal of R let E be the ideal

$$N + (N \cap I)x + (N \cap I^2)x^2 + \ldots$$

of S, i.e. E consists of all polynomials

$$c_0 + c_1 x + \ldots + c_n x^n$$

such that $n \geqslant 0$, $c_0 \in N$ and $c_j \in N \cap I^j$ $(1 \leqslant j \leqslant n)$. By Hilbert's Basis Theorem the commutative ring $S = R[a_1 x, \ldots, a_s x]$ is Noetherian. Thus the ideal E is finitely generated, say by elements $f_1(x), \ldots, f_m(x)$. Define $k = \max_{1 \leqslant i \leqslant m} \deg f_i(x)$. Let $n \geqslant k$. Let $a \in N \cap I^n$. Then $ax^n \in E$ and so

$$ax^n = b_1(x) f_1(x) + \ldots + b_m(x) f_m(x)$$

for some $b_1(x), \ldots, b_m(x)$ in S. Comparing coefficients of x^n we see that

$$a \in \sum_{t=0}^{k} I^{n-t} \quad (N \cap I^t)$$

where we make the convention $I^0 = R$. But, for $0 \leqslant t \leqslant k$,

$$\begin{aligned} I^{n-t}(N \cap I^t) &= I^{n-k}\, I^{k-t}(N \cap I^t) \\ &\leqslant I^{n-k}(N \cap I^k). \end{aligned}$$

Thus $a \in (N \cap I^k) I^{n-k}$. It follows that $N \cap I^n \leqslant (N \cap I^k) I^{n-k}$. The converse is clear and so $N \cap I^n = (N \cap I^k) I^{n-k}$, and this is true for all $n \geqslant k$.

In general, let M be a finitely generated R-module. Define T to be the set of 2 × 2 "matrices"

$$\begin{bmatrix} r & m \\ 0 & r \end{bmatrix}$$

with r in R and m in M. Matrix addition and multiplication make T a ring. Moreover, T is a commutative Noetherian ring. Define

$$J = \left\{ \begin{bmatrix} a & m \\ 0 & a \end{bmatrix} : a \in I,\ m \in M \right\}$$

and, for any submodule N of M

$$E = \left\{ \begin{bmatrix} 0 & n \\ 0 & 0 \end{bmatrix} : n \in N \right\}.$$

Then E and J are ideals of T and so by the first part of the proof there exists a positive integer k such that for all $n \geq k$,

$$E \cap J^n = (E \cap J^k)J^{n-k}. \qquad (1)$$

Note that

$$J^q = \left\{ \begin{bmatrix} a & m \\ 0 & a \end{bmatrix} : a \in I^q,\ m \in MI^{q-1} \right\}$$

for all positive integers q. By comparing the (1,2)-entries of (1) we obtain

$$N \cap MI^{n-1} = (N \cap MI^{k-1})I^{n-k}$$

for all $n \geq k$ and the result follows.

1.2 Corollary. Let R be a commutative Noetherian ring and E and I be ideals of R. Then there exists a positive integer n such that $E \cap I^n \leq EI$.

Proof. In the theorem, take $M = R$, $N = E$, $n = k+1$.

1.3 Corollary. Let R be a commutative Noetherian ring and E and I be ideals of R. Then there exists an element a in I such that

$$\bigcap_{n=1}^{\infty} (E+I^n) = \{r \in R : r(1-a) \in E\}.$$

Proof. Since R is a Noetherian ring it follows that the ideal $\bigcap_{n=1}^{\infty}(E+I^n) = Rx_1 + \ldots + Rx_t$ for some positive integer t and elements $x_i (1 \leq i \leq t)$ in R. Let $x = x_i$ for some $1 \leq i \leq t$. By Corollary 1.2

there exists a positive integer n such that $(xR+E) \cap I^n \leq (xR+E)I$. Since $x \in E+I^n$ it follows that $x \in E+(xR+E)I$ and hence $x(1-a_i) \in E$ for some $a_i \in I$. Define $a \in I$ by

$$1-a = \prod_{i=1}^{t} (1-a_i).$$

Then the result follows with this element a.

1.4 Corollary. Let R be a commutative Noetherian ring with Jacobson radical J. Then $\bigcap_{n=1}^{\infty} J^n = 0$.

Proof. In Corollary 1.3, take I = J, E = 0.

This last result is essentially Krull's Intersection Theorem and it together with the earlier ones are basic to the study of ideals in commutative Noetherian rings. However, as the following simple example shows, these results do not go over to non-commutative rings, even right and left Artinian ones.

1.5 Example. Let K be a field and R the ring of 2 × 2 upper triangular matrices over K. Then R is a right and left Artinian ring. Define

$$M_1 = \left\{ \begin{bmatrix} 0 & k_1 \\ 0 & k_2 \end{bmatrix} : k_i \in K (i = 1,2) \right\}$$

and

$$M_2 = \left\{ \begin{bmatrix} k_1 & k_2 \\ 0 & 0 \end{bmatrix} : k_i \in K (i = 1,2) \right\}.$$

Then $M_1 M_2 = 0$, M_2 is idempotent and for all positive integers n,

$$M_1 \cap M_2^n = M_1 \cap M_2 \neq 0$$

so that $M_1 \cap M_2^n \nleq M_1 M_2$.

Of course the Jacobson radical of the ring R in Example 1.5 is nilpotent but the following example of Herstein (12) shows that Corollary 1.4 is not true for right Noetherian rings in general.

<u>1.6 Example.</u> Let Q be the field of rational numbers and S the subring of Q consisting of rational numbers of the form m/n with m,n integers and n odd. Let R be the subring of the ring of 2 × 2 upper triangular matrices over Q consisting of all matrices of the form

$$\begin{bmatrix} a & b \\ 0 & c \end{bmatrix}$$

with a in S, b and c in Q. Then R is a right Noetherian ring. Let J denote the Jacobson radical of R. Then

$$J = \left\{ \begin{bmatrix} a & b \\ 0 & 0 \end{bmatrix} : a \in M,\ b \in Q \right\}$$

where M is the ideal of S consisting of all rational numbers m/n with m even. For each positive integer t we have

$$J^t = \left\{ \begin{bmatrix} a & b \\ 0 & 0 \end{bmatrix} : a \in M^t,\ b \in Q \right\}$$

and thus by Corollary 1.4

$$\bigcap_{t=1}^{\infty} J^t = \left\{ \begin{bmatrix} 0 & b \\ 0 & 0 \end{bmatrix} : b \in Q \right\} \neq 0.$$

Notice that the ring R in Example 1.6 is not left Noetherian for Q is not a finitely generated S-module. Thus we ask

<u>Question 1.</u> Let R be a right and left Noetherian ring with Jacobson radical J. Is $\bigcap_{n=1}^{\infty} J^n = 0$?

The answer to this question may be "yes" but in any case we have seen that Corollary 1.2 does not go over to all right and left Noetherian rings. One would expect it to go over to some non-commutative rings and this we shall investigate. Let I be an ideal of an arbitrary ring R. Then we say that I has the (right) AR property if for each right ideal E of R there exists a positive integer n such that $E \cap I^n \leq EI$. Corollary 1.2 states that every ideal of a commutative Noetherian ring has the AR property. In the converse direction we have:

1.7 Theorem. Let R be the polynomial ring $S[x]$ for some ring S. Suppose that the ideal Rx has the AR property. Then S is a right Noetherian ring.

Proof. Let $E_0 \leq E_1 \leq E_2 \leq \ldots$ be an ascending chain of right ideals of S. Define

$$F = E_0 + E_1 x + E_2 x^2 + \ldots$$

to be the right ideal of R consisting of all polynomials

$$c_0 + c_1 x + \ldots + c_n x^n$$

with $n \geq 0$ and c_i in E_i for $0 \leq i \leq n$. There exists a positive integer m such that

$$F \cap (Rx)^m \leq Fx.$$

Thus

$$E_m x^m + E_{m+1} x^{m+1} + \ldots \leq (E_0 + E_1 x + E_2 x^2 + \ldots)x$$

and it follows that $E_{m-1} = E_m = E_{m+1} = \ldots$. Thus S is right Noetherian.

An ideal I of a ring R will be said to have the finite intersection property if

$$\bigcap_{n=1}^{\infty} (E+I^n) = \{r \in R : r(1-a) \in E \text{ for some a in } I\}$$

for every finitely generated right ideal E of R. Every ideal of a commutative Noetherian ring has the finite intersection property (Corollary 1.3).

1.8 Theorem. Let R be a commutative ring.

(i) If the prime ideals of R are finitely generated than R is Noetherian.

(ii) If R is a local ring with unique maximal ideal M such that M is finitely generated and M has the finite intersection property then R is Noetherian.

Proof. (i) This is a theorem of Cohen (7).

(ii) Suppose R is not Noetherian and let P be an ideal of R maximal with respect to not being finitely generated. Then $P \underset{\neq}{<} M$. Let $c \in M$, $c \notin P$. Then $P + Rc$ is a finitely generated ideal of R by the choice of P. There exist a positive integer n and elements p_i of P and a_i of R $(1 \leqslant i \leqslant n)$ such that $P + Rc$ is generated by the elements $p_i + a_i c (1 \leqslant i \leqslant n)$. Let $A = \{r \in R : rc \in P\}$. Then $P \leqslant A$. Note that

$$P = Rp_1 + Rp_2 + \ldots + Rp_n + Ac.$$

If $P < A$ then A is a finitely generated ideal and hence so is Ac and P. If $P = A$ then $P = Q + Pc$ where $A = \Sigma_{i=1}^{n} Rp_i$ and it follows that

$$P = Q + Pc = Q + Pc^2 = \ldots \; .$$

Thus

$$P \leqslant \bigcap_{k=1}^{\infty}(Q + Rc^k) \leqslant \bigcap_{k=1}^{\infty}(Q + M^k) = Q$$

since M has the finite intersection property. In any case, P is finitely generated, a contradiction. Thus R is a Noetherian ring.

The restriction that M have the finite intersection property in (ii) above is necessary as the following example shows. Take S,Q as in

Example 1.6. Let R be the subring of the ring of 2 × 2 upper triangular matrices over Q consisting of all matrices of the form

$$\begin{bmatrix} a & b \\ 0 & a \end{bmatrix}$$

with a in S and b in Q. Then R is commutative local ring with unique maximal ideal

$$J = \left\{ \begin{bmatrix} a & b \\ 0 & a \end{bmatrix} : a \in M,\ b \in Q \right\}$$

where M is the unique maximal ideal of S. Then J is generated by the element

$$\begin{bmatrix} 2 & 0 \\ 0 & 2 \end{bmatrix}$$

but

$$J_1 = \bigcap_{n=1}^{\infty} J^n = \left\{ \begin{bmatrix} 0 & b \\ 0 & 0 \end{bmatrix} : b \in Q \right\}.$$

Note that the ideal J_1 is not finitely generated and hence R is not a Noetherian ring.

Also if K is a field and $R = K[x_1, x_2, \ldots]$, the polynomial ring in a countably infinite number of indeterminates x_i, let I be the ideal of R generated by the elements $x_i (i \geq 1)$. Then R/I^2 is a non-Noetherian ring which has a unique maximal ideal and this ideal is nilpotent.

2. The AR property

Recall that an ideal I of a ring R has the AR property if for each right ideal E of R there exists a positive integer n such that $E \cap I^n \leq EI$. A submodule N of a module M is called <u>essential</u> if $N \cap K \neq 0$ for all non-zero submodules K of M.

2.1 Theorem. The following statements are equivalent for an ideal I of a ring R.

(i) I has the AR property.

(ii) If N is an essential submodule of a finitely generated right R-module M such that $NI = 0$ then $MI^n = 0$ for some $n \geq 1$.

(iii) If N is a submodule of a finitely generated right R-module M then there exists a positive integer n such that $N \cap MI^n \leq NI$.

Proof. (i) $\Longrightarrow$ (ii). Suppose $M = m_1R + \ldots + m_kR$ for some positive integer k. For each $1 \leq i \leq k$, $N_i = N \cap m_iR$ is essential submodule of m_iR and thus without loss of generality we can suppose that M is cyclic and $M = mR$. Let $E = \{r \in R : mr \in N\}$. Then E is a right ideal of R and $E \cap I^n \leq EI$ for some positive integer n. Thus $N \cap MI^n \leq NI = 0$ and hence $MI^n = 0$.

(ii) $\Longrightarrow$ (iii). Define

$$\mathcal{S} = \{T : T \text{ is a submodule of } M \text{ and } N \cap T = NI\}.$$

Then $NI \in \mathcal{S}$ and by Zorn's Lemma $\mathcal{S}$ contains a maximal member K. It follows that $(N+K)/K$ is an essential submodule of M/K. But $(N+K)/K \cong N/N \cap K = N/NI$ and thus $(N+K)/K$ is annihilated by I. By (ii) $(M/K)I^n = 0$ for some positive integer n, i.e. $MI^n \leq K$ so that $N \cap MI^n \leq NI$.

(iii) $\Longrightarrow$ (i). Obvious.

2.2 Theorem. Let I be an ideal of a ring R. Consider the following conditions.

(i) For every finitely generated right ideal E of R there exists a positive integer n such that $E \cap I^n \leq EI$.

(ii) I has the finite intersection property.

Then (i) implies (ii). Conversely, if the ring R/I^2 is right Artinian then (ii) implies (i).

Proof. (i) $\Longrightarrow$ (ii). Let E be a finitely generated right ideal of R. Let

$r \in \bigcap_{n=1}^{\infty}(E+I^n)$. Let $F = rR + E$. Then there exists a positive integer n such that $F \cap I^n \leq FI$. Since $r \in E+I^n$ it follows that $r \in FI + E$ and hence $r(1-a) \in E$ for some a in I. Thus, I has the finite intersection property.

(ii) $\Rightarrow$ (i). Suppose that R/I^2 is a right Artinian ring. Let E be a finitely generated right ideal of R. Then the finitely generated right R-module E/EI is Artinian. Consider the descending chain of right ideals of R:

$$E \geq E \cap I \geq (E \cap I^2) + EI \geq (E \cap I^3) + EI \geq \dots .$$

Since E/EI is Artinian it follows that there exists a positive integer n such that

$$(E \cap I^n) + EI = (E \cap I^{n+1}) + EI = \dots .$$

Since the ring R/I^2 is right Artinian it follows that the right ideal I/I^2 is finitely generated. Thus there exists a finitely generated right ideal F such that $I = F + I^2$. Now suppose $E = e_1R + \dots + e_mR$. Then $EI = e_1I + \dots + e_mI$. Let $k \geq n$. Since $I = F + I^k$ it follows that $EI \leq e_1F + \dots + e_mF + I^{k+1}$. Let $G = e_1F + \dots + e_mF$. Then G is a finitely generated right ideal and

$$E \cap I^n \leq \bigcap_{k=1}^{\infty} (G + I^k).$$

Let $x \in E \cap I^n$. By hypothesis there exists $a \in I$ such that $x(1-a) \in G$. Thus $x \in xa + G \leq EI$. It follows that $E \cap I^n \leq EI$. This proves (i).

We shall say that an ideal I has the (right) <u>fAR property</u> if I satisfies (i) in the theorem. A similar proof shows that if I has the AR property then

$$\bigcap_{n=1}^{\infty} (E+I^n) = \{r \in R : r(1-a) \in E \text{ for some } a \in I\} \qquad (2)$$

for every right ideal E of R. Conversely, if I satisfies (2) for every right ideal E and the ring R/I is right Artinian then I has the fAR property. Let us say that an ideal I has the intersection property if I satisfies (2) for every right ideal E.

Question 2. Let R be a ring and I an ideal such that R/I^2 is right Artinian and I has the intersection property. Does I have the AR property?

Note that if R/I^2 is right Artinian and I has the AR property then E/EI has finite composition length for all right ideal E. For, $E \cap I^n \leq EI$ for some positive integer n and $E/E \cap I^n \cong (E+I^n)/I^n$ which is a right ideal of the right Artinian ring R/I^n.

Let R be a ring. An element c of R is regular if whenever $r \in R$, $cr = 0$ or $rc = 0$ implies $r = 0$. If A is an ideal of R then $C(A)$ will denote the set of elements r in R such that $r + A$ is a regular element of the ring R/A. Let T be a non-empty subset of R. Then R satisfies the right Ore condition with respect to T if for all r in R and t in T there exist r' in R and t' in T such that $rt' = tr'$.

2.3 Lemma. Let I be an ideal of a ring R such that I has the finite intersection property. Let $T = \{1-a : a \in I\}$. Then R satisfies the right Ore condition with respect to T. Moreover, if $K = \{r \in R : rt = 0$ for some t in T$\}$ then K is an ideal of R and $T \leq C(K)$.

Proof. Let $r \in R$, $a \in I$. Then, if $E = (1-a)R$,

$$r = (1-a^n)r + a^n r \in E + I^n, \; n \geq 1.$$

Thus

$$r \in \bigcap_{n=1}^{\infty} (E + I^n)$$

and so there exists b in I such that $r(1-b) \in E$. It follows that R satisfies the right Ore condition with respect to T. Also, by hypothesis

$$K = \bigcap_{n=1}^{\infty} I^n$$

and so immediately K is an ideal and $T \leq C(K)$.

Let I be an ideal of a ring R such that I has the finite intersection property. Let $T = \{1-a : a \in I\}$ and $K = \{r \in R : rt = 0$ for some t in T$\}$. Let $\overline{R} = R/K$ and $\overline{T} = \{t+K : t \in T\}$. Then $\overline{R}$ satisfies the right Ore condition with respect to $\overline{T}$ and $\overline{T}$ is a multiplicatively closed set of regular elements of $\overline{R}$. We form the partial right quotient ring $\overline{R}_{\overline{T}}$ consisting of all elements $\overline{r}\,\overline{t}^{-1}$ with $\overline{r} = r + K$, $r \in R$, $\overline{t} \in \overline{T}$. We denote $\overline{R}_{\overline{T}}$ by R_I and $(I/K)R_I$ by IR_I. Note that $IR_I = \{\overline{r}\,\overline{t}^{-1} : r \in I, t \in T\}$ and hence IR_I is an ideal of R_I and is contained in the Jacobson radical of R_I.

Note that if an ideal I of R has the intersection property then

$$\bigcap_{n=1}^{\infty} MI^n = \{m \in M : m(1-a) = 0 \text{ for some } a \in I\} \qquad (3)$$

holds for every cyclic right R-module M. For, without loss of generality we can suppose that $M = R/E$ for some right ideal E. Then

$$MI^n = (I^n + E)/E\ ,\ n \geq 1$$

and (2) implies (3).

<u>2.4 Theorem.</u> The following statements are equivalent for an ideal I of a ring R.

(i) I has the intersection property.

(ii) $\bigcap_{k=1}^{\infty} MI^k = \{m \in M : m(1-a) = 0$ for some a in I$\}$ for every finitely generated right R-module M.

Proof. (ii) $\Longrightarrow$ (i) Obvious.

(i) $\Longrightarrow$ (ii) Suppose $M = m_1R + \ldots + m_nR$ for some positive integer n. Suppose that I is contained in the Jacobson radical of R. In this case we prove the result by induction on n. If n = 1 then

$$\bigcap_{n=1}^{\infty} MI^n = 0$$

because (i) holds. Suppose $n > 1$ and let $N = m_1R + \ldots + m_{n-1}R$. By induction on n we can assume

$$\bigcap_{n=1}^{\infty} NI^n = 0. \tag{4}$$

Let k be any positive integer. For every integer $s \geq k$

$$MI^s = NI^s + m_nI^s \leq NI^k + m_nI^s.$$

Let $\overline{M} = M/NI^k$, $\overline{m}_n = m_n + NI^k$. Then

$$\bigcap_{s=1}^{\infty} \overline{M}I^s = \bigcap_{s=1}^{\infty} \overline{m}_nI^s = 0$$

by (i). Thus

$$\bigcap_{s=1}^{\infty} MI^s \leq NI^k.$$

But k was arbitrarily chosen, so that

$$\bigcap_{s=1}^{\infty} MI^s \leq \bigcap_{k=1}^{\infty} NI^k = 0$$

by (4). Thus (ii) holds.

In general, we pass to the ring R_I. Let $M_I = M \otimes_R R_I$. Then M is a finitely generated right R_I-module. It can easily be checked that IR_I has the intersection property and hence

$$\bigcap_{k=1}^{\infty} M_I(IR_I)^k = 0.$$

This just means that (ii) holds. This completes the proof.

A module M is an <u>essential extension</u> of a module N if N is an essential submodule of M.

2.5 Theorem (10, p.274 Theorem 2.60). Let R be a right Noetherian ring with Jacobson J such that R/J is an Artinian ring. Then the following statements are equivalent.

(i) J has the AR property.

(ii) $\bigcap_{n=1}^{\infty} MJ^n = 0$ for every finitely generated right R-module M.

(iii) Every finitely generated essential extension of an Artinian right R-module is Artinian.

Proof. (i) $\Rightarrow$ (ii). By Theorems 2.2 and 2.4, or directly.

(ii) $\Rightarrow$ (iii). Let N be an essential submodule of the finitely generated right R-module M and suppose N is Artinian. Then the descending chain $N \cap MJ \geq N \cap MJ^2 \geq \ldots$ must terminate and so there exists k such that

$$N \cap MJ^k = N \cap MJ^{k+1} = \ldots \leq \bigcap_{n=1}^{\infty} MJ^n = 0.$$

Thus $MJ^k = 0$ and hence M is Artinian.

(iii) $\Rightarrow$ (i). Let N be an essential submodule of a finitely generated right R-module M such that NJ = 0. Then N is Artinian and hence M is Artinian by (iii). Thus M has finite length and $MJ^t = 0$ for some positive integer t. By Theorem 2.1 J has the AR property.

I.M. Musson (see([6], p.105)) has shown that there exist right and left Noetherian domains with non-Artinian cyclic essential extensions of irreducible modules.

<u>2.6 Lemma.</u> Let I and J be ideals of a ring R such that $J \leq I$ and $IJ \leq JI$. Suppose that J and I/J have the AR property. Then I has the AR property.

Proof. Let M be a finitely generated right R-module and N an essential submodule of M such that NI = 0. Then NJ = 0. By Theorem 2.1 $MJ^k = 0$ for some $k \geq 1$. We prove by induction on k that some power of I annihilates M.

If $k = 1$ then this follows because I/J has the AR property. Suppose $k > 1$ and let $V = \{x \in M : xJ^{k-1} = 0\}$. Then $VI^s = 0$ for some $s \geq 1$ by induction. But $MI^sJ \leq MJI^s \leq VI^s = 0$ so that $MI^s \leq V$ and hence $(MI^s)I^s = 0$, i.e. $MI^{2s} = 0$. By Theorem 2.1 I has the AR property.

Let R be a ring and I an ideal of R. Then I has the nAR-property if for every submodule N of a Noetherian right R-module M there exists a positive integer k such that

$$N \cap MI^k \leq NI,$$

equivalently, for every essential submodule N of a Noetherian right R-module M with $NI = 0$ there exists a positive integer k such that $MI^k = 0$ (see the proof of Theorem 2.1). It is clear that Lemma 2.6 remains true if "AR" is replaced by "nAR". The next result is due to Nouazé and Gabriel (19).

2.7 Theorem. Let I be an ideal of a ring R and c a central element of R such that $c \in I$. Then I has the nAR property if and only if I/Rc has the nAR property.

Proof. The necessity is obvious. Conversely, suppose that I/Rc has the nAR property. Let N be an essential submodule of a Noetherian right R-module M such that $NI = 0$. Then $Nc = 0$. Define $f: M \to M$ by $f(m) = mc$ $(m \in M)$. Then f is an endomorphism of the Noetherian module M and so

$$\ker f \cap \operatorname{im} f^m = 0$$

for some positive integer m. But $N \leq \ker f$ and hence $\operatorname{im} f^m = 0$, i.e. $Mc^m = 0$. Thus $J = Rc$ has the nAR property (Theorem 2.1) and the result follows by Lemma 2.6.

An ideal I of a ring R is polycentral (or has a centralizing set of generators) provided there is a finite chain of ideals

$$0 = I_0 \leq I_1 \leq \dots \leq I_n = I$$

such that for each $1 \leq j \leq n$, the ideal I_j/I_{j-1} is generated by a finite collection of central elements of R/I_{j-1}. The theorem gives at once:

2.8 Corollary. Any polycentral ideal has the nAR property. In particular polycentral ideals of right Noetherian rings have the AR property.

2.9 Theorem. For any polycentral ideal I of a ring R the following are equivalent:

(i) I has the finite intersection property.

(ii) I has the AR property.

Proof. (ii) $\Longrightarrow$ (i). See the remarks after Theorem 2.2.

(i) $\Longrightarrow$ (ii). Suppose first that I is contained in the Jacobson radical of R. We prove that I is a Noetherian right R-module. Suppose not. Let E be a right ideal of R chosen maximal with respect to the properties $E \leq I$ and E is not finitely generated. Then $E \neq I$. Without loss of generality, because I is polycentral, we can choose a central element $c \in I$ with $c \notin E$. Then $E + cR$ is a finitely generated right ideal. Let $F = \{r \in R : cr \in E\}$. Then F is a right ideal of R and $E \leq F$. Thus we can copy the proof of Theorem 1.8 (ii) to conclude that E is finitely generated, a contradiction. Thus I_R is Noetherian.

Let G be a right ideal of R. Then $G \cap I$ is a submodule of the Noetherian right R-module I and hence Corollary 2.8 gives

$$(G \cap I) \cap I.I^n \leq (G \cap I)I$$

for some positive integer n, i.e.

$$G \cap I^n \leq GI.$$

It follows that I has the AR property.

In general, pass to the ring R_I (Lemma 2.3). Since IR_I is poly-central, has the finite intersection property and is contained in the Jacobson radical of R_I it follows that IR_I has the AR property. Let H be a right ideal of R. By passing to R_I we see that there exists a positive integer m such that

$$H \cap I^m \leq \{r \in R : r(1-a) \in HI \text{ for some } a \text{ in } I\}.$$

Let $h \in H \cap I^m$. Then $h(1-b) \in HI$ for some b in I and thus $h \in hb+HI = HI$. It follows that I has the AR property.

<u>2.10 Corollary.</u> Let R be the polynomial ring $S[x]$ for some ring S. Suppose that the ideal Rx has the finite intersection property. Then S is a right Noetherian ring.

Proof. By Theorems 2.9 and 1.7.

<u>3. Group rings</u>

Let J be a ring and G a multiplicative group. Let JG denote the collection of formal sums

$$\sum_{x \in G} a_x x$$

where $a_x \in J$ and $a_x \neq 0$ for at most a finite collection of elements x in G. Define

$$\sum_x a_x x = \sum_x b_x x \text{ if and only if } a_x = b_x \ (x \in G),$$

$$\sum_x a_x x + \sum_x b_x x = \sum_x (a_x + b_x)x, \text{ and}$$

$$(\sum_x a_x x)(\sum_x b_x x) = \sum_x c_x x$$

where

$$c_x = \sum_{yz=x} a_y b_z \ (x \in G).$$

Then JG is a ring called a <u>group ring</u>.

Define a mapping $\phi : JG \to J$ by

$$\phi(\sum_x a_x x) = \sum_x a_x.$$

Then ϕ is an epimorphism with kernel

$$\underline{\underline{g}} = \{\sum_x a_x x : \sum_x a_x = 0\} = \sum_{x \in G} (x-1)JG,$$

and $\underline{\underline{g}}$ is called the augmentation ideal of JG.

If J is a commutative ring then the map $x \mapsto x^{-1}$ $(x \in G)$ of G extends to an anti-automorphism of JG. Thus JG is right Noetherian if and only if it is left Noetherian and we say simply that JG is a Noetherian ring.

A group G is polycyclic provided there exists a finite chain of subgroups

$$1 = G_0 \leqslant G_1 \leqslant \dots \leqslant G_n = G \tag{5}$$

such that for each $1 \leqslant i \leqslant n$, G_{i-1} is a normal subgroup of G_i and G_i/G_{i-1} is cyclic. If $\underline{\underline{X}}$ and $\underline{\underline{Y}}$ are group classes then an $\underline{\underline{X}}$-by-$\underline{\underline{Y}}$ group is a group G with a normal subgroup N such that N is an $\underline{\underline{X}}$-group and G/N a $\underline{\underline{Y}}$-group. Polycyclic-by-finite groups are precisely the groups G such that there exists a chain (5) with each factor G_i/G_{i-1} cyclic or finite $(1 \leqslant i \leqslant n)$. The number of factors G_i/G_{i-1} which are infinite cyclic is an invariant of the group called the Hirsch number which we shall denote by h(G). The next result is due to Hall (11, Theorem 1).

3.1 Theorem. Let R be a ring which is generated by a subring S and a polycyclic-by-finite group G such that $x^{-1}sx \in S$ for all s in S and x in G. If S is right Noetherian then so is R.

Proof. Let G have a series (5) with factors G_i/G_{i-1} cyclic or finite $(1 \leqslant i \leqslant n)$. The proof is by induction on n, the case n = 0 being clear.

Suppose $n > 0$ and let $H = G_{n-1}$. Let T be the subring of R generated by H and S. By induction on n T is a right Noetherian ring. If $[G:H] = m < \infty$ and $x_1, \ldots, x_m$ is a transversal to the cosets of H in G then

$$R = x_1T + \ldots + x_mT.$$

Thus R is a Noetherian right T-module and hence a right Noetherian ring.

Suppose that G/H is an infinite cyclic group. Let $x \in G$ such that G is generated by x and H. Then

$$R = \sum_{k=-\infty}^{\infty} x^k T$$

and a Hilbert Basis Theorem argument proves that R is right Noetherian.

__3.2 Corollary.__ Let J be a right Noetherian ring and G a polycyclic-by-finite group. Then the group ring JG is right Noetherian.

__3.3 Corollary__ (J.E. Roseblade (22)). Let J be a right Noetherian ring, G a polycyclic-by-finite group and R the group ring JG. Let N be a nilpotent normal subgroup of G. Then the ideal $\underline{\underline{n}}$ R has the AR property.

__Proof.__ Suppose first that N is Abelian. For any g in G and y in N $gyg^{-1} \in N$ and it follows that $\underline{\underline{n}}\, R = R\, \underline{\underline{n}}$ so that $\underline{\underline{n}}\, R$ is an ideal of R. Now N is finitely generated, say by the elements $x_i (1 \leq i \leq n)$. It follows that $\underline{\underline{n}}$ is generated by the elements $c_i = x_i - 1$ $(1 \leq i \leq n)$. Note that c_i is central in the ring $JN (1 \leq i \leq n)$. Let S be the subring of the polynomial ring $J[x]$ generated by J and the elements $c_i x$ $(1 \leq i \leq n)$. By Hilbert's Basis Theorem S is a right Noetherian ring and for all g in G, $g^{-1}Sg \leq S$. Let T be the subring

$$R + \underline{\underline{n}}\, R\, x + \underline{\underline{n}}^2 R\, x^2 + \ldots$$

of the polynomial ring $R[x]$. Then T is generated by S and G and by the theorem T is a right Noetherian ring. As in the proof of Theorem 1.1 $\underline{\underline{n}}$ R has the AR property.

In general suppose that N is a nilpotent normal subgroup of G of class c. We prove that $\underline{\underline{n}}$ R has the AR property by induction on c. If c = 1 then the result is proved by the first part of the proof. Suppose c > 1 and let Z denote the centre of N. By induction $\underline{\underline{n}}\, R/\underline{\underline{z}}\, R$ has the AR property. Also by the first part $\underline{\underline{z}}$ R has the AR property and clearly because Z is the centre of N

$$\underline{\underline{n}}\, R\, \underline{\underline{z}}\, R = \underline{\underline{n}}\, \underline{\underline{z}}\, R \leq \underline{\underline{z}}\, \underline{\underline{n}}\, R = \underline{\underline{z}}\, R\, \underline{\underline{n}}\, R.$$

Thus by Lemma 2.6 $\underline{\underline{n}}$ R has the AR property.

3.4 Lemma. Let J be a ring which contains the rational field Q and G a polycyclic-by-finite group such that $\bigcap_{n=1}^{\infty} \underline{\underline{g}}^n = \{r \in JG : r(1-a) = 0$ for some a in $\underline{\underline{g}}\}$. Then G is finite-by-nilpotent.

Proof. For each positive integer n define

$$D_n = \{x \in G : x-1 \in \underline{\underline{g}}^n\}.$$

Then D_n is a normal subgroup of G and

$$[D_n, G] \leq D_{n+1} \tag{6}$$

for all $n \geq 1$. Here $[D_n, G]$ denotes the subgroup generated by all commutators

$$[x,y] = x^{-1}y^{-1}xy$$

with x in D_n, y in G. To see why (6) holds observe that

$$\begin{aligned}[x,y] - 1 &= x^{-1}y^{-1}(xy-yx)\\ &= x^{-1}y^{-1}\{(x-1)(y-1) - (y-1)(x-1)\}\\ &\in \underline{\underline{g}}^{n+1}\end{aligned}$$

provided $x \in D_n$, $y \in G$. Consider the chain

$$G = D_1 \geq D_2 \geq D_3 \geq \dots .$$

Since h(G) is finite there exists a positive integer m such that D_m/D_{m+1} is a torsion group. Let $x \in D_m$. There exists $k \geq 1$ such that $x^k \in D_{m+1}$ or $(x^{-1})^k \in D_{m+1}$. Suppose $x^k \in D_{m+1}$. Then

$$(x^{k-1} + x^{k-2} + \dots + x+1)(x-1) = x^k - 1 \in \underline{\underline{g}}^{m+1}$$

and since $x - 1 \in \underline{\underline{g}}^m$ we have

$$k(x-1) \in \underline{\underline{g}}^{m+1}.$$

Thus $x - 1 \in \underline{\underline{g}}^{m+1}$ and $x \in D_{m+1}$. It follows that $D_m = D_{m+1}$. We can suppose that $D_m = D_{m+1} = \dots$. Let $y \in D_m$. Then

$$y - 1 \in \bigcap_{n=1}^{\infty} \underline{\underline{g}}^n$$

and so $(y-1)(1-a) = 0$ for some $a \in \underline{\underline{g}}$. It follows that y has finite order. Thus D_m is torsion group and hence is finite. By (6) G/D_m is nilpotent.

<u>3.5 Theorem.</u> Let K be a field of characteristic zero and G a polycyclic-by-finite group. Then the following statement are equivalent.

(i) G is finite-by-nilpotent.

(ii) $\underline{\underline{g}}$ has the AR property.

(iii) Every ideal of KG has the AR property.

(iv) $\underline{\underline{g}}$ is polycentral.

(v) Every ideal of KG is polycentral.

Proof. (iii) $\Rightarrow$ (ii), (v) $\Rightarrow$ (iv) are trivial. (ii) $\Rightarrow$ (i) follows by Lemma 3.4 and Theorem 2.2. (v) $\Rightarrow$ (iii), (iv) $\Rightarrow$ (ii) follow by Corollaries 2.8 and 3.2. Finally (i) $\Rightarrow$ (v) by (23).

Theorem 3.3 is still true if K is replaced by the ring Z of rational integers provided (i) is replaced by

(i)′ G is nilpotent.

For a group G, and prime p a subgroup H is a <u>p′-group</u> if every element has finite order coprime to p. By $O_{p'}(G)$ we shall mean the intersection of all normal subgroups N such that G/N has no non-trivial normal p′-subgroup.

<u>3.6 Theorem.</u> Let K be a field of characteristic p > 0 and G a polycyclic-by-finite group. Then the following statements are equivalent.

(i) $G/O_{p'}(G)$ is an extension of a nilpotent group by a finite p-group.

(ii) $\underline{\underline{g}}$ is polycentral.

(iii) Every ideal of KG is polycentral.

Proof. See (23).

The argument of Lemma 3.4 shows that is J is a ring of characteristic p > 0 and G a finite group such that

$$\bigcap_{n=1}^{\infty} \underline{\underline{g}}^n = \{r \in JG : r(1-a) = 0 \quad \text{for some } a \in \underline{\underline{g}}\}$$

then G is an extension of a p′-group by a p-group. A group G is called p-<u>nilpotent</u> (p a prime) if every finite homomorphic image is an extension of a p′-group by a p-group.

<u>3.5 Theorem.</u> Let K be a field of characteristic p > 0 and G a polycyclic-by-finite group. Then the following statements are equivalent.

(i) G is p-nilpotent.

(ii) $\underline{\underline{g}}$ has the AR property.

(iii) Every ideal of KG has the AR property.

Proof. (i) $\Leftrightarrow$ (ii) follows by (22), (i) $\Rightarrow$ (iii) by (24).

Note that for any polycyclic-by-finite group G, there exists a p-nilpotent normal subgroup N of finite index in G. Of course, for fields K and polycyclic-by-finite groups G, Theorem 3.1 gives that the AR and fAR properties coincide. For other groups the situation is rather different.

3.6 Theorem. Let K be a field of characteristic $p \geq 0$ and G an Abelian group. Then a necessary and sufficient condition for $\underline{g}$ to have the AR property is that either

(i) $p = 0$ and G is an extension of a finitely generated group by a torsion group, or

(ii) $p > 0$ and G is an extension of a finitely generated group by a p'-group.

This theorem can be contrasted with

3.7 Theorem. Let K be a field of characteristic $p \geq 0$ and G an Abelian group. Then a necessary and sufficient condition for $\underline{g}$ to have the fAR property is that either

(i) $p = 0$, or

(ii) $p > 0$ and for every finitely generated subgroup N of G the group G/N has no p-elements of infinite p-height.

The proofs of Theorems 3.6 and 3.7 can be found in (27) and (28), respectively. By a p-element we mean an element with finite order a power of p. An element y has infinite p-height if

$$y \in \bigcap_{n=1}^{\infty} G^{p^n}$$

where $G^{p^n} = \{x^{p^n} : x \in G\}$. Finally we note the following result.

3.8 Theorem. Let K be a field and G any group. Then $\underline{g}$ has the fAR property if and only if $\underline{g}$ has the finite intersection property.

Proof. See (28).

4. Localization

We have seen that if an ideal I of a ring R has the AR property then R satisfies the right Ore condition with respect to T where $T = \{1-a : a \in I\}$. Recall that C(I) is the set of elements c in R such that whenever $r \in R$, $cr \in I$ or $rc \in I$ implies $r \in I$. We are interested in conditions under which R satisfies the right Ore condition with respect to C(I).

4.1 Lemma. Let I be an ideal of a ring R and J an ideal such that $J \leq I$ and J has the AR property. Then R satisfies the right Ore condition with respect to C(I) if and only if R/J^n satisfies the right Ore condition with respect to $C(I/J^n)$ for all positive integers n.

Proof. The necessity is clear. Conversely, suppose that R/J^n satisfies the right Ore condition with respect to $C(I/J^n)$ for all $n \geq 1$. Let $r \in R$, $c \in C(I)$. There exists $k \geq 1$ such that

$$(cR + cR) \cap J^k \leq (rR + cR)J.$$

But there exist $r' \in R$, $c' \in C(I)$ such that $rc' - cr' \in J^k$ and so

$$rc' - cr' = ra + cb$$

for some $a,b \in J$. Then

$$r(c' - a) = c(r' + b)$$

and $c' - a \in C(I)$. It follows that R satisfies the right Ore condition with respect to C(I).

4.2 Lemma. Let I be an ideal of a right Noetherian ring R and a a central element of R such that $a \in I$. If R/Ra satisfies the right Ore condition with respect to $C(I/Ra)$ then R satisfies the right Ore condition with respect to $C(I)$.

Proof. Let $r \in R$, $c \in C(I)$. Then there exist $r_1 \in R$, $c_1 \in C(I)$ such that

$$rc_1 - cr_1 \in J$$

where $J = Ra$. Let k be a positive integer and suppose

$$rc_k - cr_k \in J^k$$

for some $r_k \in R$, $c_k \in C(I)$. Suppose $s \in R$ satisfies

$$rc_k - cr_k = sa^k.$$

There exist $s' \in R$, $c' \in C(I)$ such that

$$sc' - cs' = ta$$

for some $t \in R$. Then

$$rc_k c' - cr_k c' = sa^k c' = sc'a^k = (cs' + ta)a^k$$

and so

$$rc_{k+1} - cr_{k+1} = ta^{k+1} \in J^{k+1}$$

where $c_{k+1} = c_k c' \in C(I)$, $r_{k+1} = r_k c' + s'a^k$. The result follows by Corollary 2.8 and Lemma 4.1.

4.3 Theorem. Let Q be a polycentral semiprime ideal of a right Noetherian ring R. Then R satisfies the right Ore condition with respect to $C(Q)$.

Proof. Let $c_0 = 0, c_1, \ldots, c_n$ be a finite set of elements in Q such that

$$rc_i - c_i r \in Rc_0 + \ldots + Rc_{i-1} \quad (r \in R,\ 1 \leq i \leq n).$$

The result is proved by induction on n. If $n = 0$ apply Goldie's Theorem

(9, Theorem 4.1). Suppose $n > 1$. Then c_1 is a central element and by induction R/Rc_1 satisfies the right Ore condition with respect to $C(Q/Rc_1)$. By Lemma 4.2 R satisfies the right Ore condition with respect to $C(Q)$.

4.4 Lemma. Let R be a right Noetherian ring such that every prime ideal has the AR property. Then every ideal of R has the AR property.

Proof. Suppose the result is false and let I be an ideal chosen maximal with respect to not having the AR property. Then I is not prime and hence there exist ideals A, B, each property containing I, such that $AB \leqslant I$. Let E be a right ideal of R. By the choice of I both A and B have the AR property. Thus there exists $n \geqslant 1$ such that

$$E \cap A^n \leqslant EA$$

and there exists $m \geqslant 1$ such that

$$EA \cap B^m \leqslant (EA)B = E(AB) \leqslant EI.$$

Let $k = \max\{m,n\}$. Then $E \cap I^k \leqslant EI$. It follows that I has the AR property. This contradiction proves the result.

4.5 Example. Let K be a field and $K[[x]]$ the ring of formal power series in an indeterminate x. Let R be the subring of the ring of 2×2 upper triangular matrices over $K[[x]]$ consisting of all matrices of the form

$$\begin{bmatrix} f(0) & g(x) \\ 0 & f(x) \end{bmatrix}$$

with $f(x), g(x) \in K[[x]]$. Then R is right (but not left) Noetherian and has only two prime ideals $M > P$ where

$$M = \left\{ \begin{bmatrix} 0 & g(x) \\ 0 & xf(x) \end{bmatrix} : f(x), g(x) \in K[[x]] \right\}$$

and

$$P = \left\{ \begin{bmatrix} 0 & g(x) \\ 0 & 0 \end{bmatrix} : g(x) \in K[[x]] \right\}.$$

Note that $R/M \cong K$ and $P^2 = 0$. Then every ideal of R has the AR property but R does not satisfy the right Ore condition with respect to C(P).

To check that M has the AR property let E be a right ideal of R. Without loss of generality we can suppose that $E \leqslant M$. Let $S = K[[x]]$ and N the submodule of $S \oplus S$ defined by

$$(g(x),\ h(x)) \in N \quad \text{if and only if} \quad \begin{bmatrix} 0 & g(x) \\ 0 & h(x) \end{bmatrix} \in E.$$

Then N is an S-submodule of $S \oplus S$. But S is a Noetherian ring and hence

$$N \cap (S \oplus S)x^t \leqslant Nx$$

for some $t \geqslant 1$ (Theorem 1.1). Then

$$E \cap M^{t+1} \leqslant EM.$$

Thus M has the AR property.

Let

$$r = \begin{bmatrix} 0 & 1 \\ 0 & 0 \end{bmatrix} \quad \text{and} \quad c = \begin{bmatrix} 0 & 1 \\ 0 & x \end{bmatrix}.$$

Then $c \in C(P)$ and

$$P = \{x \in R : rx \in cR\}.$$

Thus R does not satisfy the right Ore condition with respect to C(P).

The above example is essentially due to A.W. Chatters (5). If I is an ideal of a ring R such that I has the AR property then to check whether R satisfies the right Ore condition with respect to C(I) it can be supposed that I is nilpotent (Lemma 4.1).

Question 3. Let R be a (right and left) Noetherian ring and N the maximal nilpotent ideal of R. Does R satisfy the right Ore condition with respect to C(N)?

In Question 3 we can suppose without loss of generality that $N^2 = 0$ because of the following result of Cozzens and Sandomierski (8, Theorem 2.4).

4.6 Theorem. Let Q be a semiprime ideal of a right Noetherian ring R and I an ideal such that $I \leq Q$ and R/I^2 satisfies the right Ore condition with respect to $C(Q/I^2)$. Then R/I^n satisfies the right Ore condition with respect to $C(Q/I^n)$ for all positive integers n.

4.7 Corollary. Let Q be a semiprime ideal of a right Noetherian ring R and I an ideal such that $I \leq Q$, I has the AR property and R/I^2 satisfies the right Ore condition with respect to $C(Q/I^2)$. Then R satisfies the right Ore condition with respect to C(Q).

Proof. By Lemma 4.1 and Theorem 4.6.

Let R be a right Noetherian ring and N the maximal nilpotent ideal of R. Note that N is a semiprime ideal of R Define

$$L = \{r \in R : cr \in N^2 \text{ for some } c \text{ in } C(N)\}$$

and

$$K = \{r \in R : rc \in N^2 \text{ for some } c \text{ in } C(N)\}.$$

Then K is an ideal of R. For suppose $r, r_1, r_2 \in K$ and $x \in R$. There exist $c_1, c_2, c \in C(N)$ such that

$$r_1c_1 \equiv r_2c_2 \equiv rc \equiv 0 \pmod{N^2}.$$

There exist $s_1, s \in R$ and $d_1, d_2 \in C(N)$ such that

$$c_1d_1 - c_2s_1 \in N \quad \text{and} \quad xd_2 - cs \in N$$

by Goldie's Theorem (9, Theorem 4.1). But $r_i \in N (i = 1,2)$, $r \in N$ and hence

$$(r_1 - r_2)c_1d_1 \in N^2 \quad \text{and} \quad rxd_2 \in N^2.$$

Thus $r_1 - r_2 \in K$ and $rx \in K$. It follows that K is an ideal of R.

For any element a in R denote $\{r \in R : ar = 0\}$ by r(a) and define $C'(I) = \{a \in R : r(a+I) = 0\} = \{a \in R : ar \in I, r \in R \Rightarrow r \in I\}$, for any ideal I.

<u>4.8 Lemma.</u> For all r in R and c in $C'(0)$ there exist s in R and d in C(N) such that rd = cs.

Proof. See (6, p.40 Theorem 2.3(b)).

The next result is an extension of (6, p.145, Theorem 11.11).

<u>4.9 Theorem.</u> With the above notation, R satisfies the right Ore condition with respect to C(N) if and only if $L \leq K$.

Proof. Suppose R satisfies the right Ore condition with respect to C(N). Without loss of generality we can suppose that $N^2 = 0$. Suppose $a \in L$. Then there exists c in C(N) such that ca = 0. The chain of right ideals $r(c) \leq r(c^2) \leq \ldots$ terminates and there exists a positive integer t such that $r(c^t) = r(c^{t+1})$. There exist $s \in R$, $d \in C(N)$ such that $c^t s = ad$. But ca = 0 implies $c^{t+1} s = 0$, so $c^t s = 0$ and hence ad = 0. Thus $a \in K$ and if follows that $L \leq K$

Conversely, suppose $L \leq K$. By Corollary 4.7 we can suppose $N^2 = 0$. Let $c \in C(N)$. Suppose $a \in R$ and $ca \in K$. Then cad = 0 for some $d \in C(N)$. Thus $ad \in L \leq K$ and it follows that $a \in K$. Thus $C(N) \leq C'(K)$. By Lemma 4.8 R/K satisfies the right Ore condition with respect to C(N/K) and hence R satisfies the right Ore condition with respect to C(N).

Note that in Example 4.5, L = P and K = 0.

<u>4.10 Corollary</u> (26, Proposition 3.4). Let R be a Noetherian ring such that every ideal has the right and left AR properties and let Q be a semiprime ideal of R. Then R satisfies the right and left Ore conditions

with respect to C(Q).

Proof. By Corollary 4.7 we can suppose without loss of generality that $Q^2 = 0$. Let $L = \{r \in R : cr = 0$ for some r in $R\}$. Since R/Q satisfies the left Ore condition with respect to C(Q/Q) (9, Theorem 4.1) it follows that L is an ideal of R. Choose $c \in C(Q)$ such that r(c) is maximal in $\{r(y) : y \in C(Q)\}$. Let $a \in L$. Then there exists $d \in C(Q)$ such that $da = 0$. Hence $a \in Q$. Since R/Q satisfies the left Ore condition with respect to C(Q/Q) it follows that there exist $c_1 \in R$, $d_1 \in C(Q)$ such that $c_1 d - d_1 c \in Q$. Then $d_1 ca = 0$. But $r(c) \leq r(d_1 c)$ implies $r(c) = r(d_1 c)$ by the choice of c. Hence $ca = 0$. It follows that $cL = 0$. Let $I = RcR$, so $IL = 0$. There exists $t \geq 1$ such that $L \cap I^t \leq IL = 0$ and hence $LI^t \leq IL = 0$. But $c^t \in I^t \cap C(Q)$ and we conclude $L \leq K$. By the theorem R satisfies the right Ore condition with respect to C(Q). Similarly for the left Ore condition.

Let R be a semiprime right Noetherian ring and M a right R-module. Define

$$T = \{m \in M : mc = 0 \text{ for some } c \text{ in } C(0)\}.$$

Then T is a submodule (called the <u>torsion submodule</u>) of M. We call M <u>torsion</u> if M = T and <u>torsion-free</u> if T = 0. Question 3 is equivalent to (see (30, Corollary 8)).

<u>Question 4.</u> Let R be a semiprime Noetherian ring. Does there exist a non-zero R-bimodule M such that M_R is finitely generated torsion-free but $_RM$ is finitely generated torsion?

For, suppose no such R-bimodule M exists. Let R be a Noetherian ring with maximal nilpotent ideal N. We show R satisfies the right Ore condition with respect to C(N). By Corollary 4.7 we can suppose without loss of generality that $N^2 = 0$. Define $K = \{r \in R : rc = 0$ for some $c \in C(N)\}$. Then as before K is an ideal of R and N/K is a torsion-free right R-module.

By hypothesis, N/K is a torsion-free left R-module. By Theorem 4.9 R satisfies the right Ore condition with respect to C(N).

Now suppose that there exists a semiprime Noetherian ring R and an R-bimodule M such that M_R is finitely generated torsion-free and $_RM$ is finitely generated torsion. Since M_R is finitely generated and $_RM$ is torsion there exists an ideal A of R such that AM = 0 and $A \cap C(0) \neq \emptyset$. Define $\overline{R} = R/A$ and denote r + A by $\overline{r}$ ($r \in R$). Let S be the ring of all "matrices"

$$\begin{bmatrix} \overline{r} & m \\ 0 & r \end{bmatrix}$$

such that $r \in R$, $m \in M$. Then S is a right and left Noetherian ring. Let

$$Q = \left\{ \begin{bmatrix} 0 & m \\ 0 & 0 \end{bmatrix} : m \in M \right\}.$$

Then $S/Q \cong R$ and so Q is semiprime. Also $Q^2 = 0$.

Define

$$d = \begin{bmatrix} 0 & 0 \\ 0 & c \end{bmatrix} \in S \quad \text{and} \quad v = \begin{bmatrix} 0 & m \\ 0 & 0 \end{bmatrix} \in S$$

where $c \in A \cap C(0)$ and $m \in M$, $m \neq 0$. Then

$$V = \{s \in S : vs \in dS\} = \left\{ \begin{bmatrix} \overline{r} & m_1 \\ 0 & r \end{bmatrix} : m_1 \in M,\ r \in R,\ mr = 0 \right\}$$

and so $V \cap C(Q) = \emptyset$. Thus S does not satisfy the right Ore condition with respect to C(Q).

<u>4.11 Theorem</u> (17, Theorem 3.1). Let U be the enveloping algebra of a finite dimensional Lie algebra over the complex field. Let R be a homomorphic image of U and P a prime ideal of R with the AR property. Then R

satisfies the right Ore condition with respect to C(P).

Theorem 4.11 is a consequence of a theorem of Joseph and Small (14, Theorem 2.7) giving a sufficient condition in terms of Gelfand-Kirillov dimension for a homomorphic image of U to have an Artinian classical quotient ring. A similar result to Joseph and Small's but this time in terms of Krull dimension was proved by Krause, Lenagan and Stafford (16, Theorem 8). A consequence of their result is:

4.12 Theorem (2, Theorem 3.8 with 3, Theorems 2.4 and 3.13). Let J be a commutative Noetherian ring of non-zero characteristic or the ring of rational integers and G a polycyclic-by-finite group. Let P be a prime ideal of a homomorphic image R of the group ring JG. Then R satisfies the right Ore condition with respect to C(P) provided P has the AR property.

Localization and the existence of classical quotient rings are rather closely related. This can be seen not only from the proofs of Theorems 4.11 and 4.12 but also from the following example of J.T. Stafford. Let R be a Noetherian prime ring and Q a semiprime ideal of R. Let T be the ring of all "matrices"

$$\begin{bmatrix} a & b \\ 0 & c \end{bmatrix}$$

where $a,b \in R$, $b \in R/Q$. Then T is a Noetherian ring. Moreover, T has a classical right quotient ring (i.e. T satisfies the right Ore condition with respect to C(0)) if and only if R satisfies the right Ore condition with respect to C(Q).

In addition Small and Stafford (25, Corollary 2.3) have proved that if R is a Noetherian ring then there exists a finite collection of prime ideals $P_i (1 \leq i \leq n)$ such that

$$C(0) = C(P_1) \cap \ldots \cap C(P_n).$$

(For a nice proof of this fact see (17).) If each of the prime ideals $P_i (1 \leq i \leq n)$ are such that R satisfies the right Ore condition with respect to $C(P_i)$ then R has a classical right quotient ring (see (26, Theorem 5.2)).

For a recent survey on localization see (13).

5. Homological dimension

Let R be a ring, I an ideal of R and M a right R-module. We say that M is <u>I-torsion</u> if for all m in M there exists $t \geq 1$ such that $mI^t = 0$. For any right R-module M we denote the projective dimension of M by pd M and the injective dimension of M by id M. The right global dimension of R is denoted by rgℓd R.

5.1 Lemma (1, Lemma 2). Let I be an ideal of a right Noetherian ring R such that I has the AR property and let M be an I-torsion right R-module. Then id M = $\inf\{n : \mathrm{Ext}_R^{n+1}(X,M) = 0$ for all cyclic right R/I-modules X$\}$.

Proof. Suppose $\mathrm{Ext}_R^{n+1}(X,M) = 0$ for all cyclic right R/I-modules X. We prove id M $\leq$ n by induction on n. Suppose n = 0. Let E denote the injective hull of M. Suppose M $\neq$ E. Let $e \in E$, $e \notin M$. Then $N = eR \cap M$ is Noetherian and I-torsion and hence $NI^t = 0$ for some $t \geq 1$. There exists $s \geq 1$ such that $N \cap (eR)I^s \leq NI^t = 0$ and hence $(eR)I^s = 0$ because M is an essential submodule of E. Thus there exists a submodule V of E properly containing M such that V/M is a cyclic right R/I-module (i.e. VI $\leq$ M). Consider the exact sequence

$$0 \to M \to V \to V/M \to 0. \tag{7}$$

Since M is essential in E it follows that (7) does not split and hence $\mathrm{Ext}^1(V/M,M) \neq 0$, a contradiction. Thus M = E and hence id M = 0. Now suppose $n \geq 1$. Consider the exact sequence

$$0 \to M \to E \to E/M \to 0. \qquad (8)$$

For all cyclic right R/I-modules X,

$$\mathrm{Ext}^n(X,E/M) \cong \mathrm{Ext}^{n+1}(X,M),$$

and so $\mathrm{Ext}^n(X,E/M) = 0$. Since E/M is I-torsion it follows that id (E/M) $\leqslant$ n-1 and hence by (8) id M $\leqslant$ n.

5.2 Lemma (1, Lemma 3). Let J be the Jacobson radical of a right Noetherian ring R and let A be a finitely generated right R-module. Then pd A = inf{n : $\mathrm{Ext}^{n+1}(A,M) = 0$ for all finitely generated right R/J-modules M}.

Proof. Suppose $\mathrm{Ext}^{n+1}(A,M) = 0$ for all finitely generated right R/J-modules M. We prove by induction on n that pd A $\leqslant$ n. Suppose first that n = 0. Let

$$0 \to K \xrightarrow{\phi} F \to A \to 0 \qquad (9)$$

be an exact sequence of finitely generated right R-modules with F free. By assumption $\mathrm{Ext}^1(A,K/KJ) = 0$ and so the induced map $\phi^* : \mathrm{Hom}(F,K/KJ) \to \mathrm{Hom}(K,K/KJ)$ is onto. Let $\alpha : K \to K/KJ$ be the canonical epimorphism. There exists $\beta : F \to K/KJ$ such that $\alpha = \beta\phi$. Since F is projective there exists $\gamma : F \to K$ such that $\beta = \alpha\gamma$. Let $\varepsilon = \gamma\phi : K \to K$. Then $\alpha\varepsilon = \alpha$ and so $K = \varepsilon(K) + KJ$. By Nakayama's Lemma, $\varepsilon(K) = K$ and, because K is Noetherian, ε is an isomorphism. Thus (9) splits and A is projective.

In general if n $\geqslant$ 1 then (9) gives $\mathrm{Ext}^n(K,M) = 0$ for all finitely generated right R/J-modules M and hence by induction on n, pd K $\leqslant$ n - 1. Thus pd A $\leqslant$ n, as required.

5.3 Theorem. Let I be an ideal of a right Noetherian ring R such that I has the AR property and I is contained in the Jacobson radical of R. Then

$rg\ell d\ R = \sup\{pd\ X : X$ is a cyclic right R/I-module$\}$.

Proof. It is sufficient to suppose the right hand side equals some integer $n \geq 0$ and prove $rg\ell d\ R \leq n$. But in this case, id $M \leq n$ for all right R/I-modules M (Lemma 5.1) and so $rg\ell d\ R \leq n$ (Lemma 5.2).

We apply these results to right Noetherian local rings. A ring R with Jacobson radical J is local provided R/J is a simple Artinian ring. In this case R has (up to isomorphism) a unique irreducible right R-module U(say).

5.4 Corollary. Let R be a right Noetherian local ring with Jacobson radical J and suppose J has the AR property. Then

(i) ((1) Corollary) $rg\ell d\ R = pd(R/J) = pd\ U$ where U is the unique irreducible right R-module,

(ii) $pd\ A = \inf\{n : Ext^{n+1}(A,U) = 0\}$ for any finitely generated right R-module A.

5.5 Theorem. (4, Lemma 4.1). Let R be a right Noetherian local ring with Jacobson radical J and suppose J has the AR property. Let A be a right R-module with non-zero socle. If R has finite right global dimension then $pd\ A = rg\ell d\ R$.

Proof. Suppose $rg\ell d\ R = n < \infty$. Let U be an irreducible submodule of A. Then $pd\ U = n$ (Corollary 5.4) and, because

$$0 \to U \to M \to M/U \to 0$$

is exact, $pd\ M = n$ (see 15, p.169 Theorem 2)).

If R is a commutative Noetherian local ring of finite global dimension then the converse of Theorem 5.5 is true for finitely generated modules (see (18, p.209 Theorem 27)). Under these conditions, a finitely generated module A has non-zero socle if and only if there is no monomorphism $A \to AJ$. For, suppose there is no monomorphism $A \to AJ$. Then for all x in J there exists $0 \neq a \in A$ such that $ax = 0$. Now apply (18, p.204 Proposition 6*). Conversely, suppose there exists a monomorphism $\theta: A \to AJ$. There exists a positive integer n such that $S \cap AJ^n \leq SJ = 0$ where S is the socle of A. But $\theta^n(S) \leq AJ^n \cap S$ and so $\theta^n(S) = 0$ which gives $S = 0$.

5.6 Theorem (4, Corollary 4.7). Let R be a right Noetherian ring of finite right global dimension $n \geq 1$ and Jacobson radical J such that J has the AR property. Let A be a finitely generated right R-module such that A embeds in AJ. Then pd $A \leq n-1$.

To prove this theorem we require the following lemma. Let A be a finitely generated right R-module and

$$\cdots \to P_n \xrightarrow{d_n} P_{n-1} \xrightarrow{d_{n-1}} \cdots \to P_0 \xrightarrow{d_0} A \to 0 \qquad (10)$$

a projective resolution of A with each P_i finitely generated.

5.7 Lemma. Let R be a right Noetherian ring and I an ideal of R such that I has the AR property. Let A be a finitely generated right R-module such that there exists an R-homomorphism $\phi: A \to AI$. Let (10) be a projective resolution of A. Then for each $n \geq 0$ there exists a positive integer t such that ϕ^t can be lifted to a homomorphism $\theta: P_n \to P_n$ such that $\theta(P_n) \leq P_nI$.

Proof. By induction on n. Let $K = \ker d_0$. Consider the exact sequence

$$0 \to K \to P_0 \xrightarrow{d_0} A \to 0.$$

There exists a positive integer s such that $K \cap P_0I^s \leq KI$. Consider the diagram

$$\begin{array}{ccccccccc} 0 \to & K & \to & P_0 & \xrightarrow{d_0} & A & \to 0 \\ & \downarrow \chi & & \downarrow \chi & & \downarrow \phi^s & \\ 0 \to & K \cap P_0 I^s & \to & P_0 I^s & \xrightarrow{d_0} & AI^s & \to 0 \end{array}$$

There exists $\chi: P_0 \to P_0 I^s$ such that $\phi^s d_0 = d_0 \chi$. Also, $\chi: K \to K$ and $\chi(K) \leq KI$. Now consider the projective resolution

$$\ldots \to P_n \xrightarrow{d_n} P_{n-1} \to \ldots \to P_1 \xrightarrow{d_1} K \to 0.$$

By induction on n there exists t so χ^t can be lifted to $\theta: P_n \to P_n$ such that $\theta(P_n) \leq P_n I$. Thus, ϕ^{st} can be lifted to θ as required.

Proof of Theorem 5.6. Let $\phi: A \to AJ$ be a monomorphism. We know that $pd(A/\phi(A)) = k \leq n$. Suppose $k \geq 1$. For all $i \geq 1$, $\phi^i(A)/\phi^{i+1}(A) \cong A/\phi(A)$ and hence $pd(A/\phi^i(A)) = k$ for all $i \geq 1$ (see (15, p.169 Theorem 2)). Choose c such that $\phi^c: A \to AJ$ can be lifted to $\theta: P_k \to P_k$ where $\theta(P_k) \leq P_k J$ and (10) is a projective resolution for A with each P_i finitely generated. Thus we have the commutative diagram

$$\begin{array}{ccccc} \xrightarrow{d_{k+1}} P_k & \xrightarrow{d_k} & \ldots \to P_0 \to & A & \to 0 \\ \downarrow \theta & & & \downarrow \phi^c & \\ \xrightarrow{d_{k+1}} P_k & \xrightarrow{d_k} & \ldots \to P_0 \to & A & \to 0. \end{array}$$

Let M be a finitely generated right R/J-module. The exact sequence

$$0 \to A \xrightarrow{\phi^c} A \to A/\phi^c(A) \to 0$$

gives the exact sequence

$$\mathrm{Ext}^k(A,M) \xrightarrow{\theta^*} \mathrm{Ext}^k(A,M) \to \mathrm{Ext}^{k+1}(A/\phi^c(A),M) = 0.$$

Thus θ^* is onto. We have

$$\begin{array}{ccc} \xleftarrow{d^*_{k+1}} & \mathrm{Hom}(P_k,M) & \xleftarrow{d^*_k} \\ & \uparrow \theta^* & \\ \xleftarrow{d^*_{k+1}} & \mathrm{Hom}(P_k,M) & \xleftarrow{d^*_k} \end{array}$$

Let $f \in \ker d^*_{k+1}$. Then

$$f + \mathrm{im}\, d^*_k = \theta^*(g + \mathrm{im}\, d^*_k)$$

for some $g \in \ker d^*_{k+1}$. Thus

$$f = g\theta + hd_k$$

for some $h \in \mathrm{Hom}(P_{k-1},M)$. But $\theta(P_k) \leq P_kJ$ implies $g\theta(P_k) \leq g(P_k)J \leq MJ = 0$. Hence $f = hd_k$, and so $f \in \mathrm{im}\, d^*_k$. It follows that $\ker d^*_{k+1} = \mathrm{im}\, d^*_k$ and so $\mathrm{Ext}^k(A,M) = 0$. By Lemma 5.2, pd $A \leq k-1$.

If $k = 0$ then $A/\phi(A)$ is projective and hence $A = \phi(A) \oplus P$ for some projective submodule P. Thus $A \leq P + AJ$ and $A = P$ by Nakayama's Lemma.

5.8 Theorem. Let R be a right and left Noetherian ring such that every ideal has the right and left AR properties. Let P be a prime ideal of R. Then rank $P \leq \mathrm{pd}(R/P)$.

Recall that rank P is the greatest integer $n \geq 0$ such that there exists a chain of prime ideals

$$P = P_0 > P_1 > \dots > P_n$$

if n exists, or ∞.

Proof. Suppose the right R-module R/P has finite projective dimension n. We prove the result by induction on n. If $n = 0$ then R/P is projective and hence $P = eR$ for some idempotent e. If Q is a prime ideal of R such that $Q < P$ then $(1-e)P = 0 \leq Q$ and so $1-e \in Q \leq P$, a contradiction. Thus, rank $P = 0$. Suppose the result holds for all $0 \leq n < k$ and let $n = k$. Let Q be a prime ideal of R such that $Q < P$. By Corollary 4.10 R satisfies the right and left Ore conditions with respect to $C(P) = C$ and we form the ring $S = R_C$. By (29, p.57 §3), $pd_S(S/PS) \leq pd_R(R/P) = k$. Thus $rg\ell d(S/PS) \leq k$ (Corollary 5.4(i)). Also QS is a prime ideal of S and by (9, Theorem 3.9) $QS < PS$ implies there exists $c \in PS \cap C(QS)$. But $s + QS \to cs + QS$ $(s \in S)$ is a monomorphism from S/QS into (S/QS)PS. By Theorem 5.6, $pd_S(S/QS) < k$ and by induction rank Q = rank QS $< k$. Thus rank $P \leq k$.

6. Hypercentral ideals

An ideal I of a ring R is <u>hypercentral</u> if for all ideals $A < B \leq I$ there exists b in B but not A such that $b + A$ is central in R/A. We shall call a right R-module M a <u>(1-I)-torsion module</u> if for all m in M there exists a in I such that $m(1-a) = 0$.

<u>6.1 Theorem.</u> Let I be a hypercentral ideal of a ring R and M a Noetherian right R-module such that $M = MI$. Then M is a (1-I)-torsion module.

Proof. Suppose the result is false. Let N be a submodule of M maximal with respect to the property that M/N is not a (1-I)-torsion module. By passing to M/N we can suppose without loss of generality that $N = 0$. We can further suppose that M is faithful. Let c be a non-zero central element of R contained in I. Then $Mc \neq 0$ and so M/Mc is (1-I)-torsion. Let $m \in M$. There exists a_1 in I such that $m(1-a_1) = m_1c$ for some m_1 in M. Similarly there exists b in I such that $m_1(1-b) \in Mc$, whence $m(1-a_2) \in Mc^2$ where

$1-a_2 = (1-a_1)(1-b)$ and so $a_2 \in I$. Similarly for each positive integer t there exists a_t in I such that $m(1-a_t) \in Mc^t$. By Corollary 2.8 there exists $k \geqslant 1$ such that

$$mR \cap Mc^k \leqslant mcR.$$

Thus $m(1-a_k) = mcr$ for some r in R and hence $m(1-a_k-cr) = 0$ where $a_k + cr \in I$. It follows that M is (1-I)-torsion, a contradiction. The result follows.

Let I be an ideal of a ring R and M a right R-module. Define a descending chain

$$M = M_0 \geqslant M_1 \geqslant \dots \geqslant M_\alpha \geqslant M_{\alpha+1} \geqslant \dots$$

where

$$M_{\alpha+1} = M_\alpha I$$

for all ordinals $\alpha \geqslant 0$ and

$$M_\beta = \bigcap_{0 \leqslant \alpha < \beta} M_\alpha$$

if β is a limit ordinal. There exists an ordinal γ such that $M_\gamma = M_{\gamma+1}$ and we define

$$\kappa_I(M) = M_\gamma.$$

6.2 Corollary. Let I be a hypercentral ideal of a ring R and M a Noetherian right R-module. Then $\kappa_I(M) = \{m \in M : m(1-a) = 0 \text{ for some } a \text{ in } I\}$.

Proof. By the theorem.

Robinson (21) gives an example of a ring R, an ideal J which is a sum of polycentral ideals and so is hypercentral and for which the ring R/J is commutative, an element c of R and a Noetherian right R-module M such that if I = J + cR then

$$M_\omega = \bigcap_{n=1}^{\infty} MI^n = \{m \in M : mI = 0\} \neq 0.$$

Note that I is hypercentral and if $N = \{m \in M : mI = 0\}$ then

$$N \cap MI^n = N \nleq NI$$

for all $n \geq 1$.

6.3 Theorem (21, Lemma 2). Any ideal which is a sum of polycentral ideals has the nAR property.

Proof. Let R be a ring, Λ an index set, I_λ a polycentral ideal of $R(\lambda \in \Lambda)$ and $I = \Sigma_\Lambda I_\lambda$. Let M be a Noetherian right R-module and N an essential submodule of M such that $NI = 0$. We shall show that $MI^k = 0$ for some positive integer k. There exists a finite subset Λ' of Λ such that

$$MI = \Sigma_\Lambda MI_\lambda = \Sigma_{\Lambda'} MI_\lambda.$$

Let $J = \Sigma_{\Lambda'} I_\lambda$. Then J is a polycentral ideal of R and $MI = MJ$. Suppose $J = c_1R + \dots + c_nR$ where c_1 is central in R and c_i central modulo $c_1R + \dots + c_{i-1}R$ $(2 \leq i \leq n)$. We prove the result by induction on n.

Suppose first that $n = 1$. Let $c = c_1$. Then $MI = Mc$ and hence

$$MI^2 = (MI)I = (Mc)I = (MI)c = Mc^2$$

and for any positive integer t

$$MI^t = Mc^t.$$

By Corollary 2.8 Rc has the nAR property and hence $N \cap Mc^t \leq Nc = 0$ for some $t \geq 1$. Thus $MI^t = 0$.

Now suppose $n > 1$. By Corollary 2.8 $Mc_1{}^q = 0$ for some $q \geq 1$. If $q = 1$ then $MI = Mc_2 + \dots + Mc_n$ and the result follows by induction on n. Suppose $q > 1$. Let $V = \{m \in M : mc_1{}^{q-1} = 0\}$. By induction on q, $VI^s = 0$ for some $s \geq 1$. Then $MI^s c_1 = Mc_1 I^s \leq VI^s = 0$ so that $MI^s \leq V$ and hence $MI^{2s} = 0$. The result follows.

References

1. M. Boratynski, A change of rings theorem and the Artin-Rees property, Proc. Amer. Math. Soc. 53(1975), 307-310.

2. K.A. Brown, T.H. Lenagan and J.T. Stafford, Weak ideal invariance and localisation, J. London Math. Soc. (2)21(1980), 53-61.

3. K.A. Brown, T.H. Lenagan and J.T. Stafford, K-theory and the stable structure of some Noetherian group rings, Proc. London Math. Soc. (3)42(1981), 193-230.

4. K.A. Brown, C.R. Hajarnavis and A.B. MacEacharn, Noetherian rings of finite global dimension, to appear in Proc. London Math. Soc.

5. A.W. Chatters, Three examples concerning the Ore condition in Noetherian rings, Proc. Edinburgh Math. Soc. 23(1980), 187-192.

6. A.W. Chatters and C.R. Hajarnavis, Rings with chain conditions (Pitman 1980).

7. I.S. Cohen, Rings with restricted minimum condition, Duke Math. J. 17(1950), 27-42.

8. J.H. Cozzens and F.L. Sandomierski, Localization at a semiprime ideal of a right Noetherian ring, Comm. in Algebra 5(1977), 707-726.

9. A.W. Goldie, Semi-prime rings with maximum condition, Proc. London Math. Soc. (3)10(1960), 201-220.

10. A.W. Goldie, The structure of Noetherian rings, in Lectures on rings and modules, Lecture Notes in Mathematics (Springer) 246(1972), 213-321.

11. P. Hall, Finiteness conditions for soluble groups, Proc. London Math. Soc. (3)4(1954), 419-436.

12. I.N. Herstein, A counter-example in Noetherian rings, Proc. Nat. Acad. Sci. (U.S.A.) 54(1965), 1036-1037.

13. A.V. Jategaonkar, Localization in non-commutative Noetherian rings, preprint.

14. A. Joseph and L.W. Small, An additivity principle for Goldie rank, Israel J. Math. 31(1978), 105-114.

15. I. Kaplansky, Fields and rings (University of Chicago, 1969).

16. G. Krause, T.H. Lenagan and J.T. Stafford, Ideal invariance and Artinian quotient rings, J. Algebra 55(1978), 145-154.

17. T.H. Lenagan, Gelfand-Kirillov dimension in enveloping algebras, Quart. J. Math. Oxford (2)32(1981), 69-80.

18. D.G. Northcott, An introduction to homological algebra (Cambridge University Press 1962).

19. Y. Nouazé and P. Gabriel, Idéaux premiers de l'algèbre enveloppante d'une algèbre de Lie nilpotente, J. Algebra 6(1967), 77-99.

20. D. Rees, Two classical theorems of ideal theory, Proc. Cambridge Phil. Soc. 52(1956), 155-157.

21. D.J.S. Robinson, Hypercentral ideals, Noetherian modules and a theorem of Stroud, J. Algebra 32(1974), 234-239.

22. J.E. Roseblade, Applications of the Artin-Rees lemma to group rings, Symposia Math. 17(1976), 471-478.

23. J.E. Roseblade and P.F. Smith, A note on hypercentral group rings, J. London Math. Soc. (2)13(1976), 183-190.

24. J.E. Roseblade and P.F. Smith, A note on the Artin-Rees property of certain polycyclic group algebras, Bull. London Math. Soc. 11(1979), 184-185.

25. L.W. Small and J.T. Stafford, Regularity of zero divisors, preprint.

26. P.F. Smith, Localization and the AR property, Proc. London Math. Soc. (3)22(1971), 39-68.

27. P.F. Smith, The AR property and chain conditions in group rings, Israel J. Math. 32(1979), 131-144.

28. P.F. Smith, More on the AR property and chain conditions in group rings, Israel J. Math. 35(1980), 186-204.

29. B. Stenstrom, Rings of quotients (Springer-Verlag 1975).

30. R.B. Warfield Jr., Quotient rings and localization for Noetherian rings, preprint.

Department of Mathematics,
University of Glasgow,
GLASGOW,
G12 8QW.

ETUDE D'UNE CLASSE D'ALGEBRES ARTINIENNES LOCALES NON COMMUTATIVES

par J. Louis ROQUE

Soit k un corps commutatif et n un entier naturel supérieur ou égal à 2. Nous noterons $\mathcal{X}_n$ l'algèbre sur k des matrices (n,n) triangulaires supérieures ayant des coefficients diagonaux égaux et dont les éléments non situés sur la première ligne et la dernière colonne sont nuls. Autrement dit une matrice $A \in \mathcal{X}_n$ est de la forme

$$A = \begin{pmatrix} \alpha & * & * & \cdots & * \\ & \alpha & 0 & & * \\ & & \ddots & & \vdots \\ & 0 & & \ddots & * \\ & & & & \alpha \end{pmatrix}$$

Nous prendrons pour base de $\mathcal{X}_n$ les matrices suivantes.

$$\varepsilon_1 = I_n = \sum_{i=1}^{n} E_{ii} \ , \ \varepsilon_2 = E_{12}, \ldots, \varepsilon_n = E_{1n} \ , \ \varepsilon_{n+1} = E_{2n}, \ldots, \varepsilon_{2(n-1)} = E_{n-1n}$$

où les E_{ij} sont les unités matricielles classiques. On a en particulier $\dim_k \mathcal{X}_n = 2(n-1)$. Nous poserons $\nu = 2(n-1)$.

Calculons les constantes de structure de l'algèbre $\mathcal{X}_n$. Nous pouvons écrire quels que soient $i, j \in [1,\nu]$, $\varepsilon_i . \varepsilon_j = \sum_{k=1}^{\nu} e_{ijk} \varepsilon_k$ où les e_{ijk} sont les constantes de structures de l'algèbre $\mathcal{X}_n$. On obtient pour tout $i \in [1,\nu]$ $\varepsilon_1 . \varepsilon_i = \varepsilon_i . \varepsilon_1 = \varepsilon_i$; pour tous $i,j \in [2,n]$ $\varepsilon_i . \varepsilon_j = \varepsilon_j . \varepsilon_i = 0$; pour tous $i,j \in [n+1,\nu]$ $\varepsilon_i . \varepsilon_j = \varepsilon_j . \varepsilon_i = 0$; si $j \in [n+1,\nu]$ et $i \in [2,n-1]$ on a $\varepsilon_i . \varepsilon_j = \delta_j^{n+i-1} \varepsilon_n$ et $\varepsilon_j . \varepsilon_i = 0$ d'où : $e_{ij1} = \delta_{i1} . \delta_{j1}$. Puis, pour $k \neq n$, $e_{ijk} = \delta_{i1} . \delta_{jk} + \delta_{ik} . \delta_{j1}$. Enfin, pour $k = n$,

$$e_{ijn} = \delta_{i1}.\delta_{jn} + \delta_{in}.\delta_{j1} + \sum_{h=2}^{n-1} \delta_{i.h}\,\delta_{jn+h-1}\,.$$

<u>Proposition 1 : Pour tout</u> $n \geq 2$, <u>le centre de</u> $\mathcal{K}_n$ <u>est le sous-espace vectoriel de</u> $\mathcal{K}_n$ <u>engendré par</u> ε_1 <u>et</u> ε_n. <u>Le centre de</u> $\mathcal{K}_n$ <u>est donc de dimension 2. En particulier</u> $\mathcal{K}_2$ <u>est une k-algèbre commutative</u> (c'est l'algèbre des nombres duaux).

<u>Preuve</u> : Soit $A = \sum_{i=1}^{\nu} a_i\,\varepsilon_i$ un élément du centre de $\mathcal{K}_n$. On doit avoir pour tout $h \in [2,\nu]$ $\varepsilon_h.A = A.\varepsilon_h$; donc si $h \in [2,n-1]$, on doit avoir :

$$\varepsilon_h \sum_{i=1}^{\nu} a_i\,\varepsilon_i = a_1\,\varepsilon_h + a_{n+h-1}\,\varepsilon_n = (\sum_{i=1}^{\nu} a_i\,\varepsilon_i)\varepsilon_h = a_1\varepsilon_h$$

d'où $a_{n+h-1} = 0$ quel que soit $h \in [2,n-1]$; de la même manière, en prenant $h \in [n+1,\nu]$, on trouve que A se réduit à $a_1\varepsilon_1 + a_n\varepsilon_n$ ce qui achève la démonstration.

<u>Proposition 2 : L'anneau</u> $\mathcal{K}_n$ <u>est un anneau local dont l'idéal maximal est l'idéal</u> I <u>des matrices de</u> $\mathcal{K}_n$ <u>ayant le coefficient diagonal nul. De plus on a un isomorphisme</u> $\mathcal{K}_n/I \simeq k$. <u>L'algèbre</u> $\mathcal{K}_n$ <u>est donc une k-algèbre supplémentée. L'application d'augmentation de l'algèbre</u> $\mathcal{K}_n$ <u>est l'application</u> $\rho : \mathcal{K}_n \longrightarrow k$

$$\sum_{i=1}^{\nu} a_i\varepsilon_i \longmapsto a_1$$

1°) <u>Cohomologie de l'algèbre</u> $\mathcal{K}_n$

Nous noterons $H^p(\mathcal{K}_n,\mathcal{K}_n)$ le $p^{\text{ième}}$ groupe de cohomologie de l'algèbre $\mathcal{K}_n$ à valeur dans $\mathcal{K}_n$ elle même. Il s'agit ici de la cohomologie de Hochschild de l'algèbre associative $\mathcal{K}_n$ i.e. pour tout $p \geq 0$ $H^p(\mathcal{K}_n,\mathcal{K}_n) = \mathrm{Ext}^p_{\mathcal{K}_n^e}(\mathcal{K}_n,\mathcal{K}_n)$ où $\mathcal{K}_n^e$ désigne l'algèbre enveloppante de $\mathcal{K}_n$. Nous rappelons que ces groupes de cohomologie peuvent aussi se calculer comme quotient du groupe des cocycles par les groupes des cobords dans le complexe standard associé à $\mathcal{K}_n$. Nous noterons δ l'application de cobord. Nous montrons que les H^2 et H^3 sont non nuls ainsi que les H^0 et H^1, dont nous calculons explicitement la dimension.

1) <u>p = 0</u> : Ce cas est trivial car $H^0(\mathcal{K}_n,\mathcal{K}_n)$ est égal au centre de $\mathcal{K}_n$ donc

$$\dim_k H^0(\mathcal{K}_n,\mathcal{K}_n) = 2.$$

2) <u>p = 2</u> : Nous distinguerons deux cas suivant que $n = 2$ où $n > 2$.

i) <u>$n \neq 2$</u> : Considérons la 2-cochaine f définie par $f(\varepsilon_i,\varepsilon_j) = \delta_{i\nu}\delta_{j2}\,\varepsilon_n$. Je dis que f est un 2-cocycle ; en effet

$\delta f(\varepsilon_i,\varepsilon_j,\varepsilon_k) = \varepsilon_i f(\varepsilon_j,\varepsilon_k) - f(\varepsilon_i\varepsilon_j,\varepsilon_k) + f(\varepsilon_i,\varepsilon_j\varepsilon_k) - f(\varepsilon_i,\varepsilon_j)\varepsilon_k = \varepsilon_i\delta_{j\nu}\delta_{k2}\varepsilon_n$ $- \sum_{m=1}^{\nu} e_{ijm}\delta_{m\nu}\delta_{k2}\varepsilon_n + \sum_{m=1}^{\nu} e_{jkm}\delta_{i\nu}\delta_{m2}\varepsilon_n - (\delta_{i\nu}\delta_{j2})\varepsilon_n\varepsilon_k$ mais $\varepsilon_i\varepsilon_n = \delta_{i1}\varepsilon_n$ et $\varepsilon_n\varepsilon_k = \delta_{k1}\varepsilon_n$ donc $\delta f(\varepsilon_i,\varepsilon_j,\varepsilon_k) = \delta_{j\nu}\delta_{k2}\delta_{i1}\varepsilon_n - e_{ij\nu}\delta_{k2}\varepsilon_n + e_{jk2}\delta_{i\nu}\varepsilon_n$ $- \delta_{i\nu}\delta_{j2}\delta_{k1}\varepsilon_n$. Il suffit alors de remarquer que $e_{ij\nu} = \delta_{i1}\delta_{j\nu} + \delta_{i\nu}\delta_{j1}$ et que $e_{jk2} = \delta_{j2}\delta_{k1} + \delta_{j1}\delta_{k2}$ pour achever la démonstration.

Montrons alors que f n'est pas un 2-cobord. Supposons que $f(\varepsilon_i,\varepsilon_j) = \delta g(\varepsilon_i,\varepsilon_j)$ pour une cochaîne g définie par $g(\varepsilon_i) = \sum_{k=1}^{\nu} b_{ik}\varepsilon_k$. On a $\delta g(\varepsilon_i,\varepsilon_j) = \varepsilon_i g(\varepsilon_j) - g(\varepsilon_i\varepsilon_j) + g(\varepsilon_i)\varepsilon_j$

$= \sum_m \sum_k (e_{ikm} b_{jk} - e_{ijk} b_{km} + e_{kjm} b_{ik})\varepsilon_m$ donc en particulier pour $i = \nu$ et $j = 2$ on trouve que le coefficient de ε_n dans $\delta g(\varepsilon_\nu,\varepsilon_2)$ est nul alors qu'il vaut 1 dans $f(\varepsilon_\nu,\varepsilon_2)$, d'où la contradiction cherchée.

Nous venons donc de montrer, que pour $n \geqslant 2$, $H^2(\mathcal{X}_n,\mathcal{X}_n)$ est non nul.

ii) <u>$n = 2$</u> : Considérons la cochaine f définie par $f(\varepsilon_i,\varepsilon_j) = \delta_{i2}\delta_{j2}\varepsilon_1$. On vérifie de la même manière que f est un 2-cocyle qui n'est pas un 2-cobord, d'où la proposition :

<u>Proposition 3</u> : <u>On a</u> $\dim_k H^0(\mathcal{X}_n,\mathcal{X}_n) = 2$ <u>et pour tout entier</u> $n \geqslant 2$, <u>on a</u> :

$$H^2(\mathcal{X}_n,\mathcal{X}_n) \neq 0.$$

3) <u>$p = 3$</u> : Nous allons démontrer que la 3-cochaîne définie par $g(\varepsilon_i,\varepsilon_j,\varepsilon_k) = \delta_{in}\delta_{jn}\delta_{kn}\varepsilon_n$ est un 3-cocycle qui n'est pas un 3-cobord.

Calculons $\delta g(\varepsilon_i,\varepsilon_j,\varepsilon_k,\varepsilon_m) = \varepsilon_i(\delta_{jn}\delta_{kn}\delta_{mn})\varepsilon_n - e_{ijn}\delta_{kn}\delta_{mn}\varepsilon_n + \delta_{in}\, e_{jkn}\delta_{mn}\varepsilon_n$ $- \delta_{in}\delta_{jn}\, e_{kmn}\varepsilon_n + \delta_{in}\delta_{jn}\delta_{kn}\varepsilon_n\varepsilon_m$. En utilisant $\varepsilon_i\varepsilon_n = \delta_{i1}\varepsilon_n$ et $\varepsilon_n\varepsilon_m = \delta_{m1}\varepsilon_n$ et $e_{ijn} = \delta_{i1}\delta_{jn} + \delta_{in}\delta_{j1} + \sum_{h=2}^{n-1}\delta_{ih}\delta_{jn+h-1}$ on trouve $\delta g(\varepsilon_i,\varepsilon_j,\varepsilon_k,\varepsilon_m) = 0$ donc g est un 3-cocycle. On voit aisément que g n'est pas un 3-cobord d'où :

<u>Proposition</u> 4 : <u>Pour tout</u> $n \geqslant 2$, <u>on a</u> $H^3(\mathcal{X}_n,\mathcal{X}_n) \neq 0$.

4) <u>$p = 1$</u> : <u>Calcul de la dimension</u> de H^1. Pour $p = 1$ nous allons donner un calcul explicite de la dimension sur k de $H^1(\mathcal{X}_n,\mathcal{X}_n)$; pour celà nous allons utiliser le complexe standard

$$C^0 \xrightarrow{d^0} C \xrightarrow{d'} C^2 \xrightarrow{d^2} C^3 \xrightarrow{d^3} \ldots..$$

où C^n désigne l'espace des n-cochaînes. Nous calculerons d'une part $\dim_k B^1(\mathcal{X}_n,\mathcal{X}_n)$ et d'autre part $\dim_k Z^1(\mathcal{X}_n,\mathcal{X}_n)$ où B^1 et Z^1 désignent respectivement l'espace des 1-cobords et des 1-cocycles.

<u>Calcul de</u> $\dim_k B^1(\mathcal{X}_n,\mathcal{X}_n)$: Ce calcul est facile car $\dim_k B^1(\mathcal{X}_n,\mathcal{X}_n) = \dim_k C^o - \dim_k \operatorname{Ker} d^o$ mais $\operatorname{Ker} d^o$ n'est autre que le centre de $\mathcal{X}_n$ donc

$$\dim_k B^1(\mathcal{X}_n,\mathcal{X}_n) = \nu - 2 = 2n - 4.$$

<u>Calcul de</u> $\dim_k Z^1(\mathcal{X}_n,\mathcal{X}_n)$: Nous poserons $B^i(\mathcal{X}_n,\mathcal{X}_n) = B^i$ et $Z^i(\mathcal{X}_n,\mathcal{X}_n) = Z^i$

Nous utiliserons le fait que

$$\dim_k Z^1 = \nu^2 - \dim_k B^2$$

et nous calculons $\dim_k B^2$. Le cas $n = 2$ est singulier et un calcul trivial donne $\dim_k B^2(\mathcal{X}_2,\mathcal{X}_2) = 3$, et nous supposerons jusqu'à la fin du calcul $n > 2$. Soit donc f une 2-cochaîne définie par $f(\varepsilon_i,\varepsilon_j) = \sum_m a_{ijm}\varepsilon_m$; f sera un 2-cobord si et seulement s'il existe $g \in C^1$ définie par : $g(\varepsilon_i) = \sum_k b_{ik}\varepsilon_k$, tel que $f = \delta g$.

<u>Présentation des résultats</u> : Nous présenterons dans la base ε_i les coefficients a_{ijm} de f dans ν matrices $A_1,\dots,A_\nu$, la matrice A_k ayant pour élément générique le terme a_{ijk}. Les matrices A_k ont donc ν lignes et ν colonnes.

<u>Calcul des coefficients</u> a_{ijm} : Il y a onze cas à distinguer :

(I) : $i \in [2,n-1]$ $j \in [2,n]$ on trouve

α) si $i = j$ $a_{iii} = 2b_{i1}$, $a_{iin} = b_{in+i-1}$, $a_{iim} = 0$ si $m \neq i$ et n

β) si $j = n$ $a_{ini} = b_{n1}$, $a_{inn} = b_{nn+i-1} + b_{i1}$, $a_{inm} = 0$ si $m \neq i$ et n

γ) $i \neq j \neq n$ $a_{iji} = b_{j1}$, $a_{ijj} = b_{i1}$, $a_{ijn} = b_{jn+i-1}$, $a_{ijm} = 0$ si $m \neq i,j,n$

(II) : $i = n$ $j \in [2,n]$

α) $j = n$ $a_{nnn} = 2b_{n1}$, $a_{nnm} = 0$ si $m \neq n$

β) $j \neq n$ $a_{njn} = b_{j1}$, $a_{njj} = b_{n1}$, $a_{njm} = 0$ si $m \neq j,n$

(III) : $i \in [n+1,\nu]$, $j \in [2,n]$

$$a_{iji} = b_{j1},\ a_{ijj} = b_{i1},\ a_{ijm} = 0 \text{ si } m \neq i,j$$

(IV) : $i \in [n+1, \nu]$ $\quad j \in [n+1, \nu]$

α) $i = j$ $\quad a_{iii} = 2b_{i1}$, $a_{iin} = b_{i\ i+1-n}$, $a_{iim} = 0$ si $m \neq i,n$

β) $i \neq j$ $\quad a_{iji} = b_{j1}$, $a_{ijj} = b_{i1}$, $a_{ijn} = b_{i\ j+1-n}$, $a_{ijm} = 0$
si $m \neq i,j,n$

(V) : $i = n$ $\quad j \in [n+1, \nu]$

$a_{njn} = b_{j1} + b_{n\ j+1-n}$, $a_{njj} = b_{n1}$, $a_{njm} = 0$ si $m \neq j,n$

(VI) : $i \in [2, n-1]$ $\quad j \in [n+1, \nu]$

α) $n+i-1 = j$ $\quad a_{ijn} = b_{jj} + b_{ii} - b_{nn}$, $a_{iji} = b_{j1} - b_{ni}$
$a_{ijj} = b_{i1} - b_{nj}$, $a_{ijm} = -b_{nm}$ $m \neq i,j,n$

β) $n+i-1 \neq j$ $\quad a_{ijn} = b_{jn+i-1} + b_{ij+1-n}$, $a_{iji} = b_{j1}$
$a_{ijj} = b_{i1}$, $a_{ijm} = 0$ si $m \neq i,j,n$

(VII) : $i = 1$ $\quad j \in [2, n]$

$a_{1jj} = b_{11}$, $a_{1jm} = 0$ si $m \neq j$

(VIII) : $i = 1$ $\quad j \in [n+1, \nu]$

$a_{1jj} = b_{11}$, $a_{1jn} = b_{1\ j+1-n}$, $a_{1jm} = 0$ si $m \neq j,n$

(IX) : $j = 1$ $\quad i \in [2, n-1]$

$a_{i1i} = b_{11}$, $a_{i1n} = b_{1\ n+i-1}$, $a_{i1m} = 0$ si $m \neq i,n$

(X) : $j = 1$ $\quad i \in [n, \nu]$

$a_{i1i} = b_{11}$, $a_{i1m} = 0$ si $m \neq i$

(XI) : $i = j = 1$ $\quad a_{11m} = b_{1m}$

Pour la présentation finale des résultats dans les matrices A_k nous voyons qu'il faut distinguer quatre cas : $k = 1$, $k \in [2, n-1]$, $k = n$, $k \in [n+1, \nu]$. Nous donnons maintenant les matrices A_k dans les quatre cas précités.

Matrice A_1 :

A_1 =

n^{ième} colonne

b_{11}	0	0	0
0	○	0	$-b_{n1}$, $-b_{n1}$, …, $-b_{n1}$ (diagonale) ; ○
0	0	0	0
0	○	0	○

n^{ième} ligne

Matrices A_m pour $m \in [2, n-1]$

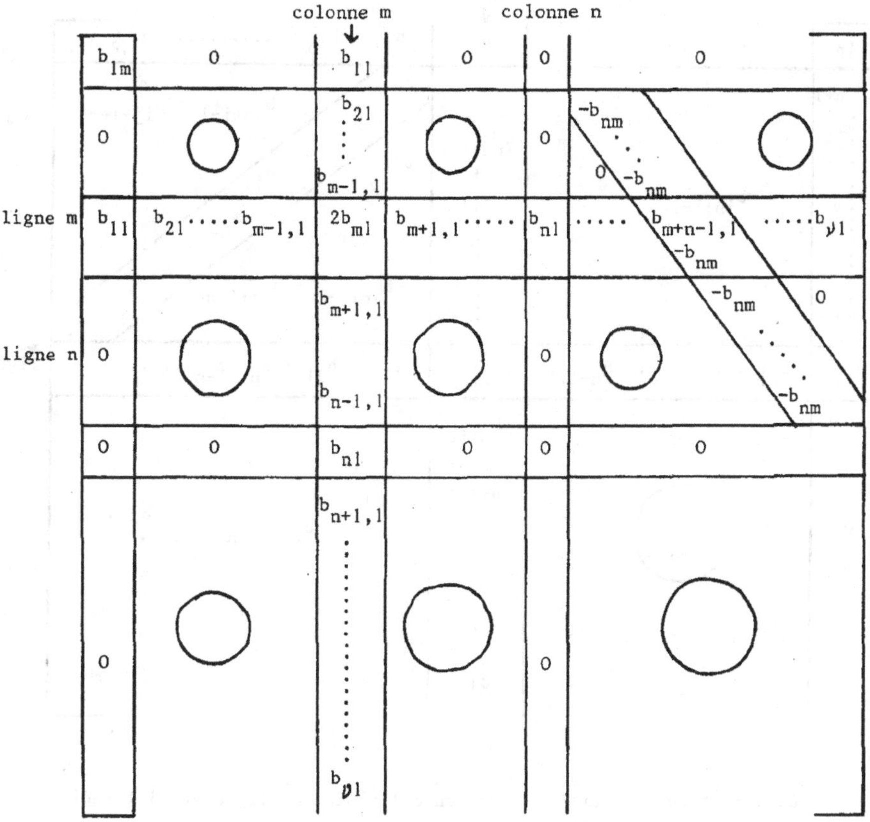

			colonne m ↓		colonne n	
	b_{1m}	0	b_{11}	0	0	0
	0	◯	b_{21} ⋮ $b_{m-1,1}$	◯	0	$-b_{nm}$ ⋱ $-b_{nm}$; 0 ; ◯
ligne m	b_{11}	$b_{21} \cdots b_{m-1,1}$	$2b_{m1}$	$b_{m+1,1} \cdots$	b_{n1}	$\cdots b_{m+n-1,1} \cdots b_{\nu 1}$; $-b_{nm}$
ligne n	0	◯	$b_{m+1,1}$ ⋮ $b_{n-1,1}$	◯	0	◯ ; $-b_{nm}$ ⋱ $-b_{nm}$; 0
	0	0	b_{n1}	0	0	0
	0	◯	$b_{n+1,1}$ ⋮ $b_{\nu 1}$	◯	0	◯

Pour chaque valeur de m la m^{ième} colonne de la matrice se déplacent et se coupent dans le carré en haut à gauche.

Matrice A_n :

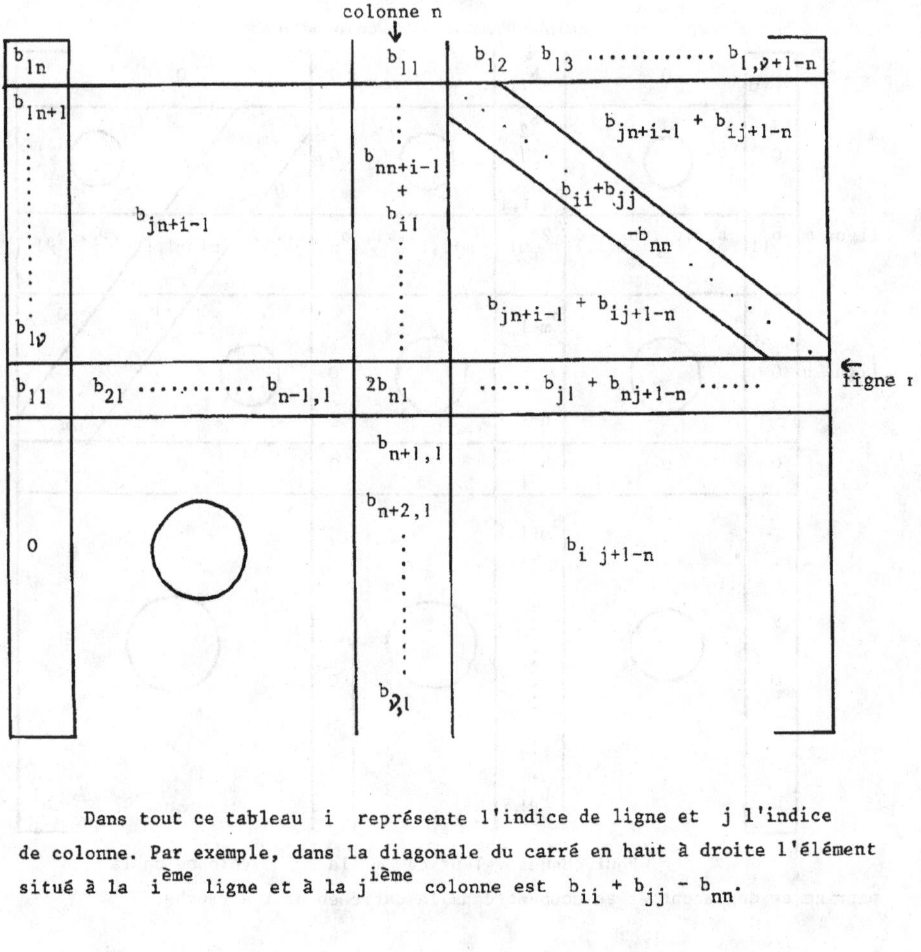

colonne n

b_{1n} | b_{11} | b_{12} b_{13} .... $b_{1,\nu+1-n}$

b_{1n+1} ... $b_{1\nu}$ | b_{jn+i-1} | $b_{nn+i-1} + b_{i1}$ | $b_{jn+i-1} + b_{ij+1-n}$; $b_{ii} + b_{jj} - b_{nn}$; $b_{jn+i-1} + b_{ij+1-n}$

b_{11} | b_{21} .... $b_{n-1,1}$ | $2b_{n1}$ | ..... $b_{j1} + b_{nj+1-n}$ ..... ligne n

0 | O | $b_{n+1,1}$ $b_{n+2,1}$ ... $b_{\nu,1}$ | $b_{i\,j+1-n}$

Dans tout ce tableau i représente l'indice de ligne et j l'indice de colonne. Par exemple, dans la diagonale du carré en haut à droite l'élément situé à la $i^{ème}$ ligne et à la $j^{ième}$ colonne est $b_{ii} + b_{jj} - b_{nn}$.

Matrices A_m pour $m \in [n+1,\nu]$

b_{1m}	0	0	0	b_{11}	0
0	$\bigcirc$	0	$-b_{nm}$ ..... $-b_{nm}$ (diagonale) ; 0 ; 0	b_{21} ... $b_{m+1-n,1}$; b_{nm} ; ... $b_{n-1,1}$	$-b_{nm}$ ..... $-b_{nm}$ (diagonale) ; 0 ; 0
0	0	0	0	b_{n1}	0
0	0	0	0	$b_{n+1,1}$	0
b_{11}	b_{12} b_{31} .... $b_{n-1,1}$	b_{n1}	$b_{n+1,1}$ ... $b_{m-1,1}$	$2b_{m,1}$	$b_{m+1,1}$ ...... $b_{\nu 1}$
0	$\bigcirc$	0	0	$\vdots$ $b_{\nu 1}$	0

Après examen de ces ν matrices A_k il ressort qu'interviennent les coefficients b_{1k} pour $k\in[1,\nu]$ et b_{k1} aussi pour $k\in[1,\nu]$ donc déjà $2\nu-1$ coefficients. Ensuite sur les diagonales des carrés supérieurs droits des matrices A_k interviennent des coefficients b_{nk} donc encore $(\nu-1)$ coefficients nouveaux (b_{n1} a déjà été compté). Enfin dans la matrice A_n on a le carré supérieur gauche faisant intervenir $(n-1)^2$ coefficients. Le carré inférieur droit fait aussi intervenir $(n-2)^2$ coefficients (tous différents des éléments déjà comptés). La diagonale du carré supérieur droit de A_n fait à nouveau intervenir $n-2$ coefficients déjà comptés et les deux triangles bordant cette diagonale en fait intervenir $(n-2)^2-(n-2)$ d'où finalement :

$$\begin{aligned}\dim_k B^2 &= 2\nu-1+(\nu-1)+3(n-2)^2-(n-2)\\ &= 2(2n-2)-1+(2n-2)-1+3n^2-12n+12-n+2\\ &= 3n^2-7n+6.\end{aligned}$$

Je donne ci-après le détail pour $n=4$ où l'on trouve 6 matrices $A_1\ A_2\ A_3 \ldots A_6$ $(\nu=6)$.

$$A_1=\begin{bmatrix} b_{11} & 0 & 0 & 0 & 0 & 0\\ 0 & 0 & 0 & 0 & -b_{41} & 0\\ 0 & 0 & 0 & 0 & 0 & -b_{41}\\ 0 & 0 & 0 & 0 & 0 & 0\\ 0 & 0 & 0 & 0 & 0 & 0\\ 0 & 0 & 0 & 0 & 0 & 0 \end{bmatrix} \qquad A_2=\begin{bmatrix} b_{12} & b_{11} & 0 & 0 & 0 & 0 & 0\\ b_{11} & 2b_{21} & b_{31} & b_{41} & b_{51} & -b_{42} & b_{61}\\ 0 & b_{31} & 0 & 0 & 0 & 0 & 0\\ 0 & b_{41} & 0 & 0 & 0 & 0 & 0\\ 0 & b_{51} & 0 & 0 & 0 & 0 & 0\\ 0 & b_{61} & 0 & 0 & 0 & 0 & 0 \end{bmatrix}$$

$$A_3=\begin{bmatrix} b_{13} & 0 & b_{11} & 0 & 0 & 0 & \\ 0 & 0 & b_{21} & 0 & -b_{43} & 0 & \\ b_{11} & b_{21} & 2b_{31} & b_{41} & b_{51} & b_{61} & -b_{43}\\ 0 & 0 & b_{41} & 0 & 0 & 0 & \\ 0 & 0 & b_{51} & 0 & 0 & 0 & \\ 0 & 0 & b_{61} & 0 & 0 & 0 & \end{bmatrix}$$

$$A_5=\begin{bmatrix} b_{15} & 0 & 0 & 0 & 0 & 0 \\ 0 & 0 & 0 & 0 & b_{21}-b_{45} & 0 \\ 0 & 0 & 0 & 0 & b_{31} & -b_{45} \\ 0 & 0 & 0 & 0 & b_{41} & 0 \\ b_{11} & b_{21} & b_{31} & b_{41} & 2b_{51} & b_{61} \\ 0 & 0 & 0 & 0 & b_{61} & 0 \end{bmatrix} \quad A_6=\begin{bmatrix} b_{16} & 0 & 0 & 0 & 0 & b_{11} \\ 0 & 0 & 0 & 0 & -b_{46} & b_{12} \\ 0 & 0 & 0 & 0 & 0 & b_{31}-b_{46} \\ 0 & 0 & 0 & 0 & 0 & b_{41} \\ 0 & 0 & 0 & 0 & 0 & b_{51} \\ b_{11} & b_{21} & b_{31} & b_{41} & b_{51} & 2b_{61} \end{bmatrix}$$

$$A_4=\begin{bmatrix} b_{14} & 0 & 0 & b_{11} & b_{12} & b_{13} \\ b_{15} & b_{25} & b_{35} & b_{45}+b_{21} & b_{22}+b_{55}-b_{44} & b_{65}+b_{23} \\ b_{16} & b_{26} & b_{36} & b_{46}+b_{31} & b_{56}+b_{32} & b_{33}+b_{66}-b_{44} \\ b_{11} & b_{21} & b_{31} & 2b_{41} & b_{51}+b_{42} & b_{61}+b_{43} \\ 0 & 0 & 0 & b_{51} & b_{52} & b_{53} \\ 0 & 0 & 0 & b_{61} & b_{62} & b_{63} \end{bmatrix}$$

$\dim_k(B^2(\mathcal{K}_4)) = 26.$

Nous avons donc montré que la dimension sur k de $B^2(\mathcal{K}_n,\mathcal{K}_n)$ est $3n^2 - 7n + 6$ (pour $n > 2$), d'où :

$$\dim_k B^2(\mathcal{K}_n,\mathcal{K}_n) = \begin{cases} 3 & \text{si } n = 2 \\ 3n^2 - 7n + 6 & \text{si } n > 2. \end{cases}$$

En conclusion : $\dim_k Z^1 = \nu^2 - \dim_k B^2$, c'est-à-dire

$$\dim_k Z^1 = \begin{cases} 1 & \text{si } n = 2 \\ n^2 - n - 2 & \text{si } n > 2 \end{cases}$$

<u>Proposition</u> 5 : <u>On a</u> $\dim_k H^1(\mathcal{K}_n,\mathcal{K}_n) = \begin{cases} 1 & \text{si } n = 2 \\ (n-1)(n-2) & \text{si } n > 2 \end{cases}$

2°) <u>Séries de Poincaré de</u> $\mathcal{K}_n$

L'idéal d'augmentation I de $\mathcal{K}_n$ est nilpotent : comme $\mathcal{K}_n/I$ est un corps, donc un anneau simple, la dimension globale de $\mathcal{K}_n$ est 0 où ∞ [1]. $\mathcal{K}_n$ n'est évidemment pas semi-simple, d'où : $\text{l.gl}\dim \mathcal{K}_n = \text{r.gl}\mathcal{K}_n = +\infty$. On a $\dim \mathcal{K}_n = +\infty$ où $\dim \mathcal{K}_n$ désigne la dimension cohomologique de $\mathcal{K}_n$ i.e. la dimension homologique de $\mathcal{K}_n$ en tant que $\mathcal{K}_n^e$-module à gauche.

L'algèbre de Lie de Heisenberg i.e. l'algèbre de Lie des matrices triangulaires de la forme $\begin{pmatrix} 0 & * & \dots & * \\ & \ddots & & * \\ & & \ddots & \vdots \\ & & & * \\ 0 & & & 0 \end{pmatrix}$ n'est autre que l'idéal I de $\mathcal{K}_n$ muni du crochet des matrices, l'injection $I \hookrightarrow \mathcal{K}_n$ préserve trivialement le crochet, ce qui prouve que l'algèbre $\mathcal{K}_n$ est un quotient de l'algèbre enveloppante d'une algèbre de Lie nilpotente (ici l'algèbre de Heisenberg).

Proposition 6 : La série de Poincaré de l'anneau local $\mathcal{K}_3$ est rationnelle et vaut :

$$P(T) = \frac{1}{(1-T)^2}$$

L'anneau local $\mathcal{K}_3$ se comporte donc du point de vue de sa série de Poincaré comme une intersection complète commutative. De plus si $H(T)$ désigne la série de Hilbert de l'anneau $\mathcal{K}_3$, on a la relation :

$$P(T).H(-T) = 1$$

où $H(T) = \sum_{p=0}^{\infty} \dim_k(I^p/I^{p+1})T^p$, où I est l'idéal maximal de $\mathcal{K}_3$.

Preuve : Nous allons construire une résolution minimale de k comme $\mathcal{K}_3$-module à gauche. Une telle résolution commence par la suite exacte $0 \longrightarrow I \longrightarrow \mathcal{K}_3 \longrightarrow k \longrightarrow 0$. Nous allons donc chercher à résoudre I. Or $\dim_k I/I^2 = 2$. On écrit donc $0 \longrightarrow \mathrm{Ker}\varphi \longrightarrow \mathcal{K}_3^2 \xrightarrow{\varphi} I \longrightarrow 0$. Calculons le noyau $\mathrm{Ker}\,\varphi$. Soit (e_1,e_2) une base de $\mathcal{K}_3^2$. Nous changerons de notation pour la base sur k de l'idéal I ; nous noterons $\varepsilon_o = \mathrm{Id}$,

$$\varepsilon_1 = \begin{pmatrix} 0 & 1 & 0 \\ 0 & 0 & 0 \\ 0 & 0 & 0 \end{pmatrix} \qquad \varepsilon_3 = \begin{pmatrix} 0 & 0 & 1 \\ 0 & 0 & 0 \\ 0 & 0 & 0 \end{pmatrix} \qquad \varepsilon_2 = \begin{pmatrix} 0 & 0 & 0 \\ 0 & 0 & 1 \\ 0 & 0 & 0 \end{pmatrix} .$$

On a la relation $\varepsilon_1\varepsilon_2 = \varepsilon_3$ donc $\varepsilon_1, \varepsilon_2$ est un système minimal de générateurs de I comme $\mathcal{K}_3$-module à gauche.

On définit alors φ par :

$$\varphi(e_1) = \varepsilon_1$$

$$\varphi(e_2) = \varepsilon_2$$

Soit $u \in \mathrm{Ker}\,\varphi$ $\quad u = x_1 e_1 + x_2 e_2$ avec $x_1 = \sum_{i=o}^{3} \lambda_{i1}\varepsilon_i$ $\quad x_2 = \sum_{i=o}^{3} \lambda_{i2}\varepsilon_i$

donc $\varphi(u) = (\sum_{i=o}^{3} \lambda_{i1}\varepsilon_i)\varepsilon_1 + (\sum_{i=o}^{3} \lambda_{i2}\varepsilon_i)\varepsilon_2$

$= \lambda_{o1}\varepsilon_1 + \lambda_{o2}\varepsilon_2 + \lambda_{12}\varepsilon_3$

donc on a $\varphi(u) = 0$ si et seulement si $\lambda_{o1} = 0 = \lambda_{o2} = \lambda_{12}$. Donc $x_1 \in I$ et $x_2 \in J$ où J est l'idéal à gauche des matrices de la forme

$\begin{pmatrix} 0 & 0 & \alpha \\ 0 & 0 & \beta \\ 0 & 0 & 0 \end{pmatrix}$: d'où $\mathrm{Ker}\,\varphi = Ie_1 \oplus Je_2$. On remarque que $I.J = I^2$ donc

$\dim_k \frac{\mathrm{Ker}\,\varphi}{I.\mathrm{Ker}\,\varphi} = 5 - 2 = 3$ donc $\mathrm{Ker}\,\varphi$ admet un système minimal de générateurs formé de 3 éléments qui peuvent être choisis comme étant $\varepsilon_1 e_1$, $\varepsilon_2 e_1$, $\varepsilon_2 e_2$.
On peut donc écrire :

$$0 \longrightarrow \mathrm{Ker}\,\varphi_2 \longrightarrow \mathcal{X}_3^3 \xrightarrow{\varphi_2} \mathrm{Ker}\,\varphi \longrightarrow 0 .$$

On note (f_1, f_2, f_3) une $\mathcal{X}_3$-base de $\mathcal{X}_3^3$; φ_2 est définie par :
$\varphi_2(f_1) = \varepsilon_1 e_1, \varphi_2(f_2) = \varepsilon_2 e_1, \varphi_2(f_3) = \varepsilon_2 e_2$. Soit $u = x_1 f_1 + x_2 f_2 + x_3 f_3$.
Si $u \in \mathrm{Ker}\,\varphi_2$ avec des notations évidentes on a :

$$(\sum_{i=o}^{3} \lambda_{i1}\varepsilon_i)\varepsilon_1 e_1 + (\sum_{i=o}^{3} \lambda_{i2}\varepsilon_i)\varepsilon_2 e_1 + (\sum_{i=o}^{3} \lambda_{i3}\varepsilon_i)\varepsilon_2 e_2 = 0$$

i.e.

$$\lambda_{o1}\varepsilon_1 e_1 + (\lambda_{o2}\varepsilon_2 + \lambda_{12}\varepsilon_3)e_1 + (\lambda_{o3}\varepsilon_2 + \lambda_{13}\varepsilon_3)e_2 = 0$$

donc $\lambda_{o1} = 0 = \lambda_{o2} = \lambda_{12} = \lambda_{o3} = \lambda_{13}$ d'où $x_1 \in I$, $x_2 \in J$, $x_3 \in J$ et $\mathrm{Ker}\,\varphi_2 = I\,f_1 \oplus J\,f_2 \oplus J\,f_3$.

Ainsi $\dim_k \frac{\mathrm{Ker}\,\varphi_2}{I\,\mathrm{Ker}\,\varphi_2} = 7 - 3 = 4$. On peut alors montrer aisément par récurrence que le nombre minimal de générateur du $\mathrm{Ker}\,\varphi_p$ est $p+2$. Comme la résolution est minimale il est connu que ces nombres minimaux de générateurs coïncident avec les nombres de Betti. Soit donc b_p le $p^{\text{ième}}$ nombre de Betti de l'anneau local $\mathcal{X}_3$. Nous avons donc montré que pour tout $p \geq 0$ $b_p = p+1$. D'où :

$$P(T) = \sum_{p=o}^{\infty} b_p\, T^p = \sum_{p=o}^{\infty} (p+1)\, T^p = \frac{1}{(1-T)^2} .$$

D'autre part la série de Hilbert d'un anneau local d'idéal maximal I est définie par : $H(T) = \sum_{p=o}^{\infty} \dim_k(I^p/I^{p+1})T^p$, donc ici, comme $I^3 = 0$, on a :

$H(T) = 1 + 2T + T^2 = (1 + T)^2$, d'où

$$H(-T) = (1-T)^2$$

ce qui achève la démonstration.

Les calculs d'une résolution minimale de k en tant que $\mathcal{X}_n$-module à gauche (par exemple), pour $n > 3$, sont nettement moins agréables que dans le cas $n=3$, que nous venons de traiter. Aussi allons-nous utiliser une autre méthode. Nous savons [(2)] que les groupes d'homologie de l'algèbre supplémentée $\mathcal{X}_n$ peuvent se calculer à l'aide des groupes d'homologie de Hochschild de l'algèbre associative $\mathcal{X}_n$. En effet , pour tout $\mathcal{X}_n$-module à droite A on a un isomorphisme

$$H_p(\mathcal{X}_n, {}_{\varepsilon}A) \simeq \mathrm{Tor}_p^{\mathcal{X}_n}(A,k) \quad \text{où}$$

${}_{\varepsilon}A$ est le $\mathcal{X}_n$-bimodule A, muni de la structure de module à droite de A et dont la structure de module à gauche est donnée par la relation :

si $\lambda \in \mathcal{X}_n$ et si $a \in A$, $\lambda . a = \mathcal{E}(\lambda)a$ où $\mathcal{E}$ est l'augmentation.

<u>Lemme</u> 1 : <u>Pour tout entier</u> $p \geq 1$ <u>on a l'isomorphisme</u>

$$\mathrm{Tor}_p^{\mathcal{X}_n}(k,k) \otimes_k I/I^2 \simeq \mathrm{Tor}_p^{\mathcal{X}_n}(I/I^2,k).$$

<u>Preuve</u> : il suffit de remarquer que I/I^2 est un $\mathcal{X}_n$-module à droite annulé par I et d'appliquer l'exercice 3 page 201 de [2].

Considérons l'inclusion $j : I^2 \hookrightarrow I$. L'application j induit pour tout p un morphisme θ_p

$$\theta_p : \mathrm{Tor}_p^{\mathcal{X}_n}(I^2,k) \longrightarrow \mathrm{Tor}_p^{\mathcal{X}_n}(I,k).$$

<u>Lemme</u> 2 : <u>Pour tout entier</u> $p \geq 0$ <u>l'application</u> θ_p <u>est l'application nulle</u>.

<u>Preuve</u> : Nous utilisons ici les groupes d'homologie de Hochschild. En effet nous savons que

$$\mathrm{Tor}_p^{\mathcal{X}_n}(I^2,k) \simeq H_p(\mathcal{X}_n, {}_{\varepsilon}I^2)$$

et

$$\mathrm{Tor}_p^{\mathcal{X}_n}(k,I) \simeq H_p(\mathcal{X}_n, {}_{\varepsilon}I).$$

Nous savons en outre que les groupes d'homologie de Hoschild peuvent se calculer avec le Bar-complexe où ce qui revient au même en remplaçant l'algèbre unitaire $\mathcal{K}_n$ par l'algèbre non unitaire I, car $\mathcal{K}_n$ n'est autre que l'algèbre non unitaire I à laquelle on a adjoint un élément unité par la méthode standard. ([2] page 181 exercice 4).

Nous noterons pour tout entier $p \geq 0$ $\quad C_p = {}_{\varepsilon}I^2 \otimes_k \underbrace{I \otimes_k \ldots \otimes I}_{p \text{ fois}}$ et

$C'_p = {}_{\varepsilon}I \otimes_k \underbrace{I \otimes_k \ldots \otimes_k I}_{p \text{ fois}}$ on a pour tout $p \geq 1$ un diagramme commutatif

$$\begin{array}{ccccc} C_{p+1} & \xrightarrow{d_{p+1}} & C_p & \xrightarrow{d_p} & C_{p-1} \\ \downarrow \varphi_{p+1} & & \downarrow \varphi_p & & \downarrow \varphi_{p-1} \\ C'_{p+1} & \xrightarrow{d'_{p+1}} & C'_p & \xrightarrow{d'_p} & C'_{p-1} \end{array}$$

où d_{p+1}, d_p sont les différentielles du Bar-complexe associé à ${}_{\varepsilon}I^2$

où d'_{p+1}, d'_p sont les différentielles du Bar-complexe associé à ${}_{\varepsilon}I$ et

où $\varphi_{p+1}, \varphi_p, \varphi_{p-1}$ sont les applications induites par l'inclusion $j : I^2 \longrightarrow I$.

Il faut donc montrer que si $z \in C_p$ est un cycle i.e. $z \in \operatorname{Ker} d_p$ alors $\varphi_p(z)$ est un bord dans C'_p i.e. $\varphi_p(z) \in \operatorname{Im} d'_{p+1}$. Nous aurons ainsi montrer que les applications induites par j sur l'homologie sont nulles.

Nous reprenons la base $\{\varepsilon_2, \varepsilon_3, \ldots, \varepsilon_\nu\}$ de I . I^2 est ainsi engendré comme k-espace vectoriel par ε_n.

Soit donc $z = \sum_{\vec{i}} \lambda_{\vec{i}}\, \varepsilon_n \otimes \varepsilon_{i1} \otimes \ldots \otimes \varepsilon_{ip}$ un élément de Ker d_p

on a $\quad 0 = d_p(z) = \sum_{\vec{i}} \lambda_{\vec{i}} \sum_{j=1}^{p-1} (-1)^j\, \varepsilon_n \otimes \varepsilon_{i_1} \otimes \ldots \otimes \varepsilon_{i_j} \varepsilon_{i_{j+1}} \otimes \ldots \otimes \varepsilon_{i_p}$

ou encore

$$\varepsilon_n \otimes \sum_{\vec{i}} \lambda_{\vec{i}} \sum_{j=1}^{p-1} (-1)^j\, \varepsilon_{i_1} \otimes \ldots \otimes \varepsilon_{i_j} \varepsilon_{i_{j+1}} \otimes \ldots \otimes \varepsilon_{i_p} = 0.$$

Comme le produit tensoriel est pris sur k, il vient :

$$(1) \qquad \sum_{\vec{i}} \lambda_{\vec{i}} \sum_{j=1}^{p-1} (-1)^j\, \varepsilon_{i_1} \otimes \ldots \otimes \varepsilon_{i_j} \varepsilon_{i_{j+1}} \otimes \ldots \otimes \varepsilon_{i_p} = 0$$

On a $\varphi_p(z) = z$ car φ_p est induite par j.

Considérons un élément $m \otimes \varepsilon_h \otimes \sum_{\vec{i}} \lambda_{\vec{i}} \varepsilon_{i_1} \otimes \ldots \otimes \varepsilon_{ip}$ de C'_{p+1} où $m \in I$. Je dis qu'il est possible de choisir m dans I et h dans $[2,\nu]$ tels que $d'_{p+1}(m \otimes \varepsilon_h \otimes \sum_{\vec{i}} \lambda_{\vec{i}} \varepsilon_{i_1} \otimes \varepsilon_{i_2} \otimes \ldots \otimes \varepsilon_{i_p}) = z$ ce qui achèvera la preuve du lemme 2.

Or $d'_{p+1}(m \otimes \varepsilon_h \otimes \sum_{\vec{i}} \lambda_{\vec{i}} \varepsilon_{i_1} \otimes \varepsilon_{i_2} \otimes \ldots \otimes \varepsilon_{i_p}) = \sum_{\vec{i}} \lambda_{\vec{i}} (m\varepsilon_h \otimes \varepsilon_{i_1} \otimes \ldots \otimes \varepsilon_{i_p}$

$- m \otimes \varepsilon_h \varepsilon_{i_1} \otimes \varepsilon_{i_2} \otimes \ldots \otimes \varepsilon_{i_p} - m \otimes \varepsilon_h \otimes \sum_{j=1}^{p-1} (-1)^j \varepsilon_{i_1} \otimes \ldots \otimes \varepsilon_{i_j} \varepsilon_{i_{j+1}} \otimes \ldots \otimes \varepsilon_{i_p})$

et par conséquent compte tenu de notre hypothèse :

$d'_{p+1}(m \otimes \varepsilon_h \otimes \sum_{\vec{i}} \lambda_{\vec{i}} \varepsilon_{i_1} \otimes \ldots \otimes \varepsilon_{i_p}) = \sum_{\vec{i}} \lambda_{\vec{i}} (m\varepsilon_h \otimes \varepsilon_{i_1} \otimes \ldots \otimes \varepsilon_{i_p} - m \otimes \varepsilon_h \varepsilon_{i_1} \otimes \ldots \otimes \varepsilon_{i_p})$

Il suffit de choisir alors $m = \varepsilon_2$ et $h = n+1$ et l'on aura $m\varepsilon_h = \varepsilon_n$ et $\varepsilon_{n+1} \varepsilon_{i_1} = 0$ ce qui achève la démonstration.

<u>Lemme</u> 3 : <u>Pour tout</u> $p \geq 1$ <u>on a un isomorphisme</u>

$$\mathrm{Tor}_{p+1}^{\mathcal{H}_n}(k,k) \simeq \mathrm{Tor}_p^{\mathcal{H}_n}(I,k) \ .$$

<u>Preuve</u> : On a la suite exacte de $\mathcal{H}_n$-module à droite

$$0 \longrightarrow I \longrightarrow \mathcal{H}_n \longrightarrow k \longrightarrow 0.$$

On en déduit la longue suite exacte d'homologie

$$\longrightarrow \mathrm{Tor}_1^{\mathcal{H}_n}(\mathcal{H}_n,k) \longrightarrow \mathrm{Tor}_1^{\mathcal{H}_n}(k,k) \longrightarrow I \otimes k \longrightarrow \mathcal{H}_n \otimes k \longrightarrow k \otimes k \longrightarrow 0$$

$$\ldots \mathrm{Tor}_2^{\mathcal{H}_n}(I,k) \longrightarrow \mathrm{Tor}_2^{\mathcal{H}_n}(\mathcal{H}_n,k) \longrightarrow \mathrm{Tor}_2^{\mathcal{H}_n}(k,k) \longrightarrow \mathrm{Tor}_1^{\mathcal{H}_n}(I,k) \longrightarrow$$

on sait que pour tout $j > 0$ on a $\mathrm{Tor}_j^{\mathcal{H}_n}(\mathcal{H}_n,k) = 0$, ce qui achève la démonstration.

<u>Lemme</u> 4 : I^2 <u>est isomorphe à</u> k <u>en tant que</u> $\mathcal{H}_n$<u>-module à droite ou à gauche.</u>

<u>Preuve</u> : Rappelons la démonstration bien connue de ce lemme. L'application $f : k \longrightarrow I^2$, qui à λ fait correspondre $\lambda\varepsilon_n$, est un isomorphisme de k-espaces vectoriels ; de plus si $x \in \mathcal{H}_n$ on a :

$$f(x\lambda) = f(\varepsilon(x)\lambda) = \varepsilon(x)\lambda\varepsilon_n = x\lambda\varepsilon_n$$

car $x - \varepsilon(x) \in I$ et $\varepsilon_n \in I^2$ donc $(x - \varepsilon(x))\varepsilon_n \in I^3$. Mais $I^3 = 0$. Ce qui achève la démonstration.

<u>Lemme</u> 5 : <u>Soit</u> b_p <u>de</u> $p^{\text{ième}}$ <u>nombre de Betti de l'anneau</u> $\mathcal{H}_n$, <u>c'est-à-dire</u>

$b_p = \dim_k \mathrm{Tor}_p^{\mathscr{X}_n}(k,k)$. Alors, pour tout $p \geq 1$ on a $b_{p+1} + b_{p-1} = (2n-4)b_p$.

Preuve : On a la suite exacte de $\mathscr{X}_n$-modules à droite

$$0 \longrightarrow I^2 \longrightarrow I \longrightarrow I/I^2 \longrightarrow 0.$$

La longue suite exacte d'homologie donne

$$\longrightarrow \mathrm{Tor}_1^{\mathscr{X}_n}(I/I^2,k) \longrightarrow I^2 \otimes k \longrightarrow I \otimes k \longrightarrow I/I^2 \otimes k \longrightarrow 0$$

$$\longrightarrow \mathrm{Tor}_2^{\mathscr{X}_n}(I,k) \longrightarrow \mathrm{Tor}_2^{\mathscr{X}_n}(I/I^2,k) \longrightarrow \mathrm{Tor}_1^{\mathscr{X}_n}(I^2,k) \longrightarrow \mathrm{Tor}_1^{\mathscr{X}_n}(I,k) \longrightarrow$$

$$\longrightarrow \mathrm{Tor}_p^{\mathscr{X}_n}(I/I^2,k) \longrightarrow \mathrm{Tor}_{p-1}^{\mathscr{X}_n}(I^2,k) \longrightarrow \mathrm{Tor}_{p-1}^{\mathscr{X}_n}(I,k) \longrightarrow \cdots\cdots$$

$$\cdots\cdots \longrightarrow \mathrm{Tor}_p^{\mathscr{X}_n}(I^2,k) \longrightarrow \mathrm{Tor}_p^{\mathscr{X}_n}(I,k) \longrightarrow$$

On a en particulier pour tout $p \geq 1$ une suite exacte

$$\mathrm{Tor}_p^{\mathscr{X}_n}(I^2,k) \xrightarrow{\theta_p} \mathrm{Tor}_p^{\mathscr{X}_n}(I,k) \longrightarrow \mathrm{Tor}_p^{\mathscr{X}_n}(I/I^2,k) \longrightarrow \mathrm{Tor}_{p-1}^{\mathscr{X}_n}(I^2,k) \xrightarrow{\theta_{p-1}} \mathrm{Tor}_{p-1}^{\mathscr{X}_n}(I,k)$$

et θ_p et θ_{p-1} sont les applications définies dans le lemme 2. On a alors :

(2) $\quad \dim_k \mathrm{Tor}_p^{\mathscr{X}_n}(I,k) + \dim_k \mathrm{Tor}_{p-1}^{\mathscr{X}_n}(I^2,k) - \dim_k \mathrm{Tor}_p^{\mathscr{X}_n}(I/I^2,k) =$

$\dim_k \mathrm{Tor}_p^{\mathscr{X}_n}(I^2,k) - \dim_k \mathrm{Ker}\,\theta_p + \dim \mathrm{Im}\,\theta_{p-1} = 0$ car θ_p et θ_{p-1} sont nulles d'après le lemme 2.

D'autre part $\mathrm{Tor}_p^{\mathscr{X}_n}(I,k) \simeq \mathrm{Tor}_{p+1}^{\mathscr{X}_n}(k,k)$ d'après lemme 3.

$\mathrm{Tor}_{p-1}^{\mathscr{X}_n}(I^2,k) \simeq \mathrm{Tor}_{p-1}^{\mathscr{X}_n}(k,k)$ d'après le lemme 4.

et $\mathrm{Tor}_p^{\mathscr{X}_n}(I/I^2,k) \simeq \mathrm{Tor}_p^{\mathscr{X}_n}(k,k) \otimes_k I/I^2$ d'après le lemme 1.

La relation (2) devient $b_{p+1} + b_{p-1} - (2n-4)\,b_p = 0$ ce qui achève la démonstration.

Cette relation permet de calculer par récurrence les nombres de Betti de l'anneau $\mathscr{X}_n$. En effet il est clair que $b_o = 1$ et $b_1 = (2n-4)$ ($b_1 = \dim_k I/I^2$).

De manière plus précise on a le :

Théorème 1 : La série de Poincaré-Betti de l'anneau $\mathscr{X}_n$ ($n > 2$) est rationnelle et vaut :

$$P(T) = \frac{1}{1-(2n-4)T+T^2} .$$

<u>D'autre part si</u> H <u>désigne la série de Hilbert de l'anneau</u> $\mathcal{H}_n$ <u>on a</u> :

$$P(T)\ H(-T) = 1.$$

<u>Preuve</u> : Considérons la série de Hilbert de l'anneau $\mathcal{H}_n$ on a : $H(T) = \sum_{p=0}^{\infty} \dim_k (I^p/I^{p+1})\ T^p$. Comme $I^3 = 0$, que $\dim_k I = 2n-3$ et que $\dim_k I^2 = 1$ on a : $H(T) = 1 + (2n-4)T + T^2$. Soit $Q(T)$ la série formelle inverse de $H(-T)$ on a : $Q(T) = \sum_{p=0}^{\infty} q_p T^p$ où les q_p sont le coefficients de Q. On a donc $(q_o + q_1 T + q_2 T^2 + \ldots)(1-(2n-4)T + T^2) = 1$ d'où : $q_o = 1$ $q_1 - (2n-4)q_o = 0$ ou encore $q_1 = 2n-4$, et en identifiant les termes en T^{p+1} :

$$q_{p+1} + q_{p-1} - (2n-4)q_p = 0.$$

Donc la suite $(q_p)_{p\in\mathbb{N}}$ vérifie la même relation de récurrence que la suite $(b_p)_{p\in\mathbb{N}}$ des nombres de Betti (lemme 5). De plus $q_o = b_o$ et $q_1 = b_1$; donc pour tout $p\in\mathbb{N}$. On a $q_b = b_p$ ce qui achève la démonstration.

Ce théorème est à rapprocher d'un théorème d'algèbre commutative. On démontre en effet dans (4) le résultat suivant :

<u>Théorème</u> : <u>Soit</u> A <u>un anneau local noethérien commutatif qui est le quotient d'un anneau local régulier</u> B <u>par un idéal engendré par des monômes de degré deux en un système minimal de générateurs de l'idéal maximal de</u> B. <u>Alors la série</u> P <u>de Poincaré-Betti de l'anneau</u> A <u>est rationnelle et si</u> H <u>est la série de Hilbert de</u> A <u>on a</u>

$$P(T)\ H(-T) = 1.$$

Or nous avons déjà remarqué que $\mathcal{H}_n$ est un quotient de l'algèbre enveloppante $U(\mathfrak{h}_n)$ de l'algèbre de Lie de Heisenberg $\mathfrak{h}_n$. Il suffit alors de localiser $U(\mathfrak{h}_n)$ par rapport à l'idéal d'augmentation pour obtenir un anneau local noethérien non commutatif qui est régulier au sens de Walker (5). On peut montrer que $\mathcal{H}_n$ est un quotient de ce localisé par un idéal engendré par des monômes de degré deux en un système minimal de générateurs (centralisant) de l'idéal maximal.

Références

(1) M. Auslander : On the dimension of modules and algebras III. Nagoya Math. J. (9) 1955 (67-77).

(2) H. Cartan et S. Eilenberg : Homological algebra, Academic Press.

(3) S. Piper : Algebras of matrices under deformation J. Diff. Geometry 5 - 1971 (437-449)

(4) R. Fröberg : Determination of a class of Poincaré Series, Math. Scand (37) 1975 (29-39)

(5) R. Walker : Local rings and normalizing sets of elements Proc. London Math. Soc (3) 24 (1972) p.27-45.

DECOMPOSITIONS REVISITED

Robert M. FOSSUM
Institute for Algebraic Meditation

0 - Introduction

Let p be a prime integer and $q = p^e$ a power of p . If the cyclic group $\mathbb{Z}/q\mathbb{Z}$ is acting on a finite dimensional vector space in characteristic zero, all of the information regarding this action can be quite explicity written down, especially in terms of the characters of the group. However the situation is not so well known in case $\mathbb{Z}/q\mathbb{Z}$ is acting in characteristic p . In particular one encounters such actions when studying deformation theory in characteristic p , when studying invariant theory, or modular representations.

For some time Almkvist and I have been studying the following problem. Let V_n be an n-dimensional vector space over a field of characteristic p on which $\mathbb{Z}/q\mathbb{Z}$ acts Then the action induces an action on the symmetric powers $S^r(V_n)$ and exterior powers $\Lambda^r(V_n)$. We want to determine :

a) the decompositions of these into the indecomposables

and

b) the rings of invariant $S^{\cdot}(V_n)^{\mathbb{Z}/q\mathbb{Z}}$ and $\Lambda^{\cdot}(V_n)^{\mathbb{Z}/q\mathbb{Z}}$.

These problems are connected with classical invariant theory, representations of the symmetric groups in characteristic p , modular representation theory, combinatorics, among other subjects. At the present time the solutions in general are beyond our reach. However some progress has been made since I last spoke on this subject in this seminar, and this paper reports some of this progress.

At this point I thank my wife Barbara for her inspiration and the United States National Science Foundation for financial support.

§ 1 - Representation Ring of a Formal Group ; Definition

Let k be a commutative ring. A formal power series

$$F(X,Y) \in k[[X,Y]]$$

is a formal group law on k if :

$$F(X,Y) = X + Y + \sum_{i,j\geq 1} a_{ij} X^i Y^j$$

and

$$F(X,F(Y,Z)) = F(F(X,Y),Z) \quad \text{in } k[[X,Y,Z]] .$$

The formal group law is commutative if

$$F(X,Y) = F(Y,X) .$$

A comprehensive reference is [Hazewinkel] . Two examples are :

$$a(X,Y) = X + Y \; ; \qquad \text{the additive law.}$$

$$m(X,Y) = X + Y + XY \; ; \qquad \text{the multiplicative law.}$$

If k is a field of characteristic zero, then any formal group law is isomorphic to $a(X,Y)$. (cf. [Hazewinkel]). Let F be a formal group law on k .

Let $\mathcal{N}$ denote the category of pairs (M,α) where M is a finitely generated k-module and $\alpha : M \to M$ is a nilpotent k-endomorphism.

A morphism in $\mathcal{N}$ is a k-homomorphism on the underlying k-module that commutes with the endomorphisms ... the diagram

$$\begin{array}{ccc} M & \xrightarrow{\alpha} & M \\ f\downarrow & & \downarrow f \\ N & \xrightarrow{\beta} & N \end{array}$$

should commute. The category has direct sums.

Let $G(\mathcal{N})$ denote the abelian group obtained by taking the free abelian group on isomorphism classes in $\mathcal{N}$ modulo the relations :

$$[(M,\alpha) \oplus (N,\beta)] \quad - \quad [(M,\alpha)] \quad - \quad [(N,\beta)] .$$

(Since I will work for the most over a field, I will not go into further details about this group. If k is a field, then $\mathcal{N}$ is the category of $k[[X]]$ - modules of finite length).

We use the formal group law to induce a tensor product on $\mathcal{N}$, namely, if $(M,\alpha),(N,\beta)$ are objects in $\mathcal{N}$, then we define :

$$(M,\alpha) \otimes (N,\beta) := (M \underset{k}{\otimes} N,\ F(\alpha\otimes 1,\ 1 \otimes \beta)) .$$

Since α and β are nilpotent (we assume of finite order), the formal power series $F(\alpha\otimes 1, 1\otimes\beta)$ is finite and it is also nilpotent.
Hence there is induced a product on $G(\mathcal{N})$, since this tensor product commutes with direct sums. The object $(k,0)$ acts as an identity, and the product is commutative and associative (because F is associative). Let $R_F(k)$ denote this commutative ring.

<u>Theorem</u> 1.1 <u>If</u> k <u>is a field of characteristic zero, then</u>

$$R_F(k) = \mathbb{Z}[V] ,$$

(V an indeterminate).

<u>Outline of proof</u> : For the first, any formal group law is isomorphic to the additive group law. And any indecomposable pair (M,α) is determined by $rk_k M$, with $\alpha = \begin{pmatrix} 0 & 0 \\ I & 0 \end{pmatrix}$. Let S,T be indeterminates and set $P = k[S,T]$. Then P is a graded ring with $rk\, S = 1 = rk\, T$. Let P_n denote the homogeneous polynomials of degree n . Then $rk_k P_n = n+1$. Define $D : P_n \to P_n$ by

$$D(f(S,T)) := \frac{\partial f}{\partial T} S .$$

With respect to the basis $T^n, ST^{n-1}, \dots, S^{n-1}T, S^n$ the matrix of D is

$$\begin{pmatrix} 0 & 0 \dots\dots 0 & 0 \\ n & 0 \quad\quad 0 & 0 \\ 0 & n-1 \quad . & . \\ . & 0 \quad\quad . & . \\ . & \quad . \quad\quad . & . \\ 0 & 0 \dots\dots 1 & 0 \end{pmatrix}$$

Let $V := (P_1, D)$.

Lemma 1.2 (a) If (M,α) is indecomposable, then $(M,\alpha) \simeq (P_r, D)$ where $r+1 = rk_k M$.

(b) $(P_r,D) \otimes (P_s,D) \cong (P_{r+s},D) \oplus (P_{r+s-2},D) \oplus \dots \oplus (P_{|r-s|}D)$.

Proof (a) This is clear. (b) This is established by induction on $\min(r,s)$. If $r = 0$, then $(P_o,D) = (k,0)$ which is the identity.

In general, note that there is a D-equivariant homomorphism

$$P_r \otimes P_s \longrightarrow P_{r+s}$$

given by multiplication in P. For the formal group law $X + Y$ gives

$$(D \otimes 1 + 1 \otimes D)(f \otimes g) = Df \otimes g + g \otimes Dg$$

and D is a derivation, so

$$D(fg) = (Df)g + fDg.$$

Suppose $r = 1$. Then consider the exact sequence

$$0 \longrightarrow W \xrightarrow{\alpha} P_1 \otimes P_s \xrightarrow{m} P_{1+s} \longrightarrow 0$$

where $\alpha : W \longrightarrow \text{Ker } m$ is D-equivariant The problem is to show that this is split exact in $\mathcal{M}$ and to identify W and α.

A general element in $P_1 \otimes P_s$ can be written $T \otimes f_1 + S \otimes f_2$ where $f_1, f_2 \in P_s$. Then $m(T \otimes f_1 + S \otimes f_2) = Tf_1 + Sf_2$. Hence $m(T \otimes f_1 + S \otimes f_2) = 0$

implies $f_1 = Sg$ and $f_2 = -Tg$ where $g \in P_{s-1}$. Therefore

$$\text{Ker } m = \{T \otimes Sg - S \otimes Tg : g \in P_{s-1}\} .$$

Define $\alpha : P_{s-1} \longrightarrow P_1 \otimes P_s$ by

$$\alpha(g) = T \otimes Sg - S \otimes Tg .$$

Since $(D \otimes 1 + 1 \otimes D)(T \otimes Sg - S \otimes Tg)$

$$= (S \otimes Sg - 0 \otimes Tg) + T \otimes SDg - S \otimes Sg - S \otimes TDg$$

$$= T \otimes SDg - S \otimes TDg$$

$$= \alpha(Dg) ,$$

it follows that α is D-equivariant, that α is injective and maps P_{s-1} onto the kernel of m .

Now I construct a splitting $\beta : P_1 \otimes P_s \longrightarrow P_{s-1}$ of α by defining

$$\beta'(T \otimes f + S \otimes g) = \frac{\partial f}{\partial S} - \frac{\partial g}{\partial T} .$$

It follows that

$$\beta'(T \otimes Sg - S \otimes Tg) = \frac{\partial Sg}{\partial S} + \frac{\partial Tg}{\partial T}$$

$$= g + \frac{\partial g}{\partial S} S + g + \frac{\partial g}{\partial T} T$$

$$= 2g + (s-1)\, g = (s+1)\, g .$$

(Since $f \in P_{s-1}$ it follows that

$$(s-1)g = \frac{\partial g}{\partial S} S + \frac{\partial g}{\partial T} T) .$$

Hence $(\beta' \circ \alpha)(g) = (s+1)g$. Since $\text{char}(A) = 0$, and $s \geqslant 2$, it follows that :

$$\beta := \frac{1}{s+1} \beta'$$

splits α . Also :

$$D \left(\frac{\partial f}{\partial S} - \frac{\partial g}{\partial T}\right) = \frac{\partial}{\partial T} \left(\frac{\partial f}{\partial S} - \frac{\partial g}{\partial T}\right) S ,$$

while $\beta'((D\otimes 1 + 1\otimes D)(T\otimes f + S\otimes g)) =$

$$\beta'(T\otimes Df + S\otimes(f+Dg)) =$$

$$\frac{\partial}{\partial S}\left(\frac{\partial f}{\partial T}S\right) - \frac{\partial}{\partial T}\left(f + \frac{\partial g}{\partial T}S\right) =$$

$$\frac{\partial^2 f}{\partial S\partial T}S + \frac{\partial f}{\partial T} - \frac{\partial f}{\partial T} - \frac{\partial^2 g}{\partial T^2}S =$$

$$D\left(\frac{\partial f}{\partial S} - \frac{\partial g}{\partial T}\right)$$

Hence β' and therefore β is D-equivariant. The general formula in the lemma now follows from the formula

$$(P_1\otimes P_s, D) \cong (P_{s+1}, D) \oplus (P_{s-1}, D) \qquad \text{q.e.d.}$$

<u>Chebyshev polynomials</u> : We take a little side step. Define polynomials $V_{n+1}(X) \in \mathbb{Z}[X]$ by the generating function

$$\frac{1}{1-XT+T^2} = \sum_{n=o}^{\infty} V_{n+1}(X)\,T^n \qquad \text{in} \qquad \mathbb{Z}[X][[T]] .$$

Then
$$V_1(X) = 1 ,$$
$$V_2(X) = X ,$$
$$V_3(X) = X^2-1 ,$$
$$\ldots$$

In general $V_2(X)\,V_n(X) = V_{n+1}(X) + V_{n-1}(X)$.

Hence, in $R_a(k)$ we have the formula

$$[P_r, D] = V_{r+1}([P_1, D]) ,$$

where $[M,\alpha]$ denotes the class of (M,α) in $R_a(k)$. Since the classes $[P_r, D]$ generate $R_a(k)$ additively, we obtain a ring homomorphism

$$\mathbb{Z}[V] \longrightarrow R_a(k)$$

by $$f(V) \longrightarrow f([P_1, D])$$

which is clearly a surjection. Considering both rings as abelian groups, then $Z[V]$ is a free $\mathbb{Z}$-module with basis given by the polynomials

$$\{V_1(X), V_2(X), ..\},$$

while $R_a(A)$ is a free $\mathbb{Z}$-module with basis :

$$\{[P_o,D], [P_1,D], ... \}.$$

Hence the map is a bijection. q.e.d.

It should be remarked that, as k-modules :

$$P_r \simeq S^r(P_1)$$

and that this isomorphism is D-equivariant. Indeed the formal group law induces an action on the symmetric and exterior powers of each object in $\mathcal{N}$ via :

$$S^r(D)(x_1...x_r) := \sum_{j=1}^{r} x_1...x_{j-1}(Dx_j)\, x_{j+1}...x_r$$

and

$$\Lambda^r(D)(x_1\Lambda...\Lambda x_r) := \sum_{j=1}^{r} x_1\Lambda...\Lambda Dx_j\Lambda...\Lambda x_r .$$

In general let $F(X_1,...X_n) = F(X_1,F(X_2,F(...,F(X_{n-1},X_n))$, which is unambiguous by the associativity of F . Then given (M,α) in $\mathcal{N}$, we get

$$S^r(\alpha) : S^r(M) \longrightarrow S^r(M)$$

via

$$S^r(\alpha) = F(\alpha\otimes 1...\otimes 1,\ 1\otimes\alpha\otimes 1,...),\quad \text{etc...}$$

Thus we obtain λ-operations on $R_F(k)$. In the case that A is a field containing $\mathbb{Q}$ these can be explicated on $\mathbb{Z}[V]$. (The definition of a λ-operation appears in section 5).

It is easier to consider the quadratic extension of $\mathbb{Z}[V]$ given by :

$$\sigma^2 - \sigma V + 1 = 0 ,$$

so

$$V = \sigma + \sigma^{-1} .$$

Let $B := \mathbb{Z}[V][\sigma]$ where $\sigma^2 - \sigma V + 1 = 0$. It follows that

$$V_{n+1}(V) = \frac{\sigma^{n+1} - \sigma^{-(n+1)}}{\sigma - \sigma^{-1}}$$

$$= \sigma^n + \sigma^{n-2} + \ldots + \sigma^{-(n)} .$$

in B. Let $a_1(\xi_1,\ldots,\xi_r)$, $a_2(\xi_1,\ldots,\xi_r)$, ... be the elementary symmetric functions and $h_1(\xi_1,\ldots)$, $h_2(\xi_1,\ldots),\ldots,$ the complete symmetric functions. Then in B define

$$\Lambda^r(V_{n+1}(V)) := a_r(\sigma^n,\sigma^{n-2},\ldots\sigma^{-n})$$

and

$$S^r(V_{n+1}(V)) := h_r(\sigma^n,\sigma^{n-1},\ldots,\sigma^{-n}).$$

It follows that these are elements in $\mathbb{Z}[V]$. More particularly, it can be shown (see [Almkvist-Fossum]), that the elements

$$\frac{V_n(V)\, V_{n-1}(V) \ldots V_{n-r+1}(V)}{V_r(V)\, V_{r-1}(V) \ldots V_1(V)} = \Lambda^r(V_n(V))$$

and

$$\frac{V_{n+1}(V)\, V_{n+2}(V) \ldots V_{n+r}(V)}{V_1(V)\, V_2(V) \ldots V_r(V)} = S^r(V_{n+1}(V))$$

in the field of fractions of $\mathbb{Z}[V]$ are in fact in $\mathbb{Z}[V]$.

<u>Proposition 1.3</u> <u>The following equations hold</u>

$$S^r(P_n) = \frac{P_n \boxtimes P_{n+1} \boxtimes \ldots \boxtimes P_{n+r-1}}{P_o \boxtimes P_1 \boxtimes P_2 \boxtimes \ldots \boxtimes P_{r-1}}$$

<u>and</u>

$$\Lambda^r(P_n) = \frac{P_n \boxtimes P_{n-1} \boxtimes \ldots \boxtimes P_{n-r+1}}{P_o \boxtimes P_1 \boxtimes \ldots \boxtimes P_{r-1}}$$

<u>in</u> $\mathbb{Z}[V] = R_a(k)$. q.e.d.

Let $\lambda_T(W) := \sum_{r=o}^{\infty} [\Lambda^r(W)]T^r$ in $R_a(k)[[T]]$.

Then $\lambda_T(P_n) = \prod_{r=o}^{n} (1+\sigma^{n-2r}T)$ in $B[[T]]$.

(A short remark about Chebyshev polynomials. Define polynomials $U_n(X)$ by $U_n(\cos\theta) = \sin(n+1)\theta / \sin\theta$. Then it follows that :

$$(1 - 2XT + T^2)^{-1} = \sum_{r=u}^{\infty} U_r(X)\, T^r.$$

Therefore $V_{r+1}(X) = U_r(X/2)$. It seems interesting that the rational functions :

$$\binom{V_m(X)}{V_n(X)} := \frac{V_m(X)\, V_{m-1}(X) \ldots V_{m-n+1}(X)}{V_n(X)\, V_{n-1}(X) \ldots V_1(X)}$$

are indeed polynomials in $\mathbb{Z}$. It follows immediately from the definition that

$$V_m(2) = m \quad \text{for all } m .$$

Which is the reason for the indexing !

Hence $\binom{V_m(2)}{V_n(2)} = \binom{m}{n}$. It is possible to show by induction that :

$$V_m(X)\, V_n(X) = \sum_{r=1}^{\inf(m,n)} V_{n+m-2r+1}(X) .$$

This gives another proof that :

$P_r \otimes P_s = P_{r+s} \oplus \ldots \oplus P_{|r-s|}$. But this also gives possibilities for finding the decomposition of $\Lambda^r(P_m)$ and $S^r(P_m)$ by hand.

For example :

$$\Lambda^3(S^7(P_1)) = P_{15} \oplus P_{11} \oplus P_9 \oplus P_7 \oplus P_5 \oplus P_3 = S^3(P_5) .$$

It also follows that :

$$S^r(S^q(P_1)) \cong S\,(S^q(P_1)) \cong \Lambda^r(S^{q-1}(P_1)), \text{ etc...})$$

§ 2 Calculation and properties for a field of positive characteristic.

In this section we study $R_m(k)$ where m is the multiplicative formal group law, i.e.

$$m(X,Y) = X + Y + X\,Y$$

and k is a field of charasteristic $p > 0$. (We note that most of the formulas that appear below hold in case $p = 2$, but to avoid having to write the special cases when $p = 2$, we delay the discussion of this case to the end of this paper).

Let $A = k[[X]]$ and set $A_q := k[[X]]/(X^q)$ where $q = p^e$. Then m induces a ring homomorphism

$$A_q \to A_q \otimes A_q$$

given by $$X \mapsto X \otimes 1 + 1 \otimes X + X \otimes X.$$ Since char $k = p$, we get

$$(X \otimes 1 + 1 \otimes X + X \otimes X)^q = 0,$$

so this map is well defined. In A_q let $S = X + 1$, then $X = S - 1$, and $(S-1)^q = S^q - 1 = 0$. It follows that :

$$S \mapsto S \otimes S.$$

Hence we conclude that A_q is the group ring $k[\mathbf{Z}/q\mathbf{Z}]$ where $\mathbf{Z}/q\mathbf{Z}$ is written multiplicatively with generator σ, and the map m is just the Hopf algebra map for a group ring.

Let $\mathcal{M}_q$ denote the (full) subcategory of $\mathcal{M}$ consisting of those pairs (M,α) such that $\alpha^q = 0$. Then $\mathcal{M}_q$ is just the category of A_q-modules of finite type. Just as before we construct the representation ring of A_q and denote it by $R^q_m(k)$ or better $R^e_m(k)$ (where $q = p^e$). We want to find a presentation of $R^e(k)$ as a ring. This is solved essentially in [Almkvist-Fossum] and we review the results here. Much of the formalities is similar to the characteristic zero case.

For the next few pages we work only with A_q-modules. Let $V_r := A_q/(X^r) = k[[X]]/(X^r)$ for $r = 1,2,\dots q$. These form a complete set of representatives for the isomorphism classes of indecomposable A_q-modules. Hence as an abelian group $R^q(k)$ is free on the generators $[V_1], \dots, [V_q]$. Green (cf.[Almkvist-Fossum]) and others have determined the multiplication tables for these representatives. It is seen that the virtual modules

$$[V_2]\ ,\ [V_{p+1}] - [V_{p-1}]\ ,\ \ldots\ ,\ [V_{p^{e-1}+1}] - [V_{p^{e-1}-1}]$$

genrate $R^e(k)$ as a $\mathbb{Z}$-algebra. Thus in $R^e(k)$ (or simply R^e) let

$$x_o = [V_2] \quad \text{and} \quad x_i = [V_{q+1}] - [V_{q-1}] \quad \text{when} \quad q = p^i .$$

Capital letters $X_o, \ldots$ will denote variables with corresponding small letters the images in R^e. For each $i > 0$ set $w_i = [V_{p^i}] - [V_{p^{i-1}}]$, with $w_o = 1$. It follows from Green's formulas that $w_i^2 = 1$ in R^e. The results in [Almkvist-Fossum] are given below.

<u>Proposition</u> 2.1 <u>The map</u> $\mathbb{Z}[X_o, \ldots, X_{e-1}] \longrightarrow R^e(k)$ <u>given by</u> $X_i \mapsto x_i$ <u>has kernel generated by the polynomials</u>

$$F_i(X_o, \ldots, X_i) := (X_i - 2W_i)\, V_p(X_i)$$

<u>where</u> $W_i := W_{i-1} V_p(X_{i-1}) - V_{p-1}(X_{i-1})$ <u>defines the polynomials</u> W_i <u>inductively</u>. (The polynomials $V_p(X)$ have been defined in the previous section).

We want to find "better" generators and relations. Note that the polynomials $V_r(X)$ are even if r is odd. Hence if Z is a 2-unipotent ($Z^2 = 1$), then $V_r(ZX) = V_r(X)$. The isomorphism above implies that

$$R^e(k) = R^{e-1}(k)\ [X_{e-1}]\ /(X_{e-1} - 2w_{e-1})\, V_p(X_{e-1})\ ,$$

(and note that $w_{e-1} \in R^{e-1}(k)$). Let $Y_{e-1} = w_{e-1} X_{e-1}$ as a variable over R^{e-1}. Then

$$(X_{e-1} - 2w_{e-1})\, V_p(X_{e-1}) = w_{e-1}(Y_{e-1} - 2)\, V_p(Y_{e-1}).$$

But w_{e-1} is a unit, so

$$R^e(k) = R^{e-1}(k)\ [\,Y_{e-1}\,]\ /(Y_{e-1} - 2)\, V_p(Y_{e-1})\ .$$

Corollary 2.2. There is a ring homomorphism

$$\mathbf{Z}[Y_o, Z_1, \ldots, Y_{e-1}] \longrightarrow R^e(k)$$

which is a surjection with kernel generated by the elements

$$(Y_i - 2)\ V_p(Y_i) \quad \underline{\text{for}} \quad i = 0,1,\ldots,e-1.$$

Now set $W(X) = V_p(X) - V_{p-1}(X)$ and let $u_i = W(y_i)$ in $R^e(k)$.

Lemma 2.3 The elements $w_i = w(y_o)\, w(y_1) \ldots w(y_{i-1})$ for $i \geqslant 1$.

The elements : $[V_p r] = V_p(y_o)\, V_p(y_1) \ldots V_p(y_{r-1})$

The main result of this section follows from Corollary 2.2.

Theorem 2.4 The ring $R^e(k) \cong R^1(k) \underset{\mathbf{Z}}{\otimes} \ldots \underset{\mathbf{Z}}{\otimes} R^1(k)$ (e times).

Proof $R^e(k) = \mathbf{Z}[Y_o,\ldots,Y_{e-1}] / ((Y_o-2)\ V_p(Y_o),\ldots(Y_{e-1}-2)\ V_p(Y_{e-1}))$.

Remark These calculations hold for the multiplicative group law $m(X,Y)$. If F is any group law on k, where char $k = p$ and $q = p^e$, then F induces a ring homomorphism (denoted by F)

$$F : k[[X]]/_{(X^q)} \longrightarrow k[[X]]/_{(X^q,Y^q)} = k[[X]]/_{(X^q)} \otimes k[[Y]]/_{(Y^q)} .$$

Hence the catorgory $\mathcal{N}_q$ is also closed under the tensor product, so we may construct the ring $R^e_F(k)$ for F just as was done in the case $F = m$. It seems likely that these rings are independent of F. I hope to return to this in a later paper. See also Section 5.

(Aded in proof : These rings are independent of F, a result that will appear Later. However the λ-operations depend of F, as is seen is Section 5).

§ 3 Consequences

In this section we draw consequences of the main result in the last section.

Theorem 3.1 *Let C be a commutative $\mathbb{Z}$-algebra. Then*

$$\mathrm{Hom}_{Alg}(R^e(k),C) = \{(r_o,\dots,r_{e-1}) \in C^e : (r_e-2)\,V_p(r_i) = 0\}.$$

Proof This follows immediately since $R^e = (R^1)^{\otimes e}$ and

$$\mathrm{Hom}_{Alg}(A\otimes B,C) = \mathrm{Hom}_{Alg}(A,C) \times \mathrm{Hom}_{Alg}(B,C)\,.$$

Theorem 3.2 *Suppose C is a reduced $\mathbb{Z}/p\mathbb{Z}$- algebra. Then*

$$\mathrm{Hom}_{Alg}(R^e(k), C) = \{c \in C : c^2 = 4\}^e$$
$$= \{(c_o,\dots,c_{e-1}) : c_i^2 = 4\}\,.$$

Proof We first need an expansion of the polynomials $V_r(X)$ about $X = 2$.

Lemma 3.3. *The Taylor series expansion of $V_r(X)$ about $X = 2$ is given by*

$$V_r(X) = \sum_{n=o}^{r-1} \binom{n+r}{2n+1}(X-2)^n\,.$$

Proof Since $(1-XT+T^2)^{-1} = \sum_{r=o}^{\infty} V_{r+1}(X)\,T^r$ we get

$$1 - XT + T^2 = (1-T)^2 - (X-2)T = (1-T)^2\left(1-(X-2)\frac{T}{(1-T)^2}\right).$$

Hence

$$(1-XT+T^2)^{-1} = (1-T)^{-2}\left(1-(X-2)\frac{T}{(1-T)^2}\right)^{-1}$$

$$= (1-T)^{-2}\sum_{n=o}^{\infty}(X-2)^n\frac{T^n}{(1-T)^{2n}} = \sum_{n=o}^{\infty}(X-2)^n\frac{T^n}{(1-T)^{2(n+1)}}$$

Now expand the terms involving T to get

$$(1-XT+T^2)^{-1} = \sum_{r=o}^{\infty}\left(\sum_{n=o}^{\infty}(X-2)^n\binom{2n+r+1}{r}T^{n+r}\right)$$

and then change the order of summation, summing over $n+r$ to get

$$(1-XT+T^2)^{-1} = \sum_{s=o}^{\infty} \left(\sum_{n=o}^{\infty} \binom{n+s+1}{2n+1} (X-2)^n \right) T^s .$$

Another formula follows by setting $X = \sigma + \sigma^{-1}$. Then

$(1-XT+T^2) = (1-\sigma T)(1-\sigma^{-1}T)$. Hence :

<u>Lemma</u> 3.4 $V_{r+1}(\sigma+\sigma^{-1}) = \dfrac{\sigma^{r+1} - \sigma^{-(r+1)}}{\sigma - \sigma^{-1}}$

<u>Corollary</u> 3.5 $V_p(X) \equiv (X-2)^{\frac{p-1}{2}} (X+2)^{\frac{p-1}{2}} = (X^2-4)^{\frac{p-1}{2}} \pmod p$.

<u>Proof</u> From Lemma 3.4 we get

$$V_p(\sigma+\sigma^{-1}) \equiv \frac{(\sigma-\sigma^{-1})^p}{\sigma-\sigma^{-1}} \equiv (\sigma-\sigma^{-1})^{p-1}$$

Now $X = \sigma+\sigma^{-1}$. Therefore

$$X^2 - 4 = (\sigma+\sigma^{-1})^2 - 4 = (\sigma-\sigma^{-1})^2 .$$

This congruence I first found in [Renaud], where some of the results of this section are suggested, in particular he proves theorem 3.2 for $R^e(k) \otimes \mathbb{Z}/p\mathbb{Z}$ in place of $R^e(k)$.

Now theorem 3.2 follows from theorem 3.1 and Corollary 3.5.

Just as in [Renaud] it is possible to study the idempotents of $R^e(k) \otimes \mathbb{Z}[\frac{1}{2}]$ (i.e. invert 2). It should be mentioned that in $R^1(k) = \mathbb{Z}[X]/(X-2)V_p(X)$ we have the $w_1 = [V_p] - [V_{p-1}]$. So in $R^1(k) \otimes \mathbb{Z}[\frac{1}{2}]$ we get two idempotents $e_2 = \frac{1}{2}(1+w_1)$ and $e_1 = \frac{1}{2}(1-w_1)$. Then letting :

$$R^1(k) \otimes \mathbb{Z}[\tfrac{1}{2}] =: R^1_{(2)}$$

we get $R^1_{(2)} = R^1_{(2)} e_1 \times R^1_{(2)} e_2$ where, of course

$$R^1_{(2)} e_1 = R^1_{(2)}/e_2 R^1_{(2)} , \quad R^1_{(2)} e_2 = R^1_{(2)}/e_1 R^1_{(2)} .$$

(The subscript (2) denotes the base change to $\mathbb{Z}[\frac{1}{2}]$). So we want to consider the rings :

$$\mathbb{Z}[X]/((X-2)\,V_p(X),\ 1+V_p(X)-V_{p-1}(X)) \qquad \text{and}$$

$$\mathbb{Z}[X]/((X-2)\,V_p(X),\ 1+V_{p-1}(X)-V_p(X)).$$

Lemma 3.6 The ideals

$$((X-2)\,V_p(X),\ 1+V_p(X)-V_{p-1}(X)) = (X-2)(V_{\frac{p+1}{2}}(X)+V_{\frac{p-1}{2}}(X))$$

$$((X-2)\,V_p(X),\ 1+V_{p-1}(X)-V_p(X)) = (X,2)(V_{\frac{p+1}{2}}(X)-V_{\frac{p-1}{2}}(X)).$$

Proof This is proved by using relations involving the elements $V_r(X) \pm V_{r-1}(X)$.

Corollary 3.7 $R^1_{(2)}e_1 \cong \mathbb{Z}[\frac{1}{2},X]\ /((X-2)\ (V_{\frac{p+1}{2}}(X)+V_{\frac{p-1}{2}}(X))$

$$R^1_{(2)}e_2 \cong \mathbb{Z}[\tfrac{1}{2},X]\ /(V_{\frac{p+1}{2}}(X)-V_{\frac{p-1}{2}}(X)).$$

Proof The ideal (X,2) in $\mathbb{Z}[\frac{1}{2},X]$ is the whole ring.

Note that $(X-2,\ V_{\frac{p+1}{2}}(X)+V_{\frac{p-1}{2}}(X)) = (X-2,p)$. Hence we get the cartesien square :

$$\begin{array}{ccc} R^1_{(2)}\,e_1 & \longrightarrow & \mathbb{Z}[\frac{1}{2},X]\ /\ (V_{\frac{p+1}{2}}(X)+V_{\frac{p-1}{2}}(X)) \\ \downarrow & & \downarrow \\ \mathbb{Z}[\frac{1}{2}] & \longrightarrow & \mathbb{Z}/p\,\mathbb{Z}\ . \end{array}$$

Examples : Let p = 3 . Then

$$R^1_{(2)} = (\mathbb{Z}[X]/(X-2)(X+1))_{(2)} \times \mathbb{Z}_{(2)}.$$

Let $p = 7$. Then

$$R^1_{(2)} = (\ \mathbb{Z}[X]/(X-2)(X^3+X^2-2X-1)_2 \times (\mathbb{Z}[X]/(X^3-X^2-2X+1))_2 .$$

Modulo p , the elements look like :

$$V_{\frac{p+1}{2}}(X) + V_{\frac{p-1}{2}}(X) \equiv (X-2)^{\frac{p-1}{2}} \qquad (\text{mod } p)$$

$$V_{\frac{p+1}{2}}(X) - V_{\frac{p-1}{2}}(X) \equiv (X+2)^{\frac{p-1}{2}} \qquad (\text{mod } p).$$

<u>Corollary</u> 3.8 $\quad R^1 \underset{\mathbb{Z}}{\otimes} \mathbb{Z}/p\mathbb{Z} \cong (\mathbb{Z}/p\mathbb{Z}\,[X]/(X-2)^{\frac{p+1}{2}}) \times (\mathbb{Z}/p\mathbb{Z}[X]/(X+2)^{\frac{p-1}{2}})$

We now apply this to R^e, using theorem 3.2 and the relation

$$(A \times B) \otimes C = (A \otimes C) \times (B \otimes C).$$

<u>Corollary</u> 3.9 $\quad R^e(k) \underset{\mathbb{Z}}{\otimes} (\mathbb{Z}/p\mathbb{Z}) \cong$

$$\overset{e}{\underset{r=o}{\times}} \binom{e}{r}\ \mathbb{Z}/p\,\mathbb{Z}\,[X_o,X_1,\ldots,X_e]/(X_o-2)^{\frac{p+1}{2}},\ldots,(X_{e-r}-2)^{\frac{p+1}{2}},(X_{e-r}+2)^{\frac{p-1}{2}},\ldots,(X_{e-1}+2)^{\frac{p-1}{2}})$$

(This means $\binom{e}{r}$ copies of the ring that follows).

<u>Corollary</u> 3.10 [Renaud] . $R^e(k) \underset{\mathbb{Z}}{\otimes} \mathbb{Z}/p\,\mathbb{Z}$ <u>is the ring product of</u> 2^e <u>local rings of the form</u>

$$\mathbb{Z}/p\mathbb{Z}\ ([U_o,U_1,\ldots U_{e-1}]/U_o^{r_o},\ldots,U_{e-1}^{r_{e-1}})$$

(truncated polynomial rings), where $U_i = X_i \pm 2$.

<u>Example</u> char $k = 5$, $q = 5^3$. Then

$$\mathbb{Z}/5\mathbb{Z} \otimes R^3(k) = \mathbb{Z}/5\mathbb{Z}\,[U_o,U_1,U_2]/U_o^3,U_1^3,U_2^3) \times 3(\,\mathbb{Z}/5\mathbb{Z}[\,U_o^3,U_1^3,U_2^2]\,)$$

$$\times\ 3(\,\mathbb{Z}/5\mathbb{Z}[\,U_o^3,U_1^2,U_2^2]) \times \mathbb{Z}/5\mathbb{Z}\,[U_o,U_1,U_2]/(U_o^2,U_1^2,U_2^2).$$

§ 4 Induction, Restriction and Inclusion

In this section we assume A is a field of characteristic $p > 0$. As noted before, $\mathcal{N}$ is the category of $A[[X]]$-modules of finite length. Let $e \in \mathbb{N}_0$ and set $\mathcal{N}^e = \{M \in \mathcal{N} : X^{p^e} M = 0\}$. Then $\mathcal{N}^e$ is isomorphic to the category of $A[[X]]/(X^{p^e})$-modules. Let $f : A[[X]]/(X^{p^e}) \longrightarrow A[[X]]/(X^{p^{e+1}})$ denote the Frobenius map given by :

$$f(\alpha(X)) := \alpha(X^p).$$

This is an injection. It is clear that the diagram

$$\begin{array}{ccc} A[[X]]/(X^q) & \xrightarrow{f} & A[[X]]/(X^{pq}) \\ \Big\downarrow F & & \Big\downarrow F \\ A[[X,Y]]/(X^q,Y^q) & \xrightarrow{f \otimes f} & A[[X,Y]]/(X^{pq},Y^{pq}) \end{array}$$

is commutative.

We get induced maps on the representation rings

$$\text{Res} : R_F^{e+1}(A) \longrightarrow R_F^e(A)$$

and

$$\text{Ind} : R_F^e(A) \longrightarrow R_F^{e+1}(A)$$

defined by $\text{Res}(M) = M$ considered as a module through f and $\text{Ind}(M) = C \otimes_B M$, where $C = A[[X]]/(X^{pq})$, $B = A[[X]]/(X^q)$ and where C is considered to be a B-module through f and $C \otimes_B M$ is a C-module through C.

There is also a quotient map $A[[X]]/(X^{pq}) \longrightarrow A[[X]]/(X^q)$ which defines the inclusion $\mathcal{N}^e \hookrightarrow \mathcal{N}^{e+1}$ and the inclusion $\text{Inc} : R_F^e(A) \rightarrow R_F^{e+1}(A)$, defining R^e as a subring of R^{e+1}. This function has been considered before.

For $F = m$ and the presentation given in § 2, the maps Ind and Res can be explicated.

<u>Proposition</u> 4.1 <u>The maps</u> Res <u>and</u> Ind <u>are given on elements by</u>

$$\text{Res}(r(x_o,x_1,\dots x_e)) = r(2,x_o,x_1,\dots,x_{e-1})$$

and $$\mathrm{Ind}\,(r(x_o,\ldots,x_{e-1})) = V_p(x_o)\, r(x_1,\ldots,x_e).$$

In particular

$$\mathrm{Res}\ \mathrm{Ind} = p.\ \mathrm{Id}.$$

Proof Consider Res first. It is clearly a ring homomorphism, so it is necessary to determine its action on the generators x_i of the algebra. Let $x \in A[[X]]/(X^q)$ with image x^p. Suppose $V_n = A[[X]]/(X^n)$, considered as an $A[[X]]/(X^{pq})$-module The action of x through f is by x^p. Let $e_o, e_1,\ldots,e_{n-1}$ be the elements $1, x,\ldots x^{n-1}$ modulo x^n. Then $x^p e_i = e_{i-p}$. Apply this to the modules V_{p^i+1} and V_{p^i-1} to get

$$\mathrm{Res}\,(V_{p^i+1}) = V_{p^{i-1}+1} \oplus (p-1)\, V_{p^i} \quad \text{while}$$

$$\mathrm{Res}\,(V_{p^i-1}) = V_{p^{i-1}-1} \oplus (p-1)\, V_{p^i}\,.$$

Hence $\mathrm{Res}\,(x_i) = x_{i-1}$. Clearly $\mathrm{Res}\,(V_2) = 2\,V_1$ so $\mathrm{Res}\,(x_o) = 2$. (The restriction map can be made quite explicit :

Let $s = s_o + s_1 p$ for $0 \leqslant s_o < p$. Then

$$\mathrm{Res}\,(V_s) = (p-s_o)\,V_{s_1} \oplus s_o\,V_{s_1+1}\,.$$

It follows that

$$\mathrm{Res}\,\Big(\bigoplus_{j=o}^{p^{k+1}} a_j V_j\Big) =$$

$$= \bigoplus_{i=o}^{p^k-1} (a_{ip+1} + 2a_{ip+2} + \ldots + (p-1)\,a_{(i+1)p-1} + pa_{(i+1)p} + (p-1)a_{(i+1)p+1} + \ldots$$
$$\ldots + 2a_{3p-2} + a_{3p-1})\,V_{i+1}$$

$$= \bigoplus_{i=o}^{p^k-1} \Big(\sum_{\ell=1}^{p} \ell\, a_{ip+e} + \sum_{\ell=p+1}^{2p-1} (2p-\ell)\,a_{ip+\ell}\Big)\,V_{i+1}\Big)$$. To get the formula for Ind it is necessary to consider the formula for $\mathrm{Ind}(V_r)$. In fact

$$\mathrm{Ind}\,(V_r) = V_{pr} \quad \text{for each } r\,.$$

For let $A = k[[X]]/(X^q)$ and $B = k[[X]]/(X^{pq})$. Then B is free as an A-module through f. The sequence :

$$0 \longrightarrow X^r A \longrightarrow A \longrightarrow V_r \longrightarrow 0$$

is exact as A-modules. Tensor this with B to get :

$$0 \longrightarrow B \otimes_A X^r A \longrightarrow B \longrightarrow \mathrm{Ind}\, V_r \longrightarrow 0 .$$

But $B \otimes_A X^r A = X^{pr} B \otimes_A A = X^{pr} B$. Hence $\mathrm{Ind}\, V_r = V_{pr}$. Thus to establish the formula it is enough to show that if

$$V_r = p(x_o, x_1, ..) \quad \text{in} \quad R(k)$$

then $$V_{pr} = V_p(x_o)\, p(x_1, x_2, ...) \quad \text{in} \quad R(k)$$. These follow from a much more general formula.

<u>Lemma</u> 4.2 <u>Suppose</u> $r \leq p^k$ <u>and</u> $p > s \geq 1$. <u>Then</u>

$$[V_{sp^k+r}] - [V_{sp^k-r}] = [V_r] \quad (V_{s+1}(x_k) - V_{s-1}(x_k)).$$

<u>Proof</u> For $r = s = 1$, we have

$$[V_{p^k+1}] - [V_{p^k-1}] = x_k \quad \text{by definition.}$$

Hence the formula is true for these values. Then the general result follows from Green's multiplication formulas.

Suppose that $p | r$. Then

$$[V_{sp^k+r}] - [V_{sp^k-r}] = V_p(x_o)\, q\,(x_1, .., x_{k-1})\; (V_{s+1}\,(x_k) - V_{s-1}(x_k))$$

by induction where $V_{r/p} = q(X_o, ..., X_{k-2})$. Hence the result follows in general, again by a mildly complication induction argument.

<u>Corollary</u> 4. 3 $[V_{p^k}] = V_p(x_o)\, V_p(x_1) \; ... \; V_p(x_{k-1})$.

<u>Proof</u> Since $V_{p^k} = \mathrm{Ind}^k(V_1)$, this follows by k applications of the result in the theorem.

Table 4.4. We list the polynomials giving the representations V_1 to V_{27} in characteristic 3 and V_1 to V_{25} in characteristic 5 .

Char k = 3

$V_1 = 1 \qquad = 1$

$V_2 = X_o \qquad = y_o$

$V_3 = X_o^2-1 \qquad = y_o^2-1 \qquad u_o = y_o^2-y_o-1 = X_o^2-X_o-1$

$V_4 = X_o+X_1 \qquad = y_o+(y_o^2-y_o-1)y_1 \qquad X_1 = U_o Y_1$

$V_5 = 1+X_oX_1 \qquad = 1+(y_o^2-2)y_1 \qquad y_1 = V_6-V_5+V_1$

$V_6 = (X_o^2-1)X_1 \qquad = (y_o^2-1)y_1$

$V_7 = X_oX_1+(X_1^2-1) \qquad = (y_o^2-2)y_1+y_1^2-1 \qquad u_1=y_1^2-y_1-1 = V_7 - V_6$

$V_8 = X_o(X_1^2-1)+X_1 \qquad = y_o(y_1^2-1)+(y_o^2-y_o-1)y_1$

$V_9 = (X_o^2-1)(X_1^2-1) \qquad =(y_o^2-1)(y_1^2-1) \qquad V_9-V_8 = u_ou_1$

$V_{10} = X_o(X_1^2-1)+X_1+X_2 \qquad = (y_o^2-1)(y_1+(y_1^2-1)y_2 + y_o(y_1^2-y_1-1)$

$V_{11} = (X_o^2-X_o-1)(X_1^2-1) + (X_o-1)X_1 + (X_1^2-1) + X_oX_2 \qquad y_2 = u_ou_1X_2$

$V_{12} = (X_o^2-1)(X_1+X_2)$

$V_{13} = (X_o+X_1)X_2 + X_oX_1+1$

[The remaining relations for y_o,y_1,y_2 are left for masochists]

$V_{14} = (1+X_oX_1)X_2 + X_o+X_1$

$V_{15} = (X_o^2-1)(1+X_1X_2)$

$V_{16} = (X_oX_1+(X_1^2-1))X_2+X_o$

$V_{17} = (X_o(X_1^2-1)+X_1)X_2+1$

$V_{18} = (X_o^2-1)(X_1^2-1)X_2$

$V_{19} = (X_o(X_1^2-1)+X_1)X_2 + X_2^2-1$

$V_{20} = (X_oX_1+ (X_1^2-1))X_2+X_o+X_o(X_2^2-2)$

$$V_{21} = (X_o^2-1)(X_1X_2 + (X_2^2-1)$$

$$V_{22} = (X_o+X_1)(X_2^2-2) + (1+X_oX_1)X_2 + X_o + X_1$$

$$V_{23} = (1+X_oX_1)(X_2^2-2) + (X_o+X_1)X_2 + X_oX_1 + 1$$

$$V_{24} = (X_o^2-1)(X_1(X_2^2-1) + X_2)$$

$$V_{25} = (X_oX_1 + (X_1^2-1))(X_2^2-2) + (X_o^2-X_o)(X_1^2-1) + (X_o-1)X_1+X_oX_2$$

$$V_{26} = X_o(X_1^2-1) + X_1 + X_2 +(X_o(X_1^2-1) + X_1)(X_2^2)2)$$

$$V_{27} = (X_o^2-1)(X_1^2-1)(X_2^2-1).$$

<u>Char k = 5</u>

$$V_1 = 1$$

$$V_2 = y_o$$

$$V_3 = y_o^2-1$$

$$V_4 = y_o^3-2y_o$$

$$V_5 = y_o^4-3y_o^2+1$$

$$V_6 = V_4+u_oy_1$$

$$V_7 = V_3 + u_o\, y_o\, y_1$$

$$V_8 = V_2 + (y_o^2-1)\; u_o\; y_1$$

$$V_9 = V_1 + (y_o^3-2y_o)\; u_o\, y_1$$

$$V_{11} = V_9 + (y_1^2-2)$$

$$V_{12} = V_8 + y_o\; (y^2-2)$$

$$V_{13} = V_7 + (y_o^2-1)(y_1^2-2)$$

$$V_{14} = V_6 + (y_o^3-2y_o)(y_1^2-2)$$

$$V_{10} = (y_o^4 - 3\,y_o^2 +1)y_1$$

$$V_{15} = (y_o^4 - 3y_o^2+1)(y_1^2-1)$$

$$V_{20} = (y_o^4 - 3y_o^2 + 1)(y_1^3-2y_1)$$

$$V_{25} = (y_o^4-3y_o^2 + 1)\,(y_1^4 - 3y_1^2+1).$$

$$u_o = V_5 - V_4$$

$$= y_o^4 - y_o^3 - 3y_o^2 + 2y_o + 1$$

$$y_o\, u_o = V_5 - V_3$$

$$= y_o^4 - 4y_o^2 + 2$$

$V_{16} = V_{14} + u_o(y_1^3 - 3y_1)$

$V_{17} = V_{13} + y_o u_o(y_1^3 - 3y_1)$

$V_{18} = V_{12} + (y_o^2 - 1)\, u_o(y_1^3 - 3y_1)$

$V_{19} = V_{11} + (y_o^3 - 2y_o)\, u_o(y_1^3 - 3y_1)$

$V_{21} = V_{19} + (y_1^4 - 4y_1^2 + 2)$

$V_{22} = V_{18} + y_o(y_1^4 - 4y_1^2 + 2)$

$V_{23} = V_{17} + (y_o^2 - 1)(y_1^4 - 4y_1^2 + 2)$

$V_{25} = V_{16} + (y_o^3 - 2y_o)(y_1^4 - 4y_1^2 + 2).$

§ 5 - λ-structures on the representation rings

In this section the λ-structures on the representations rings are studied. It is seen that different formal group laws give different λ-structures.

Let A be a commutative ring. Let $W_1(A)$ denote the ring whose underlying additive abelian group is the set of formal power series of the form $1 + \sum_{i=1}^{\infty} a_i T^i \in A[[T]]$, with multiplication as the operation, and formal inverse as the negative. The operation is still denoted by juxtaposition. The multiplication in the ring is too difficult to write in general. However, denoting it by $\boxtimes$, the formula :

$$\left(\prod_{i=1}^{\infty} (1+a_iT)\right) \boxtimes \left(\prod_{j=1}^{\infty} (1+b_jT)\right) = \prod_{i,j} (1+a_ib_jT)$$

will uniquely define the operation. So in fact :

$$(1+T) \boxtimes f(T) = f(T) ,$$

and $1+T$ is the identity element.

Let A be a commutative ring. An abelian group homomorphism :

$$\lambda_T : (A,+) \longrightarrow (W_1(A),.)$$

is called a λ-<u>structure</u> on A. Since for each $a \in A$, the element

$$\lambda_T(a) = 1 + \sum_{i=1}^{\infty} \lambda_i(a)\, T^i ,$$

a λ-structure on A is given by a family of functions $\lambda_i : A \longrightarrow A$ such that

$$\lambda_o(a) = 1 \quad \text{for all} \quad a \in A .$$

$$\lambda_i(0) = 0 \quad \text{for all} \quad i > 0 .$$

$$\lambda_1(a) = a \quad \text{for all} \quad a \in A .$$

$$\lambda_n(a+b) = \sum_{p=o}^{n} \lambda_p(a)\, \lambda_{n-p}(b), \text{ for all } a,b \subset A \text{ and all } n .$$

The ring A with λ-structure λ_T is called a λ-ring if λ_T is a ring homomorphism.

<u>Example</u> Let $\lambda_T : \mathbb{Z} \longrightarrow W_1(\mathbb{Z})$ be given by

$$\lambda_T(n) = (1+T)^n .$$

Then

$$\lambda_r(n) = \binom{n}{r} .$$

Suppose that k is a field and F is a formal group law on k . Let V be a finite length $k[[X]]$-module. Define the symmetric and exterior powers of V , denoted by $\Lambda^r_F(V)$ and $S^r_F(V)$, as follows :

The underlying vector space of $\Lambda^r_F(V)$ (resp. : of $S^r_F(V)$) is the exterior power $\Lambda^r(V)$ (resp. : the symmetric power $S^r(V)$) of V as a vector space. Each is a homomorphic image of $V^{\otimes r} = V \otimes_k \overset{r}{\cdots} \otimes_k V$. The vector spaces are given a $k[[X]]$-module structure by making these surjections F-equivariant.

<u>Example</u> If F = a , then

$$X.(v_1 \wedge \ldots \wedge v_r) := \sum_{i=1}^{r} v_1 \wedge \ldots \wedge (Xv_i) \wedge \ldots \wedge v_r$$

$$X.(v_1 \ldots v_r) := \sum_{i=1}^{r} v_1 \ldots (Xv_i) \ldots v_r$$

If F = m, then :

$$X.(v_1 \wedge v_2) = (Xv_1) \wedge v_2 + v_1 \wedge (Xv_2) + (Xv_1) \wedge Xv_2 .$$

$$X.(v_1 \wedge v_2 \wedge v_2) = Xv_1 \wedge (v_2 \wedge v_3) + v_1 \wedge X.(v_2 \wedge v_3) + Xv_1 \wedge X(v_2 \wedge v_3).$$

$$\vdots$$

Proposition 5.1 Let k be a field and F a formal group law on k (assumed commutative as always). Then the maps

$$\lambda_T^F : R_F(k) \longrightarrow W_1(R_F(k))$$

$$\sigma_T^F : R_F(k) \longrightarrow W_1(R_F(k))$$

defined by
$$\lambda_T^F(V) := \sum_{n=0}^{\infty} \Lambda^n{}_F(V)T^n \quad \text{and} \quad \sigma_T^F(V) := \sum_{n=0} S_F^n(V)T^n$$

are λ-structures on $R_F(k)$. If char $k = p > 0$, then each of these restricts to a λ-structure on $R_F^e(k)$ for each $e \geq 0$.

Proof It is sufficient to show that

$$\Lambda_F^r (V \boxtimes W) = \bigoplus_{p=0}^{r} (\Lambda_F^p(V) \underset{k}{\boxtimes} \Lambda^{r-p}(W))$$

and similarly for the symmetric powers. The isomorphism is described explicitly for $r = 2$. The general case follows by the associativity of the group law F .

The map ℓ: $\Lambda^2(V \oplus W) \xrightarrow{\sim} \Lambda^2(W) \oplus V \boxtimes W \oplus \Lambda^2(V)$ is given by

$$\ell\,(v_1 + w_1) \wedge (v_2 + w_2)) = w_1 \wedge w_2 + (v_1 \boxtimes w_2 - v_2 \boxtimes w_1) + v_1 \wedge v_2 .$$

We want to show that

$$\ell\,(X.(z_1 \wedge z_2)) = X.\ \ell(z_1 \wedge z_2)$$

where $z_1 = v_1 + w_1$, $z_2 = v_2 + w_2$. So compare the two formulas (using the fact that $F(X,Y) = F(Y,X)$.). First,

$$X(z_1 \wedge z_2) = Xz_1 \wedge z_2 + z_1 \wedge Xz_2 + \sum c_{ij} X^i z_1 \wedge X^j z_2$$

Now $(Xz_1 \wedge z_2) = Xw_1 \wedge w_2 + (Xv_1 \boxtimes w_2 - v_2 \boxtimes Xw_1) + Xv_1 \wedge v_2$. Similarly for $\ell(z_1 \wedge Xz_2)$ and $\ell\,(X^i z_1 \wedge X^j z_2)$. It is seen that the terms in $\Lambda^2(w)$ are :

$$Xw_1 \wedge w_2 + w_1 \wedge Xw_2 + \sum c_{ij}\ X^i w_1 \wedge X^j w_2 \text{ , etc...}$$

(This is just to say that the decompositions

$$\Lambda^r(V \oplus W) \cong \bigoplus_{p=o}^{r} \Lambda^p(V) \underset{k}{\boxtimes} \Lambda^{r-p}(W)$$

are F-equivariant). Thus the maps given are λ-structures. It is clear that if $X^q V = 0$, then $X^q \Lambda^r{}_F(V) = 0$ and similarly for the symmetric powers. Hence the coefficients of the power series defining Λ_T^F and σ_T^F are each in $R_F^e(k)$ if the arguments are in $R_F^e(k)$. This shows that the λ-structures restrict.

In the case char k = 0 , these λ-structures have been examined in § 1. In fact they are λ-ring structures ; they satisfy the identifies :

$$\Lambda_T^F (V \underset{k}{\boxtimes} W) = \lambda_T^F(V) \boxtimes \lambda_T^F(W) .$$

$$\lambda_{-T}^F(V) \ \sigma_T^F(V) = 1 .$$

So the λ_T and σ_T structures are essentially the same (one is the inverse of the other).

The situation in case char k = p > 0 is quite different. In [Almkvist-Fossum] there is an extensive study of the λ-structures λ_T^m , σ_T^m on $R_m^1(k)$. Further papers by Akmkvist (see references) have delved more deeply into this problem for $R_m^1(k)$. In particular they are <u>not</u> λ-ring structures.

<u>Problems</u> 5.2 : Let chark = p > 0 and let F be a formal group law on k .

a) For each indecomposable V determine the decompositions :

$$\Lambda_F^r(V) = \bigoplus_{s>o} \ell(F,r,s)\, V_s$$

$$S_F^r(V) = \bigoplus_{s>o} m(F,r,s)\, V_s .$$

b) For each indecomposable V determine rational function expression :

$$\lambda_T^F(V) = \frac{p(T)}{q(T)} \qquad \text{where} \qquad q(T) \in [T] .$$

c) Determine $\lambda_{-T}^F(V)\ \sigma_T^F(V)$.

d) Determine when :

$$\lambda_T^F(V \boxtimes W) = \lambda_T^F(V) \boxtimes \lambda_T^F(W)$$

holds.

The complexity of these problems is shown by the examples in the next section. In the remainder of this section, some general results, which may help in giving a partial solution to some of these problems, are given for the multiplicative formal group law m, the problem that I first attacked.

Several preliminary results are recalled. (In what follows the indexing by F or m is usually dropped. It is almost always assumed that the group $\mathbb{Z}/q\mathbb{Z}$, generated by an element τ, is operating. The field k has characteristic $p > 0$).

<u>Koszul complexes</u> : Let V be a finite dimensional vector space over k. Consider the complexes

$$0 \longrightarrow \Lambda^r(V) \xrightarrow{e_r} \Lambda^{r-1}(V) \otimes S^1(V) \xrightarrow{e_{r-1}} \Lambda^{r-2}(V) \otimes S^2(V) \longrightarrow \dots \xrightarrow{e_1} S^r(V) \longrightarrow 0$$

and

$$0 \longrightarrow S^r(V) \xrightarrow{d_r} \Lambda^1(V) \otimes S^{r-1}(V) \xrightarrow{d_{r-1}} \dots \xrightarrow{d_1} \Lambda^r(V) \longrightarrow 0$$

with maps given by the formulas :

$$e_j(v_1 \wedge \dots \wedge v_j \otimes s) = \sum_{s=1}^{j} (-1)^{s-1} (v_1 \wedge \dots \wedge \hat{v}_s \wedge \dots \wedge v_j) \otimes v_s s$$

and

$$d_j(e \otimes v_1 \dots v_j) = \sum_{s=1}^{j} (e \wedge v_s) \otimes v_1 \dots \hat{v}_s \dots v_j .$$

(Note that these maps are F-equivariant for any F !).

These complexes are exact (they are the graded parts of the Koszul complex giving the free resolution of k as an $S^{\cdot}(V)$ -module) and they are split exact in case $(p,r) = 1$, since :

$$d_{j+2}\, e_{r-j-1} + e_{r-j}\, d_{j+1} = r.\,\mathrm{Id}$$

for each j.

<u>Proposition</u> 5.2 <u>The power series</u> :

$$\lambda_{-T}^F(V) \quad \sigma_T^F(V) \in R_F(k)\,[[T^p]].$$

<u>Proof</u> The coefficient of T^r in this power series is :

$$\sum_{j=o}^{r} (-1)^j \; (\Lambda^{r-j}(V) \underset{k}{\otimes} S^j(V))$$

and this is zero if $(p,r) = 1$.

Another useful result concerns the decompositions of the induced modules.

<u>Proposition</u> 5.3 <u>The principal ideal generated by</u> $[V_p] = [V_p(X_o)]$ <u>in</u> $R_m^e(k)$ <u>is the ideal generated by the elements</u> $[V_{pr}]$ <u>for</u> $pr \leqslant p^e$.

<u>Proof</u> This follows directly from Prop. 4. 1.

<u>Corollary</u> 5.4. <u>If</u> $r \geqslant 1$, <u>then the elements</u>

$$S^q(V_{pr}) \quad \text{and} \quad \Lambda^q(V_{pr})$$

<u>are in the ideal generated by</u> $[V_p]$ <u>provided</u> $(q,p) = 1$.

<u>Proof</u> This follows from the relation :

$$(\text{Ind } V) \underset{k}{\otimes} W = \text{Ind } (V \underset{k}{\otimes} \text{Res } W)$$

and the fact that $S^q(V_{pr})$ and $\Lambda^q(V_{pr})$ are direct summands of $V_{pr} \otimes S^{q-1}(V_{pr})$ and $V_{pr} \otimes \Lambda^{q-1}(V_{pr})$ respectively.

Another useful result relates Res with the tensor functors.

<u>Proposition</u> 5.5. For any V , the following hold :

$$\text{Res } (\Lambda^q(V)) = \Lambda^q \text{ (Res V)}$$

$$\text{Res } (S^q(V)) = S^q \text{ (Res V)} .$$

<u>Proof</u> The underlying vector spaces are the same for the modules on each side of the equalities. And the action of X is given by X^p in both cases.

<u>Corollary</u> 5.6 <u>The diagrams</u>

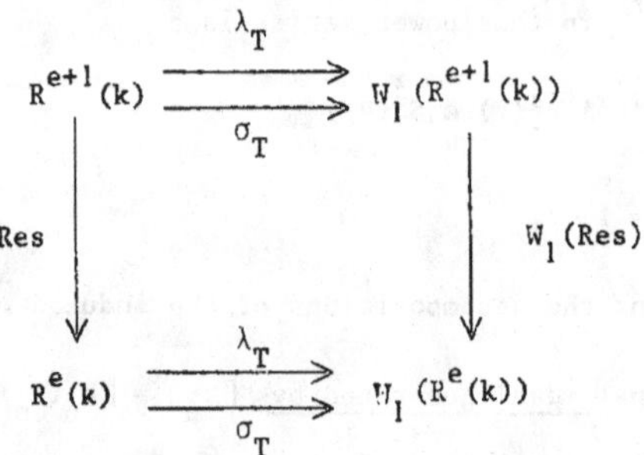

commute.

As an example, consider $\lambda_T(V_1)$ in char k = 3. Then write $\lambda_T(V_4) = \sum_{j=1}^{9} a_j(T)[V_j]$.

We conclude that:

$$\begin{aligned}\lambda_T(\mathrm{Res}\, V_4) = \; & (a_1 + 2a_2 + 3a_3 + 2a_4 + a_5)\,[V_1] \\ + \; & (a_4 + 2a_5 + 3a_6 + 2a_7 + a_8)\,[V_2] \\ + \; & (a_7 + 2a_8 + 3a_9)\,[V_3] .\end{aligned}$$

Here the $a_j(T)$ have non-negative integer coefficients. Now $\mathrm{Res}(V_4) = [V_2] + 2[V_1]$, and

$$\begin{aligned}\lambda_T(V_2 \oplus 2V_1) &= (1 + [V_2]\,T + T^2)(1+T)^2 \\ &= (1+T^2)(1+T)^2\,[V_1] + T(1+T)^2\,[V_2] .\end{aligned}$$

Hence $a_7 + 2a_8 + 3a_9 = 0$, so $a_7 = a_8 = a_9 = 0$. Then

$$a_4 + 2a_5 + 3a_6 = T(1+T)^2 .$$

Since no coefficient in $T(1+T)^2$ is as big as 3, we get $a_6 = 0$. Likewise $a_3 = 0$. Hence $a_4(T) + 2a_5(T) = T(1+T)^2$. Thus $a_4(0) = a_5(0) = 0$. So our two equations are:

$$a_1(T) + 2\,a_2(T) + 2\,a_4(T) + a_5(T) = (1+T^2)(1+T)^2$$

$$a_4(T) + 2a_5(T) = T(1+T)^2 .$$

From $a_1(0) + 2a_2(0) = 1$, we get :

$$a_1(0) = 1 \quad \text{and} \quad a_2(0) = 0 .$$

Calculation with the second equation yields :

$$a_4(T) = T + nT^2 + T^3$$

$$a_5(T) = \quad mT^2 \qquad\qquad \text{where} \quad n + 2m = 2 .$$

Then using these in the first equation yields $2n + m \leqslant 2$. Hence $n = 0, m = 1$.

So :

$$a_4(T) = T(1+T^2)$$

$$a_5(T) = \quad T^2$$

Thus $a_1(T) + 2a_2(T) = 1 + T^2 + T^4$, which implies that $a_2(T) = 0$. Hence :

$$\lambda_T(V_4) = (1+T^2+T^4)\,[V_1] + (T+T^3)\,[V_4] + T^2\,[V_5] .$$

$$= 1 + [V_4]\,T + [V_1 \oplus V_5]\,T^2 + [V_4]\,T^3 + T^4 .$$

This implies that $\Lambda^2(V_4) = V_1 \oplus V_5$.

The similar computation for V_6 yields :

$$\text{Res}\,(\lambda_T(V_6)) = \lambda_T(3V_2) = (1 + V_2T + T^2)^3 .$$

The coefficient of T^3 is $7V_2 \oplus 2\,V_3$; This implies that :

$$\Lambda^3(V_6) = a_6\,V_6 \oplus a_7\,V_7 \oplus a_8\,V_8 .$$

where $3a_6 + 2a_7 + a_8 = 7$ and $a_7 + 2a_8 = 2$.

There are two possible solutions to this set of equations :

$$a_6 = 1 ,\ a_7 = 2 ,\ a_8 = 0$$

$$a_6 = 2 ,\ a_7 = 0 ,\ a_8 = 1 .$$

A quick calculation shows :

$$\Lambda^3_m (V_6) \cong V_6 \oplus V_2 \oplus V_7 = 2V_6 \oplus V_8$$

$$\Lambda^3_a (V_6) \cong V_6 \oplus 2V_7 .$$

This shows that the λ-structures induced on $R^e(k)$ are indeed different for these two different formal group laws. And it raises more questions than it answers. We hope to return to these problems later.

Finally we derive a result that should have great possibilities in determining decompositions. Before doing so we need some additional notation. We suppose $q = p^e$ and that τ is a multiplicative generator for the group $\mathbb{Z}/pq\,\mathbb{Z}$. Let V_{pq} have a basis $X_1, X_2, \dots, X_{pq}$ on which the generator τ acts as a cycle permutation ; so that

$$\tau(X_i) = X_{i+1} \quad \text{for} \quad i < pq$$

and

$$\tau(X_{pq}) = X_1 .$$

We suppose also that $V_{pq} = k\,[[X]]/(X^{pq})$ and that the basis here is given by $e_i = X^{pq-i-1}$, so $e_{pq-1} = 1$, $e_{pq-1} = x, \dots, e_o = x^{pq-1}$. The relation between these two basis can be given by relating the operators χ and τ so that $\chi = \tau - 1$. Then $X_1 = e_{pq-1}$, $\chi X_1 = e_{pq-2} = X_2 - X_1$ etc... Note that $V_{pq-r-1} = V_{pq}/\langle e_o, e_1, \dots, e_r \rangle = k\,[[X]]/(X^{pq-r-1})$ as a $\mathbb{Z}/pq\,\mathbb{Z}$-module. The symmetric algebras $S^{\cdot}(V_n)$ are polynomial algebras on which the group $\mathbb{Z}/pq\,\mathbb{Z}$ acts as a group of algebra automorphisms. In particular :

$$S^{\cdot}(V_{pq}) = k\,[X_1, \dots, X_{pq}] = k\,[e_o, e_1, \dots, e_{pq-1}] .$$

The surjections $V_{pq} \longrightarrow V_{pq-r-1}$ yield $\mathbb{Z}/pq\,\mathbb{Z}$-equivariant homomorphisms (surjections) :

$$S^{\cdot}(V_{pq}) \longrightarrow S^{\cdot}(V_{pq-r-1}) .$$

and each has kernel generated by the linear forms $(e_o,\dots,e_r)$. Call this ideal I_{r+1}.

Let $Y_i := X_i X_{i+q} \dots$ for $i = 1,2,\dots,q$.

So $Y_i = \prod_{r=o}^{p-1} \tau^{qr}(X_i)$. Note that :

$$\tau Y_i = Y_2 ,\ \tau Y_2 = Y_3, \dots, \tau Y_{q-1} = Y_q \quad \text{and} \quad \tau Y_q = Y_1 .$$

Hence the submodule of $S^p(V_{pq})$ spanned by the $Y_1,\dots,Y_q$ is isomorphic to V_q. Let $k[Y_1,\dots,Y_q]$ denote the subalgebra spanned by these elements.

<u>Proposition</u> 5.7 a) <u>If</u> $1+r \leqslant (p-1)q$, <u>then</u>

$$I_{r+1} \cap k[Y_1,\dots,Y_q] = (0) .$$

b) <u>The elements</u> $Y_1,\dots,Y_q$ <u>are algebraically independent and form a regular sequence on</u> $S^{\cdot}(V_{pq-r-1})$ for $r+1 \leqslant (p-1)q$.

<u>Proof</u> Consider the $\mathbb{Z}/pq\,\mathbb{Z}$-equivariant homomorphisms

$$\alpha_q : k[X_1,\dots,X_{pq}] \longrightarrow k[Z_1,\dots,Z_q] , \quad Z_i \text{ variables} ,$$

given by $X_{i+rq} \longrightarrow Z_i$ for $1 \leqslant i \leqslant q$ and $0 \leqslant r \leqslant p-1$.

Since $I_o \supset I_1 \supset \dots \supset I_{(p-1)q}$, it is sufficient to show that $I_{(p-1)q} \cap k[Y_1,\dots,Y_q] = (0)$. The image of Y_i under this map is Z_i^p. Hence the composite

$$k[Y_1,\dots,Y_q] \hookrightarrow k[X_1,\dots,X_{pq}] \longrightarrow k[Z_1,\dots,Z_q]$$

is the Frobenius : $f(Y_1,\dots,Y_q) \to f(Z_1^p,\dots,Z_q^p)$.

which is an injection. So it is sufficient to show that :

$$\operatorname{Ker} \alpha_q = I_{(p-1)q} .$$

The one basis relation is

$$X_{r+1} = \sum_{j=o}^{r} \binom{r}{j} e_{pq-j-1} ,$$

since $\tau(e_s) = e_s + e_{s+1}$ for $s > 0$.

It follows that :

$$X_{r+tq} - X_r \in I_{(p-1)q}$$

and hence $\text{Ker}\ \alpha_q \subset I_{(p-1)q}$. Since both ideals are prime ideals of the same height (by a Krull dimension argument), they must be equal. This proves (a) .

Statement (b) follows immediately, since the length of :

$$k[X_1,\ldots,X_{pq}] / (I_{(p-1)q}, Y_1,\ldots,Y_4)$$

is finite (it is isomorphic to $k[Z_1,\ldots,Z_q]/(Z_1^p,\ldots,Z_q^p))$ and since this ideal is generated by pq homogeneous elements, they form a regular $k[X_1,\ldots,X_{pq}]$ -sequence. But any regular sequence of homogeneous elements remains regular under a permutation. So $(Y_1,\ldots,Y_q)$ is a regular sequence on $k[X_1,\ldots,X_{pq}]$.

There are two consequences to draw from this result.

__Corollary__ 5.8 __The algebra__ $k[X_1,\ldots,X_{pq}]$ __regarded as a graded module over__ $k[Y_1,\ldots,Y_q]$ __is free__ (but of infinite rank).

__Proof__ This follows immediately from [Bourbaki Alg. de Lie,...].

__Corollary__ 5.9 __Suppose__ $n > q$. __Then__ $S^{pr}(V_n)$ __contains a__ $\mathbb{Z}/pq\mathbb{Z}$ __direct summand isomorphic to__ $S^r(V_q)$. __Or in other words__ :

$$S^{pr}(V_n) \cong S^r(V_q) \oplus \ ?$$

__as a direct sum decomposition.__

__Proof__ The module $S^{pr}(V_n)$ is the $pr^{\underline{th}}$ homogeneous component of $S^{\cdot}(V_n)$. It was shown in [Almkvist-Fossum] that $S^r(V_q)$ is a direct summand of $S^{pr}(V_{pq})$, where

$S^r(V_q)$ was identified as the $r^{\underline{th}}$ homogeneous component of $k[Y_1,\ldots,Y_q]$ (which is graded in two ways ; with deg $Y_i = 1$ when considered as an $\mathbb{Z}/q\mathbb{Z}$-algebra and deg $Y_i = p$ when considered as a subring of $S^\cdot(V_{pq})$). The sequence :

$$0 \longrightarrow I_s \cap S^{pr}(V_{pq}) \longrightarrow S^{pr}(V_{pq}) \longrightarrow S^{pr}(V_n) \longrightarrow 0$$

is exact as $\mathbb{Z}/pq\,\mathbb{Z}$-modules (where I_s is the kernel of $S^\cdot(V_{pq}) \to S^\cdot(V_n)$). Since

$$(I_s \cap S^{pr}(V_{pq})) \quad \cap \; S^r(V_q) = 0$$

the component $S^r(V_q)$ lives in $S^{pr}(V_n)$ as a direct summand.

Of course the modules $S^r(V_q)$ have decompositions that have been explicity described in [Almkvist-Fossum] . I feel that these techniques should be extremely useful, but have not been able to exploit them fully.

§ 7 - Characteristic $p = 2$

In this section, some of the calculations from Sections 2 and 3 that do not apply in case char (k) = 2 are made. In any case the multiplications of Green apply and the elements $[V_{2^i+1}] - [V_{2^i-1}]$ still generate the algebras. So we can state the next result.

Proposition 7.1 Suppose char k = 2 . Then

$$R^e_m(k) \cong \mathbb{Z}[X_o,\ldots,X_{e-1}]/((X_o-2)X_o,(X_1-2W_1)X_1,\ldots)$$

(where the W_i are the preimages of $[V_{2^i}] - [V_{2^i-1}]$).

The elements $w_i = [V_{2^i}]-[V_{2^i-1}]$ are 2-unipotents so the generators can be changed to the $y_i = x_iw_i$ just as before. Then $R^1(k) = \mathbb{Z}[X_o]/((X_o-2)X_o)$ and we get the next result, also just as before.

<u>Proposition</u> 7.2 <u>In case</u> char (k) = 2 , then

$$R^e(k) = \mathbb{Z}[Y_o,\dots,Y_{e-1}]/((Y_o-2)Y_o,\dots,(Y_{e-1}-2)Y_{e-1})$$

$$\cong R^1(k)^{\oplus e}.$$

The next corollary also follows immediately.

<u>Corollary</u> 7.3 <u>Let</u> C <u>be a commutative ring.</u>

<u>Then</u> $\mathrm{Hom}_{Alg}(R^e(k),C) = \{(c_o,\dots c_{e-1}) : c_i^2 = 2c_i\}$.

If $1/2 \in C$, then this set is exactly the set of idempotents in C . If $2 = 0$ in C , then this is just the 2-nilpotents in C .

<u>Corollary</u> 7.4 <u>Suppose</u> C <u>is a</u> $\mathbb{Z}/2\mathbb{Z}$ - <u>algebra.</u> <u>Then</u>

$$R^e(k) \underset{\mathbb{Z}}{\otimes} C \cong C[X_o,\dots,X_{e-1}]/(X_o^2,\dots,X_{e-1}^2).$$

<u>In particular if</u> C <u>is a local ring with</u> 2 = 0 <u>in</u> C , <u>then</u> $R^e(k) \underset{\mathbb{Z}}{\otimes} C$ <u>is a local ring.</u> (Compare [Renaud] and his references) .

Since $(\mathbb{Z}[X]/((X-2)X)) \underset{\mathbb{Z}}{\otimes} \mathbb{Z}[\frac{1}{2}] = \mathbb{Z}[\frac{1}{2}] \times \mathbb{Z}[\frac{1}{2}]$, it follows that

$$R^e(k) \underset{\mathbb{Z}}{\otimes} \mathbb{Z}[\tfrac{1}{2}] = (\mathbb{Z}[\tfrac{1}{2}])^{2^e}.$$

In general the square

$$\begin{array}{ccc} R^1(k) & \longrightarrow & \mathbb{Z} \\ {\scriptstyle X\mapsto 2}\downarrow & {\scriptstyle X\mapsto o} & \downarrow \\ \mathbb{Z} & \longrightarrow & \mathbb{Z}/2\mathbb{Z} \end{array}$$

is cartesian.

References

ALMKVIST G. and R. FOSSUM : This seminaire. Lecture Notes in Mathematics n° 641, Springer.

HAZEWINKEL M. : Formal groups and Applications New York, San Francisco, London London : Academic Press 1978.

RENAUD J.C. : The characters and structure of a class of modular representations algebras of cyclic p-groups. J. Austral. Math. Soc. (Ser A) 26, 410 - 418 (1978).

ROBERT FOSSUM
University of Illinois
Department of Mathematics
1409 W. Green St.
URBANA, ILLINOIS 61801, USA

Mai 1981.

CLASSES CARACTERISTIQUES POUR LES REPRESENTATIONS DE GROUPES DISCRETS

par

Guido Mislin

Introduction

Soient $K \subset \mathbb{C}$ un corps de nombres et $\rho : G \to GL(K)$ une K-représentation d'un groupe fini G ($GL(K)$ est la réunion des groupes $GL_j(K)$, $j \geq 1$, pour les inclusions habituelles). Les classes de Chern $c_m(\rho) \in H^{2m}(G;\mathbb{Z})$ sont definies comme classes de Chern du fibré associé à ρ, qui est une fibré vectoriel complexe plat sur BG (l'espace classifiant du groupe discret G). Nous désignons par $E_K(m)$ la meilleure borne pour l'ordre des $c_m(\rho)$, ρ parcourant toutes les K-représentations de tous les groupes finis G. Ces bornes $E_K(m)$ permettent d'obtenir l'ordre précis des classes de Chern universelles $c_m(\mathcal{O}) \in H^{2m}(GL(\mathcal{O});Z)$ dans le cas où $\mathcal{O} = \mathcal{O}(K)$ est l'anneau des entiers dans un corps de nombres qui n'est pas formellement réel. Les valeurs explicites données ici pour les nombres $E_K(m)$ ont été obtenues en collaboration avec B. Eckmann; on trouve les détails concernant les sections 2 et 3 dans [5]. Dans la section 4 nous considérons la conjecture suivante concernant la fonction zêta ζ_K du corps de nombres K.

<u>Conjecture:</u> $E_K(m)\ \zeta_K(1-m) \in \mathbb{Z}$ pour tout entier $m > 1$.

((*): Voir la note à la fin de cet exposé.)

§ 1 Rappels sur les fibrés plats

Soient X un CW-complexe connexe et ξ un fibré vectoriel complexe sur X . Alors ξ est classifié par une classe d'homotopie

$$f_\xi : X \to BGL_j(\mathbb{C})$$

où j note la dimension des fibres. Le fibré ξ est dit <u>plat</u> s'il est associé à un fibré principal à groupe structural discret; cela peut s'exprimer par une factorisation de f_ξ (à une homotopie près) de la manière suivante:

$$\begin{array}{ccc} X & \xrightarrow{f_\xi} & BGL_j(\mathbb{C}) \\ & {\scriptstyle can} \searrow \quad \nearrow {\scriptstyle B\rho} & \\ & B\pi_1(X) & \end{array}$$

(où $\rho : \pi_1(X) \to GL_j(\mathbb{C})$ est une représentation.)

Si on désigne par $\hat{c}_n$ la n-ième classe de Chern du fibré universel sur $BGL_j(\mathbb{C})$, on a $c_n(\xi) := f_\xi^*(\hat{c}_n)$ et, si ξ est plat, $c_n(\xi) = can^*\ c_n(\rho)$, où $c_n(\rho) := (B\rho)^*\ \hat{c}_n \in H^{2n}(\pi_1(X);\mathbb{Z})$ est la n-ième classe de Chern de la représentation ρ ; on note que $c_n(\rho)$ est la n-ième classe de Chern d'un fibré plat, induit par $B\rho$ sur $B\pi_1(X)$.

D'après un théorème classique (le "Théorème de Chern-Weil"), les

les classes de Chern réelles d'un fibré ξ plat, $c_n(\xi)_{\mathbb{R}} \in H^{2n}(X;\mathbb{R})$, sont toutes nulles pour $n > 0$. Le théorème a tout d'abord été demontré pour une variété differentiable X et un fibré différentiable ξ admettant une connexion à courbure nulle-condition qui, dans ce contexte est équivalente à la platitude de ξ (voir [7]). C'est aussi une conséquence d'un résultat de Deligne-Sullivan [3]: Ils ont montré que pour tout fibré complexe plat sur un CW-complexe fini on peut trouver un revêtement fini $\overline{X} \to X$ de X tel que le fibré induit sur $\overline{X}$ soit trivial; cela entraîne en particulier que les classes de Chern d'un tel fibré sont annulées par le degré du revêtement $\overline{X} \to X$.

Soient $K \subset \mathbb{C}$ un corps de nombres et $\mathcal{O} = \mathcal{O}(K)$ l'anneau des entiers de K . L'inclusion $GL_j(\mathcal{O}) \subset GL_j(\mathbb{C})$ définit un fibré canonique sur $BGL_j(\mathcal{O})$ dont nous notons les classes de Chern

$$c_n(\mathcal{O},j) \in H^{2n}(GL_j(\mathcal{O});\mathbb{Z})$$

Il est bien connu que l'espace $BGL_j(\mathcal{O})$ a le type d'homotopie d'un complexe ayant pour squelettes des complexes finis. D'après le théorème de Chern-Weil les classes $c_n(\mathcal{O},j)$, $n > 0$, sont alors de torsion. Comme on sait d'après Charney [1] que la restriction

$$\mathrm{res} : H^n(GL(\mathcal{O});\mathbb{Z}) \to H^n(GL_j(\mathcal{O});\mathbb{Z})$$

est un isomorphisme pour $j >> n$, l'ordre $|c_m(\mathcal{O},j)|$ est indépendant de j pour $j >> m$, et cet ordre est égal à l'ordre

$|c_m(\mathcal{O})|$ de la classe universelle $c_m(\mathcal{O}) \in H^{2m}(GL(\mathcal{O}); \mathbb{Z})$. Une borne supérieure pour $|c_m(\mathcal{O})|$ a été calculée par Grothendieck [6], et une borne inférieure par Soulé [10]. Dans la section 2 nous donnons une borne inférieure plus précise.

§ 2 La meilleure borne $E_K(m)$

On désigne par G un groupe fini, par $K \subset \mathbb{C}$ un corps de nombres et par $\rho : G \to GL(K)$ une K-représentation, à classes de Chern $c_m(\rho) \in H^{2m}(G; \mathbb{Z})$, $m \geqslant 0$. A l'aide du caractère de la représentation ρ on peut montrer que si τ est un automorphisme de $\mathbb{C}$ qui laisse K fixe, on a

$$c_m(\rho) = j^m c_m(\rho) \quad ,$$

où j est un entier tel que τ opère sur les $|G|$-ièmes racines ω d'unité par $\tau\omega = \omega^j$. L'analyse des relations de ce type pour tout τ fournit un entier $\overline{E_K(m)}$ avec la propriété que

$$\overline{E_K(m)}\ c_m(\rho) = 0$$

pour toute K-représentation ρ d'un groupe fini arbitraire, c.à.d. qu'on obtient une borne supérieure pour l'ordre des $c_m(\rho)$. Cette borne peut être décrite de la manière suivante.

$$\overline{E_K(m)} = \text{pp cm } \{n \mid m \equiv 0 \mod \exp(\mathrm{Gal}(K_n/K))\}$$

où $K_n = K(\zeta_n)$, ζ_n une racine primitive n-ième de l'unité, et où $\exp(\mathrm{Gal}(K_n/K))$ désigne l'exposant du groupe de Galois de K_n sur K .

Pour obtenir une borne inférieure pour l'ordre des $c_m(\rho)$, on peut considérer les classes de Chern de certaines représentations de groupes cycliques. Soit $n > 0$ tel que $m \equiv 0 \mod (K_n : K)$; posons $m = i \cdot (K_n : K)$. Pour la représentation évidente du groupe cyclique C_n d'ordre n dans $(K_n)^i \cong K^m$, un calcule simple montre que la classe de Chern c_m est d'ordre n (maximum possible) dans $H^{2m}(C_n ; \mathbb{Z})$. Si on pose

$$\underline{E_K}(m) = \mathrm{pp\,cm}\{n | m \equiv 0 \mod (K_n : K)\} \quad \text{et}$$

si on définit la "meilleure borne" par

$$E_K(m) = \min\{n > 0 | n\, c_m(\rho) = 0 \text{ pour tout } \rho : G \to GL(K) \text{ et tout groupe fini } G\}$$

alors

$$\underline{E_K}(m) \leq E_K(m) \leq \overline{E_K(m)}$$

Il est facile à voir que la borne supérieure $\overline{E_K(m)}$ est la même que celle déterminée par Grothendieck [6] dans un cadre plus général. De même, la borne inférieure correspond à une borne obtenue par Soulé [10]. Ces deux bornes diffèrent au plus d'un facteur 2. Pour $E_K(m)$ nous avons le résultat suivant.

<u>Théorème 1:</u> Soit $K \subset \mathbb{C}$ un corps de nombres. La meilleure borne $E_K(m)$ pour l'ordre des classes de Chern $c_m(\rho)$ des K-représentations $\rho : G \to GL(K)$ est donné par

$$E_K(m) = \begin{cases} \overline{E_K(m)}\,, & \text{si } m \text{ est impair ou si } K \\ & \text{n'est pas formellement réel} \\ \underline{E_K(m)} = \frac{1}{2}\,\overline{E_K(m)}\,, & \text{si } m \text{ est pair et} \\ & K \text{ formellement réel.} \end{cases}$$

La démonstration dérive d'une analyse des K-représentations des 2-groupes, en particulier des groupes cycliques, diédraux, semi-diédraux et quaternioniens.

§ 3 Applications à la cohomologie des groupes arithmétiques

Soient comme avant G un groupe fini et $\rho : G \to GL_j(K)$ une K-représentation, $K \subset \mathbb{C}$ un corps de nombres. Il est facile à voir (en utilisant que $\mathcal{O}(K) = \mathcal{O}$ est un anneau de Dedekind) que $\rho \oplus 1 : G \to GL_{j+1}(K)$ est équivalente à une représentation ρ' qui se factorise par $GL_{j+1}(\mathcal{O})$. Comme $c_m(\rho) = c_m(\rho')$ pour tout m, on en déduit que $E_K(m)$ est un diviseur de $|c_m(\mathcal{O})|$, l'ordre de la classe universelle $c_m(\mathcal{O}) \in H^{2m}(GL(\mathcal{O});\mathbb{Z})$. D'autre part $|c_m(\mathcal{O})|$ divise $\overline{E_K(m)}$ d'après [6]. En utilisant le Théorème 1 on a:

<u>Théorème 2:</u> L'ordre de $c_m(\mathcal{O}) \in H^{2m}(GL(\mathcal{O});\mathbb{Z})$, $m > 0$, est soit

$E_K(m)$ soit $\overline{E_K(m)}$; il est $= \overline{E_K(m)}$ si m est impair ou si le corps des fractions K de $\mathcal{O}$ n'est pas formellement réel.

Soit p un nombre premier. Il est clair que la partie p-primaire $E_K(m)_p$ est un multiple de $E_{\mathbb{Q}}(m)_p$. Le lemme suivant donne une condition suffisante pour l'égalité de ces deux nombres. On désigne par $\mathbb{Q}_{p^\infty}$ la réunion de tous les corps cyclotomiques $\mathbb{Q}_{p^\alpha}$.

<u>Lemme 3:</u> Si $K \cap \mathbb{Q}_{p^\infty} = \mathbb{Q}$, alors la partie p-primaire $E_K(m)_p$ est donnée par

<u>cas p impair:</u> $E_K(m)_p = E_{\mathbb{Q}}(m)_p$

<u>cas p = 2 :</u> $E_K(m)_2 = E_{\mathbb{Q}}(m)_2$ si K est formellement réel ou si m est impair; et $E_K(m)_2 = 2E_{\mathbb{Q}}(m)_2$ si m est pair et K n'est pas formellement réel.

Ce lemme se déduit du fait que $K \cap \mathbb{Q}_p = \mathbb{Q}$ entraîne $K_{p^\alpha} = K \otimes_{\mathbb{Q}} \mathbb{Q}_{p^\alpha}$, d'où $\mathrm{Gal}(K_{p^\alpha}/K) = \mathrm{Gal}(\mathbb{Q}_{p^\alpha}/\mathbb{Q})$ pour tout $\alpha > 0$.

On observe que si le nombre premier p ne divise le discriminant de K la condition $K \cap \mathbb{Q}_{p^\alpha} = \mathbb{Q}$ est satisfait, car le discriminant de $\mathbb{Q}_{p^\alpha}$ est, au signe près, une puissance de p .

Considérons comme exemple le cas d'un corps quadratique K . L'intersection $K \cap \mathbb{Q}_{p^\infty}$ est alors K ou $\mathbb{Q}$. Lorsque $K \cap \mathbb{Q}_{p^\infty} = K$, K est un sous-corps quadratique de $\mathbb{Q}_{p^\infty}$. Rappelons que les sous-corps quadratiques de $\mathbb{Q}_{p^\infty}$ sont donnés par

<u>cas p impair:</u> $\mathbb{Q}(\sqrt{(-1)^{p-1/2}\,p})$, (corps de discriminant $(-1)^{p-1/2}\,p$) .

<u>cas p = 2 :</u> $\mathbb{Q}(\sqrt{-1})$, $\mathbb{Q}(\sqrt{-2})$, $\mathbb{Q}(\sqrt{2})$, (corps de discriminants -4, -8 et 8 respectivement).

<u>Corollaire 4:</u> Soit K un corps quadratique différent de $Q(\sqrt{-1})$, $Q(\sqrt{-2})$, $Q(\sqrt{2})$ et $Q(\sqrt{(-1)^{p-1/2}\,p})$, p étant un nombre premier. Alors

$$E_K(m) = \begin{cases} E_{\mathbb{Q}}(m) , & \text{si } m \text{ est impair ou } K \text{ réel} \\ 2E_{\mathbb{Q}}(m) , & \text{si } m \text{ est pair et } K \text{ imaginaire} \end{cases}$$

On peut aussi sans peine calculer $E_K(m)$ pour les corps quadratiques exclus dans ce corollaire. Le théorème suivant, qui est plus précis que le résultat de Ch. Thomas dans [11], est une conséquence du calcul de $E_K(m)$ pour tout corps quadratique K ([5]) et des formules établies dans [4]: $E_{\mathbb{Q}}(m) = 2$, pour m impair, et $=$ dén(B_m/m) , pour $m > 0$ pair (le dénominateur de B_m/m , où les B_m sont les nombres de Bernoulli: $B_2 = 1/6$, $B_4 = 1/30$ etc.).

Théorème 5: Soient $K \subset \mathbb{C}$ un corps quadratique imaginaire et $\mathcal{O} \subset K$ l'anneau des entiers de K. L'ordre des classes de Chern universelles $c_m(\mathcal{O}) \in H^{2m}(GL(\mathcal{O}); \mathbb{Z})$ est alors comme suit:

(a) : si m est impair, $c_m(\mathcal{O})$ est d'ordre 2 avec les exceptions suivantes:

$(a)_1$: si $K = \mathbb{Q}(\sqrt{-1})$, $c_{2n+1}(\mathcal{O})$ est d'ordre 4 pour tout n.

$(a)_2$: si $K = \mathbb{Q}(\sqrt{-p})$, p un nombre premier $\equiv 3(4)$ et $2m \equiv 0(p-1)$, l'ordre de $c_m(\mathcal{O})$ est $2pm_p$ (où m_p désigne la partie p-primaire de m).

(b) : si m est pair $\neq 0$, l'ordre de $c_m(\mathcal{O})$ est $=$ dén$(B_m/2m)$, (c.à.d. 24 pour $m = 2$, 240 pour $m = 4$ etc.).

§ 4 Relations avec la fonction ζ_K

Soient $K \subset \mathbb{C}$ un corps de nombres et ζ_K sa fonction zêta. Si K n'est pas totalement réel, on sait que $\zeta_K(1-n) = 0$ pour tout entier $n > 1$. Dans le cas où K est totalement réel, $\zeta_K(1-n) = 0$ pour n impair > 1, et $\zeta_K(1-n) \in \mathbb{Q}$ pour un entier pair $n > 0$ ("Théorème de Siegel"). Par un théorème classique on a pour $K = \mathbb{Q}$ et $m > 0$ pair

$$\zeta_{\mathbb{Q}}(1-m) = (-1)^{\frac{m}{2}}(B_m/m) .$$

Il suit alors de notre déscription de $E_{\mathbb{Q}}(m)$ que

$$E_{\mathbb{Q}}(m)\ \zeta_{\mathbb{Q}}(1-m) \in \mathbb{Z} \quad \text{pour tout entier} \quad m > 1 .$$

<u>Théorème 6:</u> Soit $K \subset \mathbb{C}$ un corps de nombres. Alors

$$E_K(2)\ \zeta_K(-1) \in \mathbb{Z}$$

<u>Démonstration:</u> Il suffit de considérer le cas où K est totalement réel. Dans ce cas $E_K(2) = \underline{E_K(2)}$ et, par conséquence

$$E_K(2)_p = \max_{\alpha}\{p^{\alpha} \mid (K_{p^{\alpha}} : K) \leq 2\} .$$

Si on pose $\alpha(p) = \max\{\alpha \mid (K_{p^{\alpha}} : K) \leq 2\}$

on trouve $E_K(2) = \Pi p^{\alpha(p)}$. Serre a montré dans [8, § 3] que $2\Pi p^{\alpha(p)}\ \zeta_K(-1)$ est toujours un entier $\equiv 0 \bmod 2^r$, où $r = (K : \mathbb{Q})$. On en déduit que $E_K(2)\ \zeta_K(-1) \in \mathbb{Z}$.

La relation entre le dénominateur de $\zeta_K(-1)$ et $E_K(2)$ devient plus claire si on rappelle le resultat de Serre [8] (pour un corps K de nombres totalement réel):

$$\zeta_K(-1) = \chi(SL_2(\mathcal{O}(K)))$$

Pour les autres valeurs de $\zeta_K(1-m)$ la situation reste assez mystérieuse. Rappelons tout d'abord un autre résultat de Serre [9] (nous désignons par $\mathbb{Z}_{(p)}$ les entiers localisés en (p)):

Lemma 7: Soit K un corps de nombres totalement réel. Si $m > 0$ est pair et si $r = (K : \mathbb{Q})$, on a

$$rm\ 2^{2-r}\ \zeta_K(1-m) \in \mathbb{Z}_{(2)}$$

Corollaire 8: Soit $K \subset \mathbb{C}$ un corps de nombres. Alors, pour tout entier $m > 1$,

$$E_K(m)\ \zeta_K(1-m) \in \mathbb{Z}_{(2)}$$

Démonstration: Comme on l'a vu, on peut supposer que K est totalement réel et m pair. Dans ce cas $E_K(m)_2 = 2^{\gamma} m_2$ où $\gamma = \max\{\alpha \mid K_4 = K_{2^{\gamma+1}}\} \geq 1$ (cf. [5]).
Si on pose $r = (K : \mathbb{Q})$, on a $r_2\ 2^{2-r} \leq 2$. Par conséquence, $E_K(m)_2$ est un multiple entier de la partie 2-primaire de $rm\ 2^{2-r}$, d'où le résultat, d'après le Lemme 7 .

Dans le cas où l'extension K de $\mathbb{Q}$ est abélienne, on peut utiliser le théorème suivant de Coates et Lichtenbaum [2].

Théorème 9: Soit K un corps de nombres totalement réel qui est une extension abélienne de $\mathbb{Q}$. Si $m > 0$ est pair et si p désigne un nombre premier impair, alors

$$w_m^{(p)}(K)\ \zeta_K(1-m) \in \mathbb{Z}_{(p)} .$$

Les nombres $w_m^{(p)}(K)$ sont definis comme suit ([2]):

Soit K un corps de nombres arbitraire et $\overline{K}$ une clôture algébrique de K. Soit $W_m^{(p)}$ K le group des racines de l'unité d'ordre une puissance de p (p un nombre premier arbitraire). On considère $W_m^{(p)}$ comme $Gal(\overline{K}/K)$-module par l'action $\sigma * x = \sigma^m(x)$, $\sigma \in Gal(\overline{K}/K)$ et $x \in W_m^{(p)}$. Le nombre des éléments de $W_m^{(p)}$ fixes par cette action est noté $w_m^{(p)}(K)$.

Un calcule simple montre que

$$\prod_p w_m^{(p)}(K) = \overline{E_K(m)}$$

pour tout corps de nombre K (on compare la description de $\overline{E_K(m)}$ dans [5] avec celle de $w_m^{(p)}(K)$ dans [10]).

En particulier, si p est un nombre premier impair, on a $w_m^{(p)}(K) = E_K(m)_p$. En utilisant le Théorème 9 et le Corollaire 8 on en déduit:

<u>Théorème 10:</u> Si $K \subset \mathbb{C}$ est une extension abélienne finie de $\mathbb{Q}$, alors

$$E_K(m)\ \zeta_K(1-m) \in \mathbb{Z}$$

pour tout entier $m > 0$.

Bibliographie

1 Charney, R.M.: Homology stability of GL_n of a Dedekind domain. Bull. Amer. Math. Soc. 1(2), 428-431, 1979

2 Coates, J. et Lichtenbaum, S.: On ℓ-adic zêta functions. Annals of Math. 98, 498-550 (1973)

3 Deligne, P. et Sullivan, D.: Fibrés vectoriels complexes à groupe structural discret. C.R. Acad. Sc. Paris, t. 281, Série A, 1081-1083 (1975)

4 Eckmann, B. et Mislin, G.: Rational representations of finite groups and their Euler class. Math. Ann. 245, 45-54 (1979)

5 Eckmann, B. et Mislin, G.: Chern classes of representations of finite groups (à paraître)

6 Grothendieck, A.: Classes de Chern et représentations linéaires des groupes discrets. Dans: Dix exposés sur la cohomologie des schémas. Amsterdam: North-Holland 1968

7 Milnor, J.W. et Stasheff, J.D.: Characteristic classes. Annals of Math. Studies 76 (1974)

8 Serre, J.P.: Cohomologie des groupes discrets. Annals of Math. Studies 70, 77-169 (1971)

9 Serre, J.P.: Congruence formes modulaires (d'après H.P.F. Swinnerton-Dyer). Séminaire Bourbaki 1971/1972; Springer Lecture Notes in Math. Vol. 317, 319-338

10 Soulé, C.: Classes de torsion dans la cohomologie des groupes arithmétiques. C.R. Acad. Sc. Paris, t. 284, Serié A, 1009-1011 (1977)

11 Thomas, Ch.: Characteristic classes of respresentations over imaginary quadratic fields. Springer Lecture Notes in Math. Vol. 788, 471-481

Eidg. Techn. Hochschule
Mathematikdepartement
CH - 8092 Zürich
SUISSE

(*) : Le Théorème 9 à été démontré par P. Cassou-Noguès pour K un corps de nombres totalement réel arbitraire (voir: "Valeurs aux entiers négatifs des fonctions zêta et fonctions zêta p-adiques"; Inventiones math. 51, 29-59 (1979), Théorème 17). Il suit alors de notre Corollaire 8 que $E_K(m)\ \zeta_K(1-m)$ est un entier pour tout entier $m > 1$.

AUTOMORPHISMES DE SCHEMAS ET DE GROUPES DE TYPE FINI

par Hyman Bass*

1. INTRODUCTION.

Commençons avec un problème auquel on peut appliquer les méthodes décrites ici. Soit Σ une surface compacte orientable de genre g. Le groupe :

$$M_g = \pi_0(\text{Homéom}(\Sigma))$$

des classes d'isotopie des homéomorphismes : $\Sigma \to \Sigma$ joue un rôle important dans la topologie des variétés de dimension 3 , aussi bien que dans la théorie des surfaces de Riemann (cf. [B]). Son étude est d'une surprenante difficulté. On sait que M_g est de présentation finie, et on a des renseignements assez précis sur ses sous-groupes finis. Récemment E. Grossman [E.G] a montré que M_g est résiduellement fini, autrement dit que l'intersection de ses sous-groupes d'indice fini est triviale. Sa démonstration repose sur des arguments combinatoires assez pénibles. Nous allons présenter une autre méthode pour démontrer que M_g est résiduellement fini et, en même temps, que M_g est virtuellement sans torsion, autrement dit qu'il possède un sous-groupe d'indice fini sans torsion.

Soit Γ_g le groupe fondamental de Σ. Il admet une présentation :

$$\Gamma_g = \langle a_1, b_1, \ldots, a_g, b_g \mid [a_1,b_1]\cdot[a_2,b_2]\cdots[a_g,b_g] = 1\rangle$$

---

* Il s'agit d'un travail [B-L] fait en collaboration avec Alex Lubotzky.

Un homéomorphisme $h : \sum \to \sum$ détermine une classe d'automorphismes modulo les automorphismes intérieurs de Γ_g, c'est-à-dire un élément de :

$$\mathrm{Out}\,(\Gamma_g) = \mathrm{Aut}\,(\Gamma_g)/_{\mathrm{I\,Aut}(\Gamma_g)} \,.$$

Cet élément ne dépend que de la classe d'isotopie de h, d'où un homomorphisme:

$$\alpha : M_g \to \mathrm{Out}\,(\Gamma_g) \,.$$

D'après un théorème classique de Nielsen (cf. [B], Th. 1.4), α est un isomorphisme. Il nous suffit donc de démontrer que $\mathrm{Out}\,(\Gamma_g)$ est résiduellement fini et virtuellement sans torsion.

Considérons plus généralement, un groupe Γ de type fini quelconque. Ses représentations $\mathrm{Hom}\,(\Gamma, GL_n(\mathbb{C}))$ forment, de façon naturelle, une variété algébrique affine $R_n(\Gamma)$, sur laquelle opère le groupe $GL_n(\mathbb{C})$ par conjugaison. Soit $S_n(\Gamma)$ le quotient algébrique de $R_n(\Gamma)$ par $GL_n(\mathbb{C})$. C'est une variété affine où on peut distinguer les classes des représentations semi-simples de Γ. L'opération naturelle de $\mathrm{Aut}\,(\Gamma)$ sur $R_n(\Gamma)$ définit, par passage au quotient, une opération de $\mathrm{Out}\,(\Gamma)$ sur $S_n(\Gamma)$, à laquelle on peut appliquer le théorème suivant :

<u>Théorème</u>. - <u>Soient</u> V <u>une variété algébrique sur</u> $\mathbb{C}$ <u>et</u> G <u>un groupe de type fini d'automorphismes de</u> V. <u>Alors</u> G <u>est résiduellement fini et virtuellement sans torsion</u>.

<u>Corollaire</u>. - <u>Avec les notations ci-dessus, si</u> $\mathrm{Out}\,(\Gamma)$ <u>opère fidèlement sur</u> $S_n(\Gamma)$, <u>alors tout sous-groupe de type fini de</u> $\mathrm{Out}\,(\Gamma)$ <u>est résiduellement fini et virtuellement sans torsion</u>.

C'est à l'aide de ce corollaire, que nous montrons que $M_g (\cong \mathrm{Out}\,(\Gamma_g))$ possède ces propriétés.

Dans le n°2 nous démontrons le théorème, sous une forme nettement plus générale.

2. AUTOMORPHISMES DE SCHEMAS DE TYPE FINI.

Soient k un anneau commutatif , V un k-schéma de présentation finie, et $\mathrm{Aut}_k(V)$ le groupe des automorphismes du k-schéma V .

Théorème. - Supposons que $k = \mathbb{Z}$.

a) $\mathrm{Aut}_k\, V$ est résiduellement fini.

b) Si V est plat sur $\mathbb{Z}$, $\mathrm{Aut}_k(V)$ est virtuellement sans torsion.

Corollaire. - Soit k quelconque et soit Γ un sous-groupe de type fini de $\mathrm{Aut}_k(V)$.

a) Γ est résiduellement fini.

b) Si V est plat sur $\mathbb{Z}$, Γ est virtuellement sans torsion.

Démonstration du Corollaire. A l'aide des résultats de Grothendieck ([A.G] , (8.8.2) et (8.9.1)) on déduit l'existence d'une sous-$\mathbb{Z}$-algèbre de type fini k_o de k , d'un k_o-schéma V_o de type fini, d'un isomorphisme : $k \otimes_{k_o} V_o \to V$ de k-schémas, que nous allons considérer comme une identification, et d'un sous-groupe de type fini Γ_o de $\mathrm{Aut}_{k_o}(V_o)$ tel que $\Gamma = k \otimes_{k_o} \Gamma_o$. Soit W la fermeture schématique de l'image du morphisme $V \to V_o$. Alors il est clair que W est Γ_o-invariant ; notons $\Gamma_W \subset \mathrm{Aut}_{k_o}(W)$ la restriction de Γ_o à W . Alors on déduit un diagramme commutatif d'homomorphismes surjectifs :

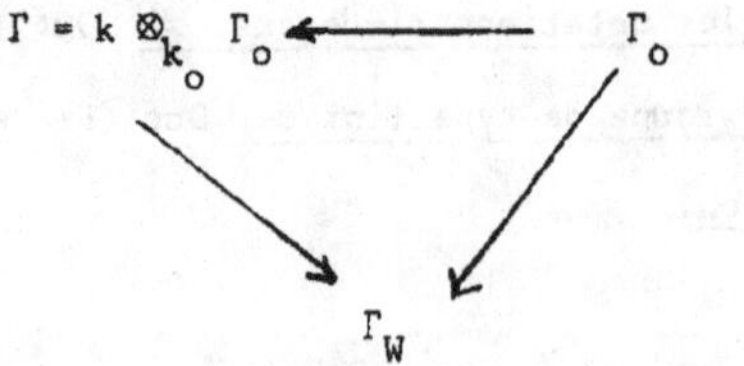

et il est facile de voir que $\Gamma \to \Gamma_W$ est bijectif. De plus on vérifie facilement que si V est plat sur $\mathbb{Z}$, alors W l'est aussi. Puisque W est de type fini sur k_o , et k_o est de type fini sur $\mathbb{Z}$, on peut appliquer le théorème à W et à Γ_W ($\cong \Gamma$) dans $\mathrm{Aut}_{\mathbb{Z}}(W)$, d'où le corollaire.

Démonstration du Théorème. - Posons $\Gamma = \mathrm{Aut}(V)$. Pour tout anneau commutatif F , Γ opère sur l'ensemble :

$$V(F) = \mathrm{Hom}\,(\mathrm{Spec}(F),V)$$

des points de V à valeurs dans F . Si F est fini , $V(F)$ est aussi fini.

Pour montrer que Γ est résiduellement fini, il suffit de montrer que si $s \in \Gamma$ et si s opère trivialement sur $V(F)$, pour tout anneau fini F , alors $s = 1$. Pour tout $x \in V$ posons $A_x = \mathcal{O}_{V,x}$, $\mathfrak{m}_x = \mathrm{rad}(A_x)$, et $k(x) = A_x/\mathfrak{m}_x$. Si x est un point fermé, alors $k(x)$ est un corps fini. L'action triviale de s sur $V(k(x))$ entraîne que s fixe x , donc opère sur A_x . Puisque s opère trivialement sur $V(A_x/\mathfrak{m}_x^r)$, il opère trivialement sur $A_x/\mathfrak{m}_x^r$. Or $\bigcap_r \mathfrak{m}_x^r = 0$, donc s opère trivialement sur A_x , ceci pour tout point fermé x de V . Donc $s = 1$.

Supposons maintenant que V soit plat sur $\mathbb{Z}$. Soient $U_i = \mathrm{Spec}(A_i)$ $(i = 1,\ldots,n)$ des ouverts affines de V qui recouvrent V . Soit $X = \bigcup_i X_i$ où , pour tout $i = 1,\ldots,n$, X_i est un ensemble fini de points fermés de U_i tel que tout $\mathfrak{p} \in \mathrm{Ass}(A_i)$ soit contenu dans $\mathfrak{m}_x$ pour au moins un $x \in X_i$. Cette conditions entraîne que l'application : $A_i \to \prod_{x \in X_i} A_x$ est injective $(i = 1,\ldots n)$.

La $\mathbb{Z}$-platitude des A_i , entraîne que, pour tout $\mathfrak{p} \in \mathrm{Ass}(A_i)$, $A_i/\mathfrak{p}$ est une $\mathbb{Z}$-algèbre plate de type fini. Elle a donc des corps de restes finis de toutes caractéristiques, sauf un nombre fini de caractéristiques premières. Par suite, on peut choisir encore un ensemble $X' = \bigcup_i X'_i$, du même genre que X , et tel que :

(*) $\quad$ $\mathrm{car.}(k(x)) \neq \mathrm{car.}(k(x'))$ pour tout $(x,x') \in X \times X'$.

Soient $A = \prod_{x \in X} A_x$, $J = \mathrm{rad}(A) = \prod_{x \in X} \mathfrak{m}_x$, et $\Gamma_X = \ker(\Gamma \to \mathrm{Aut}(V(A/J^2)))$.

Alors Γ_X est d'indice fini dans Γ , il fixe tout $x \in X$, et donc il opère sur A , trivialement modulo J^2 . Par conséquent Γ_X opère trivialement sur :

$$gr(A) = \bigoplus_{r \geq 0} J^r/_{J^{r+1}} = (A/J)\ [J/_{J^2}] .$$

Il en résulte que l'opération de Γ_X sur $A/_{J^r}$ est unipotente. Donc l'image de Γ_X dans $Aut(A/_{J^r})$ est de "(A/J) - torsion", i.e. d'ordre divisant une puissance de card(A/J) . La nature de X entraîne que l'opération de Γ_X sur A est fidèle. On déduit donc, puisque $\bigcap_r J^r = 0$, que tout sous-groupe fini de Γ_X est un groupe de (A/J)-torsion.

Maintenant prenons $A' = \prod_{x' \in X'} A_{x'}$, $J' = rad(A')$ et $\Gamma_{X'} = \ker(\Gamma \to Aut(V(A'/_{J'^2})))$. On déduit, comme précédemment, que $\Gamma_{X'}$ est d'indice fini dans Γ et que tout sous-groupe fini de $\Gamma_{X'}$ est un groupe de (A'/J')-torsion. Or $\Gamma_X \cap \Gamma_{X'}$ est d'indice fini dans Γ et tout sous-groupe fini de $\Gamma_X \cap \Gamma_{X'}$ est de (A/J)-torsion et de (A'/J')-torsion, donc trivial d'après (★) . D'où le théorème.

3. AUTOMORPHISMES DE GROUPES DE TYPE FINI.

Soit Γ un groupe. Pour $x, y \in \Gamma$, on pose : $ad(x)(y) = x\,y\,x^{-1}$, d'où la suite exacte :

$$1 \to Z(\Gamma) \to \Gamma \xrightarrow{ad} Aut(\Gamma) \to Out(\Gamma) \to 1 .$$

Si Γ est un groupe topologique, on note $Aut^c(\Gamma)$ son groupe d'automorphismes continus, et $Out^c(\Gamma) = Aut^c(\Gamma)/I\ Aut(\Gamma)$.

<u>Proposition</u> (J. Smith [J.S]). - <u>Soit</u> Γ <u>un groupe profini</u> (topologiquement) <u>de type fini. Alors</u> $Aut^c(\Gamma)$, <u>et donc aussi</u> $Out^c(\Gamma)$, <u>sont des groupes profinis</u>.

Pour tout entier q , l'intersection Γ_q des sous-groupes ouverts d'indice $\leqslant q$ est un sous-groupe ouvert et caractéristique. On voit facilement que

$\Gamma = \varprojlim \Gamma/\Gamma_q$ et que $\mathrm{Aut}^c\, \Gamma = \varprojlim_q \mathrm{Aut}(\Gamma/\Gamma_q)$.

Corollaire (Baumslag). - Soit Γ un groupe de type fini. Si Γ est résiduellement fini, $\mathrm{Aut}(\Gamma)$ l'est aussi.

En effet $\mathrm{Aut}(\Gamma) \to \mathrm{Aut}^c(\hat{\Gamma})$ est injectif.

Malheureusement ce raisonnement ne suffit pas pour montrer que $\mathrm{Out}(\Gamma)$ est résiduellement fini, car : $\mathrm{Out}(\Gamma) \to \mathrm{Out}^c(\hat{\Gamma})$ n'est pas nécessairement injectif.

Proposition. - Soient Γ un groupe et $x,y \in \Gamma$. Considérons les conditions suivantes :

(a) x et y deviennent conjugués dans tout quotient fini de Γ ;

(b) x et y deviennent conjugués dans $\hat{\Gamma}$,

(c) il existe un corps F , algébriquement clos de caractéristique zéro tel que que, pour toute représentation $\rho : \Gamma \to GL_n(F)$, on ait : $\chi_\rho(x) = \chi_\rho(y)$;

(d) Pour tout anneau commutatif k et pour toute représentation $\rho : \Gamma \to GL_n(k)$, on a : $\chi_\rho(x) = \chi_\rho(y)$;

Alors : (a) $\Longleftrightarrow$ (b) $\Leftarrow$ (c) $\Leftarrow$ (d) . Si Γ est de type fini, les conditions sont toutes équivalentes.

(b) $\Rightarrow$ (a) est triviale et (a) $\Rightarrow$ (b) résulte d'un argument standard de compacité

(c) $\Rightarrow$ (a) car les caractères des représentations sur F séparent les classes de conjugaison de tout groupe fini.

(d) $\Rightarrow$ (c) est trivial.

(a) $\Rightarrow$ (d) lorsque Γ est de type fini. En effet soit $\rho : \Gamma \to GL_n(k)$ une représentation telle que la différence $a = \chi_\rho(x) - \chi_\rho(y)$ soit $\neq 0$. Soit A une sous-$\mathbb{Z}$-algèbre de type fini de k , telle que $\rho(\Gamma) \subset GL_n(A)$, et soit $\mathfrak{a}$ un idéal d'indice fini de A ne contenant pas a . Le composé : $\Gamma \xrightarrow{\rho} GL_n(A) \to GL_n(A/\mathfrak{a})$ est une représentation σ telle que $\chi_\sigma(x) \neq \chi_\sigma(y)$. Donc $\sigma(x)$ et $\sigma(y)$ ne sont pas conjugués dans le quotient fini $\sigma(\Gamma)$ de Γ .

Définitions. - Soit Γ un groupe et soient $x,y \in \Gamma$. On dit que x et y sont trace-équivalents (en notation : $x \underset{T}{\sim} y$) dans Γ, s'ils satisfont à la condition (d), et donc à toutes les conditions de la Proposition. On pose :

$$T\,\mathrm{Aut}(\Gamma) = \{\alpha \in \mathrm{Aut}(\Gamma) \mid \alpha(x) \underset{T}{\sim} x \text{ quel que soit } x \in \Gamma\} .$$

C'est un sous-groupe distingué de $\mathrm{Aut}(\Gamma)$ contenant $I\,\mathrm{Aut}(\Gamma)$. On dit que Γ est de type (TI) si $T\,\mathrm{Aut}(\Gamma) = I\,\mathrm{Aut}(\Gamma)$.

Soit $R\,I\,\mathrm{Aut}(\Gamma) = \{\alpha \in \mathrm{Aut}(\Gamma) \mid \hat{\alpha} \in I\,\mathrm{Aut}(\hat{\Gamma})\}$, de sorte que :

$$\frac{R\,I\,\mathrm{Aut}(\Gamma)}{I\,\mathrm{Aut}(\Gamma)} = \ker\,(\mathrm{Out}(\Gamma) \to \mathrm{Out}^c(\hat{\Gamma})) .$$

D'après la condition (b) de la proposition on a :

$$I\,\mathrm{Aut}(\Gamma) \subset R\,I\,\mathrm{Aut}(\Gamma) \subset T\,\mathrm{Aut}(\Gamma)$$

lorsque Γ est de type fini.

Corollaire. - Soient Γ un groupe de type fini et F un corps algébriquement clos de caractéristique zéro. Les conditions suivantes sont équivalentes :

(a) Γ est de type (TI) ;

(b) Si $\alpha \in \mathrm{Aut}(\Gamma)$ et si : $\rho \circ \alpha \cong \rho$ pour toute représentation irréductible $\rho : \Gamma \to GL_n(F)$, alors : $\alpha \in I\,\mathrm{Aut}(\Gamma)$;

(c) Si $\alpha \in \mathrm{Aut}(\Gamma)$ et si $\chi_\rho \circ \alpha = \chi_\rho$ pour toute représentation $\rho : \Gamma \to GL_n(F)$, alors : $\alpha \in I\,\mathrm{Aut}(\Gamma)$.

Corollaire. - Soit Γ un groupe de type fini de type (TI).

1) $\mathrm{Out}(\Gamma) \to \mathrm{Out}^c(\hat{\Gamma})$ est injectif ; et donc $\mathrm{Out}(\Gamma)$ est résiduellement fini. Si Γ est résiduellement fini, on a : $N_{\hat{\Gamma}}(\Gamma) = \Gamma.Z_{\hat{\Gamma}}(\Gamma)$.

2) Soit k un anneau commutatif contenant $\mathbb{Z}$. Alors : $\mathrm{Out}(\Gamma) \to \mathrm{Out}(k\Gamma)$ est injectif et $N_{(k\Gamma)^x}(\Gamma) = \Gamma.Z(k\Gamma)^x$.

L'assertion 1) résulte des remarques ci-dessus, et 2) se déduit du critère (b) du corollaire précédent.

4. VARIETES DE REPRESENTATIONS.

Soit Γ un groupe de type fini. Pour tout $n \geqslant 1$, posons :

$$R_n(\Gamma) = \mathrm{Hom}(\Gamma, GL_n(\mathbb{C})) .$$

Si S est un système fini de générateurs de Γ, alors $R_n(\Gamma)$ s'identifie à une sous-variété fermé de $GL_n(\mathbb{C})^S$. Le groupe $GL_n(\mathbb{C})$ opère par conjugaison sur $R_n(\Gamma)$. Notons $S_n(\Gamma)$ le quotient algébrique de $R_n(\Gamma)$ par $GL_n(\mathbb{C})$. Ainsi, si $R_n(\Gamma) = \mathrm{Spec}(A)$, alors $S_n(\Gamma)$ est la variété affine $\mathrm{Spec}(A^{GL_n(\mathbb{C})})$. Notons $\rho \to (\rho)$ l'application canonique $R_n(\Gamma) \to S_n(\Gamma)$. On sait que, pour $\rho, \rho' \in R_n(\Gamma)$, les conditions suivantes sont équivalentes :

(a) $(\rho) = (\rho')$;

(b) ρ et ρ' ont les mêmes quotients de Jordan-Hölder, à isomorphisme et permutation près ;

(c) $\chi_\rho = \chi_{\rho'}$.

Le groupe $\mathrm{Aut}(\Gamma)$ opère de façon naturelle sur $R_n(\Gamma)$ et cette opération commute avec celle de $GL_n(\mathbb{C})$. Donc, par passage aux quotients, $\mathrm{Aut}(\Gamma)$ opère sur $S_n(\Gamma)$. Notons $T_n\,\mathrm{Aut}(\Gamma)$ le noyau de : $\mathrm{Aut}(\Gamma) \to \mathrm{Aut}(S_n(\Gamma))$. Ainsi :

$$T_n\,\mathrm{Aut}(\Gamma) = \bigcap_{\rho \in R_n(\Gamma)} \mathrm{Aut}(\Gamma)_{(\rho)} ,$$

où

$$\mathrm{Aut}(\Gamma)_{(\rho)} = \{\alpha \in \mathrm{Aut}(\Gamma) \mid \chi_\rho \circ \alpha = \chi_\rho\} .$$

Remarquons que si $\rho \in R_n(\Gamma)$ est semi-simple, on a : $\chi_\rho \circ \alpha = \chi_\rho$ si et seulement si $\rho \circ \alpha \cong \rho$.

Evidemment on a $T\,\mathrm{Aut}(\Gamma) = \bigcap_n T_n\,\mathrm{Aut}(\Gamma)$, et $\mathrm{Out}(\Gamma)$ opère sur $S_n(\Gamma)$ avec noyau $T_n\,\mathrm{Aut}(\Gamma)/_{I\,\mathrm{Aut}(\Gamma)}$. D'après le corollaire du théorème du n°1, on a :

Proposition. - Tout sous-groupe de type fini de $\mathrm{Aut}(\Gamma)/T_n\,\mathrm{Aut}(\Gamma)$ est résiduellement fini et virtuellement sans torsion.

Remarque. - En fait, les variétés $R_n(\Gamma)$ et $S_n(\Gamma)$ proviennent, par changement de base $\mathbb{Z} \to \mathbb{C}$, des schémas $R_n^{\mathbb{Z}}(\Gamma)$ et $S_n^{\mathbb{Z}}(\Gamma)$ qui sont plats sur $\mathbb{Z}$. De plus, les actions de $\mathrm{Aut}(\Gamma)$ se déduisent des actions sur $R_n^{\mathbb{Z}}(\Gamma)$ et $S_n^{\mathbb{Z}}(\Gamma)$ par le même changement de base. Enfin le schéma $R_n^{\mathbb{Z}}(\Gamma)$ est de type fini sur $\mathbb{Z}$, et il est bien probable que $S_n^{\mathbb{Z}}(\Gamma)$ l'est aussi. Dans ce cas, le théorème du n°1 impliquerait que $\mathrm{Aut}(\Gamma)/T_n\,\mathrm{Aut}(\Gamma)$ tout entier (et non seulement ses sous-groupes de type fini) est résiduellement fini et virtuellement sans torsion.

La proposition suivante explicite, dans certains cas, les stabilisateurs $\mathrm{Aut}(\Gamma)_{(\rho)}$.

Proposition. - Soit $\rho \in R_n(\Gamma)$ une représentation irréductible. Il existe un homomorphisme π_ρ qui rend commutatif le diagramme suivant :

$$\begin{array}{ccc} \Gamma & \xrightarrow{\quad\rho\quad} & GL_n(\mathbb{C}) \\ \mathrm{ad}\downarrow & & \downarrow\mathrm{ad} \\ \mathrm{Aut}(\Gamma)_{(\rho)} & \xrightarrow[\pi_\rho]{\quad} & PGL_n(\mathbb{C}) \end{array} .$$

On a : $\pi_\rho(\mathrm{Aut}(\Gamma)_{(\rho)}) \subset \mathrm{ad}\,(N_{GL_n(\mathbb{C})}(\rho\Gamma)) \subset N_{PGL_n(\mathbb{C})}(\mathrm{ad}\rho\Gamma)$.

Si ρ est injectif, alors π_ρ est aussi injectif, d'image égale à $\mathrm{ad}(N_{GL_n(\mathbb{C})}(\rho\Gamma))$.

En effet, soit $\alpha \in \mathrm{Aut}(\Gamma)_{(\rho)}$. L'isomorphisme : $\rho \circ \alpha \cong \rho$ est réalisé par un élément $\sigma \in GL_n(\mathbb{C})$ tel que $\sigma\,\rho(a)\,\sigma^{-1} = \rho\,\alpha\,a)$ pour tout $x \in \Gamma$. Le lemme de Schur entraîne que : $\pi_\rho(\alpha) = \mathrm{ad}(\sigma) \in PGL_n(\mathbb{C})$ est bien défini. Les assertions de la proposition se vérifient sans difficulté.

Corollaire. - Supposons qu'il existe une représentation fidèle irréductible

$\rho : \Gamma \to GL_n(\mathbb{C})$ telle que :

$$N_{GL_n(\mathbb{C})}(\rho\Gamma) = (\rho\Gamma).\mathbb{C}^x ,$$

<u>(par exemple telle que :</u> $N_{PGL_n(\mathbb{C})}(ad\rho\Gamma) = ad\rho\Gamma)$. <u>Alors</u> $Out(\Gamma)$ <u>opère fidèlement sur la variété</u> $S_n(\Gamma)$. <u>Tout sous-groupe de type fini de</u> $Out(\Gamma)$ <u>est résiduellement fini et virtuellement sans torsion.</u>

<u>Exemple</u>. - Soit Γ_g , comme dans l'introduction, le groupe fondamentale d'une surface compacte orientable de genre $g \geqslant 0$. On a $\Gamma_o = \{1\}$ et $\Gamma_1 \cong \mathbb{Z}^2$, de sorte que $Out(\Gamma_1) = Aut(\Gamma_1) \cong GL_2(\mathbb{Z})$.

Supposons $g \geqslant 2$. Notons $R_2^o(\Gamma)$ la partie de $R_2(\Gamma)$ formée des ρ qui sont injectifs et d'image un sous-groupe discret co-compact de $SL_2(\mathbb{R})$. On note $S_2^o(\Gamma)$ l'image de $R_2^o(\Gamma)$ dans $S_2(\Gamma)$. Evidemment $R_2^o(\Gamma)$ et $S_2^o(\Gamma)$ sont invariants par $Aut(\Gamma_g)$.

L'application : $\rho \longmapsto ad \circ \rho : \Gamma_g \to PGL_2(\mathbb{C})$ définit une application surjective: $R_2^o(\Gamma) \to ad\, R_2^o(\Gamma_g)$ où $ad\, R_2^o(\Gamma_g)$ est formé de <u>tous</u> les homomorphismes injectifs : $\Gamma_g \to PSL_2(\mathbb{R})$ d'images discrètes et cocompactes (cf. [P]) . Le quotient $ad\, S_2^o(\Gamma_g)$ de $ad\, R_2^o(\Gamma_g)$ par l'action de conjugaison de $PGL_2(\mathbb{R})$ est l'image de $S_2^o(\Gamma_g)$ par l'application : $(\rho) \longmapsto (ad \circ \rho)$. L'espace $ad\, S_2^o(\Gamma_g)$ paramétrise les surfaces de Riemann "marquée" de genre g . A $\rho \in R_2^o(\Gamma_g)$ correspond $\sum_g = H/_{ad\rho\Gamma_g}$, et l'on sait que :

$$Aut(\textstyle\sum_\rho) = N_{PSL_2(\mathbb{R})}(ad\rho\Gamma_g / ad\rho\Gamma_g)$$

qui est un groupe fini d'ordre $\leqslant 84(g-1)$ (cf. [L.G.]) . En fait on sait que pour un élément "général" ρ de $R_2^o(\Gamma_g)$ on a $Aut(\sum_\rho) = \{1\}$ si $g \geqslant 3$ et $Aut(\sum_\rho)$ est d'ordre 2 si $g = 2$.

Si $\rho \in R_2^o(\Gamma_g)$, il est facile de voir que : $N_{PGL_2(\mathbb{C})}(ad\rho\Gamma) = N_{PGL_2(\mathbb{R})}(ad\rho\Gamma)$. Mac beath et Singermann [M.S] ont démontré que l'indice $[N_{PGL_2(\mathbb{R})}(ad\rho\Gamma) : ad\rho\Gamma))]$ est "en général" égal à 1 pour $g \geqslant 3$ et à 2 pour $g = 2$. Il s'ensuit que $Out(\Gamma_g)$

opère fidèlement sur $S_2^o(\Gamma_g)$ pour $g \geqslant 3$, et avec un noyau N d'ordre 2 pour $g = 2$. D'après le Corollaire du théorème du n°1, $\mathrm{Out}(\Gamma_g)$ est résiduellement fini et virtuellement sans torsion si $g \geqslant 3$. Pour $g = 2$ on peut invoquer le résultat de Grossman, disant que $\mathrm{Out}(\Gamma_2)$ est résiduellement fini, pour trouver un sous-groupe distingué M d'indice fini tel que $M \cap N = \{1\}$. Alors M opère fidèlement sur $S_2(\Gamma)$; donc M, et aussi $\mathrm{Out}(\Gamma_g)$, sont virtuellement sans torsion.

L. Bers m'a signalé la démonstration suivante du fait que $\mathrm{Out}(\Gamma_g) = M_g$ est virtuellement sans torsion. D'après un théorème de Nielsen, tout élément d'ordre fini de M_g fixe au moins un point de $\mathrm{ad}\, S_2^o(\Gamma_g)$. Il suffit donc de produire un sous-groupe G d'indice fini de M_g qui opère librement sur $\mathrm{ad}\, S_2^o(\Gamma_g)$. On prend:

$$G = \ker\left(\mathrm{Out}(\Gamma_g) \to \mathrm{Aut}(H_1(\Gamma_g, \mathbb{Z}/_{3\mathbb{Z}}))\right) .$$

Si $s \in G$ fixe un point $(\mathrm{ad}\rho)$ de $\mathrm{ad}\, S_2^o(\Gamma_g)$, alors s définit un automorphisme de la surface de Riemann $\sum_\rho$, et s opère trivialement sur les points d'ordre 3 de la Jacobienne de $\sum_\rho$. D'après un résultat de Serre [J.-P.S.], un tel s est l'identité.

REFERENCES.

[B-L] H. Bass and Alex Lubotzky, Automorphisms of groups and of schemes of finite type, à paraître.

[B] J.S. Birman, The algebraic structure of mapping class groups, in Discrete groups and automorphic functions, Ed. W.J. Harvey, Acad. Press (1977) 163-198.

[L.G.] L. Greenberg, Finiteness theorems for Fuchsian and Kleinian groups, in Discrete groups and automorphic functions, Ed. W.J. Harvey, Acad. Press (1977).

[E.G.] E. Grossman, On the residual finiteness of certain mapping class groups, Jour. London Math. Soc. 9 (1974) 160-164.

[A.G.] A. Grothendieck (avec la collaboration de J. Dieudonné) Eléments de géométrie algébrique IV (Troisième partie), Publ. I.H.E.S. 28 (1966).

[M.S.] A.M. Macbeath and D. Singerman, Spaces of subgroups and Teichmüller space, Proc London Math. Soc. 31 (1975) 211-256.

[P.] S.J. Patterson, On the cohomology of Fuchsian groups, Glasgow Math. Jour. 16 (1975) 123-140.

[J.-P.S.] J.-P. Serre, Rigidité du foncteur de Jacobi d'échelon $n \geq 3$, Appendice à l'exposé 17 de A. Grothendieck, Sém. H. Cartan 13(1960/61).

[J.S.] J. Smith, On products of profinite groups, Illinois Jour. Math. 13 (1969) 680-688 .

=-=-=-=-=-=-=-=-=-=-=-=-=

SUR LES TRAVAUX DE V.K. KHARCHENKO

par J.M. Goursaud, J.L. Pascaud et J. Valette

-:-

Le but de cet exposé est de donner une nouvelle présentation des principaux résultats obtenus par V.K. Kharchenko dans la théorie des actions de groupes sur des anneaux semi-premiers. Si G est un groupe fini d'automorphismes d'un anneau semi-premier R, dans [5] V.K. Kharchenko introduit la notion d'automorphisme intérieur généralisé à l'aide de l'anneau de quotients de Martindale S de centre C , et associe au groupe une C-algèbre de type fini B , lui permettant ainsi de définir une trace dans le cas où B est semi-première, ce qui arrive dans le cas où R est sans |G|-torsion, et dans le cas où R est réduit. Dans une première partie on décompose l'anneau S en produit d'anneaux stables par G sur lesquels les automorphismes intérieurs généralisés de G sont des automorphismes intérieurs au sens classique. Dans une seconde partie on montre que la connaissance de l'auto-injectivité de B et l'étude de $B \otimes_C \overset{\circ}{B}$ facilite la définition de traces de S dans S^G .
Enfin dans la dernière partie, on présente les théorèmes de V.K. Kharchenko concernant les relations entre R et R^G en proposant des démonstrations plus concises.

Sur ce sujet on pourra également consulter les exposés de J. Fisher et J. Osterburg [4] , S. Montgomery [9] , A. Page [10].

I.– NOTATIONS ET DEFINITIONS.

1) Pour un anneau unitaire R , on note

- $Z(R)$ le centre de R ,
- $j(R)$ l'idéal singulier à gauche de R ,
- pour une partie X de R , $C_R(X)$ le centralisateur dans R de X .

2) Soit G un groupe d'automorphismes de l'anneau R .

- pour $a \in R$ et $g \in G$, a^g désigne l'image de a par g ,
- pour une partie X de R et un sous-groupe H de G on définit

$$X^H = \{x \in X \ ; \ \forall \ g \in H \quad x^g = x\}$$

- une partie X de R est H-invariante si pour tout $x \in X$ et tout $h \in H$, $x^h \in X$,
- $|X|$ désigne le cardinal de X ,
- pour $a \in R$ on appelle trace de a l'élément $\mathrm{tr}\, a = \sum_{g \in G} a^g \in R^G$.

II.– L'ANNEAU DE QUOTIENTS DE MARTINDALE

Dans toute la suite on supposera que R est un anneau semi-premier

Rappelons pour commencer la définition de l'anneau maximal de quotients à gauche de R , noté $Q_{Max}(R)$. Pour cela considérons $E = E({}_R R)$ l'enveloppe injective à gauche de l'anneau R , $A = \mathrm{Hom}_R(E,E)$; E devient ainsi un A-R-bimodule; on a alors la

Définition : $Q_{Max}(R) = \mathrm{Hom}_A({}_A E, {}_A E)$.

Des renseignements plus précis sur $Q_{Max}(R)$ sont fournis par la proposition suivante.

PROPOSITION 2.1.– *a)* $Q_{Max}(R)$ *se plonge dans* E *par l'application* ϕ *définie par* $\phi(q) = q(1)$.

b) R *s'identifie à un sous-anneau de* $Q_{Max}(R)$.

c) $Q_{Max}(R) \simeq \{x \in E \ , \ \forall\ b \in A \ , \ Rb = 0 \Longrightarrow xb = 0\}$.

En fait on utilise aussi fréquemment une autre caractérisation de $Q_{Max}(A)$ nécessitant la définition suivante :

Définition : Un sous-R-module D de $Q_{Max}(R)$ est dit dense si :

$\forall\ b \in A \ , \ Db = 0 \Longrightarrow Rb = 0$.

Si $\mathfrak{D}$ désigne l'ensemble des idéaux à gauche denses de R on a :

THEOREME 2.2.– $Q_{Max}(A) \simeq \varinjlim_{D \in \mathfrak{D}} \mathrm{Hom}(D,R)$

Pour plus de renseignements sur $Q_{Max}(R)$ on pourra se reporter à [7] .

De même si on considère l'ensemble $\mathfrak{I}$ des idéaux bilatères de R d'annulateur nul (c'est-à-dire essentiels), on peut considérer l'anneau de quotients de Martindale à gauche $S = \varinjlim_{I \in \mathfrak{I}} \mathrm{Hom}(I,R)$ [1] .

PROPOSITION 2.3.– *On a à un isomorphisme près* $R \subset S \subset Q_{Max}(R)$ *et* $S \simeq \{x \in Q_{Max}(R)\ ,\ \exists\ I \in \mathfrak{F}\quad Ix \subset R\}$.

Cet anneau possède une propriété faible d'injectivité donnée par le

LEMME 2.4.– *Soit* M *un sous-R-R-bimodule de* S , $\phi : M \to S$ *un homomorphisme de* R*-module à gauche tel que* $\phi^{-1}(R) \cap R$ *soit un sous-R-R-bimodule de* R ; *alors il existe un élément* $s \in S$ *tel que :* $\phi(b) = bs$ *pour tout élément* b *de* M .

On suppose $\phi(M)$ non nul. Soient b un élément de M tel que $\phi(b)$ soit non nul, $I \in \mathfrak{F}$ tel que $I\ \phi(b) \subseteq R$ et $Ib \subseteq R$; comme l'annulateur dans S de I est nul, on a $T = \phi^{-1}(R) \cap R \neq 0$. Alors $T \oplus \mathrm{Ann}_R\ T \in \mathfrak{F}$; définissons un morphisme ψ de $T \oplus \mathrm{Ann}_R\ T$ par $\psi(t+u) = \phi(t)$ si : $t \in T$ et $u \in \mathrm{Ann}_R\ T$. D'après la définition de S , il existe un élément s de S tel que : $\psi(\alpha) = \alpha s$ pour $\alpha \in T \oplus \mathrm{Ann}_R\ T$. Montrons que ϕ coïncide avec la multiplication à droite par s sur M . Soit b un élément de M et I un élément de $\mathfrak{F}$ tels que : $Ib \subset R$ et $I\phi(b) \subset R$.

Alors $\forall\ i \in I \qquad ib \in \phi^{-1}(R) \cap R = T$

et $\quad i\phi(b) = ibs$.

Par conséquent $I(\phi(b)-bs) = 0$ et $\phi(b) = bs$.

∎

Dorénavant on notera C le centre de S appelé centroïde de R . Les éléments de C proviennent des homomorphismes de R-R-bimodules d'un idéal I de $\mathfrak{F}$ dans R .

De plus à chaque sous-R-R-bimodule M de S, est associé un idempotent central e_M de C défini par :

$$e_M(x+y) = x \quad \text{si} \quad x \in M \quad \text{et} \quad y \in \mathrm{Ann}_R\, M .$$

On vérifie immédiatement que e_M est le plus petit idempotent central vérifiant : $\forall\; x \in M \quad e_M x = x$.

Pour tout $s \in S$, on pose $e_s = e_{RsR}$.

On obtient alors comme conséquences :

PROPOSITION 2.5.— *a)* C *est régulier de Von Neumann* [1].

b) Le sous-C*-module singulier* $j_C(S)$ *de* S *est nul.*

c) C *est auto-injectif* [2].

d) C *est un corps si et seulement si* A *est premier.*

a) Soit c un élément de $C-\{0\}$. Alors $M = Rc^2$ est un sous-R-R- bimodule de S et l'application ϕ définie par $\phi(c^2) = c$ est bien définie car $(\ell_S(c^2)c)^2 = 0$ montre que $\ell_S(c^2) \subset \ell_S(c)$; donc d'après le lemme précédent ϕ est donné par la multiplication par un élément c' qui vérifie $c = c^2c'$. On vérifie facilement que $c' \in C$.

b) Soient $s \in j_C(S)$ et $I = S\ell_C(s)$. Par définition de e_I, $e_I s = 0$ et $\ell_C(s)(1-e_I) = 0$ donc $e_I = 1$ et $s = 0$.

c) C étant régulier il suffit de montrer que pour tout idéal essentiel de C de la forme $J = \oplus\, Ce_i \; (e_i^2 = e_i)$, tout morphisme $f : J \to C$ se prolonge à C.

Puisque $f(e_i) = e_i f(e_i)$ on peut définir une application $\phi : \oplus\, Se_i \to S$ par $\phi(e_i) = f(e_i)$; elle vérifie les hypothèses du lemme 2.4. Il existe donc $s \in S$ tel que pour tout $x \in S$, $f(e_i x) = e_i xs = e_i sx$. Il en résulte $(\oplus Ce_i)(xs-sx) = 0$ et $xs = sx$ (d'après b)) : s appartient à C.

d) si R est premier, S est premier donc C est un corps.

Si R n'est pas premier, il contient un idéal bilatère J d'annulateur non nul ; e_J est un idempotent non trivial de C. ∎

III.– SOUS-ANNEAUX DE S .

Soient C un anneau commutatif régulier injectif, B un sur anneau de C tels que

1) C est contenu dans $Z(B)$

2) B est un C-module de type fini engendré par $(x_i)_{1\leq i\leq n}$ et à sous-C-module singulier nul.

3) B est semi-premier.

Il est facile de voir que B est un module projectif de type fini.

LEMME 3.1.– B *est un anneau régulier.*

Montrons que pour tout idéal maximal m de C, $B/m\,B$ est un anneau régulier. $B/m\,B$ est un C/m - espace vectoriel de dimension finie engendré par les classes des éléments x_i $(1\leq i\leq n)$. $B/m\,B$ est donc un anneau artinien. Si $B/m\,B$ n'est pas régulier il contient un idéal J de carré nul. On a alors $J = \sum_{i=1}^{p} C/m\ \overline{y_i}$ avec $\overline{y_i}\ \overline{y_j} = 0$ $\forall\ (i,j)$. Comme C est régulier il existe un idempotent e de C n'appartenant pas à m tel que :

$$\forall\ i,j,k,l \qquad e x_k\, y_i\, x_\ell\, y_j = 0\ ;$$

par conséquent l'idéal $\sum_i B e y_i$ est un idéal de carré nul, ce qui contredit notre hypothèse sur B.

LEMME 3.2.– *Si* I *est un idéal à gauche essentiel de* B, $I \cap C$ *est un idéal essentiel de* C.

Pour démontrer ce résultat il suffit de montrer que $I \cap C \neq 0$. Soit m un idéal maximal de C tel que la codimention ℓ du C/m-espace vectoriel $I_m = \frac{I+mB}{mB}$ soit minimale. On suppose ℓ non nul. On a alors $B/mB = I_m \oplus B/mB\,(1-\overline{f})$ où $f = f^2$.
Soit $\overline{z}_1,\ldots,\overline{z}_\ell$ une base du C/m-espace vectoriel $B/mB\,(1-\overline{f})$.
Il existe alors un idempotent central non nul e n'appartenant pas à m tel que les éléments $ez_1,\ldots,ez_\ell$ engendrent $B(1-f)$ en tant que C-module et $ef \in I$.
I étant essentiel dans B il existe un élément $x = x(1-f)e$ non nul appartenant à I. Soient $x = \lambda_1 ez_1+\ldots+\lambda_\ell ez_\ell$ avec $\lambda_1 e \neq 0$, e_1 un idempotent tel que $Ce_1 = C\lambda_1 e$ et m_1 un idéal maximal de C ne contenant pas e_1.
Alors dans B/m_1B, $B/m_1B\,(1-\overline{f})$ est engendré par $\overline{z}_1,\ldots,\overline{z}_\ell$ comme C/m_1-espace vectoriel et $B/m_1B\overline{f}$ est contenu dans I_{m_1}. Par définition de ℓ, $I_{m_1} = B/m_1B\overline{f}$. Il en résulte $\overline{x} = 0$ ce qui contredit le choix de e_1 et m_1.
Donc $\ell = 0$, $f = 1$ et e appartient à I.

∎

PROPOSITION 3.3.— B *est un anneau-injectif à droite et à gauche. C'est une* Z(B)*-algèbre d'Azumaya.*

a) Montrons que B est auto-injectif à gauche. B étant régulier, il suffit de montrer que tout morphisme f d'un idéal à gauche essentiel I de B dans B se prolonge. D'après le lemme précédent $I \cap C$ est un idéal essentiel de C donc contient un idéal essentiel de la forme $\oplus\, Ce_\alpha$ $(e_\alpha^2 = e_\alpha)$; la restriction de f à $\oplus\, Ce_\alpha$ est un homomorphisme de C-module qui se prolonge à C en un endomorphisme g de C dans B .
En effet B est un C-module de type fini à sous-module singulier nul et C est auto-injectif donc B est C-injectif.
Alors si x appartient à I , $f(e_\alpha x) = e_\alpha x g(1)$ et $(\oplus\, Ce_\alpha)[f(x)-xg(1)] = 0$.
D'où $f(x) = xg(1)$.

b) D'après [11, théorème 5.5.7] , il suffit de vérifier que pour tout idéal maximal m de Z(B) , B/mB a pour centre Z/m : B étant engendré par $(x_i)_{1<i\leq n}$ soit $y \in B$ tel que

$$\forall\, i \qquad yx_i - x_i y \in mB .$$

Il existe un idempotent $e \in Z\backslash m$ tel que pour tout i , $e(yx_i - x_i y) = 0$.
ey appartient à Z donc $\overline{y}$ appartient à Z/m .

∎

LEMME 3.4.— *Soit* $(e_i)_{1\leq i\leq n}$ *une famille d'idempotents centraux d'un anneau* R . *Il existe une partie finie* J *de* $\mathbb{N}$ *et des idempotents centraux orthogonaux* $(f_j)_{j\in J}$ *de* R *tels que*

$$\cdot \sum_{i=1}^{n} Re_i = \bigoplus_{j\in J} Rf_j .$$

$$\forall\, (i,j) \qquad e_i f_j = 0 \text{ ou } e_i f_j = f_j .$$

La démonstration se fait par récurrence.

Pour $n = 2$, on a $Re_1 + Re_2 = R(e_1-e_1e_2) \oplus Re_1e_2 \oplus R(e_2-e_1e_2)$.

Supposons la propriété vraie à l'ordre $n-1$ et posons :

$$\sum_{i=1}^{n-1} Re_i = \bigoplus_{j\in J} Rf_j = Re \quad (e = e^2) .$$

On a alors $\sum_{i=1}^{n} Re_i = \bigoplus_{j\in J} Rf_j(1-e_n) \oplus (\bigoplus_{j\in J} Rf_je_n) \oplus R(e_n-ee_n)$.

Les idempotents non nuls apparaissant dans le second membre donnent le résultat.

∎

<u>LEMME 3.5</u>.– *Il existe une famille d'idempotents orthogonaux* $(f_j)_{1\leq j\leq m}$ *de* C *telle que*

a) $1 = \sum_{j=1}^{m} f_j$

b) Bf_j *soit un* Cf_j*-module libre.*

Comme B est C-projectif de type fini,il existe une décomposition de la forme $B = \bigoplus_{i=1}^{n} Cb_i$. On pose $\ell_C(b_i) = C(1-e_i)$. Puisque $C \subseteq B$ on a $C = \sum_{i=1}^{n} Ce_i$ et d'après le lemme 3.4, il existe des idempotents orthogonaux $(f_j)_{1\leq j\leq m}$ de C vérifiant a) et pour tout (i,j) $e_if_j = 0$ ou $e_if_j = f_j$. Soit $I_j = \{i \ ; \ f_jb_i \neq 0\}$; il est facile de vérifier que Bf_j est un Cf_j-module libre de base $(f_jb_i)_{i\in I_j}$.

∎

LEMME 3.6.– *Soient* $C \subseteq Z$ *deux anneaux commutatifs réduits tels que* $Z = C[x]$ *où* x *désigne un élément entier sur* C . *On suppose que* C *est régulier injectif et que le sous-module singulier* $j_C(Z)$ *de* Z *est nul. Alors il existe des idempotents orthogonaux* $e_o,\ldots,e_n$ *de* C *vérifiant* $1 = e_o+\ldots+e_n$ *et tels que pour chaque* i $(o \leq i \leq n)$ *le noyau de l'homomorphisme canonique*

$$Ce_i[X] \to C[e_ix] \to 0$$

soit engendré par un polynôme unitaire.

x vérifie $x^n = \alpha_{n-1} x^{n-1}+\ldots+\alpha_o (\alpha_i \in C)$. Pour $m \leq n$ on suppose définis les idempotents orthogonaux $e_o,\ldots,e_{m-1}$ de C $(1-e = e_o+\ldots+e_{m-1})$ de sorte que le degré minimum d'un polynôme unitaire de $Ce[X]$ annulant x soit supérieur ou égal à m . On considère une famille maximale (e_α) d'idempotents orthogonaux de Ce telle que pour tout i le degré minimum d'un polynôme unitaire de $Ce_\alpha[X]$ annulant x soit égal à m et on désigne par $P_{e_\alpha} = \sum_{k=1}^{m} \gamma_{\alpha k} X^k$ un tel polynôme.

C étant régulier injectif il existe $e_m = e_m^2 \in C$ tel que $e_m = \sup e_\alpha$ et des éléments $\gamma_o,\ldots,\gamma_m$ de Ce_m tels que :

$$\forall \alpha \qquad \forall k \qquad e_\alpha \gamma_k = e_\alpha \gamma_{\alpha k} .$$

Il est clair que le polynôme unitaire $P = \sum_{k=1}^{m} \gamma_k X^k \in Ce[X]$ annule x (car $j_C(Z) = 0$) . P engendre le noyau de $Ce_m[X] \to C[e_m x]$: en effet si $Q \in Ce_m[X]$ annule x et si $Q = PR+S$ $(d^oS < d^oP)$ alors $eS(x) = 0$. Or si eS n'était pas nul il existerait $f = f^2 \in Ce$ tel que fS soit unitaire et α tel que $e_\alpha fS$ ne soit pas nul ce qui contredirait la définition de $P_{e_\alpha} = e_\alpha P$.

∎

On rappelle le résultat suivant dû à P. Martin [8, théorème 1.7].

Soit A *un anneau et* P *un polynôme unitaire à coefficients dans le centre de* A . A[X]/(P) *est auto-injectif à droite si et seulement si* A *est auto-injectif à droite.*

Avec les notations du lemme précédent si Z est contenu dans le centre d'un anneau auto-injectif A , des isomorphismes

$$A \underset{C}{\otimes} Z \tilde{\to} \prod_{i=o}^{n} A \underset{C}{\otimes} Ce_i[X] / (P_{e_i}) \tilde{\to} \prod_{i=o}^{n} Ae_i[X] / (P_{e_i})$$

où P_{e_i} désigne un polynôme unitaire de $Ce_i[X]$, il résulte que $A \underset{C}{\otimes} Z$ est auto-injectif.

<u>LEMME 3.7</u>.– $B \underset{C}{\otimes} Z(B)$ *est auto-injectif.*

B étant régulier auto-injectif son centre Z est régulier auto-injectif donc facteur direct de B . Z est donc un C-module de type fini et tout élément de Z est entier sur C . Pour tout $(z_1,\ldots,z_n) \in Z^n$, $C[z_1,\ldots,z_n]$ est donc un C-module de type fini et d'après 3.2 et 3.3, $C[z_1,\ldots,z_{n-1}] \subseteq C[z_1,\ldots,z_n]$ vérifient les hypothèses du lemme 3.6. On obtient le résultat par récurrence.

∎

<u>THEOREME 3.8</u>.– $B \underset{C}{\otimes} \overset{\circ}{B}$ *est un anneau auto-injectif à droite et à gauche.*

D'après le lemmme 3.5 il existe une famille (f_i) d'idempotents orthogonaux de Z(B) tels que $1 = f_1+\ldots+f_n$ et tels que Bf_i soit un Zf_i-module libre. Il suffit de montrer que $B \underset{C}{\otimes} \overset{\circ}{B}f_i$ est auto-injectif.

On peut donc supposer que B est Z-libre de rang r . B étant une Z-algèbre d'Azumaya, on considère les isomorphismes

$$B \underset{C}{\otimes} \overset{\circ}{B} \underset{Z}{\otimes} B \overset{\sim}{\to} B \underset{C}{\otimes} \mathrm{End}_Z B \overset{\sim}{\to} B \underset{C}{\otimes} M_r(Z) \overset{\sim}{\to} M_r(B \underset{C}{\otimes} Z)$$

B étant Z-libre, l'auto-injectivité de $(B \underset{C}{\otimes} \overset{\circ}{B}) \underset{Z}{\otimes} B$ implique celle de $B \underset{C}{\otimes} \overset{\circ}{B}$.

∎

Soit G un groupe d'automorphismes de R , alors tout élément g de G définit par prolongement un unique automorphisme de S (respectivement $Q_{Max}(R)$) . On peut donc supposer que G est un groupe d'automorphismes de S laissant R invariant.

Dans [6] Kharchenko introduit la notion d'automorphismes X-intérieurs :

DEFINITION 3.9.– Un élément g de G est dit X-intérieur si

$$\phi_g = \{x \in S \ ; \ \forall\, s \in S \quad sx = xs^g\} \quad \text{est non nul.}$$

On vérifie que $\phi_g\, \phi_h \subseteq \phi_{hg}$ et $\phi_g^h = \phi_{hgh^{-1}}$.

On notera G_{inn} l'ensemble des éléments X-intérieurs de G .

LEMME 3.10.– *ϕ_g est un C-module monogène engendré par un élément x_g inversible dans l'anneau Se_g , où e_g est l'idempotent associé à l'idéal bilatère $S\phi_g$ et g coïncide sur Se_g avec l'automorphisme intérieur défini par x_g .*

Si x est un élément de ϕ_g , Sx est un idéal bilatère de S , donc on lui associe un idempotent central que l'on notera e_x . Soit alors $(x_i)_{i\in I}$ une famille d'éléments telle que la somme $\sum_{i\in I} Se_{x_i}$ soit directe et maximale pour la propriété. Notons e l'idempotent central associé à l'idéal $\sum_{i\in I} Se_{x_i}$. On définit alors une application ϕ de $\sum Se_{x_i}$ dans S par $\phi(e_{x_i}) = x_i$; d'après le lemme 2.4., ϕ est donné par la multiplication à droite par un élément x de S et on vérifie facilement que ex appartient à ϕ_g. Montrons que ex engendre ϕ_g. D'après la maximalité de la somme directe, pour tout élément t de ϕ_g on a $t = e\,t$; il suffit donc de montrer que ex est inversible dans Se ; pour cela on considère l'application ψ de Se définie par $\psi(e) = ex$, c'est un isomorphisme.

Les éléments X-intérieurs conduisent à la définition suivante :

DEFINITION 3.11.– Soit G un groupe d'automorphismes de S. On notera $B(R;G) = \sum_{g\in G_i} \phi_g$; c'est un sous-anneau de S.

DEFINITION 3.12.– Soit G un groupe d'automorphismes de S. On dira que G est un (*)-groupe si les conditions suivantes sont réalisées.

a) G est extension finie d'un sous-groupe normal N formé d'automorphismes intérieurs au sens classique.

b) B est un C-module de type fini.

c) B est semi-premier.

DEFINITION 3.13.– Soient $(a_1,\dots,a_n) \in S^n$ et $(g_1,\dots,g_n) \in (\mathrm{Aut}\ S)^n$. On dira que a_1 est linéairement indépendant de $a_2,\dots,a_n$ modulo $g_1,\dots,g_n$ si

$$a_1 \notin \sum_{i>1} a_i\, \phi_{g_1 g_i^{-1}}$$

LEMME 3.14.– *Soient* I *un sous-R-bimodule de* S , $g \in \mathrm{Aut}\, R$ *et* $f : I \to S$ *tel que*

$$r\ ,\ r' \in R\ ,\ a \in I \implies f(rar') = rf(a)g(r')\ .$$

Alors il existe $x \in \phi_g$ *tel que*

$$a \in I \implies f(a) = ax\ .$$

On prolonge f à $I \oplus \ell(I)$ par 0 sur $\ell(I)$; d'après 2.4 il existe $x \in S$ tel que pour $a \in I \oplus \ell(I)$, $f(a) = ax$. Soit $r \in R$, on a

$$a \in \ell(I) \implies a(rx-xg(r)) = 0\ ,$$

et si $a \in I$, $f(ar) = arx = f(a)g(r) = axg(r)$: $a(rx-xg(r)) = 0$. Finalement $[I \oplus \ell(I)](rx-xr^g) = 0$ et $rx-xr^g = 0$.

PROPOSITION 3.15[5] *Soient* $a_1,\dots,a_n \in S$, $g_1,\dots,g_n \in \mathrm{Aut}\, R$. *Alors* a_1 *est linéairement indépendant de* $a_2,\dots,a_n$ *modulo* g_1 , $g_2,\dots g_r$ *si et seulement si il existe* $v_1,\dots,v_m$, $t_1,\dots,t_m \in R$ *tels que*

$$\sum_{j=1}^{m} v_j\, a_1\, t_j^{g_1} \neq 0 \quad \textit{et} \quad \sum_{j=1}^{m} v_j\, a_i\, t_j^{g_i} = 0 \quad \textit{pour} \quad 2 \le i \le n\ .$$

On raisonne par contradiction : Si $n = 2$ et si $\sum_{j=1}^{m} v_j\, a_2\, t_j^{g_2} = 0 \implies \sum_{j=1}^{m} v_j\, a_1\, t_j^{g_1} = 0$, on définit $f : R a_2 R \to S$ en posant

$$f(\Sigma\, u_j\, a_2\, w_j) = \Sigma\, u_j\, a_1\, w_j\, g_1\, g_2^{-1}$$

f vérifie les conditions du lemme avec $g = g_1\, g_2^{-1}$. On peut donc trouver $x \in \phi_{g_1 g_2^{-1}}$ tel que $a_1 = f(a_2) = a_2 x$: $a_1 \in a_2\, \phi_{g_1 g_2^{-1}}$.

On suppose la propriété vraie à l'ordre $n-1$. Soient $a_1,\dots,a_n$, $g_1,\dots,g_n$ tels que

$$\Sigma\, v_j a_i t_j^{g_i} = 0 \quad \text{pour} \quad i = 2,\dots,n \Longrightarrow \Sigma\, v_j a_1 t_j^{g_1} = 0$$

On pose $I = \{\Sigma\, v_j a_n t_j^{g_n} \;;\; \forall\, i = 2,\dots,n-1 \;,\; \Sigma\, v_j a_i t_j^{g_i} = 0\}$.

I est un sous-R-bimodule de S et

$$f : \Sigma\, v_j a_n t_j^{g_n} \in I \longrightarrow \Sigma\, v_j a_1 t_j^{g_1} \in S \;,$$

vérifie les conditions du lemme avec $g = g_1 g_n^{-1}$. On a donc

$\Sigma\, v_j a_i t_j^{g_i} = 0$ pour $i = 2,\dots,n-1 \Longrightarrow \Sigma\, v_j a_1 t_j^{g_1} = (\Sigma\, v_j a_n t_j^{g_n})\, x$,

$x \in \phi_{g_1 g_n^{-1}}$.

La propriété étant vraie à l'ordre $n-1$:

$$a_1 - a_n x \in \sum_{i=2}^{n-1} a_i \phi_{g_1 g_i^{-1}} \;, \quad \text{d'où} \quad a_1 \in \sum_{i=2}^{n} a_i \phi_{g_1 g_i^{-1}} \;.$$

∎

IV.— DECOMPOSITION DE S ET REDUCTIONS.

G désigne un (*)-groupe d'automorphismes de R et on note E l'ensemble des idempotents centraux non nuls de S .

Soit $\varepsilon \in E$. On note $H = st(\varepsilon)$ le stabilisateur de ε dans G ; il contient N donc il est d'indice fini : soient H , $g_2 H,\dots,g_n H$ les classes à droite de H dans G (on notera $g_1 = 1$) .

LEMME 4.1.— *Quel que soit* $e \in E^G$ *il existe* $\varepsilon \in E$, $\varepsilon \le e$ *tel que*

(i) $\varepsilon\, , \varepsilon^{g_2},\dots,\varepsilon^{g_n}$ *soient deux à deux orthogonaux et* $\varepsilon^G = \sum_{i=1}^{n} \varepsilon^{g_i}$ *appartienne à* E^G .

(ii) tout $f \in E$, $f \le \varepsilon$ *soit invariant par* H .

On remarque d'abord que : a) si dans E on a $e \leq e^g$ (i.e. $e = ee^g$) alors on a $e \leq e^g \leq e^{g^2} \leq \ldots \leq e^{g^m} = e$ (m est l'ordre de $\bar{g}$ dans G/N) donc $e = e^g$; b) si $f \in E$ et $f \neq f^g$ alors il existe $f_1 \leq f$, $f_1 \in E$ tel que $f_1 f_1^g = 0$ ($f_1 = f - ff^g$ convient).

Dans l'ensemble des idempotents de E inférieurs à e et dont le nombre k $(\leq |G/N|)$ de conjugués deux à deux orthogonaux est maximum, on considère un élément ε de stabilisateur H maximal et on peut supposer que ε , $\varepsilon^{g_2}, \ldots, \varepsilon^{g_k}$ sont deux à deux orthogonaux.

(ii) Soit $f \in E$, $f \leq \varepsilon$. S'il existait $h \in H$ tel que $f \neq f^h$ d'après b) il existerait $f_1 \leq f$ tel que $f_1 f_1^h = 0$ et puisque

$$f_1 \leq \varepsilon \quad f_1^h \leq \varepsilon \quad f_1^{g_2} \leq \varepsilon^{g_2} \quad \ldots \quad f_1^{g_k} \leq \varepsilon^{g_k}$$

$f_1, f_1^h, f_1^{g_2}, \ldots, f_1^{g_k}$ seraient deux à deux orthogonaux ce qui contredirait le définition de k . En conséquence $H = st(f)$.

(i) On remarque que $1, g_2, \ldots, g_k$ ne sont pas congrus modulo H et on suppose l'existence de $g \in G \setminus \bigcup_{i=1}^{k} g_i H$.

Puisque $g \notin H$ d'après b) il existe $f \in E$, $f \leq \varepsilon$ vérifiant $ff^g = 0$. On considère $f \in E$, $f \leq \varepsilon$ tel que $f, f^{g_2}, \ldots, f^{g_\ell}$ soient orthogonaux à f^g et on suppose ℓ maximum. Si on avait $k > \ell$ on en déduirait $f \neq f^{g_{\ell+1}^{-1}}$ car $H = st(f)$ et d'après a) $f_1 = f(1 - f^{g_{\ell+1}^{-1} g})$ appartiendrait à E .

Or les inégalités

$$- f_1 \leq f \leq \varepsilon$$

$$- f_1^g . f_1^{g_i} \leq f^g f^{g_i} = 0 \quad , \quad 1 \leq i \leq \ell$$

$$- f_1^g . f_1^{g_{\ell+1}} \leq f^g . (1-f^{g_{\ell+1}^{-1} g})^{g_{\ell+1}} \leq f^g (1-f^g) = 0$$

montreraient que $f_1, f_1^{g_2}, \ldots, f_1^{g_{\ell+1}}$ sont distincts et orthogonaux à f_1^g ce qui contredirait la définition de ℓ .

Donc $k = \ell$ ce qui contredit la définition de k ; k est l'indice dans G de H . ∎

On conserve les notations du lemme 4.1 et on considère dans E une famille $(\varepsilon_i)_{i \in I}$ telle que :

- pour tout i , les conjugués distincts de ε_i soient deux à deux orthogonaux et pour tout $f \in E$, $f \leq \varepsilon_i$, $st(\varepsilon_i) = st(f)$.

- pour tout $i \neq j$ on ait $\varepsilon_i^G . \varepsilon_j^G = 0$.

Le lemme 4.1 permet d'affirmer l'existence d'une telle famille avec $1 = \operatorname{Sup}_i \varepsilon_i^G$.

Pour H fixé contenant N on pose $I_H = \{i \in I \; ; \; H = st(\varepsilon_i)\}$ et $\varepsilon_H = \sup\{\varepsilon_i \; ; \; i \in I_H\}$ $(\varepsilon_H = 0$ si $I_H = \phi)$.

On remarque que si H et K sont des sous-groupes distincts de G contenant N pour tout $i \in I_H$, $j \in I_K$ et $g \in G$ on a $\varepsilon_i \varepsilon_j^g \leq \varepsilon_i^G \varepsilon_j^G = 0$ ce qui donne $\varepsilon_H . \varepsilon_K^g = 0$ et aussi $\varepsilon_H^G \varepsilon_K^G = 0$. D'autre part si $f \in E$, $f \leq \varepsilon_H$ et $g \notin H$ pour tout i et j de I_H on a $(f\varepsilon_i)(f\varepsilon_j)^g \leq \varepsilon_i \varepsilon_j^g$.

Or $\varepsilon_i \varepsilon_j^g = 0$ car si $i \neq j$, $\varepsilon_i \varepsilon_j^g \leq \varepsilon_i^G \varepsilon_j^G = 0$ et si $i = j$ on utilise $g \notin H$. On en déduit $(f\varepsilon_H)(f\varepsilon_H)^g = ff^g = 0$.

Enfin si $h \in H$, pour tout $i \in I_H$ $(f-f^h)\varepsilon_i = 0$ donc $(f-f^h)\varepsilon_H = f-f^h = 0$.
Il résulte de ces remarques que pour tout $f \in E$, $f \leq \varepsilon_H$, H est le stabilisateur de f et que ses conjugués sont deux à deux orthogonaux.
On a donc démontré le

LEMME 4.2.– *Il existe des sous-groupes* $H_1,\dots,H_s$ *de* G *contenant* N *et des idempotents centraux* $\varepsilon_{H_1},\dots,\varepsilon_{H_s}$ *de* S *tels que pour* $1 \leq i \leq s$ *et* $f \in E$, $f \leq \varepsilon_{H_i}$, H_i *soit le stabilisateur de* f *dans* G *et que les conjugués de* f *soient deux à deux orthogonaux. De plus* $1 = \varepsilon_{H_1}^G + \dots + \varepsilon_{H_s}^G$ *où les idempotents centraux invariants* $\varepsilon_{H_1}^G,\dots,\varepsilon_{H_s}^G$ *sont deux à deux orthogonaux.*

■

LEMME 4.3. [3]. *Soit* $e = e^2 \in Z(B(R;G))$ *et* $H = st(e)$. *On suppose que les conjugués distincts* e , $e^{g_2},\dots,e^{g_n}$ *de* e *sont deux à deux orthogonaux et on pose* $e^G = \sum_{i=1}^{n} e^{g_i}$. *Alors les anneaux* $(e^G S e^G)^G$ *et* $(e S e)^H$ *sont isomorphes.*

Démonstration : G opérant comme groupe de permutations sur $\{H, g_2H,\dots,g_nH\}$, il est clair que pour $x \in (e S e)^H$, $\theta(x) = x + x^{g_2} + \dots + x^{g_n}$ est invariant par G . De plus l'orthogonalité des conjugués montre que $\theta(x) = e^G \theta(x) e^G$ et que $x_1^{g_i} x_2^{g_j} = \delta_{ij}(x_1 x_2)^{g_i}$ pour x_1 et x_2 dans $e S e$

(δ_{ij} est le symbole de Kronecker) ; θ est donc un monomorphisme d'anneaux. Enfin si $y \in (e^G S e^G)^G$, $eye = ye = ey$ appartient à $(e S e)^H$ et $\theta(eye) = y$.

∎

Soit $\varepsilon \in E$. R est un anneau semi-premier et il est facile de vérifier que le filtre des idéaux bilatères de $R\varepsilon$ d'annulateur nul est identique à $\mathfrak{F}\varepsilon$ et que l'anneau des quotients $(R\varepsilon)_{\mathfrak{F}\varepsilon}$ est égal à $S\varepsilon$.

LEMME 4.4.– *Soit* $\varepsilon \in E$ *et* $H = st(\varepsilon)$. *On suppose que les conjugués distincts* $\varepsilon, \varepsilon^{g_2}, \ldots, \varepsilon^{g_n}$ *de* ε *sont deux à deux orthogonaux. Si on considère* H *comme* (*)-*groupe d'automorphismes de* εA , *on a l'équivalence*

$$g \in H_{inn} \Longleftrightarrow g \in G_{inn} \quad \textit{et} \quad \Phi_g \varepsilon \neq 0$$

De plus $B(\varepsilon R;H) = B(R;G)\varepsilon$.

Démonstration :

On considère $g \in G_{inn}$ et $y \in \Phi_g$ tel que $y\varepsilon$ ne soit pas nul. Par définition de Φ_g , on a $y\varepsilon = y\varepsilon^g$ ce qui entraîne $\varepsilon\varepsilon^g \neq 0$ et d'après l'hypothèse $g \in H \cap G_{inn} \subset H_{inn}$. L'inclusion $B(R;G)\varepsilon \subseteq B(R\varepsilon;H)$ en découle.

Réciproquement soit $g \in H_{inn}$: il existe $y\varepsilon \neq 0$ tel que, quel que soit $a \in A$, on ait $ay\varepsilon = y\varepsilon a^g$ d'où il résulte $y\varepsilon \in \Phi_g$ donc $g \in G_{inn}$ et $\Phi_g \varepsilon \neq 0$; de plus $B(R\varepsilon;H) \subseteq B(R;G)\varepsilon$.

LEMME 4.5.— *On suppose que tous les idempotents centraux de* S *sont invariants par* G . *Alors il existe des éléments* $\eta_1,\ldots,\eta_t$ *de* E *deux à deux orthogonaux tels que* $1 = \eta_1+\ldots+\eta_t$ *et tels que si on considère* G *comme groupe d'automorphismes de* $S\eta_i$, $(G)_{inn}$ *soit un sous-groupe constitué d'automorphismes intérieurs définis par des éléments inversibles de* $S\eta_i$.

Soit $g_1 = 1$, $g_2,\ldots,g_n$ un système de représentants de G modulo N . Soient $(g_\ell)_{1\le\ell\le k}$ les éléments de ce système qui appartiennent à G_{inn} . On a $C = \sum_{\ell=1}^{k} C\, e_{g_\ell}$. Le lemme 3.4 entraîne l'existence d'idempotents $(\eta_i)_{1\le i\le t}$ tels que

- $1 = \sum_{i=1}^{t} \eta_i$

- $e_{g_\ell}\eta_i = 0$ ou $e_{g_\ell}\eta_i = \eta_i$

Si $e_{g_\ell}\eta_i$ est non nul, le lemme 3.10 montre que l'automorphisme g_ℓ coïncide sur η_i avec un automorphisme intérieur.
Soit $g\in G$, tel que $\Phi_g\,\eta_i \ne 0$. Il existe $h\in N$ et $\ell(1\le\ell\le n)$ tels que $g = hg_\ell$. Sur $Se_g\,\eta_i$ on a : $g(x) = \xi x\xi^{-1}$

$$hg_\ell(x) = \xi x\xi^{-1}$$

$$g_\ell(x) = h^{-1}(\xi x\xi^{-1}) = y\xi x\xi^{-1}y^{-1} .$$

On a donc $1\le\ell\le k$. Puisque g_ℓ est intérieur au sens classique sur $S\eta_i$, g est intérieur au sens classique sur $S\eta_i$ et si on considère G comme groupe d'automorphismes de $S\eta_i$, $(G)_{inn}$ est un sous-groupe constitué d'automorphismes intérieurs au sens classique.

∎

V.– TRACES.

Ce sont les éléments de $\mathrm{Hom}({}_{SG}S_{SG}, S^G)$.

On définit sur S une structure de $B \otimes_C \overset{\circ}{B}$-module à gauche en posant :

$$(\Sigma\, a_i \otimes b_i) * s = \Sigma\, a_i\, s\, b_i\ .$$

Puisque $B(R;G) \subseteq C_S(S^G)$ et puisque, lorsque G est intérieur au sens classique, $S^G = C_S(B(R;G))$, dans ce cas on est amené à étudier l'annulateur à droite dans $B \otimes_C \overset{\circ}{B}$ de l'idéal à gauche $\sum_{b\in B} B \otimes_C \overset{\circ}{B}(1 \otimes b - b \otimes 1) =$

$\sum_{g\in G} B \otimes \overset{\circ}{B}(1 \otimes x_g - x_g \otimes 1)$ qui est le noyau de l'homomorphisme

$$B \otimes_C \overset{\circ}{B} \xrightarrow{\ \mu\ } B \longrightarrow 0\ .$$

$$\Sigma\, a_i \otimes b_i \rightsquigarrow \Sigma\, a_i\, b_i$$

1) Dans ce paragraphe B désignera un anneau régulier auto-injectif contenant dans son centre Z un anneau régulier auto-injectif C tel que B soit un C-module projectif de type fini. Notons que $B \otimes_C \overset{\circ}{B}$ étant auto-injectif et $\ker \mu$ de type fini, $\ell r(\ker \mu) = \ker \mu$ et $r(\ker \mu)$ n'est pas nul. On se propose de montrer le

THEOREME 5.1.– $r(\ker \mu)$ *est monogène.*

PROPOSITION 5.2.– $B^e = B \otimes_C \overset{\circ}{B}$ *est un anneau cohérent à droite et à gauche. En conséquence* $r(\ker \mu)$ *est de type fini.*

Soit M un sous-B^e-module de type fini d'un B^e-module à droite libre. B étant régulier et B^e étant un B-module à droite projectif de type fini, M est un B-module projectif. Donc pour toute suite exacte

$0 \to M' \to (B^e)^m \to M \to 0$ de B^e-modules, M' est un B-module de type fini donc un B^e-module de type fini.

∎

LEMME 5.3.— *Soient* ε_1 *et* ε_2 *deux idempotents orthogonaux du centre* Z *de* B *tels que* $1 = \varepsilon_1 + \varepsilon_2$. $B\varepsilon_i$ *et* $C\varepsilon_i$ $(i = 1,2)$ *vérifient les mêmes hypothèses que* B *et* C *et* $r(\ker \mu)$ *est isomorphe au produit* $r(\ker \mu_1) \times r(\ker \mu_2)$.

En effet $r(\ker \mu)$ est contenu dans $B^e(\varepsilon_1 \otimes \varepsilon_1) \oplus B^e(\varepsilon_2 \otimes \varepsilon_2)$ et $B^e(\varepsilon_i \otimes \varepsilon_i)$ est isomorphe à $B\varepsilon_i \underset{C\varepsilon_i}{\otimes} \widehat{\overset{\circ}{B\varepsilon_i}}$.

∎

LEMME 5.4.— *Il suffit de prouver le théorème lorsque* B *est commutatif.*

D'après les lemmes 3.5 et 5.3, on peut supposer que B est Z-libre. Le couple (Z,C) vérifie les mêmes hypothèses que (B,C) ; l'annulateur de $\ker(Z \underset{C}{\otimes} Z \to Z)$ est engendré par un élément u . On considère alors le morphisme naturel d'anneaux $\theta : B \underset{C}{\otimes} \overset{\circ}{B} \to B \underset{Z}{\otimes} \overset{\circ}{B}$. En utilisant le fait que $B \underset{C}{\otimes} \overset{\circ}{B}$ est un $Z \underset{C}{\otimes} Z$-module libre, on vérifie aisément que $\ker \theta = \sum_{z \in Z} B \underset{C}{\otimes} \overset{\circ}{B} (1 \otimes z - z \otimes 1)$ et que si α appartient à $r(\ker \mu)$ on a $\alpha = u\,\alpha_1$.

D'autre part B étant une Z-algèbre séparable, le noyau de $B \underset{Z}{\otimes} \overset{\circ}{B} \to B$ est engendré par un idempotent $1-e$. Soit $\tau_1 \in B \underset{C}{\otimes} \overset{\circ}{B}$ tel que $\theta(\tau_1) = e$; il en résulte $\alpha = \tau_1 \alpha = \tau_1\, u\, \alpha_1$. Enfin quel que soit $b \in B$: $(b \otimes 1 - 1 \otimes b)\tau_1 \in \ker \theta$ et $\tau_1 u \in r(\ker \mu)$.

∎

LEMME 5.5.— *Le théorème est vrai si* B *est un corps.*

On considère les extensions de degré fini $C \subseteq B_o \subseteq B$ où $C \subseteq B_o$ est séparable et $B_o \subseteq B$ radicielle. L'idéal $\sum_{b \in B_o} B^e(1 \otimes b - b \otimes 1)$ est engendré par un idempotent $1-u$ et il existe un entier n tel que $u(\ker \mu)^n = 0$ et $u(\ker \mu)^{n-1} \neq 0$. Or si $u(\ker \mu)^{n-1} = \sum_{i=1}^{n} B^e \alpha_i$, on a $\ker \mu \subseteq \mathrm{Ann}\, \alpha_1$. $\ker \mu$ étant maximal il en résulte $\ker \mu = \mathrm{Ann}\, \alpha_1$. L'auto-injectivité de B^e entraîne $\mathrm{Ann}(\ker \mu) = B^e \alpha_1$. ∎

LEMME 5.6.— *Le théorème est vrai si* B *est commutatif.*

D'après les lemmes 5.3 et 5.5, il est vérifié si B est un produit de corps.

Soit $(x_k)_{1 \leq k \leq n}$ des générateurs de $\mathrm{Ann}(\ker \mu)$. Pour tout idempotent $e \in C^*$ et tout idéal maximal m de C ne contenant pas e, il existe $y \in B^e$ dont l'image dans $B^e/m\, B^e \simeq B/m\, B \otimes_{C/m} B/m\, B$ engendre $\mathrm{Ann}(\ker \mu_m)$. Il existe donc $(e' = e'^2 \leq e, y, \lambda_k, \mu_k)$ vérifiant

(*) : $e'y = \sum_k e'\lambda_k x_k$ et $e'x_k = e'\mu_k y$. On peut alors considérer une famille $(e'_i, y_i, \lambda_{ik}, \mu_{ik})_{i \in I}$ vérifiant (*) pour tout i et telle que la somme $\sum_i C e'_i$ soit directe et essentielle. B^e étant un C-module injectif à sous-module singulier nul il existe $y, (\lambda_k, \mu_k)$ tels que pour tout i $e_i y = e_i y_i$, $e_i \lambda_k = e_i \lambda_{ik}$ et $e_i \mu_k = e_i \mu_{ik}$. On obtient $y = \sum_k \lambda_k x_k$ et $x_k = \mu_k y$ ce qui signifie que $\mathrm{Ann}(\ker \mu) = B^e y$.

∎

REMARQUES : 1) On désignera par τ un générateur de $r(\ker \mu)$. Si f et g sont des idempotents non orthogonaux de B , $f \otimes g . \tau$ n'est pas nul (sinon $f \otimes g$ appartiendrait à $\ker \mu$).

2) Si $\tau = \sum_i a_i \otimes b_i$ on a $\sum_i Bb_i = B = \sum a_i B$ (en effet si $\sum Bb_i = B(1-f)$ $(1 \otimes f).\tau = 0$ donc $f = 0$).

2) [Notations— • Soit $s \in S$ et $H = st(s)$ le stabilisateur de s dans G . $s^{G/H}$ désignera la somme des conjugués de s par un système complet de représentants des classes à droite de G modulo H . $s^{G/H}$ appartient à S^G .

• Soit $\varepsilon \in E$; on note H_ε le sous-groupe de $st(\varepsilon)$ constitué des g qui définissent sur $S\varepsilon$ des automorphismes intérieurs classiques.]

• On rappelle (4.5) qu'il existe une famille $(H_i)_{1 \le i \le s}$ de sous-groupes de G et une famille $(\varepsilon_i)_{1 \le i \le s}$ d'idempotents centraux orthogonaux telles que :

a) pour tout i et tout $f \in E$, $f \le \varepsilon_i$, H_i soit le stabilisateur de f dans G et les conjugués de f soient deux à deux orthogonaux.

b) si on considère H_i comme groupe d'automorphismes de $S\varepsilon_i$, $(H_i)_{inn}$ est formé d'automorphismes intérieurs classiques. En conséquence pour tout $f \in E$, $f \le \varepsilon_i$ on a $H_f = H_{\varepsilon_i}$ et $G_{inn} = \bigcup_{i,g} g H_{\varepsilon_i} g^{-1}$ (lemme 4.4).

c) Les idempotents (ε_i^{G/H_i}) sont deux à deux orthogonaux et

$1 = \sum_i \varepsilon_i^{G/H_i}$.

On notera E' l'ensemble des éléments non nuls de $\bigcup_{i,g} E \varepsilon_i^g$.

• Dans la suite τ désignera un générateur de l'annulateur à droite dans $B \underset{C}{\otimes} \overset{\circ}{B}$ du noyau de $B \underset{C}{\otimes} \overset{\circ}{B} \xrightarrow{\mu} B \longrightarrow 0$. Pour tout $f \in E'$ on pose :

$$\forall \quad s \in S \qquad Tr_f(s) = ((f \otimes 1).\tau * s)^{G/H_f} ,$$

Il est immédiat que les applications Tr_f sont des traces.

• Soit $0 \neq f = f^2$ un élément de B et $e \leq e_f$ $e \in E'$ tel que $(e \otimes 1)\tau$ soit de longueur minimale. On peut écrire

$$(e \otimes 1)\tau = \sum_{i=1}^{n} a_i \otimes b_i \qquad \Sigma\, a_i B = Be = \Sigma\, B\, b_i \quad ,$$

les éléments $(a_i)_{1 \leq i \leq n}$ étant Ce-libres. Si $(e \otimes 1)\tau(1 \otimes f\overset{\circ}{B})$ est nul, il en résulte que pour tout i , fBb_i est nul et donc que fe est nul ce qui est impossible. Il existe donc $\alpha \in B$ tel que $(e \otimes 1)\tau(1 \otimes f\alpha) \neq 0$. On a alors

$$0 \neq (e \otimes 1)\tau(1 \otimes f\alpha) = \sum_{i=\ell}^{m} a_i \otimes f\alpha b_i \qquad 1 \leq \ell < m \leq n \ .$$

Comme $0 \neq f\alpha b_\ell = f\alpha b_\ell e$ soit $e' \in E'$, $e' \leq e_{f\alpha b_\ell} \leq e$. Posons pour tout $s \in S$:

$$Tr_{e',f}(s) = ((e' \otimes 1)\tau(1 \otimes f\alpha) * s)^{G/H_{e'}}$$

Il est immédiat que $Tr_{e',f}$ est une trace.

<u>LEMME 5.7.</u>– *Soient* a *un élément non nul de* S *et* M *un sous-module non nul de* S . *On suppose que* $r_B(a) = (1-f_1)B$ *et* $r_B(M) = (1-f_2)B$ *où* f_1 *et* f_2 *désignent respectivement des idempotents de* B *et de* $Z(B)$ *non orthogonaux.*

Alors pour tout $e_o \in E$, $e_o \leq e_{f_1 f_2}$, *il existe* $e \in E'$, $e \leq e_o$ *tel que pour tout* $I \in \mathfrak{F}$ *on ait* $0 \neq a\, Tr_e(IM)$.

D'après la remarque 1) il existe $e \in E'$, $e \leq e_o$ tel que $(ef_1 \otimes f_2)\tau$ soit de longueur minimum $(ef_1 \otimes f_2)\tau = \sum_{k=1}^{m} ef_1 a_k \otimes b_k f_2$; pour tout $e' \in E$, $e' \leq e$ et tout k, $e'f_1 a_k$ et $b_k f_2 e'$ ne sont pas nuls. Compte-tenu des définitions de f_1 et f_2, pour tout $x \in IM$ on a :

$$aTr_e(x) = a[ef_1(\tau * x)f_2] + a[e(\tau * x)]^{G/H_e^*} = a(\sum_{k=1}^{m} ef_1 a_k x b_k f_2) + a(\sum_{k=1}^{n} ec_k x d_k)^{G/H_e^*}.$$

Par définition de m et f_1, la somme $\sum_{k=1}^{m} Caef_1 a_k$ est directe ; de plus pour tout $g \notin H_e$, $e\Phi_g$ est nul. En conséquence [3.15] il existe des éléments v_j et t_j de R tels que :

$$0 \neq \sum_j v_j a e f_1 a_1 t_j = \alpha = \alpha e \qquad 0 = \sum_j v_j a e f_1 a_k t_j \quad (1 < k \leq m)$$

$$0 = \sum_j v_j a (ef_1 c_k)^g t_j^g = 0 \quad (1 \leq k \leq n \quad \bar{g} \in G/H_e^*) .$$

Il en résulte $\sum_j v_j aTr_e(t_j x) = \alpha x b_1 f_2$ donc $\sum_j v_j aTr_e(IM) = \alpha IMb_1 f_2$.

Si $aTr_e(IM)$ était nul, on en déduirait $b_1 f_2 e_\alpha = 0$ ce qui est impossible car $e_\alpha \leq e$.

∎

COROLLAIRE 5.8.– *Soient* $\mathfrak{A}$ *un idéal à gauche (resp. à droite) invariant et* D *un idéal à gauche (resp. à droite) de* R^G *. Il existe* $e \in E'$ *et* $J \in \mathfrak{F}$ *tels que*

$$0 \neq Tr_e(J(\mathfrak{A} \cap RD)) \subset \mathfrak{A} \cap D \cap R^G$$

$$(\text{resp. } 0 \neq Tr_e((\mathfrak{A} \cap DR)J) \subset \mathfrak{A} \cap D \cap R^G).$$

Soit $0 \neq f = f^2 \in Z(B)$ tel que $r_B(\mathfrak{A} \cap RD) = (1-f)B$. Il existe $e \in E'$ tel que pour tout $J \in \mathfrak{F}$ on ait $Tr_e(J(\mathfrak{A} \cap RD)) \neq 0$. Comme $Tr_e(x) = \sum b_g x^g$ il existe $J \in \mathfrak{F}$ tel que $Tr_e(J\mathfrak{A}) \subset \mathfrak{A} \cap RG$ et $Tr_e(JD) \subset D$. ∎

VI.— LES THEOREMES.

THEOREME 6.1.— R^G *est semi-premier. De plus si* B *est* G*-simple alors* R^G *est premier.*

a) Soit $a \in R^G \setminus (0)$. D'après le lemme 5.7 appliqué à a et $M = Ra$ il existe $e \in E'$ tel que pour tout $I \in \mathfrak{F}$ on ait $0 \neq a\, Tr_e(I) a$. Il suffit de considérer $I \in \mathfrak{F}$ tel que $Tr_e(I) \subset R^G$.

b) Soient a et b deux éléments non nuls de R^G . Comme $B \subset C_S(R^G)$, $r_B(a)$ et $r_B(b)$ sont invariants donc nuls. Il suffit d'appliquer le lemme 5.7 à a et $M = Rb$. ∎

THEOREME 6.2. [5, theorem 4] . B *est le centralisateur dans* S *de* R^G.

Soit a un élément de $C_S(R^G)$.

On considère $I = \{b \in B \ ;\ ab \in B\}$ idéal à droite de B : il est essentiel dans un facteur direct $(1-f)B$ de B . Il existe donc un idéal essentiel dans C : $\oplus\, e_i C$ $(e_i \in E)$ tel que pour tout i , $e_i a(1-f) = \sum_g e_i \gamma_{ig} x_g \in B$ [3.2]. Si on considère des éléments γ_g de C tels que pour tout i , $e_i\gamma_g = e_i\gamma_{ig}$ alors $a(1-f) = \sum \gamma_g x_g \in B$ d'où il résulte $I = (1-f)B$.

On suppose $f \neq 0$; soit $e \in E'$, $e \leq e_f$ tel que $fe \otimes 1 . \tau$ soit de longueur minimum : $fe \otimes 1.\tau = \sum_{k=1}^{m} fa_k \otimes b_k$. En particulier $fa_1 \notin \sum_{k=2}^{m} fa_k C$.

Soit $J \in \mathfrak{F}$ tel que l'on ait $Tr_e(J) \subset R^G$; alors on obtient :

$$\forall\, x \in J \qquad a\, Tr_e(x) - Tr_e(x) a = 0 \ .$$

i.e. $0 = a\left(\sum_{k=1}^{m} fa_k x b_k\right) + a\left(\sum_{k=m+1}^{n} (1-f) a_k x b_k\right) - \left(\sum_{k=1}^{n} a_k x b_k\right) a + a(\tau(e \otimes 1) * x)^{G/H_e^*} - (\tau(e \otimes 1) * x)^{G/H_e^*}\, a$.

Or $afa_1 \notin \sum_{k=2}^{m} afa_kC + \sum_{k=m+1}^{n} a(1-f)a_kC + \sum_{k=1}^{n} a_kC$: dans le cas contraire, il existerait $c_2,\dots,c_m$ tels que $a(fa_1+fa_2c_2+\dots+fa_mc_m)$ appartienne à B ; on aurait $0 = fa_1+fa_2c_2+\dots+fa_mc_m$ ce qui est impossible. Comme dans la démonstration du lemme 5.7, il existe $\alpha = \alpha e \neq 0$ tel que $\alpha Jb_1 = 0$; on en déduit : $e_\alpha \leq e$ et $e_\alpha b_1 = 0$ ce qui contredit la définition de e . En conséquence $I = B$.

∎

<u>THEOREME</u> 6.3.[5,lemma 5] . *Soit* M *un sous*-R-R^G-*bimodule de* S . *Il existe* $J \in \mathfrak{F}$ *et* $f = f^2 \in B$ *tels que* $Jf \subset M$ *et* $r_B(M) = (1-f)B$.

1) Supposons d'abord que $M = RaR^G$ et $r_B(a)=(1-f)B$.

Soit $e_o \leq e_f$. On reprend les notations et la démonstration de 5.7 avec $f_1 = f$ et $f_2 = 1$:

Soit $I \in \mathfrak{F}$ tel que $Tr_e(I) \subset R^G$; il existe $\alpha_k \in R \cap Se$ tel que :

$$\alpha_k\, I\, b_k = \sum_j v_j\, a\, Tr_e(I) \subset M$$

et l'ensemble de ces idempotents e est cofinal dans $E\, e_o$. L'annulateur dans Se_o de l'idéal bilatère $L_k = \{\alpha \in R \cap Se_o \,;\, \exists\ I_\alpha \in \mathfrak{F}\ \ \alpha I_\alpha b_k \subset M\}$ est donc nul de sorte qu'il existe $0 \neq \alpha \in L_1 \cap \dots \cap L_m$ et $I \in \mathfrak{F}$ tels que $\sum_k \alpha I b_k \subset M$. Compte-tenu de la remarque 2) : $Be_of = \sum_{k=1}^{m} Bb_k$ i.e. $e_of = \sum_{k=1}^{m} \beta_k b_k$; il existe $I' \in \mathfrak{F}$ tel que $\sum_{k=1}^{m} I'\beta_k \subset I$ d'où :

$$\alpha I'f \subseteq \sum_{k=1}^{m} \alpha I'\beta_k b_k \subseteq \sum_{k=1}^{m} \alpha I b_k \subset M .$$

L'ensemble des e_o étant cofinal, l'annulateur dans Se_f de l'idéal bilatère $L = \{\alpha \in R \cap Se_f ; \exists I_\alpha \in \mathfrak{F} \quad \alpha I_\alpha f \subset M\}$ est nul. En conséquence $J = \sum_{\alpha \in L} R\alpha I_\alpha + I_o(1-e_f)$ (où $I_o \in \mathfrak{F}$ et $I_o e_f \subset R$) appartient à $\mathfrak{F}$ et vérifie $Jf \subset M$.

2) Dans le cas général on considère l'idéal $\sum_{a \in M} Bf_a$ avec $r_B(a) = (1-f_a)B$ et $J_a \in \mathfrak{F}$, $J_a f_a \subseteq R a R^G$. Cet idéal est essentiel dans un facteur direct Bf_o. On vérifie aisément que $f = f_o$. De plus d'après [3.2], il existe un idéal essentiel $\bigoplus_i Ce_i$ de C tel que pour tout i on ait $e_i f \in \sum Bf_a$. A chaque i on peut associer $J_i \in \mathfrak{F}$ tel que $J_i e_i f \subset M$: en effet il existe $\beta_1, \ldots, \beta_{n_i}$ de B tels que $e_i f = \beta_1 f_{a_1} + \ldots + \beta_{n_i} f_{a_{n_i}}$; alors si $J'_{a_n} \in \mathfrak{F}$ vérifie $J'_{a_n} \beta_n \subset J_{a_n}$ et si $J_i = J'_{a_1} \cap \ldots \cap J'_{a_{n_i}}$ on obtient $J_i e_i f \subseteq \sum J'_{a_n} \beta_n f_{a_n} \subseteq \sum J_{a_n} f_{a_n} \subset M$. Il est clair que l'on peut supposer $J_i e_i \subset R$ et que $J = \sum J_i e_i \in \mathfrak{F}$ et vérifie $Jf \subset M$.

∎

On dispose d'un théorème analogue pour les sous-R^G-R-bimodules de S.

<u>THEOREME 6.4</u>. [5,theorem 7] .*Si on désigne par* $\mathfrak{F}_1$ *le filtre des idéaux bilatères essentiels de* R^G *on a*

a) *Pour tout* $I \in \mathfrak{F}, I_1 = I \cap R^G$ *appartient à* $\mathfrak{F}_1$.

b) *Pour tout* $I_1 \in \mathfrak{F}_1$ *il existe* $J \in \mathfrak{F}$ *tel que* $J \subset RI_1$

c) $(R_{\mathfrak{F}})^G$ *est isomorphe à* $(R^G)_{\mathfrak{F}_1}$.

a) Soit $f = f^2 \in Z(B)^G$ tel que $r_B(RI_1) = (1-f)B$. Il suffit de montrer que $1-f = 0$. Dans le cas contraire il existe $e \in E'$ et $e \leq (1-f)$ et $J \in \mathfrak{F}$ tel que $0 \neq Tr_e(JI(1-f)) = Tr_e(JI)(1-f) \subset (I \cap R^G)(1-f) = 0$.

b) $r_B(I_1) = (1-f)B$ avec $f \in Z(B)$. Il existe $I \in \mathfrak{F}$ tel que $(1-f)I \subset R$. Le lemme 5.7 montre que $1-f$ est nul et il existe $I \in \mathfrak{F}$ tel que $I \subset RI_1$.

c) Soit (I_1,ϕ) un élément de $(R^G)_{\mathfrak{F}_1}$ et $\mathfrak{A}$ l'idéal à gauche $\{\Sigma\, r_i\, \phi(x_i)\ ;\ r_i \in R\ ,\ x_i \in I_1\ ,\ \Sigma\, r_i\, x_i = 0\}$. Le corollaire 5.8 montre que $\mathfrak{A}$ est nul et par suite que ϕ se prolonge à RI_1 . Comme RI_1 contient $J \in \mathfrak{F}$, (I,ϕ) définit un élément de $(R_{\mathfrak{F}})^G$.

∎

PROPOSITION 6.5.— *Si* R^G *est premier,* B *est* G*-simple.*

Comme R^G est premier et $C^G = C^{G/N}$ est régulier, le théorème 6.4 montre que C^G est un corps. Soit I un idéal essentiel de C . Comme $\bigcap_{g\in G} I^g \cap C^G = (\bigcap_{\bar g\in G/N} I^{\bar g}) \cap C^G$ est non nul, I est égal à C . Par suite C est un produit de corps et B est un anneau semi-simple. Si $\mathfrak{A}$ est un idéal de B invariant, $\mathfrak{A}$ est engendré par un idempotent invariant qui est central dans $(R^G_{\mathfrak{F}})$ et donc $\mathfrak{A} = B$.

∎

PROPOSITION 6.6.— *Si* G_{inn} *est constitué d'automorphismes définis par des éléments inversibles de* S *, alors* G/G_{inn} *est* X*-extérieur sur* $R^{G_{inn}}$.

Supposons qu'il existe $\bar g \neq \bar 1 \in G/G_{inn}$ X-extérieur sur $R^{G_{inn}}$. On peut supposer que G est engendré par G_{inn} et g . Il existe donc $0 \neq x_0 \in S^{G_{inn}}$

tel que l'on ait :

$$\forall\, s \in S^{G_{inn}} \qquad s x_o = x_o s^{\bar{g}} = x_o s^{g} .$$

En particulier x_o appartient à $C_S(S^G) = B = C_S(S^{G_{inn}})$ donc au centre de $S^{G_{inn}}$. Soit f l'idempotent central de $S^{G_{inn}}$ associé à x_o ; f est invariant et g est trivial sur $S^{G_{inn}} f$. En conséquence si $c(f)$ désigne la couverture centrale de f dans S et $e' \leq c(f)$ un élément de E' tel que la longueur m de $(e'f \otimes 1).\tau$ soit minimale, si $I \in \mathfrak{F}$ vérifie $(e'f \otimes 1).\tau * I \subset R^{G_{inn}}$, on obtient :

$$\forall\ x \in I \qquad \sum_{k=1}^{m} f a_k x b_k = \sum_{k=1}^{m} (f a_k x b_k)^g .$$

Puisque g est X-extérieur sur R et que $f a_1 \notin \sum_{k \neq 1} f a_k C$, il existe $\alpha = \alpha e \neq o$ tel que $\alpha I b_1 = o$ ce qui est impossible. ■
(cf. question 15 de [4]).

<u>PROPOSITION 6.7.</u> [5, lemma 11] . *Soit* $\mathfrak{A}$ *un idéal à gauche (à droite) essentiel dans* R , $\mathfrak{A} \cap R^G$ *est un idéal à gauche (resp. à droite) essentiel dans* R^G .

Soit D un idéal à gauche $\neq 0$ de R^G tel que $\mathfrak{A} \cap D = 0$. Soit f un idempotent de B tel que $fB \succ \sum_{M \prec \mathfrak{A}} r_B(M \cap RD)$. Il existe une famille $(e_i)_{i \in I}$ d'idempotents de C tels que

$$\oplus\, C e_i \prec C$$

$$\forall\ i \qquad e_i f \in r_B(M_i \cap RD) \qquad M_i \prec \mathfrak{A}$$

Soit $J_i \in \mathfrak{F}$ tel que $J_i e_i \in R$.

Soit $N = \Sigma J_i e_i M_i$ N est un sous-module essentiel de $\mathfrak{A}$. En effet soit $a \in \mathfrak{A}$. Il existe i $J_i e_i a \neq 0$ et soit $x \in J_i e_i a - \{o\}$.

Il existe $\alpha \in R$ $0 \neq \alpha x = \alpha x e_i \in M_i$

$$0 \neq J_i e_i \alpha x \subset J_i e_i M_i$$

On a alors $r_B(N \cap RD) = fB$. En effet

$r_B(N \cap RD) \subset fB$ d'après la définition de f

et si $x \in N \cap RD$, $x = \Sigma j_i e_i m_i$ alors $xf = \Sigma j_i m_i e_i f = 0$.

Pour tout $N' \prec N$ on a $r_B(N' \cap RD) = fB$. Soit $N' \prec N$ tel que $Tr_{e,1-f}(N') \subset N$. Le théorème 6.3 entraine l'existence de $J \in \mathfrak{F}$ tel que $I(1-f) \subset (N' \cap RD)R^G$. Soit $I_1 \in \mathfrak{F}$ tel que $Tr_{e,1-f}(I_1) \subset R^G$. On peut écrire :

$$Tr_{e,1-f}(I_1 I(1-f)) \subseteq Tr_{e,1-f}((I_1 N' \cap I_1 D)R^G) \subset (N \cap D)R^G = 0 .$$

Une démonstration analogue à celle du lemme 5.7 montre qu'alors il existe $\beta \in Se_1 R$ tel que

$$\beta I_1 I(1-f) \alpha b_\ell = 0$$

Par suite il existe $e' \leq e$ $e'(1-f)\alpha b_\ell = 0$ or $e \leq e_{(1-f)\alpha b_\ell}$.

On obtient donc une contradiction.

∎

<u>THEOREME 6.8</u> [5, theorem 9] . *Les assertions suivantes sont équivalentes :*

a) R *est de Goldie (à gauche) (resp. à droite).*

b) R^G *est de Goldie (à gauche) (resp. à droite).*

a) ⟹ b) Il suffit de montrer que si $\bigoplus_{i \in I} \mathfrak{A}_i$ est une somme directe d'idéaux à gauche de R^G la somme $\Sigma R\mathfrak{A}_i$ est directe.

Si $R\mathfrak{A}_1 \cap \sum_{i=2} R\mathfrak{A}_i$ est non nul le corollaire 5.8 entraine qu'il existe $e \in E'$ et $J \in \mathfrak{F}$ tel que

$$0 \neq Tr_e(J(R\mathfrak{A}_1 \cap \sum_{i=2}^{n} R\mathfrak{A}_i) \subset \mathfrak{A}_1 \cap \sum_{i\geq 2} \mathfrak{A}i$$

b) $\Longrightarrow$ a) L'idéal singulier de R étant un idéal invariant le corollaire 5.8 entraine qu'il est nul. Soit d un élément régulier de R^G. $\ell_R(d)$ est un idéal à gauche invariant et 5.8 entraîne qu'il est nul. Par suite tout idéal à gauche essentiel de R contient un élément de R^G régulier dans R. R est donc de Goldie.

REFERENCES

[1] S.A. AMITSUR :"*Rings of quotients*". Symposia Mathematica, vol.VIII (1972) p.149-162.

[2] K.I. BEIDAR : "*The ring of invariants under the action of a finite group of automorphisms of a rings*". Uspeki Math. Nauk 32 (1977) p.159-160.

[3] D.R. FARKAS and R.L. SNIDER :"*Noetherian fixed rings*". Pacific Journal of Math. 69(1977)p.347-353.

[4] J.W. FISHER and J. OSTERBURG :"*Finite group action on non commutative rings: a survey since 1970*". Proceedings of 1979 Oklahoma Conference (à paraître).

[5] V.K. KHARCHENKO : "*Galois theory of semi-prime rings*". Algebra and Logic 16(1976)p.208-258.

[6] V.K. KHARCHENKO : "*Fixed elements under a finite group acting on a semi-prime ring*". Algebra and Logic 14(1975)p.203-213.

[7] J. LAMBECK : "*Rings and modules*". New-York Blaisdell (1966).

[8] P. MARTIN : "*Auto-injectivité des images homomorphiques d'un anneau de polynômes*". C.R. Acad. Sc. Paris, t.285(1977)p.489-492.

[9] S. MONTGOMERY :"*Fixed rings of finite automorphisms groups of associative rings*". Lecture Notes (818) Springer Verlag.

[10] A. PAGE : "*Actions de groupes. Séminaire d'Algèbre P. Dubreil Paris (1977-1978)*". Lecture Notes in Mathematics (740) Springer Verlag.

[11] G. RENAULT : "*Algèbre non commutative*". Gauthier-Villars Paris (1975).

TRACE FUNCTIONS AND AFFINE FIXED RINGS FOR GROUPS ACTING ON NON-COMMUTATIVE RINGS

Susan Montgomery

1. Introduction

This paper is a survey of recent results extending Noether's classical theorem, on affine fixed rings for finite groups acting on commutative rings, to the noncommutative situation. It is also a survey of when the usual trace function (or an appropriate substitute) is non-trivial. As we shall see, these two topics are closely related.

More specifically, let R be an associative ring and G a group acting as automorphisms of R. We assume that G is finite (except in §8), and denote the order of G by $|G|$. The fixed ring (or ring of invariants) is $R^G = \{ r \in R \mid r^g = r, \text{ all } g \in G \}$. The trace of G on R is the R^G-bimodule homomorphism $t: R \to R^G$ given by $t(r) = \sum_{g \in G} r^g$. A ring R is affine over a commutative ring A if R is a finitely-generated A-algebra.

For completeness we state:

1.1 Theorem (E. Noether): Let R be a commutative ring which is affine over a commutative Noetherian ring A, and let G be a finite group of A-automorphisms of R. Then R^G is affine over A.

A standard proof of this theorem is to note that R is integral over R^G, and thus is a finite R^G-module, as it is affine. The Artin-Tate lemma then applies to give R^G affine.

In the non-commutative case, none of the above steps is true in general, and it is easy to construct examples of non-commutative rings such that Noether's theorem fails.

1.2 Example: Let k be a field, and $R=k\langle x,y\rangle$ the free algebra. Let g be the "exchange" automorphism; that is, g interchanges x and y, and let $G = \langle g\rangle$. Then R^G is generated by the set $\{x^n+y^n; n=1,2,...\}$ and by no finite subset. This fact is proved by G. Bergman and P.M. Cohn in [1].

E. Formanek has pointed out an easier way to look at this example. Assuming k has characteristic $\neq 2$, let $z=x+y$ and $w=x-y$; then $R=k\langle z,w\rangle$ where $z^g=z, w^g=-w$. R^G is generated by $\{z,w^2,wz^nw; n\geqslant 1\}$ and cannot be generated by a finite subset. We will return to free algebras in §6.

This example is a bit unfair, in that it is not Noetherian (a property one has automatically in Noether's theorem, by the Hilbert basis theorem). Even when R is Noetherian, though, things can go badly wrong. The next example (appearing in [18]) is due to A. Wadsworth, based on an earlier example of R. Resco.

1.3 Example: Let $A=k[x,y]$, the commutative polynomial ring over a field k of characteristic $p\neq 0$. Then $GL(2,k)$ acts on A by the usual linear action on x,y. We may choose two finite subgroups G_1, G_2 of $GL(2,k)$ such that the subgroup H generated by G_1 and G_2 is infinite. Letting $A_i = A^{G_i}, i=1,2$, we see that A is finite over $A_i, i=1,2$, but not over $A^H=A_1\cap A_2$. Moreover, A_i is not integral over $A_1\cap A_2$.

Now let t be a commuting indeterminate over A. We define $R = \begin{pmatrix} A_1+tA[t] & tA[t] \\ tA[t] & A_2+tA(t) \end{pmatrix}$ (that is, R is the set of matrices with entries in the given sets). R is a prime, Noetherian, affine k-algebra, satisfying a polynomial identity (PI). Consider $g\in Aut(R)$ given by conjugation by $\begin{pmatrix} 1 & 1 \\ 0 & 1 \end{pmatrix}$. $G=\langle g\rangle$ has order p, and

$R^G = \begin{pmatrix} A_1\cap A_2+tA[t] & tA[t] \\ 0 & A_1\cap A_2+tA[T] \end{pmatrix}$. R^G is not Noetherian, and is not affine over k. Moreover, R is not integral over R^G, and is not a finite R^G-module.

This example shows that the usual commutative arguments cannot be used. As a better model of how to proceed, we mention a result for infinite groups: the classical theorem of Hilbert.

1.4 Theorem (Hilbert). Let $R=k[x_1,...x_n]$, the polynomial ring over a field k of characteristic 0. Consider $G=SL(m,k)$ acting linearly on k (or any other group G such that every finite-dimensional characteristic 0 polynomial representation of G is completely reducible). Then R^G is affine over k.

In the lovely proof given by Mumford [19], the crucial part of the argument (for which the complete reducibility hypothesis is needed) is to construct a projection $\rho: R \to R^G$ of R onto R^G such that ρ is an R^G-module homomorphism.

When G is finite, for any ring R the trace is an R^G-module homomorphism from R to R^G. Thus we shall be concerned with situations where it behaves as much like a projection as possible.

2. Affine Noetherian rings when the order of G is invertible

When $|G|^{-1} \in R$, there exists a genuine projection $\rho: R \to R^G$, given by $\rho(x)=|G|^{-1}t(x)$. In this situation, at least when R is Noetherian, a positive result has been obtained by Montgomery and Small [18].

2.1 Theorem [18, Theorem 1]: Let R be a Noetherian ring which is affine over a commutative Noetherian ring A, and let G be a finite group of A-automorphisms such that $|G|^{-1} \in R$. Then R^G is affine over A.

The method of proof is indirect and uses the <u>skew group ring</u> R*G. That is, R*G is the free R-module with basis $\{g \mid g \in G\}$, and with multiplication given by $rg = gr^g$, all $r \in R, g \in G$. Since $|G|^{-1} \in R$, R*G contains the element $e = |G|^{-1} \sum_{g \in G} g$. One may check that $e^2 = e$ and that $eg = e$, all $g \in G$. It follows that $e(R*G)e \cong R^G$, via the mapping which sends $e(rg)e$ to $\rho(r)$.

Letting S=R*G, we see that S is affine and Noetherian also. Thus the theorem is a consequence of the next proposition.

2.2 <u>Proposition</u> [18, Corollary 1]. Let S be a Noetherian ring with 1, which is an affine algebra over a commutative Noetherian ring A, and say $e^2 = e \in S$. Then eSe is affine.

We remark that Theorem 2.1 is also true if the hypothesis that $|G|^{-1} \in R$ is replaced by the somewhat weaker hypothesis that there exists some $u \in R$ such that $t(u) = 1$. For as was observed in [9], the element $e_1 = u \sum_{g \in G} g$ is also an idempotent in R*G and $e_1(R*G)e_1 \cong R^G$. Thus Proposition 2.2 applies here also.

A different proof of Theorem 2.1 has been given by M. Lorenz (unpublished). He first proves the following lemma, by a proof similar to the usual proof of the Artin-Tate lemma.

2.3 <u>Lemma:</u> Let $S \subseteq R$ be A-algebras, such that

1) R is a finitely-generated right S-module.
2) There exists a right S-module homomorphism $\tau : R \to S$ such that $\tau(R) = S$.

Then if R is affine over A, S is affine over A.

Now, when $|G|^{-1} \in R$, it is easy to see that R Noetherian implies R^G Noetherian by using the projection ρ. Then R is a finite R^G-

module by a theorem of D. Farkas and R. Snider [7]. The theorem follows by applying Lemma 2.3 with $S=R^G$ and $t=\rho$.

3. The trace function

As the trace was so useful in §2, in this section we survey other cases in which it is known to be non-trivial. Unfortunately, the trace can be trivial even for fairly nice rings. We consider two examples:

3.1 Example (R. Snider). Let R be a division algebra of characteristic 2 which is four-dimensional over its center Z. Choose $x \in R$, $x \notin Z$ with $z^2 \in Z$. Then $(1+x)^2 = 1+x^2 \in Z$ also. Let g be conjugation by x, and let h be conjugation by 1+x. Then if $G = \langle g,h \rangle$, $G \cong Z_2 \times Z_2$ is a group of automorphisms of R with the property that $t(R) \equiv 0$.

3.2 Example [3, Example 1.1]. Let G be any non-abelian simple group, and let H be a proper subgroup of index n. Then G acts by right multiplication as permutations of the n right cosets of H in G, say $\{H, Ha_2, \ldots, Ha_n\}$. Choose any prime p dividing $|H|$, let F be a field of characteristic p, and let $R = \sum_{i=1}^{n} \oplus (F)_i$, the direct sum of n copies of F (so R is a commutative ring with no nilpotent elements). As in shown in [3], for any $r = (r_1, \ldots, r_n) \in R$, $t_G(x) = |H|\ (r,r,\ldots r)$. where $r = \sum_{i=1}^{n} r_i$. Thus $t_G(R) \equiv 0$ by our choice of p.

There are two difficulties in both of these examples: first, the characteristic of R divides $|G|$, and second, the automorphisms are inner (in Example 3.1) or at least inner on an ideal of R (in Example 3.2, each element of H fixes $(F)_1$, so is "inner" on $(F)_1$). When the first difficulty is eliminated, positive results have been obtained by G. Bergman and I.M. Isaacs [2].

3.3 <u>Theorem</u> [2, Proposition 2.3]: If R has no additive $|G|$-torsion and t(R) is nilpotent, then R is nilpotent.

Consequently, if R is semiprime with no $|G|$-torsion, then $t(R) \neq 0$, and $t(\lambda) \neq 0$ for any non-zero left (right) ideal λ of R.

For the second difficulty, we need a more general definition of inner automorphism. Let R be a semiprime ring and let $\mathfrak{F}$ be the filter of all two-sided essential ideals of R., We consider the Martindale quotient ring $Q_o(R) = \varinjlim_{I \in \mathfrak{F}} \text{Hom}(_R I, R)$ (for an exposition of $Q_o(R)$ and its properties, see Chapter 3 of [17]). It is known that $Q_o(R)$ is semiprime and contains a copy of R; the extended center C of R is the center of $Q_o(R)$; and any automorphism g of R extends uniquely to an automorphism of $Q_o(R)$. Following V.K. Kharchenko, we define

$$\phi_g = \{x \in Q_o(R) \mid xr^g = rx, \text{ all } r \in R\}.$$

We say g is <u>X-inner</u> if $\phi_g \neq 0$. For any group $G \subsetneq \text{Aut}(R)$, $G_{inn} = \{g \in G \mid g \text{ is X-inner}\}$, and G is <u>X-outer</u> if $G_{inn} = \langle 1 \rangle$. For example, in Example 3.2 above, $G_{inn} = \bigcup_i a_i^{-1} H a_i$.

When G is X-outer, the situation is again nice. It is implicit in work of Kharchenko [11, proof of Theorem 2] that when R is semiprime and G is X-outer, $t(R) \neq 0$. More generally, the present author has shown the following:

3.4 <u>Theorem</u> [15, Theorem 2.1 and 3.1]: Let R be semiprime and G an X-outer group of automorphisms of R. Then $t(\lambda) \neq 0$, for any non-zero left (right) ideal of R.

Positive results can also be obtained by restricting the nature of the group, such as considering p-groups. The next theorem is another result of Bergman and Isaacs [2].

3.5 Theorem [2, Proposition 3.3]: If R has no nilpotent elements, $pR = 0$ for some prime p, and G is a cyclic p-group acting faithfully on R, then $t(R) \not\equiv 0$.

This result was extended to prime rings by Cohen and Montgomery.

3.6 Theorem [3, Theorem 3.3] Let R be a prime ring of characteristic $p \neq 0$ and let $G = \langle g \rangle$ be a cyclic group of order p^n of X-inner automorphisms of R. Assume that g is induced by $a \in Q_o(R)$ and let k be the degree of the minimum polynomial of a over the extended center C of R. Then $t_G(R) \not\equiv 0 \Leftrightarrow p^n < 2k$.

Moreover, if G is any finite cyclic group acting on R, let H be the Sylow p-subgroup of G_{inn}. Then $t_G(R) \equiv 0$ $t_H(R) \equiv 0$, and the above criterion may be applied to H.

The above theorem has an interesting application to matrices.

3.7 Example [3, Example 3.5]: Let F be a field of characteristic p, and $R = M_K(F)$ the $k \times k$ matrices over F. For $\lambda \neq 0$ in F, let

$$A = \begin{pmatrix} \lambda & 1 & & 0 \\ & \lambda & 1 & \\ & & \ddots & \ddots \\ & & & \ddots & 1 \\ 0 & & & & \lambda \end{pmatrix},$$

an irreducible Jordan matrix, and let n be the smallest positive integer such that $p^n \geqslant k$. Then if $G = \langle g \rangle$, where g is the inner automorphism induced by conjugation by A, $|C| = p^n$ and so by the Theorem, $t_G(R) \not\equiv 0 \Leftrightarrow p^n < 2k$.

In particular if $k=4$, then $t_G(R) \not\equiv 0$ if $p=2,5$, or 7, but $t_G(R) \equiv 0$ for $p=3$ and for all $p \geqslant 11$.

The next result was proved by Faith [6] in the case of a division ring R, using non-commutative Galois theory. It was extended to arbitrary domains by Montgomery [16].

3.8 Theorem [16, Corollary 12]: Let R be a domain, and let G be any finite simple group acting non-trivially as automorphisms on R. Then $t(R) \neq 0$.

Observe that this result cannot be extended to rings with no nilpotent elements, as is seen by Example 3.2.

We have seen above a number of cases in which the trace is non-trivial. However, even when $t(R) \neq 0$, it may only be a small part of R^G, and thus will not be a suitable replacement for a projection. We therefore ask: when is the image $t(R)$ an essential ideal of R^G (that is, when does $t(R) \cap I \neq 0$, for any non-zero ideal I of R^G)?

This is true when R is semiprime with no $|G|$-torsion (by theorem 3.3) and when R is semiprime and G is X-outer on R (by Theorem 3.4).

4. Trace-like functions on rings with no nilpotent elements

The rather nice situation described above, in which $t(R)$ is essential in R^G, is one we would like to imitate for rings with no nilpotent elements. Since the trace itself can be trivial, we must look for another function. When R is a domain, an appropriate substitute exists for t, namely, a "partial trace function". That is, for a non-empty subset $\Lambda \subseteq G$, the function $t_\Lambda(x) = \sum_{g \in \Lambda} x^g$ is called a partial trace function if $t(R) \subseteq R^G$. If $t_\Lambda(R) \neq 0$, t_Λ is non-trivial. In [7], Farkas and Snider show that for a division ring R, non-trivial partial trace functions exist for any group acting on R. This was extended to domains by Montgomery [16].

4.1 Theorem [16, Theorem 13]. Let R be a domain, and G any (finite) group acting as automorphisms on R. Then there exists $\Lambda \subseteq G$ so that $0 \neq t_\Lambda(R) \subseteq R^G$.

For example, in Example 3.1, we may use the set $\Lambda = \{1,g\}$.

The proof in the domain case is essentially the same as for division rings, except that Theorem 3.8 must be used instead of the result of Faith for division rings. Moreover, the proof is constructive; we provide a sketch. Let $G=N_0 \supset N_1 \supset N_2 \ldots \supset N_k=\langle 1\rangle$ be a composition series for G. As in Galois theory consider the chain of fixed subrings $R^G \subset R^{N_1} \subset \ldots \subset R^{N_k} = R$, and note that $R^{N_{i+1}} = (R^{N_i})^{N_i/N_{i+1}}$. For each i, define $t_{\Lambda_i}: R^{N_i} \to R^{N_{i+1}}$ as follows: if N_i/N_{i+1} acts trivially on R^{N_i}, let $\Lambda_i = \{1\}$; if N_i/N_{i+1} acts faithfully on R^{N_i}, the trace is non-trivial by Theorem 3.8, so we may let Λ_i be a set of coset representatives for N_{i+1} in N_i. The theorem follows by letting $\Lambda = \Lambda_1\Lambda_2\ldots\Lambda_k$ (though in the domain case one must check non-triviality, as the t_{Λ_i}'s are not in general surjective).

As we will see in §5, such a t_Λ suffices to give an affine fixed ring result for domains. Unfortunately, however, non-trivial partial trace functions do not exist in general for rings with no nilpotent elements. The next theorem combines work of Cohen and Montgomery [3] and of Guralnick, Isaacs, and Passman [10].

4.2 <u>Theorem</u> 1) [3, Theorem 2.3] If R has no nilpotent elements and G is solvable, then non-trivial partial trace functions exist for G acting on R.

2) [10, Theorem]: If G is any non-solvable group, then a commutative ring R with no nilpotent elements can be constructed such that all partial trace functions for G on R are trivial.

Another difficulty is that even when a non-trivial partial trace function exists, $t_\Lambda(R)$ may not be essential in R^G. Thus, partial traces are not good enough for our purposes. However, in recent work of Cohen and Montgomery [4], it is shown that a trace-like function always exists, with an essential image in a ring very close to R^G.

4.3 Theorem [4, Theorem 2.8]: Let R be a ring with no nilpotent elements, let C be the extended center of R, and let E be the set of idempotents in C. Let G be any finite group acting as automorphisms on R. Then there exists a finite set $E_o \subsetneq E$ and an R^G-bimodule homomorphism $\tau : R \to (RE_o)^G$ such that $\tau(R)$ is essential in $(RE)^G$.

Moreover there exists an essential ideal K of R such that $\tau(K) \subsetneq R^G$ and is essential in R^G.

We note that in the case of a domain, the construction of τ in Theorem 4.3 specializes to the t_Λ constructed in Theorem 4.1; however the general case of Theorem 4.3 is considerably more difficult.

A different trace-like function has been constructed by Kharchenko in [12] for semiprime rings such that the algebra of the group $B = \sum_{g \in G} \Phi_g$ is semiprime (this includes the case of no nilpotent elements, as well as the cases when R has no $|G|$-torsion, or when G is X-outer). However, the image of this function is in $(RB)^G$, not $(RE)^G$, and can also be very small. Thus, although this function has been very useful in studying properties such as the Goldie conditions [13] or polynomial identities [9], it does not seem sufficient for affine fixed rings. Thus we ask:

4.4 Question : Let R be a semiprime ring, and G a (finite) group such that the algebra of the group B is semiprime. Do there exist E_o, τ, and K as in Theorem 4.3?

5. Affine fixed rings of Noetherian PI algebras

In this section, we give applications to affine fixed rings of some of the results of §4. The first such result is due to Montgomery and Small [18], and uses Theorem 4.1.

5.1 <u>Theorem</u> [18, Theorem 2] : Let R be a Noetherian domain satisfying a polynomial identity (PI) which is affine over a commutative Noetherian ring A. Let G be a finite group of A-automorphisms of R such that R^G is Noetherian. Then R^G is affine over A.

This theorem has recently been extended by Cohen and Montgomery [4] to the case of semiprime rings.

5.2 <u>Theorem</u> [4, Theorem 3.3]: Let R be a semiprime Noetherian PI ring, which is affine over a commutative Noetherian ring A. Let G be a finite group of A-automorphisms of R such that R^G is Noetherian. Then R^G is also affine over A, in any of the following cases:

1) R has no $|G|$-torsion, or
2) G is X-outer on R, or
3) R has no nilpotent elements.

There are several ingredients in the proofs of these results. First is the existence of an appropriate trace-like function: a partial trace function for Theorem 5.1, the function τ constructed in Theorem 4.3 for Theorem 5.2, part 3), and the usual trace t for Theorem 5.2, parts 1) and 2).

Second is the "trace ring" for prime PI rings studied by Schelter and by Amitsur and Small. That is, a prime PI ring R is an order in $M_n(D)$, for some division ring D finite-dimensional over its center Z. For L a maximal subfield of D, $M_n(D)\otimes_Z L = M_m(L)$, for some m. Then the <u>trace ring</u> T = T(R) is the subalgebra of $M_m(L)$ generated by the coefficients of the characteristic polynomials of all $r\otimes 1 \in M_m(L)$, for $r \in R$. In fact, $T \subsetneq Z$.

In extending Theorem 5.1 to semiprime rings, one needs to define the trace ring for semiprime Noetherian PI rings. This is done as follows: such a ring R has a classical quotient ring $Q(R) = Q_1 \oplus \cdots \oplus Q_r$, where the Q_i are simple rings. Let $\{e_1, \ldots, e_r\}$ be the set of primitive central idempotents of $Q(R)$, and let $R_i = Re_i$, for each i. Let $T_i = T(R_i)$, the trace ring of the prime PI ring R_i. Now define the <u>trace ring</u> of R to be $\tilde{T} = \sum_{i=1}^{r} \oplus T_i$.

We note that the extended center C of R is precisely the center of $Q(R)$; thus the set $E = E(R)$ as in Theorem 4.3 is just the set generated by the $\{e_i\}$ above. Let $R' = \sum_{i=1}^{r} \oplus Re_i = RE$. We are now able to state the crucial proposition for proving Theorem 5.2 (a simpler version of this was implicit in the proof of Theorem 5.1):

5.3 <u>Proposition</u> [4, Proposition 3.2]: Let R be a semiprime Noetherian PI ring which is an affine algebra over a commutative Noetherian ring A. Let G be a finite group of A-automorphisms of R such that

1) R^G is Noetherian, and
2) there exists a $\tilde{T}^G R^G$-bimodule homomorphims $\tau : \tilde{T}^G R \to (\tilde{T}^G R')^G$, and an essential ideal K of R such that $\tau(K) \subsetneq R^G$ and $\tau(K)$ contains a regular element of $\tilde{T}^G R^G$.

Then R^G is affine.

5.4 <u>Remark</u>: If Question 4.4 were answered in the affirmative, Theorem 5.2 would also be true when B is semiprime, by using Proposition 5.3 and known facts about R and R^G in this case.

Finally, we note that the hypothesis in Theorems 5.1 and 5.2 that R^G be Noetherian may not be necessary. We ask:

5.5 <u>Question</u>: Does there exist an affine Noetherian domain R and a finite group G acting on R such that R^G is not Noetherian?

6. Affine fixed rings of free algebras

Here we discuss a result of Formanek concerning fixed subrings of free algebras. Let $R=k\langle x_1,\ldots,x_n\rangle$, the free k-algebra on $x_1,\ldots,x_n$, k a field. Let G be a finite group of <u>linear</u> automorphisms: that is, for each $g \in G$, $x_i^g = \sum_{j=1}^{n} \alpha_{ijg} x_j$, some $\alpha_{ijg} \in k$, $i=1,\ldots,n$, so g corresponds to the matrix $(\alpha_{ijg}) \in GL(n,k)$. A beautiful theorem of Kharchenko [14] says that for G as above, R^G is again a free algebra, and moreover there is a one to one correspondence between free intermediate algebras T, $R^G \subset T \subset R$, and subgroups H of G, via the usual Galois correspondence.

Motivated by this result, the present author asked in [20] when R^G is affine. As noted in Example 1.2, R^G does not have to be affine, even if $|G| = 2$. If G consists of scalar matrices, on the other hand, it is not difficult to see that R^G is finitely generated, by all monomials of degree m = exponent of G. Formanek has settled the question by showing that this last case is the only type of situation in which R^G can be finitely generated.

6.1 <u>Theorem</u> [8]: Let G be a finite group of linear automorphisms of $R = k\langle x_1,\ldots,x_n\rangle$. Then R^G is finitely generated $\Leftrightarrow$ G consists of scalar matrices.

Thus the situation for free algebras is a great contrast to that of commutative polynomial rings; for in the commutative case, the fixed ring is always affine by Noether's theorem, but is a free commutative algebra (that is, a polynomial ring) if and only if the group G is generated by pseudo-reflections, by the theorems of Shepard and Todd, Chevalley, and Serre.

We remark that in the case of two variables, $R = k\langle x,y\rangle$ and k of characteristic 0, Formanek's theorem is true for any finite group G, for in that case it is known that any finite group is conjugate to a linear group action.

7. Affine fixed rings of generic matrix rings

Considering Formanek's theorem, it seems natural to ask the analogous question for finite linear groups acting on generic matrix rings. However, here the situation seems much more complex.

Let $S = k\langle X_1,\dots,X_n\rangle$, the ring of n $m\times m$ generic matrices, and let $R = k\langle x_1,\dots,x_n\rangle$, the free algebra as in §6. Now $S = R/T_m$, where T_m is the T-ideal of identities of $m\times m$ matrices. Thus for a group G acting linearly on R, there is an induced G-action on S since T_m is G-stable. Moreover if $|G|^{-1} \in k$, then S^G is the image of R^G. It is clear that if G consists of scalar matrices, then S^G is affine since R^G is affine. If G does not consist of scalar matrices, however, S^G can be affine or not; we shall give examples of both kinds of behavior.

We first note that any affine prime PI algebra A over a field k, with a group of automorphisms G such that $|G|^{-1} \in k$, may be obtained as the image of a generic matrix ring S on which G acts such that A^G is the image of S^G. For, say that A is generated by $\{a_1,\dots,a_t\}$ and that A satisfies a PI of degree 2m. Then A is the image of $S = k\langle X_{i,g} \mid i=1,\dots,t;\ g\in G\rangle$, a ring of $t|G|$ $m\times m$ generic matrices, by sending $X_{i,g} \to a_i^g$. Clearly G acts on S via $X_{i,g}^h = X_{i,gh}$, and A^G is the image of S^G. Thus if A^G is not affine, S^G is not affine. In particular, as there exist affine prime PI algebras of chracteristic 0, with non-affine fixed rings for finite G (see example 2 of ([18]), S^G is not always affine.

A more direct example has been suggested by Formanek. His

argument has been somewhat simplified by D.S. Passman.

7.1 Example: Let $S = k\langle X,Y\rangle$, the ring of 2 $m\times m$ generic matrices, $m > 1$, over a field k of characteristic not 2. Define $g \in \mathrm{Aut}(S)$ by $X^g = X$, $Y^g = -Y'$ so g corresponds to the matrix $\begin{pmatrix} 1 & 0 \\ 0 & -1 \end{pmatrix}$, and $G = \langle g\rangle$ has order 2. We claim that S^G is not finitely-generated.

In $R = k\langle x,y\rangle$ with the corresponding action, it was noted in Example 1.2 that R^G is generated by $\{x, y^2, yx^ny;\ n=1,2,\dots\}$, and thus S^G is generated by $\{X, Y^2, YX^nY;\ n=1,2,\dots\}$, the images of the generators of R^G. Say that some finite subset $\{X, Y^2, \dots, YX^rY\}$ generates S^G. This says that for some polynomial $p\in R$, $yx^{r+1}y - p(x,y^2,\dots,yx^ry) \in T_m$, the T-ideal. Since T_m is a homogeneous ideal, we must only consider monomials in p of degree r+1 in x and degree 2 in y. Thus

(*) $$yx^{r+1}y - \sum_{j=r+1} a_{ij}x^i yx^j yx^{r+1-i-j}$$

is an identity for $m\times m$ matrices.

However, make the following specialization (assuming m=2 for simplicity): $x \to A = \begin{pmatrix} 0 & 0 \\ 0 & 1 \end{pmatrix}$, $y \to B = \begin{pmatrix} 0 & 1 \\ 1 & 0 \end{pmatrix}$. Then for all $j \geqslant 1$, $C = BA^jB = \begin{pmatrix} 1 & 0 \\ 0 & 0 \end{pmatrix}$. But then $AC = 0 = CA$. It would then follow that $0 = BA^{r+1}B = \sum_{j=r+1} a_{ij}A^iB\,A^jBA^{r+1-i-j} = 0$, a contradiction. Thus (*) is not an identity, so no finite subset generates S^G.

The next example shows that S^G can be affine when G does not consist of scalar matrices. It was obtained in discussion with D.S. Passman.

7.2 Example: Let $S = k\langle X,Y\rangle$, where X and Y are 2×2 generic matrices, and assume k is a field which contains a primitive cube root of 1, say ω. Define $g \in \mathrm{Aut}(S)$ by $X^g = \omega X$, $Y^g = \omega^2 Y$; so g corresponds to the matrix $\begin{pmatrix} \omega & 0 \\ 0 & \omega^2 \end{pmatrix}$, and $G = \langle g\rangle$ has order 3. We claim that S^G is finitely-generated.

To see this, consider $R = k\langle x,y\rangle$ with the corresponding action. Then R^G is generated by

$$\{x(xy)^n y,\ x(xy)^n x^2,\ y(yx)^n x,\ y(yx)^n y^2;\ n = 0,1,2,\ldots\}.$$

For, in any fixed monomial $m=m(x,y)$, we must have $\deg_x m + 2\deg_y m = 3k$, for some k. Thus the above monomials are all fixed, and any other fixed monomial contains a factor of x^2 or y^2, so can be written as a product of the above monomials (since for any monomial m, one of m, mx, or mx^2 is fixed). We claim that R^G is also generated by the set

$$\{x(xy-yx)^n y,\ x(xy-yx)^n x^2,\ y(xy-yx)^n x,\ y(xy-yx)^n y^2;\ n=0,1,\ldots\}.$$

For let $T \subsetneq R^G$ be the subring generated by this new set. We show by induction on n that $R^G \subsetneq T$.

For $n=0$, xy, x^3, yx, and $y^3 \in T$ trivially. By induction, assume that we have shown that all $x(xy)^k y$, $x(xy)^k x^2$, $y(yx)^k x$, $y(yx)^k y^2 \in T$, all $k < n$.

Consider $x(xy)^n y$. For any $k < n$, using $xy = xy-yx+yx$,

$$\begin{aligned} x(xy)^{n-k}(xy-yx)^k y &= x(xy)^{n-k-1}(xy-yx)^{k+1}y + (\underbrace{x(xy)^{n-k-1}y}_{\in T})(\underbrace{x(xy-yx)^k y}_{\in T}) \\ &\equiv x(xy)^{n-k-1}(xy-yx)^{k+1}y \pmod T \\ &\ \ \vdots \\ &\equiv x(xy-yx)^n y \pmod T \\ &\equiv 0 \pmod T \end{aligned}$$

But then $x(xy)^n y = x(xy)^{n-1}(xy-yx)y + x(xy)^{n-1}y\, x(xy-yx)y \in T$. A similar argument works for the other generators, and thus $T = R^G$, proving the claim.

Now, return to S^G. Since X and Y are 2×2 matrices, $(XY-YX)^2$ is in the center of S, so even powers of $XY-YX$ may be "pulled out" of the images of the above generators in S. Thus S^G has $\leqslant 8$ generators, given by

$$\{XY,\ X^3, YX,\ Y^3,\ X(XY-YX)Y,\ X(XY-YX)X^2,\ Y(XY-YX)X,\ Y(XY-YX)Y^2\}.$$

Though it looks very difficult, we formally ask:

7.3 <u>Question</u>: Let G be a finite group acting linearly as automorphisms of a generic matrix ring $S = k\langle X_1,\ldots,X_n\rangle$. Can necessary and sufficient conditions on G be found to ensure that S^G is affine?

8. Affine fixed rings of enveloping algebras

In this last section we offer some elementary observations on affine fixed rings of the enveloping algebra U(g) of a finite-dimensional Lie algebra g over a field k of characteristic 0. Of course, if G is a finite group of automorphisms of U(g), then Theorem 2.1 applies to show $U(g)^G$ is affine, since U(g) is Noetherian. When G is infinite, something can still be said if G is the extension of automorphisms of the Lie algebra g. As we will see, in that case the question of $U(g)^G$ being affine reduces to the case of a commutative polynomial ring.

Consider a basis for g over k, say $\{x_1,\ldots,x_m\}$. Then U(g) has the canonical filtration given by $U(g) = \bigcup_{n\geq 0} U_n$, where $U_0 = k\,1$ and U_n is the k-subspace spanned by all monomials in the $\{x_i\}$ of "degree" $\leq n$. The associated graded algebra of U(g) may be identified with the symmetric algebra S(g) (which is just isomorphic to the commutative polynomial ring $k[x_1,\ldots,x_m]$). Now for each n, $U_n = U_{n-1} \oplus U^n$, where U^n is the set of elements of U(g) which are symmetric homogeneous of degree n [5, 2.4]. Letting $S^n(g)$ be the homogeneous polynomials of degree n, there is a canonical bijection w_n of $S^n(g)$ onto U^n, given by $w_n(y_1y_2\cdots y_n) = \frac{1}{n!}\sum_{\pi\in S_n} y_{\pi(1)}y_{\pi(2)}\cdots y_{\pi(n)}$ where $y_1,\ldots,y_n \in g$. We will use the "symmetrized basis" for U(g): that is, the image under w of monomials in $\{x_i\}$ in S(g).

Now consider a group $G \subseteq \mathrm{Aut}_k(g)$; G extends naturally to automorphisms of U(g). Not only does G preserve each U_n, but since G is acting "linearly" on the $\{x_i\}$, U^n is a G-submodule of U_n. Thus there is an induced action of G on S(g), in which the invariants of G on $S^n(g)$ come from invariants on U^n, and conversely. Now $U_n = U^0 \oplus U^1 \oplus \ldots \oplus U^n$, a direct sum of G-submodules; thus by an argument similar to the standard proof that U(g) is Noetherian, we have the following proposition.

8.1 Proposition: Let g be a finite dimensional Lie algebra over a field k of characteristic 0, and let $G \subsetneq \mathrm{Aut}(g)$. Consider G acting on U(g) and on S(g) as above. Then $U(g)^G$ is affine $\Leftrightarrow S(g)^G$ is affine.

In particular, Hilbert's theorem will apply to this situation. Finally, we note that similar arguments may be used with the Weyl algebra A_n, by using its associated graded ring.

References

1. G. Bergman and P.M. Cohn, "Symmetric elements in free powers of rings", Journal London Math Soc. 1 (second series) 1969, 525-534.
2. G.Bergman and I.M. Isaacs, "Rings with fixed-point-free group actions", Proc. London Math. Soc. 27 (1973), 69-87.
3. M. Cohen and S. Montgomery, "Trace functions for finite automorphism groups of rings", Archiv der Math. 35 (1980), 516-527.
4. ————, "Trace-like functions on rings with no nilpotent elements", Transactions AMS to appear.
5. J. Dixmier, Enveloping Algebras, North-Holland, Amsterdam, 1977 (English edition).
6. C. Faith, "Galois subrings of Ore domains are Ore domains", Bull. Amer. Math. Soc. 78 (1972), 1077-1080.
7. D. Farkas and R. Snider, "Noetherian fixed rings", Pacific J.Math 69, 1977, 347-353.
8. E. Formanek, "Poincare series and a problem of S. Montgomery", to appear.

9. J. Goursaud, J. Osterburg, J.L. Pascaud, and J. Valette, "Points fixes des anneaux reguliers auto-injectifs a gauch", Comm. Alg., to appear.

10. R. Guralnick, I.M. Isaacs, and D.S. Passman, "Non-existence of partial traces for group actions", Rocky Mountain J., to appear.

11. V.K. Kharchenko, "Generalized identities with automorphisms", Algebra i Logika 14 (1975), 215-237 (English transl. 1976, 132-148).

12. ———, Fixed elements under a finite group acting on a semiprime ring", Algebra i Logika 14 (1975), 328-344 (English transl 1976, 203-213).

13. ———, "Galois theory of semiprime rings", Algebra i Logika 16 (1977), 313-363 (English trans. 1978, 208-258).

14. ———, "Algebras of invariant of free algebras", Algebra i Logika 17 (1978), 478-487 (English trans. 1979, 316-321).

15. S. Montgomery, "Outer automorphisms of semiprime rings", Journal London Math. Soc. 18 (1978), 209-221.

16. ———, "Automorphism groups of rings with no nilpotent elements", J. Algebra 60 (1979), 238-248.

17. ———, Fixed rings of finite automorphism groups of associative rings, Lecture Notes in Math. vol 818, Springer-Verlag, 1980.

18. S. Montgomery and L. W. Small, "Fixed rings in Noetherian rings", Bull. London Math. Soc. 13 (1981), 33-38.

19. D. Mumford, "Hilbert's Fourteenth problem", in Mathematical developments arising from Hilbert's problems, Proc. Sym. Pure Math vol. 28(1976), American Math. Soc. (ed. by F.E. Browder), 431-444.

20. Noetherian rings and rings with polynomial identities, Proceedings of the Durham Conference, 1979 (University of Leeds).

University of Southern California, Los Angeles, CA 90007, USA

INVARIANTS D'UN GROUPE FINI ENGENDRE PAR DES PSEUDO-REFLEXIONS, OPERANT SUR UN ANNEAU LOCAL

Luchézar L. Avramov

INTRODUCTION.

Etant donné un anneau R, supposé noethérien, unitaire et local d'idéal maximal $\mathfrak{m}$, on se propose d'étudier le passage de R à l'anneau des invariants $R^G = \{x \in R \mid g(x) = x$ pour tout $g \in G\}$, où G est un groupe fini d'automorphismes de R. L'anneau R étant entier sur R^G, le seul idéal maximal de R^G est $\mathfrak{m}^G = \mathfrak{m} \cap R^G$, et l'inversibilité dans R de l'ordre du groupe d'inertie H de $\mathfrak{m}$ est suffisante pour assurer la descente de la propriété noethérienne.

Avec cette hypothèse, on considère l'extension $R^G \hookrightarrow R$ du point de vue de certains invariants numériques, caractérisant la singularité de R, et du point de vue de la ramification de $X = \operatorname{Spec} R$ sur $X/G = \operatorname{Spec} R^G$. Dans les deux cas on obtient, sous une condition générique peu restrictive, une information complète en supposant H engendré par des pseudo-réflexions. Par définition, un élément h de H est appelé pseudo-reflexion, si son image par l'homomorphisme canonique $\varepsilon : H \longrightarrow Gl_{R/\mathfrak{m}}(\mathfrak{m}/\mathfrak{m}^2)$ laisse fixe chaque point d'un hyperplan, et $\varepsilon(h) \neq 1$; on sait d'ailleurs que ε est injectif. Du point de vue des "singularités" le théorème de Serre [10], disant que pour R régulier (et G = H, cas auquel on peut toujours se ramener), R^G est régulier si et seulement si H est engendré par des pseudo-réflexions, a donné à plusieurs auteurs la motivation d'étudier la descente de propriétés d'être de Cohen-Macaulay, de Gorenstein, d'intersection complète, factoriel, etc... Nous renvoyons à [2,(5)] pour une comparaïson de nos résultats à ce sujet à ceux obtenus antérieurement. Du point de vue de la "ramification", le théorème (2.2) donne une extension au cas singulier du classique calcul de la différente dans le cas polynômial, en tant que produit de vecteurs correspondants aux valeurs propres $\neq 1$ des pseudo-reflexions [3]. Quelques remarques sur l'action de groupes finis sur les anneaux de Buchsbaum permettent de donner, en fin de l'exposé, un exemple d'anneau qui n'admet pas d'action non-triviale par un groupe engendré par des pseudo-reflexions.

1. SINGULARITES .

Dans ce paragraphe on considère les caractéristiques suivantes de l'anneau local R de corps résiduel $k = R/\mathfrak{m}$:

(i) codepth R = dim R - depth R, où dim désigne la dimension de Krull, et depth la longueur maximale d'une R-suite contenue dans $\mathfrak{m}$; on sait que codepth $R \geq 0$, l'égalité définissant les anneaux de Cohen-Macaulay ;

(ii) $d(R) = \mu(a) - ht(a)$, où $\hat{R} = \tilde{R}/a$ est une représentation de Cohen du complété de R en tant que quotient de l'anneau local régulier $\tilde{R}$, μ désigne le nombre minimal de générateurs, et ht la hauteur ; on sait que d(R) ne dépend pas du choix de $\tilde{R}$, et que $d(R) \geq 0$, l'égalité caractérisant les intersections complètes ;

(iii) la série de Bass de R :

$$I_R(t) = \sum_{i \geq 0} \dim_k \operatorname{Ext}_R^i(k,R)\, t^i \in \mathbb{Z}[[t]] ;$$

si R est de Cohen-Macauly, le premier coefficient non-nul apparait pour i = dim R, et sa valeur est appelée le type t(R) de R, la condition t(R) = 1 caractérisant les anneaux de Gorenstein ;

(iv) la série de Poincaré de R :

$$P_R(t) = \sum_{i \geq 0} \dim_k \operatorname{Tor}_i^R(k,k)\, t^i \in \mathbb{Z}[[t]] ,$$

qui est un polynôme si et seulement si R est régulier ;

(v) la série de Hilbert-Samuel de R :

$$H_R^{(1)}(t) = \sum_{i \geq 0} \ell_R(R/\mathfrak{m}^i)\, t^i$$

qui représente une fonction rationnelle de dénominateur $(1-t)^{\dim R+1}$; en particulier, la fonction $n \longmapsto \ell(R/\mathfrak{m}^n)$ est représentée pour n assez grand par un polynôme, et (dim R)! fois le coefficient de son terme principal donne la multiplicité e(R) de R.

Nous dirons que l'action de G sur R est génériquement sans inertie, si pour tout idéal premier P associé à R ; le groupe d'inertie $G^T(P) = \{g \in G \mid g(x) - x \in P$ pour tout x de $R\}$ est trivial. Notons que cette condition est satisfaite automatiquement lorsque R est intègre.

Théorème (1.1). On suppose $H = G^T(\mathfrak{m})$ engendré par des pseudo-reflexions, $|H|$ premier à la caractéristique de k et G opérant génériquement sans inertie sur l'anneau local R. Alors on a :

(i) codepth (R^G) = codepth(R) ; dans le cas de Cohen-Macaulay, on a en plus $t(R^G) = t(R)$;

(ii) $d(R^G) = d(R)$.

(iii) $I_{R^G}(t) = I_R(t)$.

(iv) $P_{R^G}(t) = P_R(t).(1-t)^{\mathrm{edim}R - \mathrm{edim}\, R^G}$, où $\mathrm{edim}R = \dim_k(\mathfrak{m}/\mathfrak{m}^2)$ désigne la dimension de plongement .

(v) $H^{(1)}_{R^G}(t) \ll H^{(1)}_R(t)$, où $\ll$ dénote une inégalité coefficient à coefficient, et $e(R^G) \leq e(R)$.

Le théorème montre en particulier que R^G est de Cohen-Macaulay de type t (resp. de Gorenstein, d'intersection complète) si et seulement si R l'est. Aussi, on voit que R régulier implique R^G régulier, mais la réciproque n'est pas vraie : avec $R = k[\![X^2, XY, Y^2]\!] \subset k[\![X,Y]\!]$ et G le groupe cyclique d'ordre 2, engendré par h : h(X) = -X, h(Y) = Y, on a $R^G = k[\![X^2, Y^2]\!]$ en caractéristique différente de 2 ; du coup, l'inégalité de (v) devient stricte en tout degré positif, et $e(R^G) = 1 < 2 = e(R)$.

Pour obtenir les relations numériques du théorème, il suffit de montrer que l'extension $R^G \hookrightarrow R$ est plate avec un fibre $\bar{R} = R/\mathfrak{m}^G R$ "convenable". Dans chacun des cas ce terme se spécialise comme suit : "Cohen-Macaulay" pour (i) [8] ; "Gorenstein de dimension zéro" pour (iii) [5] ; "intersection complète" pour (ii) [2,(3.6)] ; "intersection complète de dimension zéro" pour (iv) [2,(1.2)] et pour (v) [7, Remarque 4, p.87]. Donc tout ce qui est nécessaire est contenu dans l'énoncé ci-dessous. Rappelons qu'un anneau local est dit être d'intersection complète stricte si son gradué associé, pour l'idéal maximal est isomorphe à une algèbre de polynômes, factorisée par une suite régulière d'éléments homogènes ; on sait que cette condition implique la propriété d'intersection complète pour l'anneau lui-même.

Théorème (1.2). Sous les hypothèses du théorème précédent, on a :

(i) R a une base normale sur R^G, c'est-à-dire R est un module libre de rang 1 sur l'anneau de groupe $R^G[G]$; en particulier, R est R^G-libre de rang $|G|$;

(ii) la fibre $\bar{R} = R/\mathfrak{m}^G R$ est d'intersection complète stricte.

Le lemme suivant établi dans [2, (10)] joue un rôle essentiel dans la démonstration.

Lemme (1.3). Soit S l'algèbre symétrique sur k de l'espace vectoriel $\mathfrak{m}/\mathfrak{m}^2$, munie de sa graduation habituelle et de l'action de H par automorphismes de k-algèbres, induite par ε. Si $|H|$ n'est pas divisible par la caractéristique de k, on a une surjection naturelle de k-algèbres graduées :

$$gr_{\mathfrak{m}}(\bar{R}) \longleftarrow S/S_+^H S ,$$

qui est compatible avec l'action induite de H (on désigne par S_+^H l'idéal maximal irrélevant de S^H).

Nous en déduirons le théorème dans le cas où $G = H$:

Remarquons d'abord que $k[H]$ étant semi-simple par le théorème de Maschke, l'homomorphisme du lemme donne lieu à un épimorphisme de $k[H]$-modules $S/S_+^H S \longrightarrow \bar{R}$. D'autre part, par un résultat de Chevalley (cf. [4, Théorème B] en caractéristique nulle et [3, Théorème 4] en général), $S/S_+^H S$ est isomorphe à $k[H]$ en tant que $k[H]$-modules. On a donc une surjection $k[H]$-linéaire $\varphi : k[H] \longrightarrow \bar{R}$. Considérons maintenant le diagramme commutatif de $R^H[H]$-modules :

$$\begin{array}{ccc} R^H[H] & \xrightarrow{\ \phi\ } & R \\ \downarrow \pi & & \downarrow \pi' \\ k[H] & \xrightarrow{\ \varphi\ } & \bar{R} \end{array}$$

où π et π' sont les épimorphismes canoniques de noyaux $\mathfrak{m}^H R^H[H]$ et $\mathfrak{m}^H R$ respectivement, et ϕ est un relèvement de $\varphi \circ \pi$. Notons que ϕ est surjectif par Nakayama. Pour établir (i) il suffit donc de démontrer que $\operatorname{Ker} \phi = 0$.

Soit U l'ensemble des non-diviseurs de zéro dans R, $V = \{N_H(u) \mid u \in U\}$, avec $N_H(u) = \prod_{g \in H} g(u)$. Comme U est H-stable, $U^{-1}R$ reçoit une H-action induite, et on connait les isomorphismes $V^{-1}R \simeq U^{-1}R$, $V^{-1}(R^H) \simeq (U^{-1}R)^H$. On montre que l'hypothèse sur la trivialité des groupes d'inertie des idéaux associés est équivalente au fait que $U^{-1}R$ est $V^{-1}R^H$-libre de rang $|H|$. On voit donc que $V^{-1}\phi$ est un homomorphisme surjectif de $V^{-1}R^H$-modules libres de même rang, donc $V^{-1}(\operatorname{Ker} \phi) = 0$. Comme V ne contient que des éléments réguliers, et $\operatorname{Ass}_{R^H}(\operatorname{Ker} \phi) \subset \operatorname{Ass}_{R^H}(R^H)$, ceci n'est possible que dans le cas où $\operatorname{Ker} \phi = 0$.

Pour démontrer (ii) il suffit de remarquer que l'épimorphisme du lemme devient, sous nos hypothèses, bijectif, pour des raisons de dimension. Le théorème de Chevalley-Shephard-Todd [4, Théorème A],[3, Théorème 4], montre maintenant que tout système de générateurs homogènes de la k-algèbre S^H est algébriquement indépendant sur k, donc forme une suite S-régulière.

Je voudrais remercier H. Bass d'avoir suggéré de considérer un morphisme ϕ H-linéaire, ce qui dans le cas G = H a permis la simplification exposée ci-dessus de la démonstration originale, donnée dans [2].

2. RAMIFICATION .

On rappelle qu'un idéal premier P de R est dit être <u>non-ramifié</u> sur R^G, si pour l'idéal $\underline{p} = R^G \cap P$ les deux conditions suivantes sont satisfaites : (a) $\underline{p}R_P = PR_P$, et (b) l'extension de corps $R^G_{\underline{p}}/\underline{p}R^G_{\underline{p}} \hookrightarrow R_P/PR_P$ est séparable. Dans le cas contraire, on dit que P est ramifié, et on sait que l'ensemble des idéaux premiers de R ramifiés sur R^G est le fermé de Spec R, égal au support du module des différentielles de Kähler Ω_{R/R^G}. En particulier, un idéal définissant cet ensemble est la différente $\underline{D}(R/R^G)$, définie comme étant l'idéal de Fitting des mineurs maximaux d'une présentation de Ω_{R/R^G}. (Pour une exposition détaillée de la théorie locale de la ramification nous renvoyons au cours de Scheja et Storch [11]).

Dans ce numéro on se contente d'énoncer le principal résultat de [2]. On a besoin d'un lemme, qui dans le cas complet est une conséquence d'un résultat plus précis, démontré dans [10, pp.8-9].

<u>Lemme (2.1)</u> [2 ,(12)] <u>Soit h un automorphisme de R, d'ordre fini premier à la caractéristique de k ; on suppose en plus que h induit l'identité sur k. Alors h est une pseudo-reflexion si et seulement si l'idéal $\underline{a}_h$, engendré par les éléments h(x)-x, quand x parcourt R, est principal et non-nul.</u>

<u>Théorème</u> (2.2) ([2, (4)]) <u>On suppose H engendré par des pseudo-reflexsions, H premier à la caractéristique de k, et G opérant génériquement sans inertie sur l'anneau local R. Soit $\mathcal{P}$ l'ensemble des pseudo-reflexions contenues dans G, et $\mathcal{S}$ l'ensemble des idéaux de R, qui sont H-stables et ne sont pas contenus dans $\mathfrak{m}^H R$.</u>

<u>Alors on a les égalités</u> :

$$\underline{D}(R/R^G) = \bigcap_{\underline{b}\in\mathcal{S}} \underline{b} = \prod_{h\in\mathcal{P}} \underline{a}_h$$

Remarque (2.3). En suivant la démonstration donnée dans [2], on voit que l'égalité de gauche reste vraie sans supposer que H est engendré par des pseudo-reflexions, si les conditions suivantes sont satisfaites : (i) $|H|$ est premier à la caractéristique de k, et G opère sur R génériquement sans inertie ; (ii) R est R^G-libre de rang $|G|$; et (iii) $R/\mathfrak{m}^H R$ est d'intersection complète.

En reprennant un exemple de [9], soit $R = k[[X^2, XY, Y^2]] \subset k[[X,Y]]$, ω une racine 6-ième de l'unité ($6 \neq 0 \in k$), g le k-automorphisme défini par $g(X) = \omega X$, $g(Y) = Y$, $H = G = \langle g \rangle$. Alors $R^G = k[[X^6, Y^2]]$ est régulier, donc R est R^G-libre de rang 6, et $R/\mathfrak{m}^G R \simeq k[[U,V]]/(U^3, V^2)$ est d'intersection complète. On voit, par la remarque précédente ou directement, que $\underline{D}(R/R^G) = (X^5 Y)R$. D'autre part, comme $\varepsilon(g)$ est donné dans la base évidente de $\mathfrak{m}/\mathfrak{m}^2$ par la matrice $\mathrm{diag}(\omega^2, \omega, 1)$, la seule pseudo-reflexion de G est l'élément $h = g^3$, et $\underline{a}_h = (XY)R$, donc l'égalité de droite n'est plus assurée par ces conditions.

3. ANNEAUX DE BUCHSBAUM.

Il est bien connu que la propriété d'être de Cohen-Macaulay descend de R à R^H, pour tout groupe fini H dont l'ordre est inversible dans k [6]. Nous allons retrouver ce résultat, dans le cadre plus général des anneaux de Buchsbaum, introduits dans [14] par la propriété que pour tout système de paramètres $x_1, \ldots, x_k$ de R, la différence $\ell_R(R/\underline{x}R) - e_R(\underline{x}, R)$ de la longueur et de la multiplicité reste constante ; cette valeur indépendante de $\underline{x}$ est appelée l'invariant i(R) de R, et on a $i(R) \geq 0$ avec l'égalité caractérisant les anneaux de Cohen-Macaulay.

Proposition (3.1) Soit G un groupe fini d'automorphismes de l'anneau local R, tel que l'ordre de $H = G^T(\mathfrak{m})$ soit premier à la caractéristique de k.

(i) Si R est un anneau de Buchsbaum, R^G l'est aussi, et $i(R^G) \leq i(R)$;

(ii) Si H est engendré par des pseudo-reflexions, et G opère sur R génériquement sans inertie, on a

$$i(R) = |H| i(R^G).$$

Démonstration. (i) Soit $x_1, \ldots, x_d$ un système arbitraire de paramètres de R^H. D'après [12], pour voir que R^H est de Buchsbaum, il suffit de montrer que les groupes d'homologie $H_i(K)$ du complexe de Koszul de R^H sur $x_1, \ldots, x_d$ sont annulés par $\mathfrak{m}^H$ pour tout $i \geq 1$. Comme $K \otimes_{R^H} R$ est un complexe de Koszul sur un système de pramètres de R, on a déjà $\mathfrak{m} H_i(K \otimes_{R^H} R) = 0$ pour $i \geq 1$.

Or l'opérateur de Reynolds $\rho : R \longrightarrow R^H$ $(\rho(x) = |H|^{-1} \sum_{g \in H} g(x))$ s'étend à une section de l'inclusion canonique $K \hookrightarrow K \otimes_{R^H} R$, qui commute aux différentielles, car elles sont données par des matrices à coefficients dans R^H. On obtient donc que la suite exacte de R^H-modules :

$$0 \longrightarrow R^H \longrightarrow R \longrightarrow T \longrightarrow 0$$

donne en homologie de Koszul les isomorphismes de R^H-modules :

$$H_i(K \otimes_{R^H} R) \cong H_i(K) \oplus H_i(K \otimes_{R^H} T).$$

D'une part on voit que $\mathfrak{m}^H$ annule l'homologie de Koszul de R^H et de T, donc R^H est un anneau de Buchsbaum et T est un R^H-module de Buchsbaum. D'autre part, comme on a

$$(3.2) \quad \ell_R(R/\underline{x}R) - e(\underline{x},R) = \sum_{i=1}^{d} (-1)^{i-1} \ell_R\, H_i(K \otimes_{R^H} R) \ ,$$

et comme l'égalité $R^H/\mathfrak{m}^H = R/\mathfrak{m}$ montre que la longueur d'un R module peut être calculée sur R^H, on voit que

$$(3.3) \qquad i(R) = i(R^H) + i(T)$$

ce qui démontre (i) pour $G = H$. Pour passer au cas général il suffit, puisque R^H est R^G-libre de rang $|G/H|$ et $\mathfrak{m}^H = \mathfrak{m}^G R^H$ (e.g. [2,(7)]), d'appliquer le lemme suivant :

<u>Lemme</u> (3.4). <u>Soit</u> $(A,\mathfrak{m}) \longrightarrow (B,\mathfrak{n})$ <u>un homomorphisme local, pour lequel</u> B <u>est</u> A<u>-plat et</u> $\mathfrak{m}B = \mathfrak{n}$. <u>Alors</u> A <u>est de Buchsbaum si et seulement si</u> B <u>l'est, et dans ce cas</u> : $i(A) = i(B)$.

<u>Démonstration</u>. Soit K^A le complexe de Koszul de A sur un système minimal de générateurs de $\mathfrak{m}$, et posons $H^i(K^A) = H_i\, \mathrm{Hom}_A(K^A,A)$. D'après nos hypothèses $K^B = K^A \otimes_A B$, $H^i(K^B) = H^i(K^A) \otimes_A B = H^i(K^A) \otimes_{A/\mathfrak{m}} B/\mathfrak{n}$ et on a un diagramme commutatif :

$$\begin{array}{ccccc}
H_i(K^A) & \longrightarrow & H^i(K^A) \otimes_A B & \cong & H^i(K^B) \\
\downarrow \varphi^A & & \downarrow \varphi^A \otimes B & & \downarrow \varphi^B \\
H^i_{\mathfrak{m}}(A) & \longrightarrow & H^i_{\mathfrak{m}}(A) \otimes_A B & \cong & H^i_{\mathfrak{n}}(B)
\end{array}$$

où l'isomorphisme des modules de cohomologie locale vient de la platitude :

$H^i_{\mathfrak{m}}(A) \underset{A}{\otimes} B = \varinjlim \mathrm{Ext}^i_A(A/\mathfrak{m}^n, A) \underset{A}{\otimes} B \simeq \varinjlim \mathrm{Ext}^i_B(B/\mathfrak{m}^i B, B) = H^i_{\mathfrak{n}}(B)$. On voit que la surjectivité de φ^A est équivalente à celle de φ^B (descente fidèlement plate), et cette propriété caractérise les anneaux de Buchsbaum [13]. D'autre part, en calculant i(A) comme dans (3.2), on obtient i(A) = i(B).

(ii). D'après le lemme on peut supposer G = H, et comme R est d'après (1.2) R^H-libre de rang |H|, on a le résultat par (3.3).

Exemple (3.5). Soit R l'anneau gradué $k[X^4, X^3Y, XY^3, Y^4]$ du cône sur la quartique gauche. Il n'existe pas de groupe H engendré par des pseudo-réflexions, opérant par k-automorphismes homogènes sur R, tel que |H| soit inversible dans k. En effet, R est un domaine de Buchsbaum avec i(R) = 1, et on conclut par (3.1 ii)

Bibliographie

[1] L.L. AVRAMOV, Homology of local flat extensions and complete intersection defects, Math. Ann. 228, (1977), 27-37.

[2] L.L. AVRAMOV, Pseudo-reflection group actions on local rings, Nagoya Math. J. (to appear)

[3] N. BOURBAKI, Groupes et algèbres de Lie, chapitre V, §.5, Hermann, Paris, 1968.

[4] C. CHEVALLEY, Invariants of finite groups generated by pseudo-reflections, Amer. J. Math. 67 (1955), 778-782.

[5] H.-B. FOXBY and A. THORUP, Minimal injective resolutions under flat base change, Proc. Amer. Math. Soc. 67 (1977), 27-31.

[6] M. HOCHSTER and J.A. EAGON, Cohen-Macaulay rings, invariant theory, and the generic perfection of determinantal loci, Amer. J. Math. 93 (1971), 1020-1056

[7] C. LECH, Inequalities related to certain couples of local rings, Acta. Math. 112 (1964), 69-89.

[8] H. MATSUMURA, Commutative Algebra, Benjamin-Cummings, New-York, 1980

[9] E. PLATTE und U. STORCH, Invariante reguläre Differential-formen auf Gorenstein Algebren, Math. Z. 157 (1977), 1-11

[10] J.-P. SERRE, Groupes finis d'automorphismes d'anneaux locaux réguliers, Colloque d'Algèbre E.N.S.J.F., 1967.

[11] G. SCHEJA und U. STORCH, Lokale Venzweigungstheorie, Schriftenreiche Math. Inst. Univ. Fribourg/Suisse, n°5, 1974.

[12] P. SCHENZEL, Applications of dualizing complexes to Buchsbaum rings, Adv. in Math. (to appear).

[13] J. STÜCKRAD, Kohomologische Charakterisierung von Buchsbaum-Modules, Math. Nachr. 95 (1980), 265-272.

[14] J. STÜCKRAD und V. VOGEL, Eine Verallgemeinerung der Cohen-Macaulay Ringe und Anwendung auf ein Problem der Multiplizitätstheorie, J. Math. Kyoto Univ. 13 (1973), 513-528.

ON NONNORMALITY OF AFFINE QUASI-HOMOGENEOUS SL(2,$\mathbb{C}$)-VARIETIES

Dina Bartels

Introduction:

By an affine quasi-homogeneous Sl(2,$\mathbb{C}$)-variety we will mean an affine algebraic variety with regular Sl(2,$\mathbb{C}$)-action containing a dense orbit. The normal affine quasi-homogeneous Sl(2,$\mathbb{C}$)-varieties have been classified up to Sl(2,$\mathbb{C}$)-isomorphism by Popov [11] . We were interested to know, whether orbit-closures in simple Sl(2,$\mathbb{C}$)-modules, in particular those orbit-closures containing zero, are normal. For instance, Kraft and Procesi [8] have shown that for the adjoint module of Sl(n,$\mathbb{C}$) all orbit-closures are normal. It is known that in simple Sl(2,$\mathbb{C}$)-modules all orbit-closures of dimension less than three are normal. Special examples of non-normal three-dimensional orbit-closures were given by Popov [11] , Kraft and Hesselink [4] , using various types of arguments. The aim of this paper is to introduce a new method, which is sufficient to show in general that for simple Sl(2,$\mathbb{C}$)-modules all three-dimensional orbit-closures containing zero are not normal.

Let us mention furthermore that Luna and Vust [9] have classified up to Sl(2,$\mathbb{C}$)-isomorphism all normal quasi-homogeneous, but not necessarily affine, Sl(2,$\mathbb{C}$)-varieties.

I would like to thank W.Borho for several discussions and hints.

0. Notations

We denote by G the group $Sl(2,\mathbb{C})$ of complex 2×2 matrices $\begin{bmatrix} a & b \\ c & d \end{bmatrix}$ such that $ad-bc=1$. Throughout the paper, we use the following symbols to denote some special subgroups of G:

B is the Borel-subgroup of matrices $\begin{bmatrix} a & b \\ 0 & a^{-1} \end{bmatrix}$.

B^- is the Borel-subgroup of matrices $\begin{bmatrix} a & 0 \\ c & a^{-1} \end{bmatrix}$.

U is the unipotent subgroup of matrices $\begin{bmatrix} 1 & b \\ 0 & 1 \end{bmatrix}$.

U^- is the unipotent subgroup of matrices $\begin{bmatrix} 1 & 0 \\ c & 1 \end{bmatrix}$.

T is the subgroup of diagonal matrices $\begin{bmatrix} t & 0 \\ 0 & t^{-1} \end{bmatrix}$.

$N(T)$ is the normalizer of T in G, consisting of matrices $\begin{bmatrix} t & 0 \\ 0 & t^{-1} \end{bmatrix}$, $\begin{bmatrix} 0 & t \\ -t^{-1} & 0 \end{bmatrix}$.

$\mathfrak{C}_m$ is the cyclic subgroup of matrices $\begin{bmatrix} \zeta & 0 \\ 0 & \zeta^{-1} \end{bmatrix}$ such that $\zeta^m=1$.

U_m is the semidirect product $\mathfrak{C}_m \cdot U$.

By R_n we denote the complex vector-space of all binary n-forms with complex coefficients. For a particular such form f, we use the notation

$$f=f(X,Y)=\sum_{\nu=0}^{n} \alpha_\nu \binom{n}{\nu} X^{n-\nu}Y^\nu \qquad (\alpha_\nu \in \mathbb{C})$$

as polynomial in X and Y. The group G acts on R_n by

$$\begin{bmatrix} a & b \\ c & d \end{bmatrix}^{-1} f(X,Y)=f(aX+bY,cX+dY) \quad .$$

1. Orbit-closures in finite-dimensional G-modules

1.1 By an (affine) G-variety we will mean an affine algebraic variety on which G acts regularly. A G-module is a vector-space on which G acts linearly. The following fact is easy and well-known [10].

Theorem: For every (affine) G-variety there exists a closed G-equivariant embedding into some finite-dimensional G-module.

From representation theory of G it is well-known that

1) every finite-dimensional G-module is semisimple,
2) every simple G-module is isomorphic to some R_n .

Definition: A G-variety is called quasi-homogeneous, if it contains a dense orbit.

The subject of this paper is to study quasi-homogeneous G-varieties. As a consequence of the facts listed above, it is equivalent to study closures of G-orbits in a G-module $V=R_{n_1}\oplus\ldots\oplus R_{n_r}$ for any numbers $n_1,\ldots,n_r$.

1.2 Let us recall the structure of closures of G-orbits in finite-dimensional G-modules as described by Popov in [12]. Let Gf denote an orbit, generated by f , and let $\overline{Gf}$ denote its closure. We will distinguish six types, according to the dimension of $\overline{Gf}$ and the structure of the boundary $\overline{Gf}\setminus Gf$. These are listed in the following table. In addition, the stabilizer G_f of the generator f (up to G-isomorphism) is given in the last column, where we use the notations introduced in section 0.

type	dim $\overline{Gf}$	$\overline{Gf}\setminus Gf$	G_f
1)	0	$\emptyset$	G
2 a)	2	$\emptyset$	T or N(T)
b)	2	type 1) $=\{0\}$	U_m for some m
3 a)	3	$\emptyset$	any finite subgroup of G
b)	3	type 2a)	$\mathfrak{C}_m$ for some m
c)	3	type 2b)	$\mathfrak{C}_m$ for some m

1.3 For our subsequent considerations orbit-closures of type 3c) will be of special interest. Let us make some suitable choice for the generator f of the dense orbit in this case.

As a consequence of the Hilbert-Mumford-criterion, f can be chosen such that

$$\lim_{t\to 0}\begin{bmatrix} t & 0 \\ 0 & t^{-1}\end{bmatrix}^{-1} f = 0 \quad ,$$

as explained in [11]. Moreover, f can be chosen such that the cyclic stabilizer G_f is the cyclic group $\mathfrak{C}_m$ of order m. For an f in $R_{n_1}\oplus\ldots\oplus R_{n_r}$, say $f=f_1\oplus\ldots\oplus f_r$, these special assumptions on f amount to choosing $f_i\in R_{n_i}$ for each $i=1,\ldots,r$ as follows:

$$(*)\qquad f_i=\sum_{\nu=0}^{n_i}\alpha_\nu^{(i)}\binom{n_i}{\nu}X^{n_i-\nu}Y^{\nu}$$

$$\text{with } \alpha_\nu^{(i)}=0 \text{ if } m\nmid n_i-2\nu \text{ or if } \nu\geq\left[\frac{n_i+1}{2}\right] .$$

For each nonzero f_i we denote by k_i the maximal number ν such that $\alpha_\nu^{(i)}$ is $\neq 0$. Note that by the above assumptions $0 \leq k_i < \left[\frac{n_i+1}{2}\right]$ and that $k_i > 0$ for at least one i. (Remark: The number $\max_i \frac{k_i}{n_i}$, where $\frac{k_i}{n_i}$ is defined to be zero for $f_i = \{0\}$, is the "height" of the variety $\overline{Gf}$ in the terminology of [11]).

2. Regular functions on orbits and their closures

For an algebraic variety $\mathcal{V}$ let $\mathcal{R}(\mathcal{V})$ denote its ring of regular functions. An affine algebraic variety $\mathcal{V}$ is completely determined by $\mathcal{R}(\mathcal{V})$. So, taking the algebraic point of view here, we will consider the ring $\mathcal{R}(\overline{Gf})$ for an orbit-closure $\overline{Gf}$.

2.1 First we will state some general results concerning the regular functions on G-varieties. (All these results are valid for any reductive linear algebraic group).

Let $\mathcal{V}$ be any G-variety, then G acts on $\mathcal{R}(\mathcal{V})$ linearly by: $g(F(v)) := F(g^{-1}v)$ for every $v \in \mathcal{V}$, $F \in \mathcal{R}(\mathcal{V})$ and $g \in G$.

Proposition: If $\mathcal{V}$ is a G-variety, then $\mathcal{R}(\mathcal{V})$ is a semi-simple G-module.

This follows since G acts locally finitely on $\mathcal{R}(\mathcal{V})$, and since G is a reductive group.

It is well-known from representation theory that every simple

G-module is generated by a highest weight vector, i.e. a vector which is fixed by a maximal unipotent subgroup of G . For example, our group U of upper triangular unipotent matrices in G is such a subgroup. Therefore, by the proposition, $\mathcal{R}(\mathcal{V})$ is generated as a G-module by $\mathcal{R}(\mathcal{V})^U := \{ F \in \mathcal{R}(\mathcal{V}) \mid u.F=F \quad \forall\, u \in U \}$. The subring $\mathcal{R}(\mathcal{V})^U$ will be called the ring of U-invariant functions on $\mathcal{V}$. It will be convenient to study the ring $\mathcal{R}(\mathcal{V})^U$ instead of $\mathcal{R}(\mathcal{V})$. It has the following two remarkable properties:

Theorem: Let $\mathcal{V}$ be a G-variety. Then

a) the ring $\mathcal{R}(\mathcal{V})^U$ is finitely generated.

b) $\mathcal{R}(\mathcal{V})$ is normal (i.e. integrally closed) if and only if $\mathcal{R}(\mathcal{V})^U$ is normal.

For the proofs, we refer to Hadžiev [3] and Grosshans [2] for a) and to Luna and Vust [15] for b) .

2.2 In this section we will determine $\mathcal{R}(\mathcal{V})^U$ for the special case $\mathcal{V}=G$, with G acting on itself by: $g.v:=vg^{-1}$ for $g,v \in G$. According to 2.1 , G acts on $\mathcal{R}(G)$ by : $gF(g'):=F(g'g)$ for every $g,g' \in G$ and $F \in \mathcal{R}(G)$. It is well-known that the ring $\mathcal{R}(G)$ may be described by generators A, B, C, D and the simple relation AD-BC=1 . Here A (resp. B, C, D) identifies with the function which has value a (resp. b, c, d) on a matrix $\begin{bmatrix} a & b \\ c & d \end{bmatrix}$.

Lemma: $\mathcal{R}(G)^U = \mathbb{C}[A,C]$ is the subring generated by A and C .

Proof: $\mathcal{R}(G)^U$ is the ring of those functions $F \in \mathcal{R}(G)$ such that $u.F(g):=F(gu)=F(g)$ for all $u \in U$. So, we have to find all functions on G which are constant on the cosets gU . Let $g=\begin{bmatrix} a & b \\ c & d \end{bmatrix}$ and $g'=\begin{bmatrix} a' & b' \\ c' & d' \end{bmatrix}$ be in G . The cosets gU and $g'U$ are equal if and only if $a=a'$ and $c=c'$. Thus, the U-invariant functions on G are exactly those functions whose values on a matrix $\begin{bmatrix} a & b \\ c & d \end{bmatrix}$ depend only on a and c . These are just the polynomial functions in A and C . This proves the lemma.

2.3 Next, $\mathcal{R}(\overline{Gf})^U$ will be described as a subring of $\mathcal{R}(G)^U$. In the following we will consider only those 3-dimensional orbit-closures which contain a fixed point, i.e. those of type 3c) in the table 1.2 .

Choosing a generator f with stabilizer G_f for the dense orbit means to fix a G-equivariant morphism $G \to \overline{Gf}$, or an embedding ι_f of the quotient $G_f\backslash G$ into $\overline{Gf}$. Here ι_f maps $G_f g$ to $g^{-1}f$. This induces an embedding ι_f^* of the ring $\mathcal{R}(\overline{Gf})$ into $\mathcal{R}(G)$, or, more precisely, into the ring

$$^{G_f}\mathcal{R}(G) := \{ F \in \mathcal{R}(G) \mid hF(g) := F(hg) = F(g) \quad \forall\, h \in G_f ,\ g \in G \}$$

which is a subring and a G-submodule of $\mathcal{R}(G)$. The embedding is given by the formula

$$\iota_f^*(F)(g) := F(\iota_f(g)) \qquad \text{for all } g \in G .$$

Since ι_f is G-equivariant, this map ι_f^* is a G-module-homomorphism. So, restriction of ι_f^* to U-invariants gives an injection:

$$\mathcal{R}(\overline{Gf})^U \hookrightarrow {}^{G_f}\mathcal{R}(G)^U \subset \mathcal{R}(G)^U ,$$

which we will also denote by ι_f^* .

We choose a generator f as in 1.3 (✱) with $G_f=\mathcal{C}_m$. To determine the ring ${}^{\mathcal{C}_m}\mathcal{R}(G)^U$, we have to ask for all functions F on G with $F(hgu)=F(g)$ for all $h\in\mathcal{C}_m$, $g\in G$ and $u\in U$, i.e. those U-invariant functions F which moreover satisfy $F(hg)=F(g)$ for all $h\in\mathcal{C}_m$ and all $g\in G$. Let $h=\begin{bmatrix}\zeta & 0\\ 0 & \zeta^{-1}\end{bmatrix}$ in $\mathcal{C}_m$ and $g=\begin{bmatrix}a & b\\ c & d\end{bmatrix}$ in G, then $hg=\begin{bmatrix}\zeta a & \zeta b\\ \zeta^{-1}c & \zeta^{-1}d\end{bmatrix}$. By the lemma in 2.2 we have to consider only the action of h on the entries a and c. It is well-known that for this action of the cyclic group $\mathcal{C}_m$ the ring of invariants is the subring $\mathbb{C}[AC,A^m,C^m]$ of $\mathbb{C}[A,C]$, see for example [14]. The above considerations prove the following:

<u>Lemma</u>: <u>Let</u> f <u>be chosen as in 1.3 (✱) with stabilizer</u> $\mathcal{C}_m$. <u>Then the ring of U-invariant functions</u> $\mathcal{R}(\overline{Gf})^U$ <u>is isomorphic to a subring of</u> ${}^{\mathcal{C}_m}\mathcal{R}(G)^U=\mathbb{C}[AC,A^m,C^m]$.

<u>Convention</u>: In the sequel, we shall identify $\mathcal{R}(\overline{Gf})^U$ with a subring of $\mathbb{C}[AC,A^m,C^m]$ by means of the isomorphism mentioned in the lemma. (In performing computations, we shall frequently use that A and C are algebraically independent).

There arises the question which kinds of finitely generated subrings of $\mathbb{C}[AC,A^m,C^m]$ occur in this way. In the normal case, when $m=1$, this is solved in [7]. In section 3 we shall consider special properties of the ring $\mathcal{R}(\overline{Gf})^U$ for an orbit-closure $\overline{Gf}$.

<u>Remark</u>: Since Gf is dense in $\overline{Gf}$, the quotientfields of ${}^{\mathcal{C}_m}\mathcal{R}(G)$ and $\mathcal{R}(\overline{Gf})$ coincide. In particular, the <u>quotientfield of</u> $\mathcal{R}(\overline{Gf})^U$ <u>is equal to</u> $\mathbb{C}(AC,A^m,C^m)$.

2.4 In the preceding section we studied the relation between $\mathcal{R}(\overline{Gf})$ and $\mathcal{R}(G)$. On the other hand, the G-equivariant embedding of $\overline{Gf}$ into a finite-dimensional G-module V gives rise to a surjective ring-homomorphism

$$\mathcal{R}(V) \longrightarrow \mathcal{R}(\overline{Gf})$$

which is given by restricting the functions on V to $\overline{Gf}$.
By complete reducibility (2.1), its restriction to U-invariants

$$\mathcal{R}(V)^U \longrightarrow \mathcal{R}(\overline{Gf})^U$$

is also surjective. Clearly, this is a homomorphism of G-modules.

Notation: The composed ring-homomorphism

$$\mathcal{R}(V)^U \longrightarrow \mathcal{R}(\overline{Gf})^U \hookrightarrow \mathbb{C}\left[AC, A^m, C^m\right]$$

(cf. 2.3) shall be denoted by Φ_f.

It will be useful to consider first $\mathcal{R}(V)^U$ more closely.
It is an old problem in invarianttheory to describe the structure of $\mathcal{R}(V)^U$, or equivalently the structure of the "ring of covariants". For instance, one would like to know a minimal set of generators. Even in the case $V=R_n$, this problem is not yet solved. (Estimates are given by Jordan [6]). However, there are at least some well-known functions belonging to $\mathcal{R}(R_n)^U$, known as "Apolare" (Schur [13]). To define them, let us denote by C_i the function on R_n which takes the value α_i on a form $f=\sum_{\nu=0}^{n} \alpha_\nu \binom{n}{\nu} X^{n-\nu} Y^\nu$. Note that with this notation, the functions $C_0, C_1, \ldots, C_n$ are algebraically independent generators of $\mathcal{R}(R_n)$.

Proposition 1: The function $Q_j := \sum_{\mu=0}^{2j} (-1)^\mu \binom{2j}{\mu} C_{2j-\mu} C_\mu$ is a U-invariant function on R_n for every $0 \le j \le \left[\frac{n}{2}\right]$.

In the classical terminology, for which we refer for example to Grace-Young [1] or to Springer [14], the function Q_j is called "Apolare", or the leading term of the "2j-th transvectant" of the groundform f .

The ring $\mathcal{R}(R_n)$ is graded by the subspaces $\mathcal{R}(R_n)_i := \mathbb{C}[c_0,..c_n]_i$ of homogeneous polynomials of degree i in $c_0,...,c_n$. Since each $\mathcal{R}(R_n)_i$ is a G-submodule, this gives also a graduation on the ring of U-invariants:

$$\mathcal{R}(R_n)^U = \bigoplus_{i=0}^{\infty} \mathcal{R}(R_n)_i^U \quad .$$

<u>Proposition 2</u>:

a) $\mathcal{R}(R_n)_1^U = \mathbb{C}\, c_0$

b) $\mathcal{R}(R_n)_2^U = \bigoplus_{j=0}^{\left[\frac{n}{2}\right]} \mathbb{C}\, Q_j \quad .$

<u>Proof</u>: Obviously c_0 is a U-invariant function. Then the proposition is a consequence of the theorem of Cayley-Sylvester, see Springer [14] .

3. A Nonnormality theorem

<u>Definition</u>: An affine algebraic variety $\mathcal{V}$ is said to be <u>normal</u>, if the ring $\mathcal{R}(\mathcal{V})$ is normal, i.e. integrally closed in its quotientfield.

3.1

Theorem: Let V be a simple G-module, then every 3-dimensional orbit-closure containing zero is not normal.

The essential part of the proof is the following:

Lemma 1: Let $f \in V$ be chosen as in 1.3 (✱), with $G_f = \mathcal{C}_m$, then there exists a positive integer l such that A^{lm} belongs to $\mathcal{R}(\overline{Gf})^U$ (considered as a subring of $\mathbb{C}[AC, A^m, C^m]$, cf. 2.3).

On the other hand, we will show that

Lemma 2: A^m does not belong to $\mathcal{R}(\overline{GF})^U$.

Proof of the theorem: By lemma 1, A^m is an element of the integral closure of $\mathcal{R}(\overline{Gf})^U$ in its quotientfield $\mathbb{C}(AC, A^m, C^m)$, (2.3) . Hence lemma 2 shows that $\mathcal{R}(\overline{Gf})^U$ is not normal. Consequently, $\mathcal{R}(\overline{Gf})$ is not normal (2.1 theorem b)).

Proof of lemma 1: Let $V = R_n$ for some n . The choice of a generator f as in 1.3 (✱) means:

$$f = \sum_{\nu=0}^{k} \alpha_\nu \binom{n}{\nu} X^{n-\nu} Y^\nu$$

with $0 < k < \left[\frac{n+1}{2}\right]$, $\alpha_k \neq 0$, and $\alpha_\nu \neq 0$ only if $n-2\nu$ is divisible by m . In particular, $n-2k$ is divisible by m , say

$n-2k = sm$, where $s \in \mathbb{N}$.

Recall the ring-homomorphism

$$\Phi_f : \mathcal{R}(R_n)^U \longrightarrow \mathbb{C}[AC, A^m, C^m]$$

defined in 2.4 . It is enough to find a function F in $\mathcal{R}(R_n)^U$

with $\Phi_f(F)=A^{lm}$ for some $l\in\mathbb{N}$, since this implies lemma 1 . It turns out that the U-invariant function Q_k introduced in 2.4 does the job with $l=2s$.

<u>Assertion</u>: $\Phi_f(Q_k)=\gamma A^{2sm}$ where $0\neq\gamma\in\mathbb{C}$.

<u>Proof</u>: From the theory of linear algebraic groups (see for example Humphreys [5]) it is well-known that B^-U , the "big cell", is dense in G . Therefore any function on $\overline{Gf}$ is completely determined by its values on the dense subset B^-Uf . Moreover, a function $F\in\mathcal{R}(\overline{Gf})^U$ is even determined by its values on B^-f , since such a function satisfies

$$F((\mathcal{b}u)^{-1}f) = F(u^{-1}\mathcal{b}^{-1}f) = F(\mathcal{b}^{-1}f) \quad \text{for all } u\in U,\ \mathcal{b}\in B^- .$$

Therefore, it is enough to evaluate the function Q_k on B^-f in order to determine $\Phi_f(Q_k)$. Let $\mathcal{b}=\begin{bmatrix} a & 0 \\ c & a^{-1}\end{bmatrix}\in B^-$, then

$$\mathcal{b}^{-1}f=\sum_{\nu=0}^{k}\alpha_\nu'\binom{n}{\nu}X^{n-\nu}Y^\nu \quad , \text{ where for all } \nu=0,\dots,k$$

$$(**)\qquad \alpha_\nu'=a^{n-2k}\sum_{\lambda=\nu}^{k}\alpha_\nu\binom{n-\nu}{\lambda-\nu}(ac)^{\lambda-\nu}a^{2(k-\lambda)}$$

In particular

$$(***)\qquad \alpha_k'=\alpha_k a^{n-2k} \qquad \text{and} \qquad \alpha_\nu'=0 \quad \text{if} \quad \nu>k .$$

Now for every $\mathcal{b}\in B^-$ as above, we compute

$$Q_k(\mathcal{b}^{-1}f) = \sum_{\mu=0}^{2k}(-1)^\mu\binom{2k}{\mu}\alpha_\mu'\alpha_{2k-\mu}' .$$

Observe that in the sum all summands with $\mu\neq k$ vanish, since either α_μ' or $\alpha_{2k-\mu}'$ is 0 by (***) . Therefore, only the summand $\mu=k$ contributes to the sum, and (***) gives

$$Q_k(\mathcal{b}^{-1}f) = (-1)^k\binom{2k}{k}\alpha_k{}^2a^{2(n-2k)} .$$

We conclude that

$$\Phi_f(Q_k) = (-1)^k \binom{2k}{k} \alpha_k{}^2 A^{2(n-2k)} \quad .$$

Since α_k was $\neq 0$ by the choice of our notations, this gives our assertion, and finishes the proof of lemma 1 .

<u>Proof of lemma 2:</u> Let $0 \neq F \in \mathcal{R}(R_n)^U$ be a U-invariant function on R_n of degree ≥ 2 . From (✱✱) we conclude that $\Phi_f(F)$ belongs to $A^{2(n-2k)}\mathbb{C}\left[AC, A^m, C^m\right]$ or, in other words, has degree $\geq 2(n-2k) \geq 2m$ as a polynomial in A . According to 2.4 ,

$$\mathcal{R}(R_n)^U_1 = \mathbb{C}\, C_0 \quad \text{and}$$

$$\Phi_f(C_0) = A^{n-2k} \sum_{\lambda=0}^{k} \alpha_\lambda \binom{n}{\lambda} (AC)^\lambda A^{2(k-\lambda)} \qquad \text{by 3.1 (✱✱)}$$

which has degree at least $n-k > n-2k \geq m$ in A (since $k>0$) . We conclude that all functions in $\mathcal{R}(\overline{Gf})^U$ except the scalars have degree greater than m as polynomials in A . This proves $A^m \notin \mathcal{R}(\overline{Gf})^U$, as claimed in lemma 2 .

<u>3.2</u>

<u>Corollary</u>: <u>Let</u> $V = \bigoplus_{i=1}^{n} R_{n_i}$ <u>and</u> $f = f_1 \oplus \ldots \oplus f_r \in V$ <u>as in 1.3 (✱)</u> <u>with</u> $G_f = \mathscr{C}_m$. <u>Then for</u> $\overline{Gf}$ <u>to be normal it is necessary that</u> $m = n_i$ <u>and</u> $f_i = \alpha X^m$ <u>for some</u> i <u>, where</u> $0 \neq \alpha \in \mathbb{C}$.

<u>Proof</u>: Generalizing the considerations about $\mathcal{R}(R_n)^U$ in 2.4 , we get:

$$\mathcal{R}(V) = \mathbb{C}\left[C_0^{(1)}, \ldots, C_{n_1}^{(1)}, C_o^{(2)}, \ldots, C_{n_2}^{(2)}, \ldots, C_0^{(r)}, \ldots, C_{n_r}^{(r)}\right]$$

where $C_j^{(i)}(f_i) = \alpha_j$ and $C_j^{(i)}(f_\varkappa) = 0$ for $\varkappa \neq i$.

A graduation on $\mathcal{R}(V)$ is given by:

$$\mathcal{R}(V) = \bigoplus_{\eta=0}^{\infty} \mathcal{R}(V)_\eta \qquad \text{with} \quad \mathcal{R}(V)_\eta := \mathbb{C}\left[C_0^{(1)}, \quad , C_{n_r}^{(r)}\right]_\eta$$

and likewise

$$\mathcal{R}(V)^U = \bigoplus_{\eta=0}^{\infty} \mathcal{R}(V)^U_\eta \quad .$$

According to proposition 2 in 2.4 we have:

$$\mathcal{R}(V)^U_1 = \bigoplus_{i=1}^{r} C\,\mathbb{C}_0^{(i)} \quad .$$

Let

$$Q_j^{(i)} := \sum_{\mu=0}^{2j} (-1)^\mu \binom{2j}{\mu} C_\mu^{(i)} C_{2j-\mu}^{(i)} \quad .$$

Note that our map Φ_f (2.4) is defined in the present situation.

Assertion: For each $i=1,\ldots,r$, such that $f_i \neq \{0\}$, we have

$$\Phi_f(Q_{k_i}^{(i)}) = \gamma_i A^{l_i m}$$

for a nonzero scalar γ_i , and a positive integer l_i .

The proof of the assertion is the same as for that in 3.1 .

We conclude again that A^m is integral over $\mathcal{R}(\overline{Gf})^U$. On the other hand, (✱✱) shows that $\Phi_f(F) \in A^{2m}\mathbb{C}[AC, A^m, C^m]$ for every $F \in \mathcal{R}(V)^U$ of degree ≥ 2 , since m divides n_i-2k_i for every i , and that $\Phi_f(C_0^{(i)}) \neq \text{const.} \cdot A^m$ unless $k_i=0$ and $m=n_i$. We conclude again, that all non-constant functions in $\mathcal{R}(\overline{Gf})^U$ have degree $\geq m+1$ as polynomials in A , excepting only the case that $k_i=0$ and $m=n_i$ for some i . Therefore, $A^m \notin \mathcal{R}(\overline{Gf})^U$ unless $f_i = \alpha X^m$ for an i . This proves the corollary.

References:

[1] Grace, J.H. and Young, A. , The algebra of invariants, Cambridge Univ. Press, (1903) .

[2] Grosshans, F. , Observable groups and Hilbert's fourteenth problem, Amer. J. Math. 95 (1973), 229-253 .

[3] Hadžiev, D. , Some questions in the theory of vector invariants, Math. USSR Sbornik 1 (1967), 383-396.

[4] Hesselink, W. , Desingularizations of varieties of null-forms, Invent. Math. 55 (1979), 141-163 .

[5] Humphreys, J.E. , Linear algebraic groups, Springer GTM 21 (1975) .

[6] Jordan, C. , Memoire sur les covariants des formes binaires, Journal de Math. (3) 2 (1876), 177-233 and (3) 5 (1879), 345-378 .

[7] Kraft, H. , Geometrische Methoden in der Invariantentheorie Vorlesungsausarbeitung, Bonn WS 77/78 .

[8] Kraft, H. and Procesi, C. , Closures of conjugacy classes of matrices are normal, Invent. Math. 53 (1979), 227-247 .

[9] Luna, D. and Vust, Th. , Plongements d'espaces homogenes, preprint (1981) .

[10] Mumford, D. , Geometric invariant theory, Erg. der Math. 34 Springer Verlag (1970) .

[11] Popov, V.L. , Quasihomogeneous affine algebraic varieties of the group Sl(2), Math. USSR Izv. 7 No. 4, (1973), 793-831 .

[12] Popov, V.L. , Structure of the closure of orbits in spaces of finite-dimensional linear Sl(2)-représentations, Math. Notes 16 No. 6 (1974), 1159-1162 .

[13] Schur, I. , Vorlesungen über Invariantentheorie, Grundl. Math. Wiss. 143, Springer Verlag (1968) .

[14] Springer, T.A. , Invariant theory, Lecture Notes in Math. 585, Springer Verlag (1977) .

[15] Vust, Th. , Sur la theorie des invariants des groupes classiques, Ann. Inst. Fourier 26 2 (1976), 1-31 .

Dina Bartels
Fachbereich Mathematik
Gesamthochschule Wuppertal
Postfach 10 01 27
D 5600 Wuppertal 1

SEMINORMALITY and PROJECTIVE MODULES

by Douglas L. COSTA

This is an account of recent developments in the study of seminormality in commutative rings. In 1970 Traverso [20], inspired by Andreotti, Bombieri, and Salmon, defined seminormality and studied its relation to the Picard group. Let $A \subseteq B$ be rings with B integral over A The <u>seminormalization of A in B</u> was defined by Traverso to be ${}_{B}^{+}A = \{x \in B \mid x/1 \in A_p + J(B_p) \text{ for all } P \in \text{Spec}(A)\}$, where J denotes the Jacobson radical. Equivalently, ${}_{B}^{+}A$ is the largest subring C of B containing A such that (i) Spec(C) $\to$ Spec(A) is injective and (ii) for all $Q \in \text{Spec}(C)$ the canonical map of residue class fields $k(Q \cap A) \to k(Q)$ is an isomorphism. A is <u>seminormal in B</u> if $A = {}_{B}^{+}A$, and A is <u>seminormal</u> if it is seminormal in its integral closure. Thus seminormality, as here defined, describes a relation between a ring and its total quotient ring.

By using methods of Bass-Murthy [5], Traverso was able to prove the following theorem, generalizing results of Endo [10].

<u>Theorem 1</u>. Let A be a reduced Noetherian ring with integral closure $\bar{A}$ a finitely generated A-module. Let $X_1, \ldots, X_n$ be indeterminates over A. The canonical monomorphism $\text{Pic}(A) \to \text{Pic}(A[X_1, \ldots, X_n])$ is an isomorphism if and only if A is seminormal.

In current terminology the conclusion of Theorem 1 may be restated as follows : A is seminormal if and only if every rank one projective $A[X_1, \ldots, X_n]$ -module is extended from A. [If $A \subseteq B$, then a B-module M is <u>extended</u> from A if there exists an A-module M_o such that $M \cong M_o \otimes_A B$.] This establishes a connection between seminormality and projective modules over polynomial rings. It is upon this connection that this report is centered.

I - Projective modules over polynomial rings.

In 1976 Quillen [17] and simultaneously Suslin, stunned the world of commutative algebra by settling the Serre conjecture. Equally stunning was Quillen's use of a methodology not in vogue at the time. Quillen's starting point was the following "localization theorem".

Theorem 2. Let A be a commutative ring, $X_1, \ldots, X_n$ indeterminates over A, and M a finitely presented $A[X_1, \ldots, X_n]$-module. M is extended from A if and only if for every maximal ideal $\mathfrak{m}$ of A, $M_{\mathfrak{m}}$ is extended from $A_{\mathfrak{m}}$.

Via a theorem of Horrocks, Quillen then obtained the following corollary in which X is an indeterminate and $A\langle X\rangle$ is the localization of the polynomial ring $A[X]$ at the set of monic polynomials.

Theorem 3. If P is a finitely generated projective $A[X]$-module for which $P \otimes_{A[X]} A\langle X\rangle$ is free, then P is free.

An easy induction argument based on Theorem 3 now shows that the following methodology is valid.

Methodology. Let $\mathcal{C}$ be a class of commutative rings such that

(Q.1) $A \in \mathcal{C} \Rightarrow A\langle X\rangle \in \mathcal{C}$, and

(Q.2) $A \in \mathcal{C} \Rightarrow$ fin. gen. proj. , A-modules are free.

Then for any $n \geq 1$ and any $A \in \mathcal{C}$, fin. gen. proj. $A[X_1, \ldots, X_n]$ - modules are free.

Quillen then showed that the class of P.I.D.'s satisfies axioms (Q.1) and (Q.2), proving the Serre conjecture. It follows immediately from Theorem 2 that fin. gen. proj. $A[X_1, \ldots, X_n]$ -modules are extended from A if A is a Dedekind domain. Recently Lindel [14] has extended this result to a wide clas of regular noetherian rings.

Now a striking feature of the methodology above is that it presupposes no particular conditions on the class $\mathcal{C}$ of rings, and thus one is lead to suspect that it is applicable to classes of possibly non-Noetherian rings. This was the approach taken in [6] , where the following theorem was proved.

Theorem 4. Let A be an integral domain, not a field. Then $A\langle X\rangle$ is a Bezout domain if and only if A is a one-dimensional Bezout domain. In case these conditions hold, then $A\langle X\rangle$ is also one-dimensional.

Recalling that Bezout domains are precisely those domains for which finitely generated submodules of free modules are free, we see that axiom (Q,2) is satisfied by the class of all Bezout domains. But theorem 4 tells us that axiom (Q,1) fails unless we restrict ourselves to the class of one-dimensional Bezout domains. The methodology above thus yields the validity of the Serre conjecture for one-dimensional Bezout domains, while at the same time we see that the methodology is limited by dimension considerations.

This obstacle was overcome by Lequain and Simis [13] , who cleverly reformulated the axioms to obtain a new methodology.

Methodology. Let $\mathcal{C}$ be a class of commutative rings such that

(LS.0) $A\in\mathcal{C}\Rightarrow$ every non-maximal prime ideal of A has finite height,

(LS.1) $A\in\mathcal{C}\Rightarrow A_P\in\mathcal{C}$ for every prime ideal P of A,

(LS. 2) $A\in\mathcal{C}\Rightarrow A[X]_{P[X]}\in\mathcal{C}$ for every prime ideal P of A, and

(LS.3) $A\in\mathcal{C}$ and A quasi-local $\Rightarrow$ fin. gen. proj. $A[X]$-modules are free.

Then for any $n\geq 1$ and $A\in\mathcal{C}$, fin. gen. proj. $A[X_1,\ldots,X_n]$-modules are extended from A.

This methodology enabled Lequain and Simis to prove the following remarkably general version of the Serre conjecture.

Theorem 5. If A is a Prüfer domain, then finitely generated projective $A[X_1,\ldots, X_n]$-modules are extended from A.

Sketch of proof. Recall that a Prüfer domain is an integral domain which is locally a valuation ring. By theorem 2 it suffices to prove the theorem for valuation rings. If P is a finitely generated projective $A[X_1,\ldots, X_n]$-module, then P is determined by an idempotent matrix (f_{ij}) whose entries are polynomials in $X_1,\ldots, X_n$ with coefficients in the valuation ring A. Let $\{C_1,\ldots, C_t\}$ be the finite set of these coefficients, and let K be the prime subfield of the quotient field of A. Then $B = K(C_1,\ldots, C_t) \cap A$ is a valuation ring of the field $K(C_1,\ldots, C_t)$, and since $K(C_1,\ldots, C_t)$ has finite transcendence degree over K, B is finite dimensional. Now the idempotent matrix (f_{ij}) determines a projective module P_o over $B[X_1,\ldots, X_n]$ such that $P \cong P_o \otimes A[X_1,\ldots, X_n]$, so P will be free (*i.e.* extended) if we can show that P_o is free. Thus it suffices to prove the theorem for finite dimensional valuation rings.

Let $\mathcal{C}$ be the class of finite dimensional valuation rings. It is well-known that $\mathcal{C}$ satisfies (LS.0), (LS.1), and (LS.2). If $A \in \mathcal{C}$, then Spec $(A[X])$ is a finite union of one-dimensional subspaces, and a theorem of Serre [4] shows that every finitely generated projective $A[X]$-module has the form $I \oplus F$, where I is an ideal and F is free. Since A is a valuation ring, $A[X]$ is a GCD-domain and hence invertible (projective) ideals are principal in $A[X]$. Thus every finitely generated projective $A[X]$-module is free, and (LS.3) holds for $\mathcal{C}$. By the Lequain-Simis methodology the theorem is proved.

II - Seminormality and the Picard group.

In theorem 1 we have seen how traverso established a connection between seminormality and the Picard group. Earlier Bass [3] had shown that for a noetherian integrally closed domain A, $\mathrm{Pic}(A) \cong \mathrm{Pic}(A[X])$. And Endo [10] had given versions of theorem 1 for n = 1,2 and A a one-dimensional noetherian domain with finite integral closure, using a notion he called "weak normality". To give evidence for the necessity of some kind of normality condition on A in order that $\mathrm{Pic}(A) \cong \mathrm{Pic}(A[X])$, Bass cited the following important example due to Schanuel.

<u>Schanuel's example</u>. Let A be an integral domain with quotient field K and let $a \in K$ be an element such that $a^2, a^3 \in A$ but $a \notin A$. Then there is a rank one projective $A[X]$-module which is not extended (i.e. $\mathrm{Pic}\, A \ncong \mathrm{Pic}(A[X])$).

Proof. Consider the fractional ideals $I = (a^2, 1+aX)$ and $J = (a^2, 1-aX)$ of $A[X]$. $IJ = (a^4, a^2 + a^3X, a^2 - a^3X, 1-a^2X^2) \subseteq A[X]$. Now $X^4a^4 + (1+a^2X^2)(1 - a^2X^2) = 1$, so $IJ = A[X]$, and I and J are invertible with $I^{-1} = J$. To see that I is not extended from A, we may localize A at a prime ideal P such that $a \notin A_p$ and so we may assume that A is quasi-local. If I were extended, then, it would be principal, say $I = (f)$. Since we have $IK[X] = K[X]$, $f \in K$. Then since $1 - aX \in J$, $f - faX \in A[X]$, so $f \in A$. This gives $I = (f) \subseteq A[X]$, contradicting $a \notin A$.

As in [7] we say that a ring A with total quotient ring Q(A) is (2,3) - closed if $a \in Q(A)$, $a^2, a^3 \in A$ imply $a \in A$. Schanuel's example tells us that for integral domains, at least, (2,3)-closure is a necessary condition for $\mathrm{Pic}(A) \rightarrow \mathrm{Pic}(A[X])$ to be an isomorphism. Part of the next theorem says that it is also sufficient. Hence in a very real sense Schanuel's example is <u>the</u> example of a non-extended invertible ideal.

Also part of the next theorem is the equivalence for suitable rings A, of seminormality with (2,3)-closure and $\mathrm{Pic}(A) \to \mathrm{Pic}(A[X])$ an isomorphism. This strengthens Traverso's result and brings the relationship between seminormality and Pic into sharper focus.

<u>Theorem 6</u>. Let A be a ring with total quotient ring Q(A), and integral closure $\bar{A}$. Let $X_1, \ldots, X_n, \ldots$ be indeterminates over A. Then for A an integral domain (Brewer-Costa [7], Gilmer-Heitmann [11]), a reduced noetherian ring (Gilmer-Heitmann [11]), or a reduced ring which has a finite number of minimal primes (Rush [18]), the following are equivalent.

(1) $\mathrm{Pic}(A) \to \mathrm{Pic}(A[X_1, \ldots, X_n])$ is an isomorphism for $n \geq 1$.

(2) $\mathrm{Pic}(A) \to \mathrm{Pic}(A[X_1])$ is an isomorphism.

(3) A is seminormal.

(4) For $a \in \bar{A} \setminus A$ the conductor of A in $A[a]$ is a radical ideal of $A[a]$.

(5) A is (2,3)-closed.

(6) For $a \in Q(A)$, $a^n \in A$ for all sufficienthy large n implies $a \in A$.

<u>Sketch of proof for the case of A an integral domain.</u>

(5) $\Leftrightarrow$ (6). Easy.

(1) $\Rightarrow$ (2). Clear.

(2) $\Rightarrow$ (5). Schanuel's example.

(5) $\Rightarrow$ (4). Suppose $a \in \bar{A} \setminus A$ is chosen so that the conductor $C = (A : A[a])$ of A in $A[a]$ is not a radical ideal of $A[a]$. Since C is the largest common ideal of A and $A[a]$ we may choose $b \in \sqrt{C} \setminus A$. Say $b^n \in C$. Since C is an ideal of $A[a]$, $b^m \in C \subseteq A$ for all $m \geq n$.

Let n_o be the smallest integer such that $b^m \in A$ for $m \geq n_o$. Then $n_o > 1$ and $d = b^{n_o - 1} \notin A$, but d^2, $d^3 \in A$ contradicting (5).

(4) ⇒ (3). If B is a subring of $\bar{A}$ containing A, then $+_B A = +_{\bar{A}} A \cap B$. suppose A is not seminormal and let $a \in +_{\bar{A}} A \setminus A$.

Let $B = A[a]$. Then $+_B A = B$, so for each prime ideal P of A there is a unique prime ideal Q of B lying over P and the map $k(P) \to k(Q)$ is an isomorphism. We shall now show that the conductor $C = (A:B)$ is not a radical ideal in B, contradicting (4).

$C \neq 0$ since B is a finite A-module, and $C \neq A$ since $A \neq B$. Let P be a prime ideal of A minimal over C, and let Q be the prime ideal of B lying over P. Then $CA_P = (A_P : B_P)$ is the conductor of A_P in B_P and since $CA_P \neq A_P$, $A_P \neq B_P$. But we also have that $B_P = B_Q$. Now if C is a radical ideal, $CA_P = PA_P$ and since QB_Q is the radical of PB_Q, $CA_P = QB_Q$ also. Thus $PA_P = QB_Q$. We now have from the isomorphism $k(P) \to k(Q)$, $A_P/PA_P \to B_Q/PA_P$ an isomorphism, and hence $A_P = B_P$, which is false.

(3) ⇒ (1) (After Gilmer-Heitmann). If P is a rank one projective $A[X_1, \ldots, X_n]$-module, then P is determined by an idempotent matrix of polynomials. Let $\{c_1, \ldots, c_t\}$ be the finite set of coefficients occurring in the matrix and let B be the subring of A generated by $c_1, \ldots, c_t$. Since B is a finitely generated $\mathbb{Z}$-algebra, it is noetherian with finite integral closure $\bar{B}$, and hence the same is true of the seminormalization $+_{\bar{B}} B$. Let $A_o = +_{\bar{B}} B$. It is possible to see that $A_o \subseteq A$, since A is seminormal. The matrix which defines P also defines a projective $A_o[X_1, \ldots, X_n]$ -module P_o such that $P \cong P_o \otimes A[X_1, \ldots, X_n]$. P_o is of rank 1 and A_o satisfies the hypotheses of theorem 1, so P_o is extended from A_o and therefore P is extended from A.

III - Seminormality and projective modules over polynomial rings.

Theorem 6 allied with the methodology of Lequain-Simis now permits us to extend the Serre conjecture a bit farther, and in a way which brings seminormality into play.

Theorem 7. [7, Thm.2] Let A be an integral domain such that

(i) $\bar{A}$ is a Prüfer domain, and

(ii) Spec (A_p) is finite for all $P \in$ Spec(A).

Then finitely generated projective A $[X_1, \ldots, X_n]$-modules are extended, $n \geq 1$, if and only if A is seminormal.

Sketch of proof. If finitely generated projective A $[X_1, \ldots, X_n]$-modules are extended, then A is seminormal by theorem 6. To show the converse we apply the Lequain-Simis methodology to the class $\mathcal{C}$ of integral domains A satisfying (i),(ii) and (iii) A is seminormal. It is clear from (ii) that (LS .0) holds in $\mathcal{C}$. (LS.1) is easy to verify. (LS.2) is less easy, but follows from well-known facts about Prüfer domains.

Let us now verify (LS.3) : if $A \in \mathcal{C}$ is quasi-local, then finitely generated projective A [X]-modules are free. By (ii) Spec (A) is finite and as in the proof of theorem 5 we have by the theorem of Serre that every projective A [X]-module has the form $I \oplus F$ with I an ideal and F free. But A is seminormal, so by theorem 6 I is extended and hence is principal.

The following immediate corollary to theorem 7 seems to have been known to several authors.

Corollary. If A is a one-dimensional Noetherian domain, then finitely generated projective A $[X_1, \ldots, X_n]$-modules are extended for $n \geq 1$ if and only if A is seminormal.

This result applies to algebraic curves.(For a precise interpretation of seminormality in algebraic curves the reader should consult Davis's enlightening paper [9].) The result is limited to curves, however, since Pedrini [15] has shown that the two-dimensional affine domain $A = K[X,Y,Z] / (Y^3 + Z^2 - X^2 YZ)$ has $Pic(A) \to Pic(A[X_1])$ an isomorphism, so A is seminormal, but $NK_o(A) \neq 0$. ($NK_o(A)$ is the cokernel of the canonical injection $K_o(A) \to K_o(A[X_1])$).

IV - <u>Seminormality and Pic : Swan completes the picture</u>.

To be sure, theorem 6 provides us with a refinement of Traverso's original result. But with the apparently disparate circumstances in which theorem 6 holds, it is a result which muddies the waters more than it clears them. This situation has recently been resolved by Swan [19] in a remarkable way : by redefining seminormality.

Let us henceforth refer to Traverso's seminormality as T-seminormality. In addition to theorem 6, Gilmer and Heitmann gave further information on the relation between T-seminormality and Pic. They showed that for any reduced ring A, T-seminormality is a necessary condition for $Pic(A) \to Pic(A[X_1,\ldots, X_n])$ to be an isomorphism. In addition they constructed a reduced ring A equal to its own total quotient ring, and hence T-seminormal, with $Pic(A) \to Pic(A[X_1,\ldots,X_n])$ not surjective. Thus for reduced rings T-seminormality is formally weaker than the condition on the Picard groups.

Swan gives the following elegant definition of seminormality.

A commutative ring A is <u>seminormal</u> if whenever $b, c \in A$ satisfy $b^3 = c^2$ there exists $a \in A$ such that $a^2 = b$ and $a^3 = c$.

Let us first note that A seminormal implies A reduced. For suppose $b \in A$ and $b^2 = 0$. Then $b^2 = b^3 = 0$ and so there is an $a \in A$ with $a^2 = b$,

$a^3=b$. Then $b=a^3=ab$ which implies that $b=a^2b=b^2=0$.

It is fairly easy to see that if Q(A) is a product of fields, then A is seminormal if and only if it is T-seminormal. Thus the two definitions agree in all situations hypothesized in theorem 6, i.e. in all situations of consequence vis-à-vis the Picard group. And so the following theorem of Swan [19] provides the ultimate version of Traverso's theorem. (A_{red} denotes A modulo its nilradical).

Theorem 8. Let A be a commutative ring. The following are equivalent.

1) $Pic(A) \to Pic(A[X_1,..,X_n])$ is an isomorphism for $n \geq 1$

2) $Pic(A) \to Pic(A[X_1])$ is an isomorphism.

3) A_{red} is seminormal.

V - Related results.

Here we shall compile a list of results related to those above.

Anderson [1] has studied the question of when, for a graded integral domain $A=A_0 \oplus A_1 \oplus \ldots$, the natural monomorphism $Pic(A_0) \to Pic(A)$ is an isomorphism. He has obtained the following result.

Theorem 9. For $A=A_0 \oplus A_1 \oplus \ldots$ a graded integral domain, A is seminormal if and only if A_0 is seminormal and $Pic(A_0) \to Pic(A)$ is an isomorphism.

This has the interesting corollary that for A_0 a field, A is seminormal if and only if $Pic(A) = 0$. Anderson also gives an example with $Pic(A_0) \to Pic(A)$ an isomorphism, but neither A nor A_0 seminormal.

By using the theory of divisorial ideals, Querré [16] has proved the next theorem.

Theorem 10. Let A be an integral domain. Then A is integrally closed if and only if every rank one reflexive $A[X]$-module is extended

This puts theorem 6 in greater perspective.

It is an interesting formal consequence of theorem 8 that if A is seminormal, then so is A [X] . In [8] a direct proof of this result was given in the case when Q(A) is absolutely flat. That proof stemmed from the following quite general result.

Theorem 11. Let $A \subsetneq B$ be rings. If A is (2,3)-closed in B, then A [X] is (2,3)-closed in B [X] .

(In Swan's formulation of seminormality, A(2,3)-closed in B is equivalent to A seminormal in B [19, Thm 2.5] , so theorem 11 shows that "A seminormal in B" extends to polynomial rings.)

Proof. Suppose that $A \subsetneq B$ are commutative rings with A (2,3)-closed in B, and suppose also that A [X] is not (2,3)-closed in B [X]. We shall refer to any polynomial $f(X) \in B[X]$ such that $f^2, f^3 \in A[X]$ but $f \notin A[X]$ as a counterexample.

A minimal counter example is a counter-example $f = a_0 + \dots + a_n X^n$ such that (i) f has minimal degree n among all counter examples and (ii) among all counter examples of degree n, f has the longest initial string of coefficients in A (i.e. $a_0, \dots, a_{i-1} \in A$, $a_i \notin A$). Note that every counterexample at least has $a_0 \in A$ since A is (2,3)-closed in B. By our assumption that A [X] is not (2,3)-closed in B [X] , minimal counter examples exist.

Let $f = a_0 + \dots + a_n X^n$ be a minimal counterexample with $a_0, \dots, a_{i-1} \in A$ and $a_i \notin A$. If $r \in A$ and $ra_i \in A$, then $rf \in A[X]$. For otherwise rf would again be a counterexample and would violate the minimality of f.

The coefficient of X^i in f^2 is $2a_0 a_i +$ terms in A and hence $2a_0 a_i \in A$. It follows from the preceding paragraph that $2a_0 f \in A[X]$. likewise, the coefficientof X^i in f^3 is $3a_0^2 a_i$ + terms in A and so

$3a_0^2 f \in A[X]$. Now consider the polynomial. $g(X) \in B[X]$ defined by $X_g(X) = f(X) - a_o$. $X^2 g(X)^2 = f(X)^2 - 2a_o f(X) + a_o^2 \in A[X]$ and $X^3 g(X)^3 = f(X)^3 - 3a_o f(X)^2 + 3a_o^2 f(X) - a_o^3 \in A[X]$, and hence g^2, $g^3 \in A[X]$. But $\deg g(X) < \deg f(X)$, so $g(X) \in A[X]$. Then $f = a_o + Xg(X) \in A[X]$, a contradiction.

We say that A is <u>n-root closed</u> in B if $A \subseteq B$ and $b \in B$, $b^n \in A$ implies $b \in A$. In much the same way that theorem 11 was proved one can show that A n-root closed in B implies A $[X]$ n-root closed in B $[X]$. This was done in [8]. Recently Watkins [21] has studied the stability of such closure conditions under formation of formal power series rings. He has shown that $\mathbb{Z}[[X]]$ is not n-root closed in $\mathbb{Q}[[X]]$, but that for $A \subseteq B$ with A absolutely flat, A n-root or (2,3)-closed implies the same for A $[[X]]$ in B $[[X]]$.

Finally, we remark that in [19] Swan also defined a notion called p-seminormality, for each integer p. He then generalized result of Hamann [12] and Asanuma [2] by proving that if A is any reduced ring, then A $[X]$ is A-invariant if and only if A is p-seminormal for all primes p. The reader is referred to [19] for a discussion this result.

<u>References</u>

[1] D.F. Anderson, Seminormal graded rings, to appear.

[2] T. Asanuma, D-algebras which are D-stably equivalent to D $[X]$, preprint.

[3] H. Bass, Torsion free and projective modules, Trans. A.M.S. 102(1962), 319-327.

[4] H. Bass, Algebraic K-Theory, Benjamin, N.Y., 1968.

[5] H. Bass and M.P. Murthy, Grothendieck groups and Picard groups of

abelian group rings, Ann. of Math. 86 (1967), 16-73.

[6] J. Brewer and D. Costa, Projective modules over some non-Notherian polynomial rings, J. Pure App. Alg. 13 (1978), 157-163.

[7] J. Brewer and D. Costa, Seminormality and projective modules over polynomial rings, J. Algebra 58 (1979), 208-216.

[8] J. Brewer, D. Costa and K. Mc Crimmon, Seminormality and root closure in polynomial rings and algebraic curves, J. Algebra 58 (1979), 217-226.

[9] E. Davis, On the geometric interpretation of seminormality, Proc. A.M.S. 68 (1978), 1-5.

[10] S. Endo, Projective modules over polynomial rings, J. Math. Soc. Japan 15 (1963), 339-352.

[11] R. Gilmer and R. Heitmann, On Pic R [X] for R seminormal, J. Pure App. Alg. 16 (1980), 251-264.

[12] E. Hamann, The R-invariance of R [X] , J. Algebra 35 (1975),1-16.

[13] Y. Lequain and A. Simis, Projective modules over R $[X_1,..,X_n]$, R a Prüfer domain, J. Pure App. Alg. 18 (1980), 165-172.

[14] H. Lindel, On a conjecture of Quillen and Suslin, preprint.

[15] C. Pedrini, On the K_o of certain polynomial extensions, in <u>Alg. K-theory II</u>, Spinger Lecture Notes n° 342, 1973, 92-108.

[16] J. Querré, Idéaux divisoriels d'un anneau de polynômes, J. Algebra 64 (1980), 270-284.

[17] D. Quillen, Projective modules over polynomial rings, Inv. Math. 36 (1976), 167-171.

[18] D.E. Rush, Seminormality, to appear.

[19] R.G. Swan, On seminormality, preprint.

[20] C. Traverso, Seminormality and Picard group, Ann. Scuola Norm. Sup. Pisa, 24 (1970), 585-595.

[21] J. Watkins, Root and integral closure for R[[X]] , preprint.

On the maximal number of $\mathfrak{a}$-independent elements in ideals of noetherian rings

By Jan-Erik Björk

Introduction

Let R be a commutative noetherian ring and let $\mathfrak{a}$ be an ideal of R. In [1] G. Valla introduced the following concept:

0.1 Definition. A subset $\{a_1 \dots a_k\}$ of $\mathfrak{a}$ is called $\mathfrak{a}$-independent if every homogeneous form in $R[x_1 \dots x_k]$ vanishing at $(a_1 \dots a_k)$ has all its coefficients in $\mathfrak{a}$.

This leads to:

0.2 Definition. Put $\sup(\mathfrak{a}) = \sup\{k \geq 0 : \exists$ a k-tuple of $\mathfrak{a}$-independent elements$\}$.

Following a recent work by N.V. Trung we are going to determine $\sup(\mathfrak{a})$ for each given ideal $\mathfrak{a}$ of a commutative noetherian ring R. So except for some minor modifications the proof of the Main Theorem below is contained in [5]. Before it can be announced we need some notations. They are introduced in Section 2 below, while the proof of the Main Theorem is carried out in the subsequent sections.

0.3 Remark. In [3, p.35] it was proved that the following inequalities hold for every ideal $\mathfrak{a}$: grade $(\mathfrak{a}) \leq \sup(\mathfrak{a}) \leq \mathrm{ht}(\mathfrak{a})$ where ht denotes the height of the ideal $\mathfrak{a}$, or equivalently the Krull dimension of the R-module $\mathfrak{a}$.

1. Statement of the Main Theorem

Let us first observe that $\sup(\mathfrak{a})$ decreases under localisations. In general, if $\mathfrak{p}$ is a prime ideal of R which contains $\mathfrak{a}$ then we get the local ring $R_{\mathfrak{p}}$ where $\mathfrak{a}$ generates the ideal $\mathfrak{a}R_{\mathfrak{p}}$ and with these notations we have.

1.1 Lemma. $\sup(\mathfrak{a}) \leq \sup(\mathfrak{a}R_{\mathfrak{p}})$

Proof. Let $\{a_1 \ldots a_k\}$ be $\mathfrak{a}$-independent. It is then sufficient to prove that they are also $\mathfrak{a}R_{\mathfrak{p}}$-independent. To show this we consider some homogeneous form $\sum s_{\alpha} x^{\alpha}$ in $R_{\mathfrak{p}}[x_1 \ldots x_k]$ - so here $\alpha = (\alpha_1 \ldots \alpha_k)$ are multi-indices and $|\alpha| = \alpha_1 + \ldots + \alpha_k = m$ for some fixed integer m.

Suppose now that $\sum s_{\alpha} a^{\alpha} = 0$ in the ring $R_{\mathfrak{p}}$. Then we can find $t \in R \setminus \mathfrak{p}$ so that all the coefficients $ts_{\alpha} \in R$ and in addition $\sum (ts_{\alpha}) a^{\alpha} = 0$ in R.

Since $\{a_1 \ldots a_k\}$ are $\mathfrak{a}$-independent it follows that $ts_{\alpha} \in \mathfrak{a}$ for all α and then $s_{\alpha} \in \mathfrak{a}R_{\mathfrak{p}}$ for all α which proves that $\{a_1 \ldots a_k\}$ are $\mathfrak{a}R_{\mathfrak{p}}$-independent.

1.2 The completion $R^*_{\mathfrak{p}}$ of the local ring $R_{\mathfrak{p}}$ can also be introduced. We know that $R^*_{\mathfrak{p}}$ is faithfully flat over $R_{\mathfrak{p}}$ and using this fact it is easily seen that $\mathfrak{a}R_{\mathfrak{p}}$-independent elements are $\mathfrak{a}R^*_{\mathfrak{p}}$ -independent which together with Lemma 1.1 gives.

1.3 Lemma. $\sup(\mathfrak{a}) \leq \sup(\mathfrak{a}R^*_{\mathfrak{p}})$ hold for all $\mathfrak{p} \supseteq \mathfrak{a}$.

This inequality will ve used to prove the Main Theorem. Before it is announced we shall need another definition.

1.4 The ideals $U_i(\mathfrak{b})$. Given an ideal $\mathfrak{b}$ of some noetherian ring S we define the sets $U_i(\mathfrak{b})$ for each $i \geq 0$ by :

$$U_i(\mathfrak{b}) = \{x \in S : \exists t \geq 1 \text{ so that } \mathrm{Kr.dim}(\mathfrak{b}^t x) \leq i\} .$$

Of course, if $t \geq 1$ then $\mathfrak{b}^t x$ appears as an ideal of S and considered as an S-module it has a Krull dimension. This explains the definition of the sets $U_i(\mathfrak{b})$. It is easily seen that $U_i(\mathfrak{b})$ are ideals of S and of course they increase. Finally, if $\mathrm{Kr.dim}(S) = \delta$ is finite then $U_{\delta}(\mathfrak{b}) = S$ holds.

1.5 Main Theorem. For a given ideal $\mathfrak{a}$ of a ring R we have

$$\sup(\mathfrak{a}) = \inf_{\mathfrak{p}} \left\{ \inf_{i \geq 0} U_i(\mathfrak{a}R^*_{\mathfrak{p}}) \not\subseteq \mathfrak{a} : \mathfrak{p} \in \mathrm{Ass}(R/\mathfrak{a}) \right\}$$

The proof requires several steps. In Section 2 we prove the easy part, namely the inequality $\leq$, the opposite inequality is more involved and to prove it we need several preliminary results in Section 3, while the actual construction of $\mathfrak{a}$-independent elements is carried out in Section 4.

2. <u>An upper bound of</u> $\sup(\mathfrak{a})$

Given an ideal $\mathfrak{a}$ in a ring R we have already seen that $\sup(\mathfrak{a}) \leq \sup(\mathfrak{a}R^*_{\mathfrak{p}})$ for all $\mathfrak{p}$ in $\mathrm{Ass}(R/\mathfrak{a})$. Therefore the inequality $\leq$ in the Main Theorem follows from the result below- applied to the ideals $\mathfrak{a}R^*_{\mathfrak{p}}$.

2.1 <u>Proposition. Let</u> S <u>be a local ring and let</u> $\mathfrak{b} \subset S$ <u>be an ideal. Then</u> $\sup(\mathfrak{b}) \leq \inf\{i \geq 0 : U_i(\mathfrak{b}) \not\subset \mathfrak{b}\}$.

<u>Proof</u>. Let $k \geq 0$ and assume that $U_k(\mathfrak{b}) \not\subset \mathfrak{b}$. We must prove that $\sup(\mathfrak{b})$ is $\leq k$. To do this we begin with some preliminary observations.

<u>Sublemma</u> 1. $\exists\, w \geq 1$ such that $\mathrm{Kr.dim}(\mathfrak{b}^w U_k(\mathfrak{b})) \leq k$.

<u>Proof of Sublemma</u> 1. First, the ideal $U_k(\mathfrak{b})$ is finitely generated, say by $x_1 \ldots x_T$. Now we can choose w so large that $\mathrm{Kr.dim}(\mathfrak{b}^w x_t) \leq k$ for all $1 \leq t \leq T$. It follows easily that $\mathrm{Kr.dim}(\mathfrak{b}^w x) \leq k$ for all x in $U_k(\mathfrak{b})$, and then that $\mathrm{Kr.dim}(\mathfrak{b}^w U_k(\mathfrak{b})) \leq k$ where $\mathfrak{b}^w U_k(\mathfrak{b})$ = the ideal generated by all products ξx with $\xi \in \mathfrak{b}^w$ and $x \in U_k(\mathfrak{b})$.

Starting with this we now introduce:

<u>The ideal</u> $\mathcal{J} = (0 : \mathfrak{b}^w U_k(\mathfrak{b})) = \{x \in S : x\mathfrak{b}^w U_k(\mathfrak{b}) = 0\}$ and the factor rings $S/\mathcal{J} = \bar{S}$. Let also $\mathcal{M}$ be the ideal in $\bar{S}$ which is generated by the maximal ideal $\mathfrak{m}$ of S. Then we can prove:

<u>Sublemma</u> 2. If $\{a_1 \ldots a_k\}$ are $\mathfrak{b}$-independent then their images $\{\bar{a}_1 \ldots \bar{a}_k\}$ are $\mathcal{M}$-independent.

<u>Proof</u>. Suppose that $\Sigma \bar{s}_\alpha \bar{a}^\alpha = 0$ in $\bar{S}$. This means that $\Sigma s_\alpha a^\alpha$ belongs to the ideal $\mathcal{J}$ in the ring S. Since $a_1^w \in \mathfrak{b}^w$ this means that $\forall y \in U_k(\mathfrak{b})$ we get $ya_1^w \Sigma s_\alpha a^\alpha = 0$. Now we observe that $x_1^w \Sigma s_\alpha x^\alpha$ also is a homogeneous form and since $a_1 \ldots a_k$ are $\mathfrak{b}$-independent it follows that $ys_\alpha \in \mathfrak{b}$ for all α . This holds for all y in $U_k(\mathfrak{b})$ which means that the coefficients $s_\alpha \in (\mathfrak{b} : U_k(\mathfrak{b}))$ and this ideal is $\subseteq \mathfrak{m}$ because $U_k(\mathfrak{b}) \not\subset \mathfrak{b}$ was assumed. Hence $s_\alpha \in \mathfrak{m}$ for all α so that $\bar{s}_\alpha \in \mathcal{M}$ for all α and this proves that $\{\bar{a}_1 \ldots \bar{a}_k\}$ are $\mathcal{M}$-independent.

Final part of the proof Sublemma 2 shows that $\sup(\mathfrak{a}) \leq \sup(\mathcal{M})$ and here $\mathcal{M}$ is the maximal ideal of the local ring $\bar{S}$ and now we observe that $\bar{S} = S/\mathcal{J}$ where $\mathcal{J}$ was the annihilating ideal of the R-module $\mathfrak{b}^w U_k(\mathfrak{b})$ whose Krull dimension is $\leq k$. This implies that $\mathrm{Kr.dim}(\bar{S}) \leq k$ and then the inequality $\sup(\mathcal{M}) \leq k$ follows from [3, page 35].

3. Some useful criteria for $\mathfrak{a}$-independence

Following [5] we are going to establish some useful conditions which ensure $\mathfrak{a}$-independence. We shall use the following:

Notations. If $a_1 \ldots a_i$ are elements of a local ring R then $(a_1 \ldots a_i)$ denotes the ideal they generate. By V(a) we denote the family of prime ideals of R which do not contain the given element a.

Now we can announce (See also [5, Lemma 6]).

3.1 Lemma. Let $a_1 \ldots a_d$ be elements of an ideal $\mathfrak{a}$ satisfying

(i) $a_i \notin \mathfrak{p}$ for all $\mathfrak{p} \in \mathrm{Ass}(R/(a_1 \ldots a_{i-1})) \setminus V(a_d)$: $1 \leq i \leq d$

(If i = 1 we use Ass(R) above)

(ii) The ideal $J = \{x \in R : a_d^t x \in (a_1 \ldots a_{d-1})$ for some $t \geq 0\}$ is $\subseteq \mathfrak{a}$

Conclusion. $\exists w$ so that $a_1, \ldots, a_{d-1}, a_d^w$ are $\mathfrak{a}$-independent.

Proof. First we can choose an integer w so that the ideal J equals the ideal $((a_1 \ldots a_{d-1}) : a_d^{w-1})$ and now we are going to prove that $a_1 \ldots a_{d-1}$ and a_d^w are $\mathfrak{a}$-independent. For this we shall need.

The ring $S = R[z_1 \ldots z_{d-1}]$ where $z_i = a_i/a_d^{w-1}$. In other words, S appears as a subring of the localisation $R[a_d^{-1}]$. The choice of w easily gives the following:

Sublemma 1. $S/(z_1 \ldots z_{d-1}) \cong R/J$.

Besides, it is also obvious that the image $\bar{a}_d$ in the factor ring $S/(z_1 \ldots z_{d-1})$ is a non-zero divisor. Also, using the fact that $S[a_d^{-1}] = R[a_d^{-1}]$ and that primes of $R[a_d^{-1}]$ correspond to primes in R which do not contain a_d, the condition (i) of Lemma 3.1 gives.

Sublemma 2. The (d-1)-tuple $a_1 \ldots a_{d-1}$ is a regular sequence of the ring S.

At this stage we must distinguish two cases in order to finish the proof and begin with.

Case 1. Here we assume that a_d is a non-zero divisor in the ring R. Then [2, Corollary 2] applies and shows that the elements $z_1 \dots z_{d-1}$, a_d is a regular sequence of the local ring S_M, where M is the maximal ideal of S which is generated by $z_1 \dots z_{d-1}$ and the maximal ideal $\mathfrak{m}$ of the given local ring R. Then [3, Proposition 3] implies that the d-tuple $\{z_1 \dots z_{d-1}, a_d\}$ are $(z_1 \dots z_{d-1}, a_d)S_M$ -independent- in other words, regular sequences are independent in the ideal they generate.

If we now recall that $z_i = a_i/a_d^{w-1}$ it follows that $\{a_1 \dots a_{d-1}, a_d^w\}$ are $(a_1 \dots a_{d-1}, a_d^w) S_M$ -independent and at this stage their required $\mathfrak{a}$-independence follows from:

Sublemma 3. The intersection $R \cap (a_1 \dots a_{d-1}, a_d^w)S_M$ is $\subseteq \mathfrak{a}$.

We leave out the easy proof and it remains only to consider

Case 2. When a_d is a zero-divisor. In this case we first introduce the ideal $J_o = \{x \in R : a_d^t x = 0 \text{ for some } t \geq 1\}$. Then the image of a_d is a non-zero divisor on the factor ring $R' = R/J_o$ and it is easy to verify that (i) and (ii) of Lemma 3.1 hold for the images $\bar{a}_1 \dots \bar{a}_d$ in R' as well.

So by Case 1 we can conclude that $\exists w$ so that $\{\bar{a}_1 \dots \bar{a}_{d-1}, \bar{a}_d^w\}$ are $\bar{\mathfrak{a}}$-independent. But then their $\mathfrak{a}$-independence follows immediately if we observe that the ideal $J_o \subseteq J = \{x : a_d^t x \in (a_1 \dots a_{d-1}) \text{ for some } t \geq 1\}$ and that $J \subseteq \mathfrak{a}$.

3.2 Another formulation of Lemma 3.1. The condition (ii) of Lemma 3.1 is not so easily realised. It turns out that if R is complete, then (ii) can be replaced by a condition which will be easier to control during the subsequent proof of the Main Theorem in Section 4. Let us explain how this is done.

So let R be a complete local ring and suppose that $a_1 \dots a_d$ is some set of elements in its maximal ideal. Then we define.

3.3 The ideal $\mathcal{U} = \{x \in R : \forall \mathfrak{p} \in \text{Ass}(R/(a_1 \dots a_{d-1})) \setminus V(a_d)$ there exists some $s \in R \setminus \mathfrak{p}$ so that $sx = 0\}$.

Then we can prove the following.

3.4 Proposition. Assume that the d-tuple $\{a_1 \dots a_d\}$ satisfies (i) in Lemma 3.1. Then $\exists$ an integer N so that the ideal $\mathcal{U}$ equals the ideal

$$\{x \in R : a_d^t x \in (a_1^N \dots a_{d-1}^N) \text{ for some } t \geq 1\}$$

Proof. Condition (i) of Lemma 3.1 implies that $a_1 \dots a_{d-1}$ is a regular sequence in $R[a_d^{-1}]$ and this gives:

<u>Sublemma</u> 1. $\mathrm{Ass}(R/(a_1 \ldots a_{d-1})) \setminus V(a_d) = \mathrm{Ass}(R/(a_1^n \ldots a_{d-1}^n)) \setminus V(a_d)$ <u>hold for all</u> $n \geq 2$.

Next, the definition of the ideal $\mathfrak{U}$ shows that $x \in \mathfrak{U}$ if and only if its image in the localisations $R_{\mathfrak{p}}$ are zero for all primes $\mathfrak{p}$ of the set $\Gamma = \mathrm{Ass}(R/(a_1 \ldots a_{d-1})) \setminus V(a_d)$.

Besides, using Krull's Intersection Theorem in these local rings, we see that x belongs to $\mathfrak{a}$ if and only if its image $\in \bigcap_{n=1}^{\infty} (a_1^n \ldots a_{d-1}^n) R_{\mathfrak{p}}$ for all primes $\mathfrak{p}$ in Γ.

At this stage we need:

<u>Sublemma</u> 2. <u>If</u> $n \geq 1$ <u>is given then an element</u> x <u>in</u> R <u>belongs to</u> $(a_1^n \ldots a_{d-1}^n) R_{\mathfrak{p}}$ <u>for all</u> $\mathfrak{p}$ <u>in</u> Γ <u>if and only if</u> $\exists t \geq 1$ <u>so that</u> $a_d^t x$ <u>belongs to</u> $(a_1^n \ldots a_{d-1}^n)$.

<u>Proof</u>. Follows because Sublemma 1 shows that $\Gamma = \mathrm{Ass}(R/(a_1^n \ldots a_{d-1}^n)) \setminus V(a_d)$.

Summing up, we have now proved the following equality :

$$\mathfrak{a} = \bigcap_{n=1}^{\infty} \bigcup_{t=1}^{\infty} (a_1^n \ldots a_{d-1}^n : a_d^t)$$

and now [4, Theorem 10, page 270] is applied to the complete ring R and shows that there exists a positive integer N so that

$$\mathfrak{U} = \bigcup_{t=1}^{\infty} (a_1^N \ldots a_{d-1}^N : a_d^t) \quad \text{holds.}$$

3.5 <u>Another useful independence condition</u>

Using Lemma 3.1 and Proposition 3.4 we can now prove the following.

3.5 <u>Proposition</u>. <u>Let</u> R <u>be a complete local ring and let</u> $\mathfrak{a}$ <u>be an</u> $\mathfrak{m}$<u>-primary ideal of</u> R <u>and let</u> $\{a_1 \ldots a_d\}$ <u>be elements of</u> $\mathfrak{a}$ <u>satisfying</u>

(i) $a_i \notin \mathfrak{p}$ <u>for all</u> $\mathfrak{p} \in \mathrm{Ass}(R/(a_1 \ldots a_{i-1})) \setminus V(a_d) : 1 \leq i \leq d$.

(ii) $\forall \mathfrak{p} \in \mathrm{Ass}(R)$ <u>with</u> $\mathrm{Kr.dim}(R/\mathfrak{p}) \geq d$ <u>there exists some</u> $\tilde{\mathfrak{p}}$ <u>in</u> $\mathrm{Ass}(R/(a_1 \ldots a_{d-1})) \setminus V(a_d)$ <u>so that</u> $\mathfrak{p} \subseteq \tilde{\mathfrak{p}}$.

(iii) <u>The ideal</u> $U_{d-1}(\mathfrak{a}) \subseteq \mathfrak{a}$.

<u>Conclusion</u>. $\exists N$ <u>and</u> w <u>so that</u> $\{a_1^N \ldots a_{d-1}^N, a_d^w\}$ <u>are</u> $\mathfrak{a}$<u>-independent</u>.

<u>Proof</u>. Using Lemma 3.1 and Proposition 3.5 we see that it is sufficient to prove that the ideal $\mathfrak{U}$ which was defined in 3.3 is $\subseteq \mathfrak{a}$. Next, since $U_{d-1}(\mathfrak{a}) \subseteq \mathfrak{a}$ was assumed, it is sufficient to prove the following:

The inclusion $\mathfrak{U} \subseteq U_{d-1}(\mathfrak{a})$. To prove this we shall use (ii) in Proposition 3.5 which together with the definition of the ideal $\mathfrak{U}$ obviously implies:

Sublemma 1. $\mathfrak{U} \subseteq \tilde{\mathfrak{U}} = \{x \in R : \forall \mathfrak{p} \in \mathrm{Ass}(R)$ and $\mathrm{Kr.dim}(R/\mathfrak{p}) \geq d$ there exists some $s \in R \setminus \mathfrak{p}$ so that $sx = 0\}$.

Also, since $\mathfrak{a}$ is $\mathfrak{m}$-primary, which means that $\mathfrak{m}^t \subset \mathfrak{a}$ for some t it follows that $U_{d-1}(\mathfrak{a}) = U_{d-1}(\mathfrak{m})$, so it remains only to show.

Sublemma 2. $\tilde{\mathfrak{U}} \subseteq U_{d-1}(\mathfrak{m})$.

Proof of Sublemma 2. Let $x \in \tilde{\mathfrak{U}}$ be given. Consider the R-module $(x) = Rx$ and recall that $\mathrm{Ass}(Rx)$ is a subset of $\mathrm{Ass}(R)$. The definition of the ideal then shows that if $\mathfrak{p} \in \mathrm{Ass}(Rx)$ then $\mathrm{Kr.dim}(R/\mathfrak{p}) \leq d-1$ and this implies that $\mathrm{Kr.dim}(Rx) \leq d-1$ and a fortiori $\mathrm{Kr.dim}(\mathfrak{m}x) \leq d-1$ which proves that x belongs to $U_{d-1}(\mathfrak{m})$.

4. Proof of the Main Theorem

Using Proposition 3.5 we are now prepared to enter the proof of the inequality $\geq$ in the Main Theorem. For a given ideal $\mathfrak{a}$ of a ring R we define.

The integer $\delta(\mathfrak{a}) = \inf_{\mathfrak{p}} \{\inf \{i \geq 0 : U_i(\mathfrak{a}R^*_{\mathfrak{p}}) \not\subset \mathfrak{a}R^*_{\mathfrak{p}} : \mathfrak{p} \in \mathrm{Ass}(R/\mathfrak{a})\}\}$.

So in Section 2 we have proved that $\sup(\mathfrak{a}) \leq \delta(\mathfrak{a})$ and it remains to prove that $\sup(\mathfrak{a}) \geq \delta(\mathfrak{a})$. For this we shall first need the follwing result from [5, Proposition 8] where we use the notations : $V(\mathfrak{a})$ = all primes not containing $\mathfrak{a}$ and $W(\mathfrak{a})$ = all primes $\subseteq$ some prime in $\mathrm{Ass}(R/\mathfrak{a})$.

4.1 Proposition. Let $d = \delta(\mathfrak{a}) \geq 1$ and let $a_1 \dots a_d$ be a d-tuple of elements in $\mathfrak{a}$ satisfying:

(i) $a_i \notin \mathfrak{p}$ for all $\mathfrak{p} \in [\mathrm{Ass}(R/(a_1 \dots a_{i-1}) \setminus V(\mathfrak{a})] \cap W(\mathfrak{a})$

(ii) If $d > 1$ and if $\mathfrak{p} \in [\mathrm{Ass}(R/(a_1 \dots a_{i-1}) \setminus V(\mathfrak{a})] \cap W(\mathfrak{a})$ is such that $\exists \tilde{\mathfrak{p}} \in \mathrm{Ass}(R/\mathfrak{a})$ with $\mathfrak{p} \subset \tilde{\mathfrak{p}}$ and $\mathrm{ht}(\tilde{\mathfrak{p}}/\mathfrak{p}) \geq d-i+1$, then there exists some prime $\hat{\mathfrak{p}}$ so that $(\mathfrak{p}, a_i) \subset \hat{\mathfrak{p}}$ and $\hat{\mathfrak{p}} \notin V(\mathfrak{a})$.

Conclusion. $\exists N$ and w so that $\{a_1^N \dots a_{d-1}^N, a_d^w\}$ are $\mathfrak{a}$-independent.

The proof requires a preliminary observation :

4.2 Lemma. Let $\mathfrak{a} = Q_1 \cap \dots \cap Q_s$ be some primary decomposition of the ideal $\mathfrak{a}$. Then the primes $\mathfrak{p}_i = \sqrt{Q_i}$ all belong to $\mathrm{Ass}(R/\mathfrak{a})$ and if

$\{\xi_1 \dots \xi_d\}$ is a d-tuple of $\mathfrak{a}$ such that their images in $R^*_{\mathfrak{p}_i}$ are $Q_i R^*_{\mathfrak{p}_i}$-independent for each $1 \leq i \leq s$, then they are $\mathfrak{a}$-independent.

Proof. Follows easily from faithful flatness and the definition of primary ideals.

Proof of Proposition 4.1 First, by Lemma 4.2 it is sufficient to prove that if Q is a primary component of $\mathfrak{a}$ and if S is the complete local ring $R^*_{\mathfrak{p}}$ - with $\mathfrak{p} = \sqrt{Q}$, then $\{a_1^N \dots a_{d-1}^N, a_d^w\}$ are $QR^*_{\mathfrak{p}}$-independent.

Using the flatness of $R^*_{\mathfrak{p}}$ over R, it is not difficult to show that the assumptions (i) and (ii) are valid in $R^*_{\mathfrak{p}}$ too-with respect to the elements $a_1 \dots a_d$ there and with the ideal $\mathfrak{a}$ replaced by the $\mathfrak{m}$-primary ideal $QR^*_{\mathfrak{p}}$ of this complete local ring. Here $\mathfrak{m} = \mathfrak{p}R^*_{\mathfrak{p}}$ of course.

We leave out this verification to the reader. Next, using (ii) in this case, it follows by an easy induction that every prime $\mathfrak{q}$ in $\mathrm{Ass}(R^*_{\mathfrak{p}})$ for which $\mathrm{Kr.dim}(R^*_{\mathfrak{p}}/\mathfrak{q}) \geq d$, is contained in some prime which belongs to $\mathrm{Ass}(R/(a_1 \dots a_{d-1})) \setminus V(a_d)$ and then the required $QR^*_{\mathfrak{p}}$-independence follows from Proposition 3.5

4.3. The construction of $\mathfrak{a}$-independent elements

Given an ideal $\mathfrak{a}$ of a ring R it remains only to find elements in $\mathfrak{a}$ which guarantee the two conditions expressed by (i) and (ii) of Proposition 4.1. Following [5] this is achieved, using the following inductive choice :

1. The choice of a_1. Consider first the following family of primes $\Gamma_0 = \{\mathfrak{p} \in [\mathrm{Ass}(R) \setminus V(\mathfrak{a})] \cap W(\mathfrak{a})$, and then all pairs $\mathfrak{p} \subset \tilde{\mathfrak{p}}$ with $\tilde{\mathfrak{p}} \in \mathrm{Ass}(R/\mathfrak{a})$ and $\mathrm{ht}(\tilde{\mathfrak{p}}/\mathfrak{p}) \geq d$. We shall assume that $d \geq 2$ here.

For the construction we shall need

Sublemma. To each pair $\mathfrak{p} \subseteq \tilde{\mathfrak{p}}$ as above there exists a prime $\hat{\mathfrak{p}}$ so that : $\mathrm{ht}(\hat{\mathfrak{p}}/\mathfrak{p}) = 1$ and $\hat{\mathfrak{p}} \in \mathrm{Spec}(R) \setminus V(\mathfrak{a})$ and $\hat{\mathfrak{p}} \not\subset \cup \mathfrak{p} : \mathfrak{p} \in \Gamma_0$.

Proof. First, $\tilde{\mathfrak{p}}$ contains $\mathfrak{a}$ so the definition of Γ_0 shows that $\tilde{\mathfrak{p}}$ is not contained in $\cup \mathfrak{p} : \mathfrak{p} \in \Gamma_0$ and hence there exist α in $\tilde{\mathfrak{p}}$ which is outside $\cup \mathfrak{p} : \mathfrak{p} \in \Gamma_0$.

Given such an element α we find a minimal prime divisor $\hat{\mathfrak{p}}$ of the ideal $(\mathfrak{p}, \alpha)$ and then $\mathrm{ht}(\hat{\mathfrak{p}}/\mathfrak{p}) = 1$. It remains only to see that we can choose $\hat{\mathfrak{p}}$ so that $\hat{\mathfrak{p}} \in \mathrm{Spec}(R) \setminus V(\mathfrak{a})$ also holds. This follows because $\mathrm{ht}(\tilde{\mathfrak{p}}/\mathfrak{p}) \geq 2$ which shows that there exist infinitely many primes $\hat{\mathfrak{p}}$ as above and since $\mathfrak{p} \in \mathrm{Spec}(R) \setminus V(\mathfrak{a})$ we see that we can choose $\hat{\mathfrak{p}}$ from this infinite family of minimal prime divisors of $\mathfrak{p}$ so that $\hat{\mathfrak{p}}$ also $\in \mathrm{Spec}(R)\setminus V(\mathfrak{a})$.

Summing up, to all pairs $\mathfrak{p} \subset \tilde{\mathfrak{p}}$, with $\mathfrak{p} \in \Gamma_0$ and $\tilde{\mathfrak{p}} \in \mathrm{Ass}(R/\mathfrak{a})$ and $\mathrm{ht}(\tilde{\mathfrak{p}}/\mathfrak{p}) \geq d$ we find some $\hat{\mathfrak{p}} = \hat{\mathfrak{p}}(\mathfrak{p}, \tilde{\mathfrak{p}})$ so that $\mathfrak{p} \subset \hat{\mathfrak{p}} \subset \tilde{\mathfrak{p}}$ and $\mathrm{ht}(\hat{\mathfrak{p}}/\mathfrak{p}) = 1$ and $\hat{\mathfrak{p}} \in \mathrm{Spec}(R)\setminus V(\mathfrak{a})$.

Now we can choose a_1 so that a_1 belongs to the intersection of all the primes $\hat{\mathfrak{p}} = \hat{\mathfrak{p}}(\mathfrak{p}, \tilde{\mathfrak{p}})$ and in addition a_1 is outside $\cup \mathfrak{p} : \mathfrak{p} \in \Gamma_0$.

2. The inductive choice. Suppose now that $a_1 \ldots a_{i-1}$ have been constructed. Then we consider the family of primes:

$\Gamma_{i-1} = \{\mathfrak{p} \in [\mathrm{Ass}(R/(a_1 \ldots a_{i-1}))\setminus V(\mathfrak{a})] \cap W(\mathfrak{a})$ and all pairs $\mathfrak{p} \subset \tilde{\mathfrak{p}}$ with $\tilde{\mathfrak{p}} \in \mathrm{Ass}(R/\mathfrak{a})$ and $\mathrm{ht}(\tilde{\mathfrak{p}}/\mathfrak{p}) \geq d-i+1\}$. Assuming that $i \leq d-1$ we find as before primes $\hat{\mathfrak{p}} = \hat{\mathfrak{p}}(\mathfrak{p}, \tilde{\mathfrak{p}})$ for all these pairs so that $\mathrm{ht}(\hat{\mathfrak{p}}/\mathfrak{p}) = 1$ and $\hat{\mathfrak{p}}$ is outside $\cup \mathfrak{p} : \mathfrak{p} \in \Gamma_{i-1}$ and in $\mathrm{Spec}(R)\setminus V(\mathfrak{a})$.

Then $a_i \in \mathfrak{a}$ is chosen so that a_i belongs to $\cap \hat{\mathfrak{p}}$ and in addition a_i is outside $\cup \mathfrak{p} : \mathfrak{p} \in \Gamma_{i-1}$.

The construction of a_d. Having found $a_1 \ldots a_{d-1}$ we simply choose a_d so that a_d is outside all primes $\in [\mathrm{Ass}(R/(a_1 \ldots a_{d-1}))\setminus V(\mathfrak{a})] \cap W(\mathfrak{a})$.

This completes the construction in the case when $d \geq 2$.

A remark. If $d = 1$ we simply choose a_1 so that a_1 is outside the family $\cup \mathfrak{p} : \mathfrak{p} \in \Gamma_0$.

The careful choice above of the d-tuple $a_1 \ldots a_d$ ensures that (i) and (ii) hold in Proposition 4.1 and therefore Proposition 4.1 gives

4.4 Proposition $\sup(\mathfrak{a}) \geq \delta(\mathfrak{a})$ hold for all ideals $\mathfrak{a}$.

This completes the proof of the Main Theorem.

5. Final Remarks

Actually a more general result was proved in [5] which computes the integer $\mathfrak{b}$-sup $\mathfrak{a}$ = the maximal number of $\mathfrak{a}$-independent elements which belong to a given ideal $\mathfrak{b} \subset \mathfrak{a}$. In [5] there are also nice applications of the Main Theorem. In particular it gives an affirmative answer to a question raised in [6] and here is the result.

5.1 Theorem. Let R be a local ring, then the following statements are equivalent:

(i) R is unmixed (i.e. $\dim(R^*/\mathfrak{p}) = \dim(R)$ for all $\mathfrak{p} \in \mathrm{Ass}(R^*)$)

(ii) $\sup(\mathfrak{a}) = \mathrm{ht}(\mathfrak{a})$ for all ideals $\mathfrak{a}$ of R

(iii) $\sup(\mathfrak{m}^t) = \dim(R)$ <u>for all</u> $t \geq 1$.

References

[1] Valla, G., Elementi independenti rispetto ad un ideale. Rend. Sem. Mat. Univ. Padova <u>44</u> (1970), 339-354.

[2] Eisenbud, D., Herrmann, M., Vogel W., Remarks on regular sequences, Nagoya Math. Journ. <u>67</u> (1977), 117-180.

[3] Barshay, J., Generalised analytic independence, Proc. AMS <u>58</u> (1976), 32-36.

[4] Zariski, O., Samuel P., Commutative Algebra, vol.2

[5] Trung, N.V. Generalised analytic independence. To appear. Preprint from institute of Math. 208 D Dôi Can, Hanoi.

[6] Valla, G., Remarks on generalised analytic independence. Math. Proc. Cambridge Phil. Soc.

DIMENSIONS PURES DE MODULES

par Danielle SALLES

INTRODUCTION. Depuis la définition en 1961, par J.E. Roos des dérivés du foncteur $\varprojlim$, de nombreux auteurs (en particulier Barbara Osofsky (1) et Christian Jensen (2)) se sont attachés à déterminer les systèmes projectifs et/ou les anneaux pour lesquels les dérivés de $\varprojlim$ s'annulent pour des entiers inférieurs ou égaux à un entier n. Un article récent de C.U. Jensen ("Dimensions cohomologiques reliées aux foncteur $\varprojlim^{i}$" à paraître aux Proceedings du Séminaire Dubreil Malliavin, Lectures Notes in Math) montre que ces problèmes sont étroitement liés à celui de la détermination de la dimension pure globale des anneaux.

Nous répondons ici (corollaire 5) partiellement à une question orale de C.U. Jensen : Déterminer les anneaux qui ont une grande dimension pure globale (partiellement, car nous ne savons pas répondre lorsque l'anneau est Noethérien.)

Ce type de résultat est utile en Topologie Algébrique (voir par exemple les travaux de Pezennec (3)). Nous montrons ensuite que, sous certaines conditions lorsqu'on passe du foncteur Ext au foncteur Pext (le foncteur dérivé relatif aux résolutions pures) un certain nombre d'isomorphismes classiques (4) sont conservés (Propositions 9, 10, Théorème 11).

Nous donnons enfin (proposition 12) une majoration de la dimension pure injective de certains systèmes projectifs, une caractérisation des anneaux dont la dimension globale pure est ≤ 1 (proposition 13) et une amélioration (proposition 14) d'un résultat déjà énoncé (5) sur la commutativité de $\varprojlim$ et de Tor.

On utilisera les abréviations suivantes :

pp. dim : dimension pure projective
pi. dim : dimension pure injective
gl. dim : dimension globale
p.gl.dim : dimension pure-globale
W. dim : dimension faible.

Les anneaux sont commutatifs

RAPPELS. Soit A un anneau,

1. Pour tous A-modules E, F, G il existe un isomorphisme :

$$\mathrm{Hom}\,(E \otimes F,G) \leftrightarrow \mathrm{Hom}\,(E,\ \mathrm{Hom}\,(F,G)) \qquad (6)$$

2. Si G est un A-module injectif et F un module plat alors Hom (F,G) est injectif (6) .

3. Si G est un A-module injectif et E de présentation finie alors il existe un isomorphisme

$$\mathrm{Hom}\,(\mathrm{Hom}\,(E,F),G) \leftrightarrow \mathrm{Hom}\,(F,G) \otimes E \qquad (6)$$

4. Un A-module est dit <u>pur projectif</u> s'il est projectif pour les suites exactes pures.

5. Un A-module de présentation finie est pur projectif. (7)

6. Un A-module est pur-projectif si et seulement si c'est un facteur direct d'une somme directe de modules de présentation finie. (7)

7. Pour tout A-module M il existe une suite exacte pure :

$$0 \to M \to G \to T \to 0$$

où G est un module pur-injectif. (7) . Il est donc possible de définir pour tout module M une résolution pure-injective :

$$0 \to M \to G_0 \to G_1 \to \dots \to G_i \to \dots$$

L'homologie en n du complexe $\mathrm{Hom}(F, G_i)$, où F est un A module quelconque, est notée $\mathrm{Pext}^n(F,M)$.

Dualement on définit les résolutions pures-projectives des modules. On a alors :

$\mathrm{Pext}^n(F,M) = 0$ pour tout $n \in \mathbb{N}$ et pour tout A-module F si et seulement si M est pur-injectif

et

$\mathrm{Pext}^n(N,F) = 0$ pour tout $n \in \mathbb{N}$ et pour tout A-module F, si et seulement si N est pur-projectif.

PROPOSITION 1. *Soient A un anneau et G un A-module pur injectif alors $\mathrm{Hom}(M,G)$ est pur injectif pour tout A-module M.*

Preuve : Soit $0 \to R \to S \to T \to 0$ une suite exacte pure, alors la suite :

$$0 \to R \otimes M \to S \otimes M \to T \otimes M \to 0$$

est exacte pure (6) donc, G étant pur injectif la suite :

$$0 \to \mathrm{Hom}(T \otimes M, G) \to \mathrm{Hom}(S \otimes M, G) \to \mathrm{Hom}(R \otimes M, G) \to 0$$

est exacte. On utilise alors l'isomorphisme 1)

$$\mathrm{Hom}(E, \mathrm{Hom}(F,G)) \longleftrightarrow \mathrm{Hom}(E \otimes F, G)$$

qui montre que la suite :

$$0 \to \mathrm{Hom}(T, \mathrm{Hom}(M,G)) \to \mathrm{Hom}(S, \mathrm{Hom}(M,G)) \to \mathrm{Hom}(R, \mathrm{Hom}(M,G)) \to 0$$

est exacte, donc que $\mathrm{Hom}(M,G)$ est pur-injectif.

PROPOSITION 2. *Le produit tensoriel de deux modules pur-projectifs est pur-projectif.*

Preuve : Un module P_1 est pur projectif si et seulement si il est facteur direct d'une somme directe de modules de présentation finie, d'après 6).

Soient donc P_1 et P_2 deux modules pur projectifs, il existe S_1 et S_2

et deux familles $(L_1)_I$ et $(L_2)_J$ de modules de présentation finie telles que :

$$P_1 \oplus S_1 = \bigoplus_{i \in I} L_{1i} \qquad P_2 \oplus S_2 = \bigoplus_{j \in J} L_{2j}$$

le produit tensoriel de deux modules de présentation finie est un module de présentation finie ; le produit tensoriel commutant avec la somme directe ; $P_1 \otimes P_2$ est facteur direct d'une somme directe de modules de présentation finie, il est donc pur-projectif.

PROPOSITION 3. *La dimension projective d'un module plat est égale à sa dimension pure projective (en particulier tout module plat de présentation finie est projectif).*

Preuve : Soit $\ldots \to L_1 \to L_o \to F \to 0$ une résolution projective du module plat F, les modules L_i étant projectifs, sont pur-projectifs. D'autre part, si K_o est le noyau de $L_o \to F \to 0$, la suite $0 \to K_o \to L_o \to F \to 0$ est pure car F est plat ; K_o est plat car L_o est plat. Par récurrence on montre ainsi facilement que la résolution projective de F est une résolution pure projective donc $\mathrm{Ext}^n(F,M) = \mathrm{Pext}^n(F,M)$ pour tout module M ; en particulier, si F est pur projectif il est projectif.

COROLLAIRE 4. *La dimension globale d'un anneau régulier au sens de Von Neumann A est égale à sa dimension pure globale* (8) .

Preuve : Tous les A-modules sont alors plats.

COROLLAIRE 5. *La dimension globale d'un anneau A est inférieure à la somme de sa dimension globale pure et de sa dimension faible*

$$g\ell \dim A \leqslant p.\ g\ell \dim A + W \dim A.$$

Preuve : Soit r la dimension faible de A (supposée finie, si $r = \infty$ le résultat est trivial). Alors tout A-module M admet au moins une résolution de longueur r telle que L_i $(0<i<r-1)$ soit projectif et P soit plat :

$$0 \to P \to L_{r-1} \to \ldots \to L_1 \to L_o \to M \to 0.$$

Alors par décalage on a :

dimension projective M = dim proj $P+r$.

On sait que (proposition 3)

dim proj P = dim pure projective P.

Si on appelle s la dimension pure globale (supposée finie de même que précédemment) de A on obtient ainsi l'inégalité recherchée.

Ce corollaire nous permet de répondre (partiellement) à une question de C.U. Jensen : les résultats obtenus jusqu'alors dans ce domaine, concernent surtout les anneaux pour lesquels les dérivés n-ièmes de $\varprojlim$ s'annulent pour $n \leqslant 3$. Existe-t-il des anneaux "pathologiques" où ces dérivés pour certains systèmes projectifs caractéristiques, s'annulent pour un n grand $(n>3)$? On sait (8) que ces anneaux doivent avoir une "grosse" dimension pure globale. S'ils existent ce seront donc des anneaux de grande dimension globale et de petite dimension faible. Barbara Osofsky a montré (9) qu'on peut construire des anneaux de valuation ayant une aussi grande dimension globale que l'on veut. Ces anneaux sont de dimension faible égale à 1 ; ce sont donc de bons exemples d'anneaux ayant une grande dimension pure globale.

Récapitulons ci-dessous ces résultats :

	gℓ. dim	p gℓ. dim	weak dim.
Absol. plat	?	n	n
Noethérien	n	?	n
Valuation	>n	> n-1	1

PROPOSITION 6. *Soient* A *un anneau,* G *un* A*-module injectif,* F *un* A*-module quelconque et* $0 \to K \to P \to F \to 0$ *une suite exacte pure. Alors la suite :*

$$0 \to Hom(F,G) \to Hom(P,G) \to Hom(K,G) \to 0$$

est une suite scindée de modules pur-injectifs.

Montrons que la suite :

① $$0 \to \mathrm{Hom}(F,G) \to \mathrm{Hom}(P,G) \to \mathrm{Hom}(K,G) \to 0$$

est pure. Elle est exacte car G est injectif. Elle est pure si et seulement si la suite :

$$\text{②} \qquad 0 \to \mathrm{Hom}\,(F,G)\otimes R \to \mathrm{Hom}\,(P,G)\otimes R \to \mathrm{Hom}\,(K,G)\otimes R \to 0$$

est exacte pour tout A-module R de présentation finie (car les limites inductives commutent aux suites exactes et aux produits tensoriels.). La suite :

$$0 \to \mathrm{Hom}\,(R,K) \to \mathrm{Hom}\,(R,P) \to \mathrm{Hom}\,(R,F) \to 0$$

est exacte car R étant de présentation finie est pur projectif. La suite :

$$\text{③} \qquad 0 \to \mathrm{Hom}(\mathrm{Hom}(R,F),G) \to \mathrm{Hom}(\mathrm{Hom}(R,P),G) \to \mathrm{Hom}(\mathrm{Hom}(R,K),G) \to 0$$

est exacte car G est injectif.

On sait que G étant injectif et R étant de présentation finie, on a :

$$\mathrm{Hom}\,(\mathrm{Hom}\,(R,F),G) \approx \mathrm{Hom}\,(F,G)\otimes R \quad \text{pour tout } F.$$

Les suites ② et ③ sont isomorphes, donc la suite ① est pure. G étant injectif, $\mathrm{Hom}\,(F,G)$ et $\mathrm{Hom}\,(P,G)$ sont pur-injectifs (proposition 1); $\mathrm{Hom}\,(F,G)$ est pur-injectif et sous-module pur d'un pur injectif, il en est donc facteur direct et la suite ① est scindée.

PROPOSITION 7. *Soient A un anneau cohérent, F un A module injectif alors $\mathrm{Hom}\,(E,F)$ est un A module plat pour tout module E sous module pur d'un injectif.*

Preuve : Soit $0 \to E \to P \to R \to 0$ une suite exacte pure où P est injectif. Montrons tout d'abord que $\mathrm{Hom}\,(P,F)$ est plat. Il suffit de montrer qu'il est plat vis à vis des modules de présentation finie. Soit donc $0 \to D \to B \to C \to 0$ une présentation finie, où B est projectif, de C. L'anneau A étant cohérent, D est de présentation finie. Il nous suffit de montrer que la suite :

$$0 \to \mathrm{Hom}\,(P,F)\otimes D \to \mathrm{Hom}\,(P,F)\otimes B \to \mathrm{Hom}\,(P,F)\otimes C \to 0$$

est exacte, soit encore, puisque D, B et C sont de présentation finie et que F

est injectif, que la suite isomorphe :

$$0 \longrightarrow \mathrm{Hom}\,(\mathrm{Hom}(D,P),F) \longrightarrow \mathrm{Hom}\,(\mathrm{Hom}(B,P),F) \longrightarrow \mathrm{Hom}\,(\mathrm{Hom}(C,P),F) \longrightarrow 0$$

est exacte.

Le module F étant injectif, il suffit de montrer que :

$$0 \longrightarrow \mathrm{Hom}\,(C,P) \longrightarrow \mathrm{Hom}\,(B,P) \longrightarrow \mathrm{Hom}\,(A,P) \longrightarrow 0$$

est exacte, ce qui est vérifié car P est injectif ; $\mathrm{Hom}\,(P,F)$ est donc plat.

Nous avons vu (prop. 4) que si $0 \to E \to P \to R \to 0$ est une suite pure, la suite :

$$0 \longrightarrow \mathrm{Hom}\,(R,F) \longrightarrow \mathrm{Hom}\,(P,F) \longrightarrow \mathrm{Hom}\,(E,F) \longrightarrow 0$$

est scindée, donc $\mathrm{Hom}\,(E,F)$ et $\mathrm{Hom}\,(R,F)$ sont plats.

Rappelons qu'un module F est dit FP injectif s'il vérifie $\mathrm{Ext}^1\,(M,F) = 0$ pour tout module M de présentation finie.

COROLLAIRE 8. *Soient A un anneau cohérent, E un A-module FP injectif, F son enveloppe injective, alors $\mathrm{Hom}\,(E,F)$ est plat.*

<u>Preuve</u> : Les modules FP injectifs sont purs dans tous les modules qui les contiennent.

PROPOSITION 9. *Soient A un anneau, F un A-module pur-injectif, G un A-module injectif, alors :*

1) $(\mathrm{Hom}\,(F,G) \otimes E)$ est isomorphe à $\mathrm{Hom}\,(\mathrm{Hom}\,(E,F),G)$ pour tout sous module pur de type fini E d'un module de présentation finie.

2) Si, de plus, l'anneau A est cohérent, alors $\mathrm{Tor}_n(\mathrm{Hom}\,(F,G),E)$ est isomorphe à $\mathrm{Hom}\,(\mathrm{Ext}^n(E,F),G)$.

<u>Preuve</u> :

1) Rappelons que cet isomorphisme est toujours vérifié quand F est quelconque et E de présentation finie (6). Soit $0 \longrightarrow E \xrightarrow{i} P \longrightarrow R \longrightarrow 0$ une

suite exacte pure où i est l'injection canonique et P est de présentation finie. Puisque E est de type fini, R est de présentation finie. Comme F est pur injectif, la suite :

$$0 \to \mathrm{Hom}\,(R,F) \to \mathrm{Hom}\,(P,F) \to \mathrm{Hom}\,(E,F) \to 0$$

est exacte et G étant injectif, la suite :

$$0 \to \mathrm{Hom}(\mathrm{Hom}(E,F),G) \to \mathrm{Hom}(\mathrm{Hom}(P,F),G) \to \mathrm{Hom}(\mathrm{Hom}(R,F),G) \to 0$$

est exacte et isomorphe à la suite exacte :

$$0 \to \mathrm{Hom}\,(F,G) \otimes E \to \mathrm{Hom}\,(F,G) \otimes P \to \mathrm{Hom}\,(F,G) \otimes R \to 0$$

car les deux derniers termes des deux suites sont isomorphes. On a donc :

$$\mathrm{Hom}\,(\mathrm{Hom}\,(E,F),G) \approx \mathrm{Hom}\,(F,G) \otimes E.$$

2) On termine comme dans (4) en prenant une résolution projective de E et en calculant l'homologie des complexes induits. En effet E est de présentation finie comme sous-module d'un module de présentation finie car A est cohérent.

PROPOSITION 10. *Soient A un anneau, E et G deux A-modules, F un A-module pur projectif, alors pour tout $n \in \mathbb{N}$ les A-modules : $\mathrm{Pext}^n\,(E \otimes F, G)$ et $\mathrm{Pext}^n\,(E, \mathrm{Hom}\,(F,G))$ sont isomorphes.*

Preuve : Soient

$$0 \to G \to Q_o \to Q_1 \to \dots \to Q_n \to \dots$$

une résolution pure injective du module G, et

$$\dots \to L_p \to \dots \to L_1 \to L_o \to E \to 0$$

une résolution pure projective du module E. Considérons le bi-complexe :

$$\mathrm{Hom}\,(L_p \otimes F, Q_q) \text{ isomorphe à } \mathrm{Hom}\,(L_p, \mathrm{Hom}\,(F,Q_q)).$$

Calcul des suites spectrales convergentes associées à ce bi-complexe :

1) Calcul de l'homologie quand q est fixé : on obtient :

$$\mathrm{Pext}^p\,(E, \mathrm{Hom}\,(F, Q_q))$$

or Q_q étant pur injectif, Hom (F,Q_q) est pur-injectif (prop. 1) il reste donc :

$$\mathrm{Hom}\,(E,\ \mathrm{Hom}\,(F,Q_q)) \simeq \mathrm{Hom}\,(E \otimes F,\ Q_q)$$

dont l'homologie en q est $\mathrm{Pext}^q\,(E \otimes F,G)$.

2) Calcul de l'homologie quand p est fixé : on obtient :

$$\mathrm{Pext}^q\,(L_p \otimes F,G)$$

or L_p étant pur-projectif ainsi que F, $L_p \otimes F$ est un module pur-projectif (prop. 2) donc on obtient :

$$\mathrm{Hom}\,(L_p \otimes F,G) \simeq \mathrm{Hom}\,(L_p,\ \mathrm{Hom}\,(F,G))$$

dont l'homologie en p est $\mathrm{Pext}^p\,(E,\ \mathrm{Hom}\,(F,G))$. Les suites spectrales associées au bi-complexe $\mathrm{Hom}\,(L_p \otimes F,Q_q)$ dégénèrent donc en les isomorphismes :

$$\mathrm{Pext}^n\,(E \otimes F,G) \simeq \mathrm{Pext}^n\,(E,\ \mathrm{Hom}\,(F,G)).$$

COROLLAIRE 10. Bis. *Soient* F *un* A*-module pur projectif et* G *un* A*-module alors* :

$$p.i\ \dim \mathrm{Hom}\,(F,G) \leq p.i\ \dim G.$$

THEOREME 11. *Soient* A *un anneau,* G *un* A*-module,* F *un* A*-module de dimension pure-projective égale à* s*,* E *un* A*-module plat de dimension pure projective finie* r*. Alors :*

1) $\mathrm{Pext}^r\,(E,\ \mathrm{Pext}^s\,(F,G)) \simeq \mathrm{Pext}^{r+s}\,(E \otimes F,G)$ *et*

2) $pp.\ \dim E \otimes F \leq r+s$.

<u>Preuve</u> : Par récurrence sur s a) Faisons $s = 1$. Soient

$$0 \to G \to Q_o \to \dots \to Q_q \to \cdots$$

une résolution pure injective de G et

$$\to L_p \to \dots \to L_1 \to L_o \to E \to 0$$

une résolution projective de E (elle est pure projective car E est plat) (prop. 3).

F étant de dimension pure projective égale à 1, il existe une suite exacte pure :

$$\text{①} \qquad 0 \to F_1 \to F_o \to F \to 0$$

où F_o et F_1 sont pur-projectifs.

Appliquons à cette suite exacte le foncteur $\mathrm{Hom}(.,G)$ il vient :

$$0 \to \mathrm{Hom}(F,G) \to \mathrm{Hom}(F_o,G) \to \mathrm{Hom}(F_1G) \to \mathrm{Pext}^1(F,G) \to 0$$

car F_o et F_1 sont pur-projectifs.

Pour p fixé, appliquons à cette suite exacte le foncteur $\mathrm{Hom}(L_p,.)$. L_p étant projectif on obtient la suite exacte :

$$\text{②} \qquad 0 \to \mathrm{Hom}(L_p,\mathrm{Hom}(F,G)) \to \mathrm{Hom}(L_p,\mathrm{Hom}(F_o,G)) \to \ldots$$
$$\to \mathrm{Hom}(L_p,\mathrm{Hom}(F_1,G)) \to \mathrm{Hom}(L_p,\mathrm{Pext}^1(F,G)) \to 0$$

Reprenons la suite exacte ① et appliquons lui le foncteur $(L_p \otimes .)$. On obtient :

$$0 \to L_p \otimes F_1 \to L_p \otimes F_o \to L_p \otimes F \to 0.$$

Appliquons à cette suite exacte pure le foncteur $\mathrm{Hom}(.,G)$ on obtient la suite exacte :

$$(3) \qquad 0 \to \mathrm{Hom}(L_p \otimes F,G) \to \mathrm{Hom}(L_p \otimes F_o,G) \to$$
$$\to \mathrm{Hom}(L_p \otimes F_1,G) \to \mathrm{Pext}^1(L_p \otimes F,G) \to 0$$

car $L_p \otimes F_o$ et $L_p \otimes F_1$ sont pur-projectifs.

Les isomorphismes des 3 premiers termes des suites exactes ② et ③ se prolongent aux dérivés ; on a :

$$\mathrm{Hom}(L_p, \mathrm{Pext}^1(F,G)) \simeq \mathrm{Pext}^1(L_p \otimes F,G).$$

Considérons alors le bi-complexe :

$$\mathrm{Hom}(L_p \otimes F, Q_q) \simeq \mathrm{Hom}(L_p, \mathrm{Hom}(F,Q_q))$$

et calculons les suites spectrales qui lui sont associées :

1) Quand q est fixé, on obtient (comme dans la proposition précédente) :

$$\text{Pext}^p \,(E,\ \text{Hom}\,(F,Q_q))$$

car la résolution projective de E est pure projective puisque E est plat.

$\text{Hom}\,(F,Q_q)$ est pur injectif, donc il ne reste que $\text{Hom}\,(E,\ \text{Hom}\,(F,Q_q))$ qui est isomorphe à $\text{Hom}\,(E \otimes F,Q_q)$ dont l'homologie en q est :

$$E_2^{pq} = \text{Pext}^q \,(E \otimes F,G).$$

2) Quand p est fixé, on obtient :

$$\text{Pext}^q \,(L_p \otimes F,G)$$

qui ne s'annule pas nécessairement car F n'est pas pur projectif. On sait que : (voir plus haut)

$$\text{Hom}\,(L_p,\ \text{Pext}^1\,(F,G)) \simeq \text{Pext}^1\,(L_p \otimes F,G).$$

Il est d'autre part, immédiat de voir que $\text{Pext}^i\,(L_p \otimes F,G) = 0$ quand $i > 1$ et que $\text{Hom}\,(L_p,\ \text{Pext}^i\,(F,G)) = 0$ quand $i > 1$ car F est de dimension pure projective égale à 1.

Les termes restants quand p est fixé sont donc :

$$\text{Hom}\,(L_p \otimes F,G)$$

dont l'homologie en p est $\text{Pext}^p\,(E,\ \text{Hom}\,(F,G))$ et

$$\text{Pext}^1(L_p \otimes F,G) \simeq \text{Hom}\,(L_p,\ \text{Pext}^1(F,G))$$

dont l'homologie en p est :

$$\text{Pext}^p\,(E,\ \text{Pext}^1\,(F,G)).$$

<u>Récapitulation</u> Les termes de la suite spectrale de 1er terme :

$$''E_2^{pq} = \text{Pext}^p\,(E,\ \text{Pext}^q(F,G))$$

sont réduits à 0 dès que $q > 1$ et on a :

$$''E_2^{pq} = \text{Pext}^p\,(E,\ \text{Pext}^q(F,G)) \Longrightarrow \text{Pext}^n(E \otimes F,G)$$

$$''E_2^{pq} = 0 \quad \text{si} \quad q \geq 2.$$

Remarquons que puisque E est plat, il revient au même d'écrire :

$$''E_2^{pq} = \mathrm{Ext}^p\ (E,\ \mathrm{Pext}^q\ (F,G)) \Longrightarrow \mathrm{Pext}^n\ (E \otimes F,G).$$

En particulier, si E est de dimension pure projective égale à r on a lorsque $n = r+2$

$$\left.\begin{array}{l} \mathrm{Pext}^{r+1}\ (E,\ \mathrm{Pext}^1\ (F,G)) = 0 \\ \mathrm{Pext}^{r+2}\ (E,\ \mathrm{Hom}\ \ (F,G)) = 0 \\ \mathrm{Pext}^{r}\ \ (E,\ \mathrm{Pext}^2(F,G)) = 0 \end{array}\right\} \quad \begin{array}{l}\text{on a aussi pour tout } i \\ \mathrm{Pext}^i(E,\mathrm{Pext}^{r+2-i}(F,G))=0\end{array}$$

Donc $\mathrm{Pext}^{r+2}\ (E \otimes F,G) = 0$ pour tout G. On obtient ainsi

<u>Si</u> F <u>est un module de dimension pure projective égale à</u> 1 <u>et</u> E <u>un module plat de dimension pure projective finie</u> r <u>alors</u> $E \otimes F$ <u>est de dimension pure projective au plus égale à</u> $r+1$ <u>c'est-à-dire</u> :

$$\text{p.p. dim } E \otimes F \leq \text{p.p. dim } E+1.$$

La suite spectrale précédente montre de plus que

$$\mathrm{Pext}^r\ (E,\ \mathrm{Pext}^1\ (F,G)) \simeq \mathrm{Pext}^{r+1}\ (E \otimes F,G)$$

(il suffit de remarquer que lorsque $n = r+1$ E_2^{pq} n'est non nulle qu'en $\begin{array}{l}p = r\\ q = 1\end{array}$ et est égale à E_∞^{pq}.

La proposition est donc vraie en $s = 1$.

b) Supposons qu'elle est vraie en $s-1$ $(s>1)$, alors :

$$\mathrm{Pext}^r\ (E,\ \mathrm{Pext}^{s-1}\ (H,G)) \simeq \mathrm{Pext}^{r+s-1}\ (E \otimes H,G) \qquad (*)$$

pour tout module H de dimension pure-projective $s-1$.

Soit F un module de dimension pure projective s et soit

$$0 \to F_s \to F_{s-1} \to \ldots \to F_1 \to F_o \to F \to 0$$

une résolution pure projective de F. Appelons H le noyau du morphisme $F_o \to F \to 0$ alors H est de dimension pure projective $(s-1)$ et vérifie $(*)$

F_o est pur projectif donc la suite exacte pure :

$$0 \to H \to F_o \to F \to 0$$

montre que $\mathrm{Pext}^s(F,G) \simeq \mathrm{Pext}^{s-1}(H,G)$, d'autre part la suite :

$$0 \to H \otimes E \to F_o \otimes E \to F \otimes E \to 0$$

est exacte donc $\mathrm{Pext}^{r+s}(E \otimes F,G) = \mathrm{Pext}^{r+s-1}(E \otimes H,G)$, l'isomorphisme (*) devient :

$$\mathrm{Pext}^r(E, \mathrm{Pext}^s(F,G)) \simeq \mathrm{Pext}^{r+s}(E \otimes F, G)$$

ce qui termine la récurrence.

La technique de récurrence est la même pour montrer que :

$$\text{pp. dim } E \otimes F \leqslant \text{pp. dim } E + \text{pp. dim } F.$$

PROPOSITION 12. *Soient A un anneau, $(G_\alpha)_{\alpha\in\mathbb{N}}$ un système projectif de modules plats et pur-injectifs dont les morphismes intermédiaires sont surjectifs. Alors la dimension pure-injective de $\varprojlim_{\alpha\in\mathbb{N}} G_\alpha$ est inférieure ou égale à 1.*

Preuve : Soient

$$\cdots \to L_p \to \cdots \to L_1 \to L_o \to M \to 0$$

une résolution pure projective de M et

$$0 \to \varprojlim_{\alpha\in\mathbb{N}} G_\alpha \to \pi^o \to \pi^1 \to \cdots \to \pi^q \to \cdots$$

le complexe de modules introduit par Jensen dans ((2) page 3.)

Considérons le bi-complexe $\mathrm{Hom}(L_p, \pi^q)$ et étudions les suites spectrales convergentes qui lui sont associées.

1) Calcul de l'homologie quand q est fixé : on obtient $\mathrm{Pext}_p(M, \pi^q)$ puisque L_p est une résolution pure-projective de M. π^q est un produit de modules pur-injectifs, c'est un module pur-injectif et les seuls termes restant sont :

Hom (M,π^q) dont on sait (2) que l'homologie est $\varprojlim_{\alpha}^{q}$ Hom (M,G_α).

2) Calcul de l'homologie quand p est fixé : on obtient

$$\varprojlim_{\alpha}^{q} \text{Hom}\,(L_p, G_\alpha) \qquad (1)$$

Montrons que le système Hom (L_p, G_α) est flasque.

$(G_\alpha)_{\alpha \in \mathbb{N}}$ est un système projectif flasque (2), donc pour tout sous-ensemble I de $\mathbb{N}$ le morphisme s : $\varprojlim_{\alpha \in \mathbb{N}} G_\alpha \longrightarrow \varprojlim_{\alpha \in I} G_\alpha$ est surjectif.

Puisque I est fini, $\varprojlim_{I} G_\alpha$ est isomorphe à G_I qui est un module plat ; la suite :

$$0 \to K \to \varprojlim_{N} G_\alpha \xrightarrow{s} \varprojlim_{I} G_\alpha \to 0$$

où K est le noyau de s, est exacte pure et puisque L_p est pur-projectif le morphisme :

$$\text{Hom}\,(L_p, \varprojlim_{\mathbb{N}} G_\alpha) \longrightarrow \text{Hom}\,(L_p, \varprojlim_{I} G_\alpha)$$

est surjectif et isomorphe au morphisme :

$$\varprojlim_{\mathbb{N}} \text{Hom}\,(L_p, G_\alpha) \to \varprojlim_{I} \text{Hom}\,(L_p, G_\alpha)$$

ce qui montre que le système projectif :

$$(\text{Hom}\,(L_p, G_\alpha))_{\alpha \in \mathbb{N}} \quad \text{est flasque.}$$

Le terme (1) se réduit donc à :

$$\varprojlim_{\alpha} \text{Hom}\,(L_p, G_\alpha) \text{ isomorphe à } \text{Hom}\,(L_p, \varprojlim_{\alpha} G_\alpha) \;;$$

l'homologie en p est donc $\text{Pext}^p\,(M, \varprojlim_{\alpha} G_\alpha)$.

Les suites spectrales convergentes associées au bi-complexe Hom (L_p, π^q) dégénèrent donc et on a pour tout $n \in \mathbb{N}$ l'isomorphisme :

$$\varprojlim_{\alpha \in \mathbb{N}}^{n} \text{Hom}\,(M, G_\alpha) \simeq \text{Pext}^n\,(M, \varprojlim_{\alpha \in \mathbb{N}} G_\alpha) \qquad (*).$$

Le système $(\mathrm{Hom}\,(M,G_\alpha))_{\alpha\in\mathbb{N}}$ étant indexé sur $\mathbb{N}$ annule tous les dérivés de $\varprojlim$ sauf $\varprojlim^1$ et $\varprojlim^0$ (2), on obtient donc :

$$\text{pi. dim} \varprojlim G_\alpha \leq 1.$$

L'isomorphisme (*) montre de plus que $\varprojlim^1 \mathrm{Hom}\,(M,G_\alpha) \simeq \mathrm{Pext}^1(M,\varprojlim G_\alpha)$.

RAPPEL. Un anneau commutatif est dit héréditaire si :

- Tout sous-module d'un module projectif est projectif.
- Tout quotient d'un module injectif est injectif.

On montre (4) qu'un anneau commutatif est héréditaire si et seulement si sa dimension globale est inférieure ou égale à 1.

La proposition suivante montre qu'on peut définir de façon similaire des anneaux pur-héréditaires dont la dimension globale pure est inférieure ou égale à 1.

PROPOSITION 13. *Soit A un anneau, tout sous-module pur d'un pur projectif est pur projectif et tout quotient "pur" d'un pur injectif est pur injectif si et seulement si la dimension globale pure de l'anneau est* ≤ 1.

Preuve : Soit A un anneau, Q un module quelconque, alors il existe un module P pur projectif dont Q est quotient et tel que la suite exacte

$$0 \longrightarrow J \longrightarrow P \xrightarrow{s} Q \longrightarrow 0$$

soit pure (J = ker s).

Appliquons le foncteur Hom (.,L) où L est quelconque, on obtient :

$$0 \longrightarrow \mathrm{Hom}\,(Q,L) \longrightarrow \mathrm{Hom}\,(P,L) \longrightarrow \mathrm{Hom}\,(J,L) \longrightarrow$$

$$\longrightarrow \mathrm{Pext}^1(Q,L) \longrightarrow \mathrm{Pext}^1\,(P,L) \longrightarrow \mathrm{Pext}^1\,(J,L) \longrightarrow$$

$$\mathrm{Pext}^1\,(P,L) = 0$$

$$\longrightarrow \mathrm{Pext}^2(Q,L) \longrightarrow \mathrm{Pext}^2\,(P,L) \ldots$$

$$\mathrm{Pext}^2\,(P,L) = 0$$

donc :

$$\mathrm{Pext}^1\,(J,L) \simeq \mathrm{Pext}^2\,(Q,L)$$

$$\mathrm{Pext}^i\,(J,L) \simeq \mathrm{Pext}^{i+1}(Q,L)$$

ou Q est un module quelconque et J est sous-module pur d'un pur projectif, ce qui montre que J est pur projectif si et seulement si la dimension globale pure de $A \leq 1$; de même tout module J est sous module pur d'un pur injectif et on obtient l'autre partie de la proposition.

<u>EXEMPLES</u> d'anneaux <u>pur-héréditaires</u> : Jensen a montré (8) que tout anneau dénombrable est de dimension globale pure ≤ 1 ; en particulier les anneaux de groupe $Z(\pi)$ ou π est un groupe fini sont donc pur-héréditaires.

Nous avions montré dans (5) que lorsqu'un anneau A est cohérent auto FP injectif, il est possible d'obtenir des résultats sur la commutativité du foncteur Tor et du foncteur $\varprojlim$. Le théorème démontré faisait appel à une définition de dimension injective des systèmes projectifs ; cette notion est difficile à manier, très abstraite ; nous donnons maintenant, lorsque le système projectif est indexé sur $\mathbb{N}$, une démonstration n'utilisant pas cette notion. Elle n'est hélas pas, a priori, généralisable, même à $\mathbb{R}$.

PROPOSITION 14. *Soient A cohérent auto-FP injectif et* $(P_\alpha)_{\alpha \in \mathbb{N}}$ *un système projectif de modules injectifs, alors ; pour tout A-module M de présentation finie :*

1) $M \otimes \varprojlim_{\mathbb{N}}^1 (P_\alpha)$ *est isomorphe à* $\varprojlim_{\mathbb{N}}^1 (M \otimes P_\alpha)$

2) $M \otimes \varprojlim (P_\alpha)$ est isomorphe à $\varprojlim (M \otimes P_\alpha)$ ssi $\varprojlim^1 P_\alpha$ est plat.

<u>Preuve</u> : Montrons tout d'abord un <u>lemme</u> : <u>tout système projectif $(F_\alpha)_{\alpha \in \mathbb{N}}$ flasque</u> (i.e dont les morphismes intermédiaires sont surjectifs) <u>de modules injectifs a pour limite un module plat</u>. (l'anneau A comme dans le texte de la proposition 14).

On considère pour tout module M de présentation finie, la suite spectrale convergente de deuxième terme :

$$\mathrm{Ext}^p (M, \varprojlim_{\mathbb{N}}{}^q F) \Longrightarrow \varprojlim_{\mathbb{N}}{}^p (\mathrm{Ext}^q (M,F_\alpha)).$$

F_α étant injectif $\forall a \in \mathbb{N}$ et $(F_\alpha)_{\alpha \in \mathbb{N}}$ étant flasque, elle dégénère en les isomorphismes :

$$\mathrm{Ext}^n (M, \varprojlim_{\mathbb{N}} F_\alpha) \simeq \varprojlim_{\mathbb{N}}{}^n \mathrm{Hom} (M,F_\alpha) \qquad (*)$$

Le module M étant de présentation finie est pur projectif. Le système $(F_\alpha)_{\alpha\in\mathbb{N}}$ étant flasque, ses morphismes intermédiaires $f_{\alpha\beta}$ $(\beta\geq\alpha)$ sont surjectifs. Les modules F_α $(\alpha \in \mathbb{N})$ étant injectifs sont plats (5) les suites exactes :

$$0 \longrightarrow \ker (f_{\alpha\beta}) \longrightarrow F_\beta \longrightarrow F_\alpha \longrightarrow 0$$

sont donc pures et le foncteur Hom (M,.) conserve leur exactitude.

Les morphismes $\mathrm{Hom} (M,F_\beta) \longrightarrow \mathrm{Hom} (M,F_\alpha)$ déduits de $f_{\alpha f}$ $(\forall(\alpha,\beta)\in \mathbb{N}^2$ $\beta\geq\alpha)$ sont donc surjectifs, le système projectif $(\mathrm{Hom}\ M, F_\alpha)_{\alpha \in \mathbb{N}}$ est flasque et annule les dérivés de $\varprojlim$. Les isomorphismes (*) montrent que $\varprojlim_{\mathbb{N}} F_\alpha$ est FP-injectif donc plat. (5).

Soit maintenant un système projectif $(P_\alpha)_{\alpha\in\mathbb{N}}$ non nécessairement flasque de modules injectifs. D'après Jensen (4) (P_α) admet une résolution flasque de longueur au plus 1 :

$$0 \longrightarrow (P_\alpha) \longrightarrow (F^o_\alpha) \longrightarrow (F^1_\alpha) \longrightarrow 0 \qquad (1)$$

l'anneau A étant cohérent, M admet une résolution projective par des modules libres de type fini :

$$\dots \longrightarrow L_n \longrightarrow \dots \longrightarrow L_1 \longrightarrow L_o \longrightarrow M \longrightarrow 0 \qquad (2)$$

La résolution (1) étant bornée à droite et à gauche, la suite spectrale associée au bicomplexe $B^{pq} = L_{-p} \otimes \varprojlim F^q$ est convergente. Comme dans (3) on calcule les 2èmes termes de la suite spectrale en remarquant que d'après ce qui précède $\varprojlim F^o_\alpha$ et $\varprojlim F^1_\alpha$ sont des modules plats. On obtient :

1) $M \otimes \varprojlim^1 P_\alpha$ est isomorphe à $\varprojlim^1 (M \otimes P_\alpha)$ et la suite :

$$0 \longrightarrow \mathrm{Tor}_2 (M, \varprojlim_{\mathbb{N}}{}^1 P_\alpha) \longrightarrow M \otimes \varprojlim_{\mathbb{N}} P_\alpha \longrightarrow \varprojlim_{\mathbb{N}} (M \otimes P_\alpha) \longrightarrow$$

$$\longrightarrow \mathrm{Tor}_1 (M, \varprojlim_{\mathbb{N}}{}^1 P_\alpha) \longrightarrow 0 \quad \text{est exacte donc :}$$

2) $M \otimes \varprojlim_{\mathbb{N}} P_\alpha$ est isomorphe à $\varprojlim_{\mathbb{N}} (M \otimes P_\alpha)$ si et seulement si $\varprojlim^1 P_\alpha$ est plat.

BIBLIOGRAPHIE

(1) B. OSOFSKY, *The subscript of* $\aleph_n$ *, projective dimension, and the vanishing of* $\varprojlim^n$. *Bull. of the Amer. Math. Society. Vol. 80, number 1, January 1974.*

(2) C.U. JENSEN, *Les foncteurs dérivés de* $\varprojlim$, *Lecture Notes in Math. n°254.*

(3) J.L. PEZENNEC, *Propriétés topologiques de* $[X,Y]$, *Bull. Soc. Math. de Fr. 107. 1979 p. 113-126.*

(4) CARTAN, *Homological Algebra, Princeton.*

(5) D. SALLES, *Dualisation de la platitude, Lecture Notes n° 795.*

(6) BOURBAKI, *Algèbre linéaire (ch. 2).*

(7) WARFIELD, *Purity and algebraic compactness, Pac J. of Math 28 n°3 (1969) p. 699-719.*

(8) JENSEN, *Dimensions cohomologiques reliées aux foncteurs* $\varprojlim^i$ *, à paraître aux Lectures Notes in Math.*

(9) B. OSOFSKY, *Globaledimension of valuation rings, Trans Am Math Soc. (1967) 127.*

UNIVERSITE DE CAEN
U.E.R. de Sciences
Département de Mathématiques
Pures.

ANNEAUX FILTRES COMPLETS ET SUITES SPECTRALES ASSOCIEES

par Elena Wexler-Kreindler

Nous proposons dans cette note une étude de certaines propriétés homologiques des anneaux filtrés complets, à partir de moyens étroitement liés à leur structure. Ceci permet d'étendre certains des résultats exposés dans [2,ch.2] à d'autres anneaux filtrés que ceux noethériens,ainsi qu'à des modules filtrés qui ne sont pas de type fini.

Dans le §.1 nous décrivons le gradué associé au groupe filtré $\mathrm{Hom}_f(M,N)$ des f-morphismes de A-modules filtrés $M \longrightarrow N$ sur un anneau filtré complet A (avec filtration décroissante), ainsi que le groupe gradué $\mathrm{HOM}_{grA}(grM,grN)$ engendré par les morphismes gradués $grM \longrightarrow grN$ de degré p, $p\in\mathbb{N}$, sur l'anneau gradué associé grA. Le théorème 1.7 décrit les foncteurs dérivés $\mathrm{EXT}^n_{grA}(.,grN)$ du foncteur $\mathrm{HOM}_{grA}(.,grN)$.

Dans le §.2 nous établissons l'indépendance des foncteurs dérivés du foncteur $\mathrm{Hom}_f(.,N)$, défini à partir des résolutions f-projectives, définies dans [7] (théorème 2.3), ainsi que quelques unes de ses propriétés.

A toute résolution f-libre $\mathbb{L} \longrightarrow M$ d'un A-module filtré M, on peut associer le complexe filtré $\mathbb{L}^* = (\mathrm{Hom}_f(L_n,I))_{n\in\mathbb{N}}$, où I est un idéal bilatère fermé de A (§.3), auquel nous associons une suite spectrale de cohomologie, dont le premier terme est $\bigoplus_{n\in\mathbb{N}} \mathrm{EXT}^n_{grA}(grM,grI)$ (proposition 3.3). Lorsque A est linéairement compact à droite,le complexe filtré $\mathbb{L}^*$ est complet et la suite spectrale associée converge, dans le sens de [4,ch.XI,§.8] vers le module différentiel gradué $\bigoplus_{n\in\mathbb{N}} \mathrm{Ext}^n_f(M,I)$, muni d'une filtration décroissante qui est séparée,mais qui n'est pas cobornée dans le sens de [4, ch. XI, §.8], (théorème 3.5).

Une étude des suites spectrales associées aux complexes filtrés complets, à l'aide des couples dérivés de Massey, se trouve dans [3]. Des suites spectrales utilisées dans l'étude des modules de type fini sur des anneaux filtrés noethériens (avec "bonnes" filtrations croissantes) peuvent être trouvées dans [2, ch.2]. Notons que toutes ces suites spectrales proviennent de modules différentiels gradués, munis de filtrations croissantes (décroissantes) qui sont bornées (resp. cobornées) et c'est justement cette propriété qui permet, essentiellement leur convergence. Ceci n'est pas le cas, en général, dans l'étude que nous proposons.

Pour les questions concernant les anneaux et modules filtrés et gradués, nous renvoyons le lecteur à [1] et [5] et pour les questions de homologie à [4] et [6]. Les notions de modules f-libre et f-projectif, ainsi que celle de résolution f-projective, ont été introduiteset étudiées par l'auteur dans [7,8,9].

1°) Foncteur Hom_f et gradués associés.

Dans tout ce qui suit, A désigne un anneau unitaire, muni d'une filtration $\{F^pA\}_{p\in\mathbb{N}}$ décroissante, exhaustive et séparée, pour laquelle A est un anneau filtré complet. Tout A-module filtré M est muni d'une filtration $\{F^pM\}_{p\in\mathbb{N}}$ décroissante, exhaustive, compatible avec celle de A et séparée, si rien d'autre n'est mentionné. La fonction d'ordre associée à une filtration sera désignée par ω . Un sous-module $N\subseteq M$ sera muni de la filtration induite ($F^pN = N\cap F^pM$, $\forall p\in\mathbb{N}$) et le module quotient M/N de la filtration quotient ($F^p(M/N) = (F^pM+N)/N \simeq F^pM/F^pN$, $\forall p\in\mathbb{N}$) qui est séparée ssi N est un sous-module fermé de M. Si rien d'autre n'est indiqué, tous les modules sont des A-modules à gauche.

L'anneau gradué associé sera désigné par grA et le grA-module gradué associé, par grM. On utilisera les notations :

$$grA = \bigoplus_{p=o}^{\infty} gr_pA,\ gr_pA = F^pA/F^{p+1}A$$

et des notations analogues pour grM. La forme principale d'un élément $x\in M$ est la classe $\bar{x}\in grM$ de x modulo $F^{\omega(x)+1}M$.

Pour tout entier $k\in\mathbb{N}$, $M_{(k)}$ va désigner le A-module filtré F^kM, muni de la filtration induite : $F^pM_{(k)} = F^{p+k}M$, $p\in\mathbb{N}$.

Rappelons qu'une application A-linéaire $\varphi: M \longrightarrow N$, où M et N sont des A-modules filtrés, est un f-morphisme (f-morphisme strict) si pour tout

$p\in\mathbb{N}$, $\varphi(F^pM)\subseteq F^pN$ ($\varphi^pM = F^pN\cap\varphi(M)$, respectivement). Sur le groupe abélien $\mathrm{Hom}_f(M,N)$ des f-morphismes de A-modules à gauche filtrés $M \longrightarrow N$ on définit la filtration, en posant pour tout $p\in\mathbb{N}$:

$F^p\mathrm{Hom}_f(M,N) = \{\varphi\in\mathrm{Hom}_f(M,N) \mid \varphi(F^nM)\subseteq F^{n+p}N$, quel que soit $n\in\mathbb{N}\}$,

qui est séparée, puisque N l'est. Si l'on suppose en plus que N est un A-module à droite et que sa filtration est compatible à droite avec celle de A (i.e. si N est un A-<u>bimodule filtré</u>), alors $\mathrm{Hom}_f(M,N)$ est un A-module à droite filtré. Dans ces conditions, pour tout f-morphisme $\psi: M \longrightarrow M'$ de A-modules à gauche filtrés, l'application $\varphi \longmapsto \varphi\circ\psi$, $\varphi\in\mathrm{Hom}_f(M',N)$ est un f-morphisme de A-modules à droite filtrés :

$$\mathrm{Hom}_f(\psi,N) : \mathrm{Hom}_f(M',N) \longrightarrow \mathrm{Hom}_f(M,N)$$

et $\mathrm{Hom}_f(.,N)$ est un foncteur contrevariant de la catégorie (préabélienne) des A-modules à gauche filtrés dans la catégorie des A-modules à droite filtrés. En particulier, ceci est vrai pour tout idéal bilatère de A.

<u>Lemme 1.1 Soit</u> M <u>et</u> M' <u>des A-modules filtrés,</u> N <u>un</u> A-<u>bimodule filtré et</u> $\psi\in\mathrm{Hom}_f(M,M')$ <u>un f-morphisme surjectif strict. Alors</u> $\mathrm{Hom}_f(\psi,N)$ <u>est un f-morphisme injectif strict de A-modules à droite filtrés et</u> $\mathrm{Hom}_f(M',N)$ <u>est f-isomorphe à un sous-module fermé de</u> $\mathrm{Hom}_f(M,N)$.

<u>Preuve</u> : L'application injective $\mathrm{Hom}_f(\psi,N)$ associe à chaque f-morphisme $\varphi\in\mathrm{Hom}_f(M',N)$, le f-morphisme de même ordre $\varphi\circ\psi\in\mathrm{Hom}_f(M,N)$, car $\forall k\in\mathbb{N}$, $\psi(F^kM) = F^kM'$ et

$$(\varphi\circ\psi)(F^kM) = \varphi(\psi(F^kM)) = \varphi(F^kM')\subseteq F^{k+\omega(\varphi)}M'.$$

Soit $(\varphi_n)_{n\in\mathbb{N}}$ une suite d'éléments de $\mathrm{Hom}_f(M',N)$, telle que la suite $(\varphi_k\circ\psi)_{k\in\mathbb{N}}$ converge vers $\alpha\in\mathrm{Hom}_f(M,N)$. Alors pour tout $y\in M'$, la suite $(\varphi_k(y))_{k\in\mathbb{N}}$ converge dans N vers $\alpha(x)$, où $x\in M$ et $\psi(x) = y$. L'application $\varphi: M' \longrightarrow N$, $\varphi(y) = \lim_k \varphi_k(y)$, $\forall y\in M'$, est A-linéaire et $\alpha = \varphi\circ\psi$. Il reste à montrer que $\varphi\in\mathrm{Hom}_f(M,N)$. En effet, si $y\in M$ et $\varphi(y)\neq 0$, de $\omega(\varphi(y)-\varphi_k(y))\longrightarrow\infty$ on déduit qu'il existe un entier k_o, tel que pour $k\geq k_o$, $\omega(\varphi(y)-\varphi_k(y))\geq\omega(\varphi_k(y))$, car $\{\omega(\varphi_k(y)\}_{k\in\mathbb{N}}$ est une suite bornée. Par suite, pour $k\geq k_o$,

$$\omega(\varphi(y))\geq\inf\{\omega(\varphi_k(y)),\omega(\varphi(y)-\varphi_k(y)\}\geq\omega(\varphi_k(y))\geq\omega(y),$$

donc φ est un f-morphisme, ce qui achève la démonstration.

Notons que l'ordre d'un f-morphisme surjectif strict $\neq 0$ est 0.

Lemme 1.2 Soit $\{M_\lambda\}_{\lambda\in\Lambda}$ une famille de A-modules filtrés complets. Alors le produit $M = \prod_{\lambda\in\Lambda} M_\lambda$, muni de la filtration produit $F^pM = \prod_{\lambda\in\Lambda} F^pM_\lambda$, $\forall p \in \mathbb{N}$, est un A-module filtré complet.

Preuve : Vérification standard de l'isomorphisme $M \simeq \varprojlim_p M/F^pM$ en utilisant que chaque M_λ est complet.

Un A-module filtré est f-libre s'il existe un système d'éléments $\{x_\lambda\}_{\lambda\in\Lambda}$ de L appelé f-base, dont les formes principales $\{\bar{x}_\lambda\}_{\lambda\in\Lambda}$ constituent une base homogène du grA-module gradué grL. Un A-module f-projectif est un A-module filtré dont le gradué associé est projectif (v. [7]). Rappelons que tout f-morphisme $L \longrightarrow N$ est uniquement défini sur une f-base de L.

Lemme 1.3 Soit L un A-module f-libre, $\{x_\lambda\}_{\lambda\in\Lambda}$ une f-base de L et soit N un A-module filtré qui est supposé complet si Λ n'est pas fini. Le groupe abélien $Hom_f(L,N)$ est un A-module à gauche filtré si, pour $a \in A$ et $\varphi \in Hom_f(L,N)$, on définit $\forall\lambda\in\Lambda$, $(a.\varphi)(x_\lambda) = a\varphi(x_\lambda)$ et il y a un f-morphisme de A-modules à gauche filtrés

$$Hom_f(L,N) \simeq \prod_{\lambda\in\Lambda} N_{(\omega(x_\lambda))},$$

qui est aussi un f-isomorphisme de A-modules à droite filtrés, si N est en plus un A-bimodule filtré.

Preuve : Puisque N est complet, toute famille $(y_\lambda)_{\lambda\in\Lambda}$ d'éléments de N définit de manière unique un f-morphisme $\varphi \in Hom_f(L,N)$ avec $\varphi(x_\lambda) = y_\lambda$, $\forall\lambda\in\Lambda$. Le reste est vérification standard (v. [8] et [9]).

Proposition 1.4 Soit M un A-module à gauche filtré et soit N un A-bimodule filtré complet. Alors $Hom_f(M,N)$ est un A-module à droite filtré complet.

Preuve : Il existe un A-module à gauche f-libre L et un f-morphisme surjectif strict $\alpha : L \twoheadrightarrow M$ [8, corollaire de la proposition 3.5]. Par le lemme 1.1, $Hom_f(M,N)$ est f-isomorphe à un sous-module fermé de $Hom_f(L,N)$ qui est complet par le lemme 1.3 et le lemme 1.2. Alors $Hom_f(M,N)$ est complet.

Considérons maintenant deux A-modules à gauche filtrés M et N. Par $Hom_{grA}(grM,grN)$ on désigne le groupe des applications grA-linéaires $grM \longrightarrow grN$. Pour tout entier $k \geq 0$, soit :

$$HOM_{grA}(grM,grN)_p$$

le sous-groupe de $Hom_{grA}(grM,grN)$ des morphismes homogènes de degré p, i.e. des applications grA-linéaires φ: grM $\longrightarrow$ grN, telles que, pour tout $n\in\mathbb{N}$,

$$\varphi(gr_n M) \subseteq gr_{n+p} N.$$

La somme de ces sous-groupes est directe et on pose

$$HOM_{grA}(grM,grN) = \bigoplus_{p=o}^{\infty} HOM_{grA}(grM,grN)_p ,$$

qui est un groupe abélien gradué plongé dans $Hom_{grA}(grM,grN)$. On appelle les éléments de $HOM_{grA}(grM,grN)$ encore des morphismes gradués. Les isomorphismes sont des morphismes de degré O. Les foncteurs dérivés sont notés par $EXT^n_{grA}(.,grN)$.

Sur le grA-module grM on peut définir la filtration $\{\Gamma^n grM\}_{n\in\mathbb{N}}$, $\Gamma^n grM = \bigoplus_{p=n}^{\infty} gr_p M$. Posons $(grM)_{(n)} = \Gamma^n M$, qui est un groupe gradué.

Proposition 1.5 Soit M et L deux A-modules filtrés, L f-libre, $\{x_\lambda\}_{\lambda\in\Lambda}$ une f-base de L et $\bar{x}_\lambda$ la forme principale de x_λ. Alors :

a) $grL = \bigoplus_{\lambda\in\Lambda} (grA)\,\bar{x}_\lambda$.

b) L'application $\varphi \longmapsto (\varphi(\bar{x}_\lambda))_{\lambda\in\Lambda}$, $\varphi\in HOM_{grA}(grL,grM)$ est un isomorphisme de groupes gradués

$$HOM_{grA}(grL,grM) \simeq \prod_{\lambda\in\Lambda} (grM)_{\omega(x_\lambda)} .$$

c) Soit $g\in Hom_f(L,M)$ un f-morphisme et soit $\bar{g}\in gr\, Hom_f(L,M)$ la forme principale de g. Si $\varphi_{\bar{g}}\in HOM_{grA}(grL,grM)$ est le morphisme gradué, tel que

$$\forall\lambda\in\Lambda,\ \varphi_{\bar{g}}(\bar{x}_\lambda) = \widetilde{g(x_\lambda)} = \text{classe de } g(x_\lambda) \text{ modulo } F^{\omega(g)+\omega(x_\lambda)+1}M,$$

alors $\varphi_{\bar{g}}$ est un morphisme gradué homogène, dont le degré est l'ordre de g dans $Hom_f(L,M)$. L'application $\bar{g} \longmapsto \varphi_{\bar{g}}$ définit un morphisme gradué homogène de degré o et injectif de groupes gradués

$$gr(Hom_f(L,M)) \hookrightarrow HOM_{grA}(grL,grM) ,$$

qui est un isomorphisme si M est complet.

Si, en plus, on suppose que M est un A-bimodule filtré , les isomorphismes b) et c) sont des isomorphismes de grA-modules à droite gradués.

Preuve : a) C'est un corollaire du [7, théorème 2.2].

b) De a) on déduit que tout morphisme homogène $\varphi\in HOM_{grA}(grL,grM)$ est défini de manière unique par ses valeurs sur $\{\bar{x}_\lambda\}_{\lambda\in\Lambda}$. Si $\varphi(\bar{x}_\lambda)\neq o$ alors, en désignant par ∂ les fonctions degrés dans les groupes gradués respectifs, on a

$$\partial(\varphi(\bar{x}_\lambda)) = \partial(\varphi) + \partial(\bar{x}_\lambda) = \partial(\varphi) + \omega(x_\lambda)$$

d'où on déduit l'isomorphisme des groupes gradués

$$HOM_{grA}(grA\bar{x}_\lambda,\ grM) \simeq (grM)_{\omega(x_\lambda)}\ ,$$

la graduation de $grA\bar{x}_\lambda$ étant celle induite de grL. On conclut alors à l'isomorphisme énoncé.

c) Soit g,h deux éléments de $Hom_f(L,M)$, d'ordre p. Alors $\partial(\bar{g}) = \partial(\bar{h}) = p$. Supposons $\bar{g} = \bar{h}$, par suite $h = g+f$, où $\bar{f} = \bar{o}$, donc $\omega(f) \geq p+1$. Soit x_λ un élément de la f-base de L. Alors $\omega(f(x_\lambda)) \geq \omega(f) + \omega(x_\lambda)$, par suite $\widetilde{g(x_\lambda)} = \widetilde{h(x_\lambda)}$, $\forall\lambda\in\Lambda$ et $\varphi_{\bar{g}} = \varphi_{\bar{h}}$. Notons que par a), $\varphi_{\bar{g}}$ est une application linéaire bien définie si l'on pose, pour une somme finie $\Sigma\,\bar{a}_\lambda\bar{x}_\lambda$

$$\varphi_{\bar{g}}(\Sigma\,\bar{a}_\lambda\bar{x}_\lambda) = \Sigma\,\bar{a}_\lambda\,\widetilde{g(x_\lambda)},$$

où $\bar{a}_\lambda$ sont des éléments $\neq o$, homogènes de grA. Soit $\Sigma\,\bar{a}_\lambda\bar{x}_\lambda \in gr_k L$. Alors $\partial(\bar{a}_\lambda\bar{x}_\lambda) = \partial(\bar{a}_\lambda) + \partial(\bar{x}_\lambda) = k$. Supposons $\widetilde{g(x_\lambda)} \neq o$, donc $\omega(g(x_\lambda)) \leq p+\omega(x_\lambda)$ et puisque $\omega(g(x_\lambda)) \geq \omega(g) + \omega(x_\lambda) = p + \omega(x_\lambda)$, on déduit $\omega(g(x_\lambda)) = p+\omega(x_\lambda) = \partial(\widetilde{g(x_\lambda)})$. Alors $\bar{a}_\lambda.\widetilde{g(x_\lambda)} \in gr_{p+\omega(x_\lambda)+\partial(\bar{a}_\lambda)}M = gr_{p+k}M$. Par suite $\varphi_{\bar{g}}(gr_k L) \subseteq gr_{k+p}L$ et $\varphi_{\bar{g}}$ est un morphisme homogène gradué de même degré que $\bar{g}$.

On définit une application additive

$$\psi:\ gr(Hom_f(L,M)) \longrightarrow HOM_{grA}(grL,grM),$$

si l'on pose $\psi(\sum_{i=1}^{s}\bar{g}_i) = \sum_{i=1}^{s}\varphi_{\bar{g}_i}$, qui est un morphisme homogène de degré o gradué. Si $\sum_{i=1}^{s}\varphi_{\bar{g}_i} = o$, alors $\forall\lambda\in\Lambda$, $\sum_{i=1}^{s}\widetilde{g_i(x_\lambda)} = o$. Puisque $\bar{x}_\lambda$ est homogène on aura, pour tous les $\bar{g}_j$ ayant même degré, $\widetilde{\Sigma\,g_j(x_\lambda)} = o$, $\forall\lambda\in\Lambda$. Si $\bar{h} = \Sigma\,\bar{g}_j \neq o$, alors $\partial(\bar{h}) = \partial(\bar{g}_j)$ et $\varphi_{\bar{h}} = \psi(\Sigma\,\bar{g}_j)$ est de même degré que $\bar{h}$, donc n'est pas nulle, or $\varphi_{\bar{h}}(\bar{x}_\lambda) = o$, $\forall\lambda\in\Lambda$. On déduit que ψ est injective.

Soit maintenant M complet et $\varphi\in HOM_{grA}(grL,grM)$ un morphisme homogène gradué de degré k. Si $\varphi(\bar{x}_\lambda) \neq o$, alors $\partial(\varphi(\bar{x}_\lambda)) = k + \partial(\bar{x}_\lambda)$. Soit $y_\lambda\in M$, $\omega(y_\lambda) = \partial(\varphi(x_\lambda))$ dont la partie principale est $\varphi(\bar{x}_\lambda)$. Il existe $y_\lambda\in M$, tel que $\omega(y_\lambda) = \partial(\varphi(\bar{x}_\lambda))$. Puisque M est complet, il existe $g\in Hom_f(L,M)$, tel que $\forall\lambda\in\Lambda$, $g(x_\lambda) = y_\lambda$ si $\varphi(x_\lambda) \neq o$ et $g(x_\lambda) = o$ si $\varphi(x_\lambda) = o$, par [8, proposition 3.5]. On a $\widetilde{g(x_\lambda)} = \varphi(\bar{x}_\lambda)$ et par le f-isomorphisme du lemme 1.3, on obtient $\omega(g) = \inf_\lambda \omega(y_\lambda)$. Soit $\bar{g}$ la forme principale de g. Alors $\varphi_{\bar{g}} = \varphi$ et ψ est surjectif.

Les autres affirmations sont immédiates.

Soit M un A-module à gauche filtré. Il existe alors [7] une résolution f-libre strictement exacte de M :

$$(\mathbb{L} \longrightarrow M) : \dots \to L_n \xrightarrow{d_n} L_{n-1} \to \dots \to L_1 \xrightarrow{d_1} L_o \xrightarrow{\varepsilon} M \to 0 .$$

Pour tout A-bimodule filtré complet N et pour tout entier $n \in \mathbb{N}$, posons :

$$L_n^* = \mathrm{Hom}_f(L_n, N) , \ d_n^* = \mathrm{Hom}_f(d_n, N) ;$$

pour les gradués associés posons :

$$(grL_n)^* = \mathrm{HOM}_{grA}(grL_n, grN), (grd_n)^* = \mathrm{HOM}_{grA}(grd_n, grA).$$

Avec ces notations on a les résultats suivants.

<u>Proposition</u> 1.6 <u>La suite de grA-modules gradués et de morphismes de modules gradués</u>

$$gr(\mathbb{L} \to M) : \dots \to grL_n \xrightarrow{grd_n} grL_{n-1} \to \dots \to grL_1 \xrightarrow{grd_1} grL_o \xrightarrow{gr\varepsilon} grM \to 0$$

<u>est une résolution graduée libre du grA-module gradué</u> grM.

<u>Preuve</u> : Puisque L_n est f-libre, grL_n est un grA-module libre. Soit $d_n = i_n \circ \theta_n \circ p_n$ la décomposition canonique. Puisque tous les f-morphismes sont stricts, par [1, p.25, proposition 2] , on déduit que $gr\varepsilon$ est surjectif, θ_n est un f-isomorphisme, p_n un f-morphisme surjectif strict dont le noyau est $\mathrm{Ker}\, d_n = \mathrm{Im}\, d_{n+1} = d_{n+1}(L_{n+1})$ et par suite $gr(p_n)$ est surjectif, $gr(i_n)$ est injectif et $gr(\theta_n)$ un isomorphisme gradué, $grd_n = gri_n \circ gr\theta_n \circ grp_n$. On a la suite strictement exacte :

$$0 \to d_{n+1}(L_{n+1}) = \mathrm{Ker}\, p_n = \mathrm{Ker}\, d_n \xrightarrow{i_{n+1}} L_n \xrightarrow{p_n} p_n(L_n) \to 0 ,$$

où $p_n(L_n) \simeq L_n/\mathrm{Im}\, d_{n+1} \simeq d_n(L_n)$. Alors $d_{n+1}(L_{n+1})$ est muni de la filtration de L_n et $p_n(L_n)$ de la filtration quotient de $L_n/\mathrm{Im}\, d_{n+1}$, d'où l'on obtient pour tout entier n, la suite exacte graduée :

$$0 \to gr\, d_{n+1}\, L_{n+1} \xrightarrow{gr(i_n)} gr\, L_n \xrightarrow{gr\, p_n} gr\, p_n(L_n) \to 0$$

et les égalités $\mathrm{Ker}\, gr\, p_n = gr\, d_{n+1}(L_{n+1}) = \mathrm{Ker}\, gr\, d_n$. Puisque $gr\, p_n \circ gr\, i_{n+1} = 0$, on a l'inclusion

$$(grd_{n+1})(grL_{n+1}) \subseteq (grd_{n+1})(L_{n+1}).$$

Si $x \in L_{n+1}$ et $d_{n+1}(x) \in F^pL_n \setminus F^{p+1}L_n$, la forme principale $\overline{d_{n+1}(x)} \in \dfrac{F^pL_n}{F^{p+1}L_n}$.

Puisque d_{n+1} est strict, il existe $y \in F^pL_{n+1} \setminus F^{p+1}L_{n+1}$, tel que $d_{n+1}(y) = d_{n+1}(x)$, d'où $\overline{d_{n+1}(y)} = \overline{d_{n+1}(x)}$. Alors

$(\mathrm{grd}_{n+1})(\bar{y}) = \overline{d_{n+1}(y)} \in (\mathrm{grd}_{n+1})(\mathrm{grL}_{n+1})$. On déduit l'égalité $\mathrm{Kergr}_p = (\mathrm{grd}_{n+1})(\mathrm{grL}_n)$.

<u>Théorème</u> 1.7 <u>Les complexes de</u> grA-<u>modules à droite gradués sont isomorphes</u> :

$$\mathrm{gr}\mathbb{L}^* : 0 \longrightarrow \mathrm{grL}_o^* \xrightarrow{\mathrm{gr}(d_1^*)} \mathrm{grL}_1^* \xrightarrow{(\mathrm{grd}_2)^*} \mathrm{grL}_2^* \rightarrow \cdots$$

$$(\mathrm{gr}\mathbb{L})^* : 0 \longrightarrow (\mathrm{grL}_o)^* \xrightarrow{(\mathrm{grd}_1)^*} (\mathrm{grL}_1)^* \xrightarrow{(\mathrm{grd}_2)^*} (\mathrm{grL}_2^*) \longrightarrow \cdots$$

<u>et pour tout entier</u> $n \in \mathbb{N}^*$ <u>il y a un isormorphisme de</u> grA-<u>module gradués</u>

$$\mathrm{EXT}^n_{\mathrm{grA}}(\mathrm{grM},\mathrm{grN}) \simeq \frac{\mathrm{Kergrd}^*_{n+1}}{\mathrm{Imgrd}^*_n} .$$

<u>Preuve</u> : Montrons que pour tout entier $n \in \mathbb{N}^*$, les isomorphismes $j_n : \mathrm{grL}_n^* \longrightarrow (\mathrm{grL}_n)^*$ définis dans la proposition 1.5, partie c), sont tels que

$$j_n \circ \mathrm{grd}_n^* = (\mathrm{grd}_n)^* \circ j_{n-1} .$$

Soit $\varphi \in \mathrm{grL}^*_{n-1}$. Il existe un nombre fini de f-morphismes $g_i \in L_n^*$, tels que $\varphi = \sum_{i=1}^{s} \bar{g}_i$, où $\bar{g}_i$ est la forme principale de g_i. Considérons le diagramme :

$$\begin{array}{ccc}
\mathrm{grL}^*_{n-1} = \mathrm{gr}(\mathrm{Hom}_f(L_{n-1},N)) & \xrightarrow{\mathrm{gr}(d_n^*)} & \mathrm{grL}_n^* = \mathrm{gr}(\mathrm{Hom}_f(L_n,N)) \\
\Big\downarrow j_{n-1} & & \Big\downarrow j_n \\
(\mathrm{grL}_{n-1})^* = \mathrm{HOM}_{\mathrm{grA}}(\mathrm{grL}_{n-1},\mathrm{grN}) & \xrightarrow[(\mathrm{grd}_n)^*]{} & (\mathrm{grL}_n)^* = \mathrm{Hom}_{\mathrm{grA}}(\mathrm{grL}_n,\mathrm{grA})
\end{array}$$

On aura $j_{n-1}(\varphi) = \sum_{i=1}^{s} \varphi_{\bar{g}_i} \in \mathrm{HOM}_{\mathrm{grA}}(\mathrm{grL}_{n-1},\mathrm{grN})$ et

$$\alpha = ((\mathrm{grd}_n)^* \circ j_{n-1})(\varphi) = (\mathrm{grd}_n)^*(j_{n-1}(\varphi)) = j_{n-1}(\varphi) \circ \mathrm{grd}_n \in \mathrm{HOM}_{\mathrm{grA}}(\mathrm{grL}_n,\mathrm{grN}).$$

Pour tout élément x_λ d'une f-base de L_n, soit $\bar{x}_\lambda$ sa forme principale. Puisque $\bar{x}_\lambda$ est un élément homogène, on aura :

$$(\mathrm{grd}_n)(\bar{x}_\lambda) = \begin{cases} \bar{0} \text{ si } \omega(d_n(x_\lambda)) > \omega(x_\lambda) \\ \overline{d_n(x_\lambda)} \in \mathrm{gr}_{\omega(x_\lambda)} N, \text{ si } \omega(d_n(x_\lambda)) = \omega(x_\lambda). \end{cases}$$

Soit $\omega(d_n(x_\lambda)) > \omega(x_\lambda)$, alors $\alpha(\bar{x}_\lambda) = 0$. Sinon,

$$\alpha(\bar{x}_\lambda) = j_{n-1}(\varphi)(\overline{d_n(x_\lambda)}) = \sum_{i=1}^{s} \bar{g}_i . \overline{d_n(x_\lambda)} = \sum_{i=1}^{s} \widetilde{g_i(d_n(x_\lambda))}.$$

Posons

$$\beta = (j_n \circ \mathrm{gr}(d_n^*))(\Sigma \bar{g}_i) = j_n(\sum_{i=1}^{s} (\mathrm{gr}(d_n^*))(\bar{g}_i)) = \sum_{i=1}^{s} \beta_i ,$$

où $\beta_i = j_n((\mathrm{gr}(d_n^*))(\bar{g}_i))$. On aura

$$(\mathrm{gr}(d_n^*))(\bar{g}_i) = \begin{cases} \bar{0}, \text{ si } \omega(d_n^*(g_i)) = \omega(g_i \circ d_n) > \omega(g_i) , \\ \overline{g_i \circ d_n} \in \mathrm{gr}_{\omega(g_i)} L_n^* , \text{ si } \omega(g_i \circ d_n) = \omega(g_i). \end{cases}$$

Dans le premier cas, $\beta_i = o$ et $\beta_i(\bar{x}_\lambda) = o = \widetilde{(g_i \circ d)(x_\lambda)}$.

Dans le second cas, on aura $\beta_i = j_n(\overline{g_i \circ d_n})$ et $\beta_i(\bar{x}_\lambda) = \widetilde{(g_i \circ d_n)(x_\lambda)}$.

On déduit que $\alpha(\bar{x}_\lambda) = \beta(\bar{x}_\lambda)$, d'où la commutativité du diagramme, ce qui achève la démonstration.

2°) <u>Définition des foncteurs</u> Ext_f^n .

Un f-<u>complexe</u>, ou complexe filtré, est un complexe de A-modules filtrés M_n et de f-morphismes d_n, $n \in \mathbb{Z}$:

$$(\mathcal{M}) \ldots \longrightarrow M_{n+1} \longrightarrow M_n \longrightarrow M_{n-1} \longrightarrow \ldots\ldots$$

Les groupes d'homologie $H_n(\mathcal{M}) = \mathrm{Ker}\, d_n / \mathrm{Im}\, d_{n+1}$ seront munis de la filtration quotient. Un f-morphisme $\bar{f} : \mathcal{M} \longrightarrow \mathcal{M}'$ de f-complexes est une suite $(f_n)_{n \in \mathbb{N}}$ de f-morphismes $f_n : M_n \longrightarrow M'_n$ qui commutent avec les d_n : $d'_{n+1} f_{n+1} = f_n d_{n+1}$, $\forall n$. On lui associe les applications canoniques $H_n(\bar{f}) : H_n(\mathcal{M}) \longrightarrow H_n(\mathcal{M}')$, des groupes d'homologie.

<u>Lemme</u> 2.1 <u>Pour tout f-morphisme</u> $\bar{f} : \mathcal{M} \longrightarrow \mathcal{M}'$ <u>de f-complexes et tout entier</u> $n \in \mathbb{N}$, $H_n(\bar{f})$ <u>est un</u> f-<u>morphisme. Si</u> $\bar{f}$ <u>est un</u> f-<u>isomorphisme alors</u> $H_n(\bar{f})$ <u>est un</u> f-<u>isomorphisme</u>.

<u>Preuve</u> : Vérification standard.

Remarque. Par la suite on rencontrera essentiellement des complexes bornés à droite ou à gauche. On va toujours les indexer sur $\mathbb{N}$.

Pour tout A-module filtré M, nous désignons par $\hat{M}$ le complété de M et par $i_M : M \longrightarrow \hat{M}$ l'injection canonique. Pour tout f-morphisme $\varphi: M \longrightarrow N$, nous désignons par $\hat{\varphi}: \hat{M} \longrightarrow \hat{N}$, l'unique f-morphisme vérifiant $i_N \circ \hat{\varphi} = \hat{\varphi} \circ i_M$ Le complété du f-complexe $\mathcal{M} = (M_n, d_n)$ est $\hat{\mathcal{M}} = (\hat{M}_n, \hat{d}_u)$.

Soit M un A-module filtré et

$$(\mathbb{P} \longrightarrow M) \ldots P_n \xrightarrow{d_n} P_{n-1} \longrightarrow \ldots P_1 \xrightarrow{d_1} P_o \xrightarrow{\varepsilon} M \longrightarrow 0$$

une résolution f-projective strictement exacte de M[7]. Pour tout A-module filtré N, le foncteur $\mathrm{Hom}_f(\cdot, N)$ définit le f-complexe croissant séparé, qui est complet si N est complet (début du §.1 et prop. 1.4) :

$$\mathrm{Hom}_f(\mathbb{P}, N) : 0 \longrightarrow \mathrm{Hom}_f(P_o, N) \xrightarrow{d_1^*} \mathrm{Hom}_f(P_1, N) \xrightarrow{d_2^*} \mathrm{Hom}_f(P_2, N) \longrightarrow \ldots$$

où $d_n^* = \mathrm{Hom}_f(d_n, N)$, dont les groupes de cohomologie sont :

$$n \geq 1, \mathrm{Ext}_f^n(M,N) \overset{\text{not}}{=} \frac{\mathrm{Ker}\ d_{n+1}^*}{\mathrm{Im}\ d_n^*} = H_n(\mathrm{Hom}_f(\mathbb{P}, M),), \quad \mathrm{Ext}_f^o(M,N) = \mathrm{Hom}_f(M,N).$$

Lemme 2.2 Si $\mathrm{Im}\ d_n^*$ est un fermé de $\mathrm{Hom}_f(P_n, N)$, alors la filtration de $\mathrm{Ext}_f^n(M,N)$ est séparée et si, en plus, N est complet, alors $\mathrm{Ext}_f^n(M,N)$ est complet.

Remarque. Lorsque N est un A-bimodule filtré, les d_n^* étant des f-morphismes de A-modules à droite filtrés, $\mathrm{Ext}_f^n(M,N)$ sont des A-modules à droite filtrés.

Théorème 2.3 Soit N un A-module filtré complet. Pour tout entier $n \geq 1$ et tout A-module filtré M, les groupes de cohomologie $\mathrm{Ext}_f^n(M,N)$ ne dépendent pas de la résolution f-projective $\mathbb{P} \longrightarrow M$, jusqu'à un f-isomorphisme près. En plus il y a un f-isomorphisme canonique

$$\mathrm{Ext}_f^n(\hat{M}, N) \simeq \mathrm{Ext}_f^n(M,N), \quad \forall n \in \mathbb{N}^*.$$

Si N est un A-bimodule filtré, alors $\mathrm{Ext}_f^n(M,N)$ sont des A-modules à droite filtrés, qui sont séparés et complets, lorsque $\mathrm{Hom}_f(d_n, N)$ sont des f-morphismes stricts.

La démonstration fait intervenir des résultats, que nous énonçons à part. Considérons le diagramme

$$(D_1)\qquad \begin{array}{ccccccccccc} \cdots\to X_n & \xrightarrow{d_n} & X_{n-1} & \to & \cdots & \to & X_1 & \xrightarrow{d_1} & X_o & \xrightarrow{\varepsilon} M & \to O \\ \downarrow \varphi_n & & \downarrow \varphi_{n-1} & & & & \downarrow \varphi_1 & & \downarrow \varphi_o & \downarrow \varphi & \\ \cdots\to X'_n & \xrightarrow{d'_n} & X'_{n-1} & \to & \cdots & \to & X'_1 & \xrightarrow{d'_1} & X'_o & \xrightarrow{\varepsilon'} M' & \to O \end{array}$$

où les lignes sont des f-complexes. La suite $(\varphi_n)_{n\in\mathbb{N}}$ de f-morphismes <u>relève</u> de f-morphisme φ, si $(\varphi,\varphi_n)_{n\in\mathbb{N}}$ est un morphisme de f-complexes.

<u>Lemme</u> 2.4 (de comparaison). <u>Si dans le diagramme</u> (D_1) <u>tout</u> X_n <u>est</u> f-<u>projectif</u>, <u>tout</u> X'_n <u>est complet et si la deuxième ligne est strictement exacte, alors il existe au moins une suite</u> $(\varphi_n)_{n\in\mathbb{N}}$ <u>de</u> f-<u>morphismes qui relève le</u> f-<u>morphisme</u> $\varphi : M \longrightarrow M'$. <u>Deux suites de f-morphismes qui relèvent</u> φ <u>sont homotopiques et les applications d'homotopie sont des</u> f-<u>morphismes</u>.

<u>Preuve</u> : Il suffit de noter que si dans le diagramme (D_1) d'_n est un f-morphisme strict, alors l'application canonique $X'_n \twoheadrightarrow \operatorname{Im} d'_n$ est un f-morphisme surjectif strict. Ceci permet de procéder comme dans le cas des modules sans filtration (v. par exemple [6, th.6.9, p.179]), pour obtenir la suite (φ_n) et l'homotopie, en tenant compte qu'un A-module P est f-projectif si et seulement si pour tout diagramme de modules filtrés et de f-morphismes

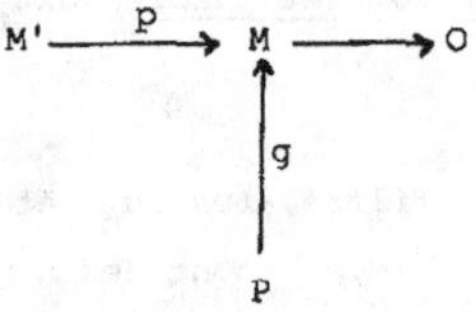

où M' est complet et la ligne est strictement exacte, il existe un f-morphisme $h : P \longrightarrow M'$, tel que $p \circ h = g$ [8].

Pour toute résolution f-projective $\mathbb{P} \longrightarrow M$ du A-module filtré M, nous considérons le f-complexe complété

$$(\hat{\mathbb{P}} \longrightarrow \hat{M}) \ : \ \cdots \to \hat{P}_n \xrightarrow{\hat{d}_n} \hat{P}_{n-1} \to \cdots \to \hat{P}_1 \xrightarrow{\hat{d}_1} \hat{P}_o \xrightarrow{\hat{\varepsilon}} \hat{M} \to O$$

et le diagramme

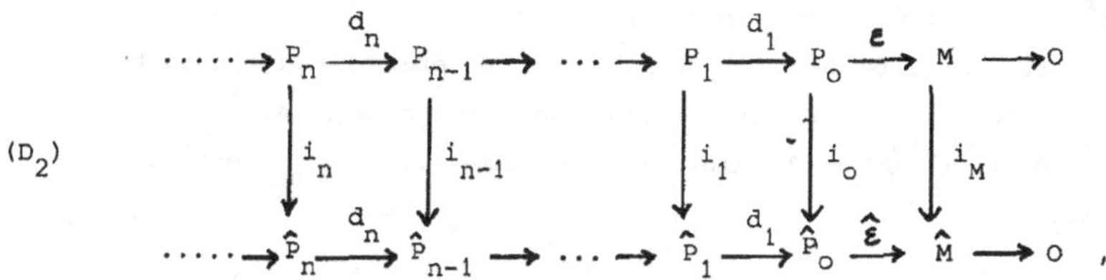

où $i_n : P_n \longrightarrow \hat{P}_n$ et $i_M : M \longrightarrow \hat{M}$ sont les inclusions canoniques. Avec ces conventions on a le résultat suivant.

Proposition 2.5 a) Si $(\mathbb{P} \longrightarrow M)$ est une résolution f-projective strictement exacte de M, alors $(\hat{\mathbb{P}} \longrightarrow \hat{M})$ est une résolution f-projective strictement exacte du complété $\hat{M}$ et la suite $(i_n)_{n\in\mathbb{N}}$ relève le f-morphisme i_M.

b) Soit N un A-module (bimodule) filtré complet, $(\varphi_n)_{n\in\mathbb{N}}$ une suite de f-morphismes qui relève i_M et $\overline{\varphi}^* = (\varphi_n^*)_{n\in\mathbb{N}^*}$ le morphisme de f-complexes

$$\overline{\varphi}^* : \mathrm{Hom}_f(\hat{\mathbb{P}},N) \longrightarrow \mathrm{Hom}_f(\mathbb{P},N) ,$$

où $\varphi_n^* = \mathrm{Hom}_f(\varphi_n,N)$, $\forall n\in\mathbb{N}^*$. Alors, pour tout entier $n \geqslant 1$, $H^n(\overline{\varphi}^*) : H_n(\mathrm{Hom}_f(\hat{\mathbb{P}},N)) \longrightarrow H_n(\mathrm{Hom}_f(\mathbb{P},N))$ est un f-isomorphisme de groupes (A-modules à droite) filtrés.

Preuve : a) Le complété d'un module f-projectif est f-projectif [8, lemme 4.3] et si la suite $M' \xrightarrow{f} M \xrightarrow{g} M''$ est strictement exacte, alors $\hat{M}' \xrightarrow{\hat{f}} \hat{M} \xrightarrow{\hat{g}} \hat{M}''$ est aussi une suite strictement exacte [1, §.2, n°12] , d'où le résultat.

b) Notons d'abord que d'après a) et le lemme 2.4 $(i_n)_{n\in\mathbb{N}^*}$ et $(\varphi_n)_{n\in\mathbb{N}^*}$ sont homotopiques et par suite $(i_n^*)_{n\in\mathbb{N}^*}$ et $(\varphi_n^*)_{n\in\mathbb{N}^*}$ le sont aussi, d'où $\forall n\in\mathbb{N}^*$, $H_n(i_n^*) = H_n(\varphi_n^*)$. Il reste à montrer que pour tout entier $n \geqslant 1$, $H_n(i_n^*)$ est un f-isomorphisme.

Montrons que i_n^* est un f-morphisme injectif strict. Soit $\varphi\in\mathrm{Hom}_f(\hat{P}_n,N)$, $i_n^*(\varphi) = \varphi \circ i_n \in \mathrm{Hom}_f(P_n,N)$. Supposons que $\varphi \in F^p(\mathrm{Hom}_f(\hat{P}_n,N))$. Alors pour tout $k\in\mathbb{N}$,

$$(\varphi \circ i_n)(F^k P_n) \subseteq \varphi(F^k \hat{P}_n) \subseteq F^{k+p} N,$$

d'où $(\varphi \circ i_n) \in F^p \mathrm{Hom}_f(P_n,N)$. Si $\omega(\varphi) = p$, il existe $k \in\mathbb{N}$, tel que $\varphi(F^k \hat{P}_n) \notin F^{k+p+1} N$. Alors il existe $x \in F^k \hat{P}_n$, tel que $\omega(\varphi(x)) < k+p+1$. Puisque $F^k \hat{P}_n$ est le complété de $F^k P_n$, il existe une suite d'éléments $(x_s)_{s\in\mathbb{N}}$, $x_s \in F^k P_n$, qui converge vers x. Si pour tout s , $\omega(\varphi(x_s)) \geqslant k+p+1$, alors $\omega(\varphi(x)) \geqslant k+p+1$, ce qui est contradictoire. Il existe alors un entier s, tel que $\omega(\varphi(x_s)) \leqslant k+p$, d'où $(\varphi \circ i_n)(F^k P_n) \notin F^{k+p+1} N$ et $\omega(\varphi \circ i_n) = p = \omega(\varphi)$. Si $i_n^*(\varphi) = 0$, alors $(\varphi \circ i_n)(P_n) = 0$. Puisque P_n est dense dans $\hat{P}_n$ et φ est continue, on déduit l'injectivité de i_n^* et par suite i_n^* est un f-morphisme injectif strict.

Pour la surjectivité de i_n^*, on note que, N étant complet, à tout f-morphisme $\varphi \in \mathrm{Hom}_f(P_n, N)$ on associe l'unique f-morphisme $\hat{\varphi}: \hat{P}_n \longrightarrow N$ qui prolonge φ et $i^*(\hat{\varphi}) = \varphi$. Alors par le lemme 2.1 $H_n'(i_n^*)$ est un f-isomorphisme.

Lemme 2.6 Soit M et N deux A-modules filtrés complets (N un A-bimodule complet) et deux résolutions f-projectives strictement exactes de M :

$(\mathbb{P} \longrightarrow M) : \ldots \longrightarrow P_n \longrightarrow \ldots \longrightarrow P_o \longrightarrow M \longrightarrow 0$,

$(\mathbb{Q} \longrightarrow M) : \ldots \longrightarrow Q_n \longrightarrow \ldots \longrightarrow Q_o \longrightarrow M \longrightarrow 0$,

où les P_n et Q_n sont des modules filtrés complets.

Alors pour tout entier $n \geq 1$, il y a un f-isomorphisme

$$H_n(\mathrm{Hom}_f(\mathbb{P},N)) \simeq H_n(\mathrm{Hom}_f(\mathbb{Q},N))$$

de groupes filtrés (A-modules à droite filtrés).

Preuve : Le lemme de comparaison 2.4 permet de relever l'identité $1_M : M \longrightarrow M$ dans les deux sens et d'appliquer l'argument canonique qui fonctionne dans le cas des modules sans filtration (v.[6]).

Preuve du thérème dans le cas général. Lorsque M n'est pas complet, on considère deux résolutions f-projectives de M, qu'on complète et on obtient le résultat en appliquant la proposition 2.5 et le lemme 2.6.

3°) Suites spectrales.

Dans cette partie nous considérons une résolution f-libre strictement exacte d'un A-module filtré M :

$$(\mathbb{L} \longrightarrow M) : \ldots \longrightarrow L_n \xrightarrow{d_n} L_{n-1} \rightarrow \ldots \rightarrow L_2 \xrightarrow{d_2} L_1 \xrightarrow{d_1} L_o \xrightarrow{\varepsilon} M \longrightarrow 0$$

et un idéal bilatère fermé I de A. Nous posons

$$L_n^* = \mathrm{Hom}_f(L_n, I) \ , \ d_n^* = \mathrm{Hom}_f(d_n, I).$$

On obtient le f-complexe séparé et complet de A-modules à droite (début du §.1 et prop. 1.4) :

$$\mathbb{L}^* : 0 \longrightarrow L_o^* \xrightarrow{d_1^*} L_1^* \xrightarrow{d_2^*} L_2^* \xrightarrow{d_3^*} \ldots \xrightarrow{d_n^*} L_n^* \xrightarrow{d_{n+1}^*} \ldots$$

ainsi que le complexe gradué associé (théorème 1.7) :

$$\mathrm{gr}\,\mathbb{L}^* : 0 \longrightarrow \mathrm{gr}\,L_o^* \xrightarrow{\mathrm{gr}\,d_1^*} \mathrm{gr}\,L_1^* \xrightarrow{\mathrm{gr}\,d_2^*} \ldots\ldots L_n^* \xrightarrow{\mathrm{gr}\,d_{n+1}^*} \ldots\ldots$$

La filtration du complexe filtré séparé et complet $\mathbb{L}^*$ étant décroissante, on peut lui associer une suite spectrale de cohomologie [4, ch. XI, §.8] de la manière suivante. A $\mathbb{L}^*$ nous associons le A-module à droite gradué,

$$\mathcal{L} = \oplus L_n^*$$

(tous les éléments de A étant de degré 0), filtré par la filtration exhaustive et séparée

$$F^p\mathcal{L} = \bigoplus_{n\in\mathbb{N}} F^p L_n^* \quad , \forall p\in\mathbb{N}$$

et muni du f-morphisme différentiel $\Delta : \mathcal{L} \longrightarrow \mathcal{L}$, défini par $\Delta(\varphi) = d_n^*(\varphi)$, pour tout $\varphi\in L_n^*$ et pour tout $n\in\mathbb{N}$. ($\Delta^2 = 0$). Le A-module à droite de cohomologie $H(\mathcal{L})$ du A-module différentiel gradué filtré $\mathcal{L}$ est muni de la filtration quotient :

$$F^p\, H(\mathcal{L}) = (F^p\, \mathrm{Ker}\,\Delta + \mathrm{Im}\,\Delta)/\mathrm{Im}\,\Delta$$

et le gradué associé à cette filtration est

$$(1) \qquad \mathrm{gr}(H(\mathcal{L})) = \bigoplus_{p\in\mathbb{N}} \frac{F^p\,\mathrm{Ker}\,\Delta + \mathrm{Im}\,\Delta}{F^{p+1}\,\mathrm{Ker}\,\Delta + \mathrm{Im}\,\Delta} ,$$

qui est un gr A-module à droite gradué et un A-module à droite bigradué.

Posons le résultat suivant évident.

<u>Lemme</u> 3.1 <u>Les</u> A-<u>modules gradués et bigradués filtrés suivants sont isomorphes</u>

$$H(\mathcal{L}) \simeq \bigoplus_{n\in\mathbb{N}} \mathrm{Ext}_f^n(M,I) \quad , \quad \mathrm{gr}\,H(\mathcal{L}) \simeq \bigoplus_{n\in\mathbb{N}} \mathrm{gr}\,\mathrm{Ext}_f^n(M,I) ,$$

<u>le deuxième isomorphisme étant aussi un isomorphisme de</u> gr A-<u>modules à droite gradués</u>.

$$\text{Notation : } \bigoplus_{n=o}^{\infty} \mathrm{Ext}_f^n(M,I) \overset{\text{not}}{=} \mathrm{Ext}_f(M,I) .$$

On associe d'autre part au module filtré $\mathcal{L}$ le gr A-module à droite gradué $\mathrm{gr}\mathcal{L}$.

<u>Proposition</u> 3.2 <u>Les</u> gr A-<u>modules à droites gradués suivants sont isomorphes</u> :

$$\mathrm{gr}\mathcal{L} \simeq \bigoplus_{n=o}^{\infty} \mathrm{gr}\, L_n^*$$

<u>et sont munis de l'application graduée différentielle</u>

$\mathrm{gr}\,\Delta = \bigoplus_{n=0}^{\infty} \mathrm{gr}\, d_n^*$. Le groupe de cohomologie est un gr A-module à droite et il y a l'isomorphisme

$$H(\mathrm{gr}\,\mathcal{L}) \simeq \bigoplus_{n=0}^{\infty} \mathrm{EXT}^n_{\mathrm{gr}A}(\mathrm{gr}M, \mathrm{gr}I).$$

Preuve : On applique le théorème 1.7.

Proposition 3.3 Il existe une suite spectrale (E_k, δ_k), $k = 0,1,2,\ldots$, où $\delta_k : E_k \longrightarrow E_k$ est une application bigraduée de bidegré $(k,-k+1)$, vérifiant les conditions suivantes :

a) $E_0 \simeq \mathrm{gr}\,\mathcal{L}$ et $\delta_0 = \mathrm{gr}\,\Delta$.

b) $E_1 \simeq H(\mathrm{gr}\,\mathcal{L})$; plus précisément, pour tout entier $p \geq 0$ et tout entier q, tel que $p+q = n \geq 0$,

$$E_1^{p,q} \simeq \frac{\mathrm{Ker}\ \mathrm{gr}_p\, d^*_{p+q+1}}{\mathrm{Im}\ \mathrm{gr}_p\, d^*_{p+q}} .$$

c) Il existe un homomorphisme injectif de grA-modules à droite gradués

$$\mathrm{gr}(H(\mathcal{L})) \hookrightarrow E_\infty ,$$

qui est défini à partir des applications A-linéaires (à droite) injectives :

$$\mathrm{gr}_p(H_{p+q}(\mathcal{L})) \hookrightarrow E_\infty^{p,q} , \ \forall p,q.$$

Preuve . Construction de la suite spectrale associée à $\mathcal{L}$. Nous utilisons de manière appropriée les notations canoniques utilisées pour les modules Z-gradués avec filtrations croissantes. Les propriétés des modules que nous introduisons sont ou bien immédiates, ou bien s'obtiennent de manière analogue au cas précédent (v. [2] et [4]).

Pour tout entier $k \in \mathbb{N}$ et tout entier $p \in \mathbb{N}$, posons

$$Z_k^p = \{ x \in F^p\mathcal{L} \mid \Delta x \in F^{p+k}\mathcal{L} \}$$

$$B_k^p = F^p\mathcal{L} \cap \Delta(F^{p-k+1}\mathcal{L}),\ Z_\infty^p = F^p\ \mathrm{Ker}\,\Delta,\ B_\infty^p = F^p\mathcal{L} \cap \mathrm{Im}\,\Delta ,$$

en faisant la convention que pour $p+1 < k$, $F^{p-k+1}\mathcal{L} = F^o\mathcal{L} = \mathcal{L}$. Pour l'indice complémentaire $q = n-p$, où $n \geq o$ $(q \geq -p)$, on pose, en omettant parfois d'écrire les indices pour d^* :

(2)
$$Z_k^{p,q} = \{ x \in F^p L^*_{p+q} \mid d^*(x) \in F^{p+k} L^*_{p+q+1} \} ,\ Z_\infty^{p,q} = F^p\ \mathrm{Ker}\, d^*_{p+q+1} ,$$

$$B_k^{p,q} = F^p L^*_{p+q} \cap d^*(F^{p-k+1} L^*_{p+q-1}) ,\ B_\infty^{p,q} = F^p L^*_{p+q} \cap d\, L^*_{p+q-1} .$$

On aura la décomposition directe

$Z_k^p = \overset{\infty}{\underset{q=-p}{\oplus}} Z_k^{p,q}$ et de même pour B_k^p, Z_∞^p et B_∞^p respectivement. Les suites des cocycles $(Z_k^{p,q})_{k\in\mathbb{N}}$, $(Z_k^p)_{k\in\mathbb{N}}$ sont décroissantes et celles des cobornes $(B^k_{p,q})_{k\in\mathbb{N}}$, $(B_k^p)_{k\in\mathbb{N}}$ croissantes. On a les inclusions

$$B_k^{p,q} \subseteq B_\infty^{p,q} \subseteq Z_\infty^{p,q} \subseteq Z_k^{p,q}$$

et les égalités :

$$B_\infty^{p,q} = Z_\infty^{p,q} \cap \operatorname{Im} d_{p+q}^* = \bigcup_{k=0}^{\infty} B_k^{p,q} = B_{p+1}^{p,q} , \tag{3}$$

la dernière, en vertu de l'égalité $B_k^{p,q} = B_{p+1}^{p,q}$, pour $k \geq p+1$. Puisque la filtration de $\mathcal{L}$ est séparée, on déduit enfin

$$Z_\infty^{p,q} = \bigcap_{k=0}^{\infty} Z_k^{p,q} . \tag{4}$$

Posons pour tout $k\in\mathbb{N}$ et tout $p\in\mathbb{N}$, $n = p+q \geq 0$,

$$E_k^{p,q} = \frac{Z_k^{p,q} + F^{p+1} L_{p+q}^*}{B_k^{p,q} + F^{p+1} L_{p+q}^*} , \quad E_k^p = \frac{Z_k^p + F^{p+1}\mathcal{L}}{B_k^p + F^{p+1}\mathcal{L}} , \quad E_k = \overset{\infty}{\underset{p=0}{\oplus}} E_k^p .$$

On aura alors

$$E_k^p = \overset{\infty}{\underset{q=-p}{\oplus}} E_k^{p,q} , \quad \forall k\in\mathbb{N},\ \forall p\in\mathbb{N}.$$

De même, pour tout p, tout k et tout q, on a

$$Z_{k+1}^{p-1} = Z_k^p \cap F^{p+1}\mathcal{L} , \quad Z_{k+1}^{p+1,q-1} = Z_k^{p,q} \cap F^{p+1} L_{p+q}^*$$

$$\Delta(Z_k^p) = B_{k+1}^{p+k} , \quad d_{p+q+1}^* (Z_k^{p,q}) = B_{k+1}^{p+k,q-k+1}$$

$$E_k^{p,q} \simeq \frac{Z_k^{p,q}}{B_k^{p,q} + Z_{k-1}^{p+1,q-1}} .$$

<u>Preuve de</u> a) : Evident, puisque $B_0^p = \Delta F^{p+1}\mathcal{L}$. On pose $\delta_0 = \operatorname{gr}\Delta$.

<u>Preuve de</u> b) On a par la proposition 3.2

$$H(\operatorname{gr}\mathcal{L}) = \frac{\operatorname{Ker} \operatorname{gr} \Delta}{\operatorname{Im} \operatorname{gr} \Delta} = \overset{\infty}{\underset{n=0}{\oplus}} \frac{\operatorname{Ker} \operatorname{gr} d_{n+1}^*}{\operatorname{Im} \operatorname{gr} d_n^*}$$

et, d'autre part,

$$E_1^{p,q} \simeq (Z_1^{p,q}/F^{p+1}L_{p+q+1}^*)/(B_1^{p,q}/F^{p+1}L_{p+q+1}^*) =$$

$$= (\operatorname{Ker} \operatorname{gr}_p d_{p+q+1}^*)/\operatorname{Im} \operatorname{gr}_p d_{p+q}^* ,$$

ce qui prouve b) et l'égalité $E_1 = H(E_o, \delta_o)$.

Définition de δ_k : Pour tout $x \in Z_k^{p,q}$, on pose

$$\bar{x} = \text{classe de } x \text{ modulo } (B_k^{p,q} + Z_{k-1}^{p+1,q-1}) \in E_k^{p,q},$$

$$\delta_k^{p,q}(\bar{x}) = \overline{d^*_{p+q+1}(x)} =$$

$$= \text{classe de } d^*_{p+q+1}(x) \text{ modulo } (B_k^{p+k,q+1-k} + Z_{k-1}^{p+k+1,q-k}),$$

$$\delta_k^{p,q}(\bar{x}) \in E_k^{p+k,q-k+1}.$$

Alors $\delta_k^{p,q}$ est bien défini, $\delta_k^{p+k,q-k+1} \circ \delta_k^{p,q} = 0$ et $\delta_k = (\delta_k^{p,q})$ est une application différentielle de E_k. Les arguments standards de modularité [2] et les propriétés énoncées au début conduisent aux isomorphismes

$$\operatorname{Ker} \delta_k^{p,q} \simeq \frac{Z_{k+1}^{p,q} + F^{p+1} L^*_{p+q}}{B_k^{p,q} + F^{p+1} L^*_{p+q}},$$

$$\operatorname{Im} \delta_k^{p-k,q+k-1} \simeq \frac{B_{k+1}^{p,q} + F^{p+1} L^*_{p,q}}{B_k^{p,q} + F^{p+1} L^*_{p,q}},$$

ce qui prouve $H(E_k, \delta_k) \simeq E_{k+1}$.

Preuve de c). On va calculer le module limite E_∞. Pour tout k, E_k est un sous-facteur de E_o, donc de $\frac{F^p \mathcal{L}}{F^{p+1}\mathcal{L}}$. Alors pour p et q, $p+q = n \geq 0$, $E_k^{p,q}$ est un sous-facteur de $\frac{F^p L^*_{p+q}}{F^{p+1} L^*_{p+q}}$. On considère la suite de sous-modules de $F^p L^*_{p+q}$:

$$B_1^{p,q} \subseteq B_2^{p,q} \subseteq B_3^{p,q} \subseteq \ldots \subseteq Z_3^{p,q} \subseteq Z_2^{p,q} \subseteq Z_1^{p,q} \subseteq F^p L^*_{p+q},$$

dont l'image dans $gr_p L^*_{p+q}$ est la suite croissante

$$\mathcal{B}_1^{p,q} \subseteq \mathcal{B}_2^{p,q} \subseteq \mathcal{B}_3^{p,q} \subseteq \ldots \subseteq \mathcal{Z}_3^{p,q} \subseteq \mathcal{Z}_2^{p,q} \subseteq \mathcal{Z}_1^{p,q} \subseteq E_o^{p,q} = gr_p L^*_{p+q},$$

où $\mathcal{B}_k^{p,q} = \dfrac{B_k^{p,q} + F^{p+1} L^*_{p+q}}{F^{p+1} L^*_{p+q}}$, $\mathcal{Z}_k^{p,q} = \dfrac{Z_k^{p,q} + F^{p+1} L^*_{p+q}}{F^{p+1} L^*_{p+q}}$.

Il y a des isomorphismes pour tout k :

$$\mathcal{Z}_k^{p,q} / \mathcal{B}_k^{p,q} \simeq E_k^{p,q}$$

et l'on a les inclusions

$$\frac{\mathcal{B}_{k+1}^{p,q}}{\mathcal{B}_k^{p,q}} \subseteq \frac{\mathcal{B}_{k+2}^{p,q}}{\mathcal{B}_k^{p,q}} \subseteq \ldots \subseteq \frac{\mathcal{Z}_{k+2}^{p,q}}{\mathcal{B}_k^{p,q}} \subseteq \frac{\mathcal{Z}_{k+1}^{p,q}}{\mathcal{B}_k^{p,q}} \subseteq E_k^{p,q}$$

et les isomorphismes

$$(\mathcal{Z}_{k+1}^{p,q}/\mathcal{B}_k^{p,q})/(\mathcal{B}_{k+1}^{p,q}/\mathcal{B}_k^{p,q}) \simeq E_{k+1}^{p,q} .$$

Ceci prouve, que si $\mathcal{B}_\infty^{p,q} = \bigcup_k \mathcal{B}_k^{p,q}$, $\mathcal{Z}_\infty^{p,q} = \bigcap_k \mathcal{Z}_k^{p,q}$,

alors
$$E_\infty^{p,q} = \frac{\mathcal{Z}_\infty^{p,q}}{\mathcal{B}_\infty^{p,q}} ,$$

ou bien

$$E_\infty^{p,q} \simeq \frac{\bigcap_k (Z_k^{p,q} + F^{p+1} L_{p+q}^*)}{B_{p+1}^{p,q} + F^{p+1} L_{p+q}^*} . \tag{5}$$

D'autre part on a, d'après (1) et (2) ,

$$gr_p(H_{p+q}\mathcal{L}) = \frac{F^p \operatorname{Ker} d_{p+q+1}^* + \operatorname{Im} d_{p+q}^*}{F^{p+1} \operatorname{Ker} d_{p+q+1}^* + \operatorname{Im} d_{p+q}^*} = \frac{Z_\infty^{p,q} + \operatorname{Im} d_{p+q}^*}{Z_\infty^{p+1,q-1} + \operatorname{Im} d_{p+q}^*} .$$

Le dernier module est isomorphe à

$$\frac{Z_\infty^{p,q}}{Z_\infty^{p,q} \cap (Z_\infty^{p+1,q-1} + \operatorname{Im} d_{p+q}^*)} = \frac{Z_\infty^{p,q}}{Z_\infty^{p+1,q-1} + (Z_\infty^{p,q} \cap \operatorname{Im} d_{p+q}^*)}$$

la dernière égalité s'obtenant par modularité. Par (3), (4) et (5), on obtient :

$$gr_p(H_{p+q}\mathcal{L}) \simeq \frac{\bigcap_{k=o}^{\infty} Z_k^{p,q}}{B_{p+1}^{p,q} + Z_\infty^{p+1,q-1}} \simeq \frac{Z_\infty^{p,q} + F^{p+1} L_{p+q}^*}{B_{p+1}^{p,q} + F^{p+1} L_{p+q}^*} \subseteq E_\infty^{p,q} , \tag{6}$$

car on a les égalités , la dernière par modularité :

$$B_{p+1}^{p,q} + Z_\infty^{p+1,q-1} = B_{p+1}^{p,q} + (F^{p+1} L_{p+q}^* \cap Z_\infty^{p,q}) = Z_\infty^{p,q} \cap (B_{p+1}^{p,q} + F^{p+1} L_{p+q}^*).$$

Ceci achève la démonstration.

Proposition 3.4 Soit M un A-module filtré et I un idéal bilatère fermé de A. Pour tout entier $n \in \mathbb{N}$, on munit le A-module à droite de cohomologie

$$\mathrm{Ext}_f^n(M,I)$$

de la filtration quotient. Alors $\mathrm{gr}\,\mathrm{Ext}_f^n(M,I)$ est un sous-facteur du grA-module gradué $\mathrm{EXT}_{grA}^n(grM,grI)$.

Preuve : On utilise les propositions 3.2 et 3.3. On en déduit que E_∞ est un sous-facteur de E_1, qui est isomorphe à $\mathrm{EXT}_{grA}^n(grM,grI)$. On conclut par le lemme 3.1 et la partie c) de la proposition 3.3.

Théorème 3.5 Soit A un anneau filtré complet, qui est linéairement compact comme A-module à droite et soit I un idéal bilatère fermé de A. Alors la filtration canonique définie sur $\mathrm{Ext}_f^n(M,I)$ est séparée pour tout $n \geq o$ et la suite spectrale (E_k,δ_k) définie dans la proposition 3.3 converge :

$$E_r^p \underset{p}{\Longrightarrow} \mathrm{Ext}_f(M,I).$$

Preuve : Puisque A est linéairement compact à droite, par [1,p.109, ex. 15d] et le lemme 1.3, pour tout A-module f-libre L, $\mathrm{Hom}_f(L,I)$ est un A-module à droite filtré linéairement compact. Par suite, les f-morphismes d_n^* du f-complexe $\mathbb{L}^*$ étant continus, les sous-modules $\mathrm{Im}\, d_n^*$ sont des fermés de L_n^* [1, p.109, exercice 16]. Alors la filtration de $\mathrm{Ext}_f^n(M,I)$ est séparée. Pour la convergence de la suite spectrale, on utilise les isomorphismes (5) et (6) et on applique [1, ex.16, p.109]. Le A-module à droite filtré L_{p+q}^* est linéairement compact et $F^{p+1}L_{p+q}^*$ est un fermé de L_{p+q}^*. La suite décroissante $\{Z_k^{p,q}\}_{k\in\mathbb{N}}$ de sous-modules de L_{p+q}^* est une base de filtre de variétés linéaires, d'où

$$\bigcap_{k=o}^{\infty}(Z_k^{p,q} + F^{p+1}L_{p+q}^*) = \bigcap_{k=o}^{\infty} Z_k^{p,q} + F^{p+1}L_{p+q}^* .$$

On conclut à la convergence.

Remarque : Si le gradué associé de l'anneau A est noethérien à gauche et à droite et si grM est de type fini, alors on obtient des résultats analogues à ceux de [2, ch.II, §.6] et la convergence de la suite spectrale définie dans la proposition 3.3.

Bibliographie

[1] N. Bourkaki, Algèbre commutative, ch.III, Herman, Paris 1961.

[2] J.E. Björk, Rings of differential operators, North-Holland, Publ. comp., 1979.

[3] S. Eilenberg, J.C. Moore, Limits and spectral sequences, Topology, 1 (1961), pp.1-23.

[4] S. Mc Lane, Homology, Springer Verlag, 1967.

[5] C. Nastasescu, Ivan Oystaeyen, Graded and Filtred Rings and Modules, Springer Verlag, 1979, L.N. 758.

[6] J.J. Rotman, An introduction to homological algebra, Ac. Press, 1979.

[7] E. Wexler-Kreindler, Sur la dimension projective des modules filtrés sur des anneaux filtrés complets, Sém. d'Algèbre, Paris 1979, Springer-Verlag, L.N. 795, pp.225-250.

[8] E. Wexler-Kreindler, Polynômes de Ore, séries formelles tordues et anneaux filtrés complets héréditaires, Comm. in algebra, 8 (4), (1980), pp.339-371.

[9] E. Wexler-Kreindler, Skew power series rings and some homological properties of filtered rings, Ring theory Antwerp 1980, Springer-Verlag, L.N. 825, pp.198-209.

Manuscrit remis en
Mai 1981

Université P. et M. Curie
Département de Mathématiques
4, Place Jussieu
75230 Paris Cedex 05

Vol. 759: R. L. Epstein, Degrees of Unsolvability: Structure and Theory. XIV, 216 pages. 1979.

Vol. 760: H.-O. Georgii, Canonical Gibbs Measures. VIII, 190 pages. 1979.

Vol. 761: K. Johannson, Homotopy Equivalences of 3-Manifolds with Boundaries. 2, 303 pages. 1979.

Vol. 762: D. H. Sattinger, Group Theoretic Methods in Bifurcation Theory. V, 241 pages. 1979.

Vol. 763: Algebraic Topology, Aarhus 1978. Proceedings, 1978. Edited by J. L. Dupont and H. Madsen. VI, 695 pages. 1979.

Vol. 764: B. Srinivasan, Representations of Finite Chevalley Groups. XI, 177 pages. 1979.

Vol. 765: Padé Approximation and its Applications. Proceedings, 1979. Edited by L. Wuytack. VI, 392 pages. 1979.

Vol. 766: T. tom Dieck, Transformation Groups and Representation Theory. VIII, 309 pages. 1979.

Vol. 767: M. Namba, Families of Meromorphic Functions on Compact Riemann Surfaces. XII, 284 pages. 1979.

Vol. 768: R. S. Doran and J. Wichmann, Approximate Identities and Factorization in Banach Modules. X, 305 pages. 1979.

Vol. 769: J. Flum, M. Ziegler, Topological Model Theory. X, 151 pages. 1980.

Vol. 770: Séminaire Bourbaki vol. 1978/79 Exposés 525–542. IV, 341 pages. 1980.

Vol. 771: Approximation Methods for Navier-Stokes Problems. Proceedings, 1979. Edited by R. Rautmann. XVI, 581 pages. 1980.

Vol. 772: J. P. Levine, Algebraic Structure of Knot Modules. XI, 104 pages. 1980.

Vol. 773: Numerical Analysis. Proceedings, 1979. Edited by G. A. Watson. X, 184 pages. 1980.

Vol. 774: R. Azencott, Y. Guivarc'h, R. F. Gundy, Ecole d'Eté de Probabilités de Saint-Flour VIII-1978. Edited by P. L. Hennequin. XIII, 334 pages. 1980.

Vol. 775: Geometric Methods in Mathematical Physics. Proceedings, 1979. Edited by G. Kaiser and J. E. Marsden. VII, 257 pages. 1980.

Vol. 776: B. Gross, Arithmetic on Elliptic Curves with Complex Multiplication. V, 95 pages. 1980.

Vol. 777: Séminaire sur les Singularités des Surfaces. Proceedings, 1976-1977. Edited by M. Demazure, H. Pinkham and B. Teissier. IX, 339 pages. 1980.

Vol. 778: SK1 von Schiefkörpern. Proceedings, 1976. Edited by P. Draxl and M. Kneser. II, 124 pages. 1980.

Vol. 779: Euclidean Harmonic Analysis. Proceedings, 1979. Edited by J. J. Benedetto. III, 177 pages. 1980.

Vol. 780: L. Schwartz, Semi-Martingales sur des Variétés, et Martingales Conformes sur des Variétés Analytiques Complexes. XV, 132 pages. 1980.

Vol. 781: Harmonic Analysis Iraklion 1978. Proceedings 1978. Edited by N. Petridis, S. K. Pichorides and N. Varopoulos. V, 213 pages. 1980.

Vol. 782: Bifurcation and Nonlinear Eigenvalue Problems. Proceedings, 1978. Edited by C. Bardos, J. M. Lasry and M. Schatzman. VIII, 296 pages. 1980.

Vol. 783: A. Dinghas, Wertverteilung meromorpher Funktionen in ein- und mehrfach zusammenhängenden Gebieten. Edited by R. Nevanlinna and C. Andreian Cazacu. XIII, 145 pages. 1980.

Vol. 784: Séminaire de Probabilités XIV. Proceedings, 1978/79. Edited by J. Azéma and M. Yor. VIII, 546 pages. 1980.

Vol. 785: W. M. Schmidt, Diophantine Approximation. X, 299 pages. 1980.

Vol. 786: I. J. Maddox, Infinite Matrices of Operators. V, 122 pages. 1980.

Vol. 787: Potential Theory, Copenhagen 1979. Proceedings, 1979. Edited by C. Berg, G. Forst and B. Fuglede. VIII, 319 pages. 1980.

Vol. 788: Topology Symposium, Siegen 1979. Proceedings, 1979. Edited by U. Koschorke and W. D. Neumann. VIII, 495 pages. 1980.

Vol. 789: J. E. Humphreys, Arithmetic Groups. VII, 158 pages. 1980.

Vol. 790: W. Dicks, Groups, Trees and Projective Modules. IX, 127 pages. 1980.

Vol. 791: K. W. Bauer and S. Ruscheweyh, Differential Operators for Partial Differential Equations and Function Theoretic Applications. V, 258 pages. 1980.

Vol. 792: Geometry and Differential Geometry. Proceedings, 1979. Edited by R. Artzy and I. Vaisman. VI, 443 pages. 1980.

Vol. 793: J. Renault, A Groupoid Approach to C*-Algebras. III, 160 pages. 1980.

Vol. 794: Measure Theory, Oberwolfach 1979. Proceedings 1979. Edited by D. Kölzow. XV, 573 pages. 1980.

Vol. 795: Séminaire d'Algèbre Paul Dubreil et Marie-Paule Malliavin. Proceedings 1979. Edited by M. P. Malliavin. V, 433 pages. 1980.

Vol. 796: C. Constantinescu, Duality in Measure Theory. IV, 197 pages. 1980.

Vol. 797: S. Mäki, The Determination of Units in Real Cyclic Sextic Fields. III, 198 pages. 1980.

Vol. 798: Analytic Functions, Kozubnik 1979. Proceedings. Edited by J. Ławrynowicz. X, 476 pages. 1980.

Vol. 799: Functional Differential Equations and Bifurcation. Proceedings 1979. Edited by A. F. Izé. XXII, 409 pages. 1980.

Vol. 800: M.-F. Vignéras, Arithmétique des Algèbres de Quaternions. VII, 169 pages. 1980.

Vol. 801: K. Floret, Weakly Compact Sets. VII, 123 pages. 1980.

Vol. 802: J. Bair, R. Fourneau, Etude Géometrique des Espaces Vectoriels II. VII, 283 pages. 1980.

Vol. 803: F.-Y. Maeda, Dirichlet Integrals on Harmonic Spaces. X, 180 pages. 1980.

Vol. 804: M. Matsuda, First Order Algebraic Differential Equations. VII, 111 pages. 1980.

Vol. 805: O. Kowalski, Generalized Symmetric Spaces. XII, 187 pages. 1980.

Vol. 806: Burnside Groups. Proceedings, 1977. Edited by J. L. Mennicke. V, 274 pages. 1980.

Vol. 807: Fonctions de Plusieurs Variables Complexes IV. Proceedings, 1979. Edited by F. Norguet. IX, 198 pages. 1980.

Vol. 808: G. Maury et J. Raynaud, Ordres Maximaux au Sens de K. Asano. VIII, 192 pages. 1980.

Vol. 809: I. Gumowski and Ch. Mira, Recurences and Discrete Dynamic Systems. VI, 272 pages. 1980.

Vol. 810: Geometrical Approaches to Differential Equations. Proceedings 1979. Edited by R. Martini. VII, 339 pages. 1980.

Vol. 811: D. Normann, Recursion on the Countable Functionals. VIII, 191 pages. 1980.

Vol. 812: Y. Namikawa, Toroidal Compactification of Siegel Spaces. VIII, 162 pages. 1980.

Vol. 813: A. Campillo, Algebroid Curves in Positive Characteristic. V, 168 pages. 1980.

Vol. 814: Séminaire de Théorie du Potentiel, Paris, No. 5. Proceedings. Edited by F. Hirsch et G. Mokobodzki. IV, 239 pages. 1980.

Vol. 815: P. J. Slodowy, Simple Singularities and Simple Algebraic Groups. XI, 175 pages. 1980.

Vol. 816: L. Stoica, Local Operators and Markov Processes. VIII, 104 pages. 1980.

Vol. 817: L. Gerritzen, M. van der Put, Schottky Groups and Mumford Curves. VIII, 317 pages. 1980.

Vol. 818: S. Montgomery, Fixed Rings of Finite Automorphism Groups of Associative Rings. VII, 126 pages. 1980.

Vol. 819: Global Theory of Dynamical Systems. Proceedings, 1979. Edited by Z. Nitecki and C. Robinson. IX, 499 pages. 1980.

Vol. 820: W. Abikoff, The Real Analytic Theory of Teichmüller Space. VII, 144 pages. 1980.

Vol. 821: Statistique non Paramétrique Asymptotique. Proceedings, 1979. Edited by J.-P. Raoult. VII, 175 pages. 1980.

Vol. 822: Séminaire Pierre Lelong-Henri Skoda, (Analyse) Années 1978/79. Proceedings. Edited by P. Lelong et H. Skoda. VIII, 356 pages, 1980.

Vol. 823: J. Král, Integral Operators in Potential Theory. III, 171 pages. 1980.

Vol. 824: D. Frank Hsu, Cyclic Neofields and Combinatorial Designs. VI, 230 pages. 1980.

Vol. 825: Ring Theory, Antwerp 1980. Proceedings. Edited by F. van Oystaeyen. VII, 209 pages. 1980.

Vol. 826: Ph. G. Ciarlet et P. Rabier, Les Equations de von Kármán. VI, 181 pages. 1980.

Vol. 827: Ordinary and Partial Differential Equations. Proceedings, 1978. Edited by W. N. Everitt. XVI, 271 pages. 1980.

Vol. 828: Probability Theory on Vector Spaces II. Proceedings, 1979. Edited by A. Weron. XIII, 324 pages. 1980.

Vol. 829: Combinatorial Mathematics VII. Proceedings, 1979. Edited by R. W. Robinson et al.. X, 256 pages. 1980.

Vol. 830: J. A. Green, Polynomial Representations of GL_n. VI, 118 pages. 1980.

Vol. 831: Representation Theory I. Proceedings, 1979. Edited by V. Dlab and P. Gabriel. XIV, 373 pages. 1980.

Vol. 832: Representation Theory II. Proceedings, 1979. Edited by V. Dlab and P. Gabriel. XIV, 673 pages. 1980.

Vol. 833: Th. Jeulin, Semi-Martingales et Grossissement d'une Filtration. IX, 142 Seiten. 1980.

Vol. 834: Model Theory of Algebra and Arithmetic. Proceedings, 1979. Edited by L. Pacholski, J. Wierzejewski, and A. J. Wilkie. VI, 410 pages. 1980.

Vol. 835: H. Zieschang, E. Vogt and H.-D. Coldewey, Surfaces and Planar Discontinuous Groups. X, 334 pages. 1980.

Vol. 836: Differential Geometrical Methods in Mathematical Physics. Proceedings, 1979. Edited by P. L. García, A. Pérez-Rendón, and J. M. Souriau. XII, 538 pages. 1980.

Vol. 837: J. Meixner, F. W. Schäfke and G. Wolf, Mathieu Functions and Spheroidal Functions and their Mathematical Foundations Further Studies. VII, 126 pages. 1980.

Vol. 838: Global Differential Geometry and Global Analysis. Proceedings 1979. Edited by D. Ferus et al. XI, 299 pages. 1981.

Vol. 839: Cabal Seminar 77 – 79. Proceedings. Edited by A. S. Kechris, D. A. Martin and Y. N. Moschovakis. V, 274 pages. 1981.

Vol. 840: D. Henry, Geometric Theory of Semilinear Parabolic Equations. IV, 348 pages. 1981.

Vol. 841: A. Haraux, Nonlinear Evolution Equations- Global Behaviour of Solutions. XII, 313 pages. 1981.

Vol. 842: Séminaire Bourbaki vol. 1979/80. Exposés 543–560. IV, 317 pages. 1981.

Vol. 843: Functional Analysis, Holomorphy, and Approximation Theory. Proceedings. Edited by S. Machado. VI, 636 pages. 1981.

Vol. 844: Groupe de Brauer. Proceedings. Edited by M. Kervaire and M. Ojanguren. VII, 274 pages. 1981.

Vol. 845: A. Tannenbaum, Invariance and System Theory: Algebraic and Geometric Aspects. X, 161 pages. 1981.

Vol. 846: Ordinary and Partial Differential Equations, Proceedings. Edited by W. N. Everitt and B. D. Sleeman. XIV, 384 pages. 1981.

Vol. 847: U. Koschorke, Vector Fields and Other Vector Bundle Morphisms – A Singularity Approach. IV, 304 pages. 1981.

Vol. 848: Algebra, Carbondale 1980. Proceedings. Ed. by R. K. Amayo. VI, 298 pages. 1981.

Vol. 849: P. Major, Multiple Wiener-Itô Integrals. VII, 127 pages. 1981.

Vol. 850: Séminaire de Probabilités XV. 1979/80. Avec table générale des exposés de 1966/67 à 1978/79. Edited by J. Azéma and M. Yor. IV, 704 pages. 1981.

Vol. 851: Stochastic Integrals. Proceedings, 1980. Edited by D. Williams. IX, 540 pages. 1981.

Vol. 852: L. Schwartz, Geometry and Probability in Banach Spaces. X, 101 pages. 1981.

Vol. 853: N. Boboc, G. Bucur, A. Cornea, Order and Convexity in Potential Theory: H-Cones. IV, 286 pages. 1981.

Vol. 854: Algebraic K-Theory. Evanston 1980. Proceedings. Edited by E. M. Friedlander and M. R. Stein. V, 517 pages. 1981.

Vol. 855: Semigroups. Proceedings 1978. Edited by H. Jürgensen, M. Petrich and H. J. Weinert. V, 221 pages. 1981.

Vol. 856: R. Lascar, Propagation des Singularités des Solutions d'Equations Pseudo-Différentielles à Caractéristiques de Multiplicités Variables. VIII, 237 pages. 1981.

Vol. 857: M. Miyanishi. Non-complete Algebraic Surfaces. XVIII, 244 pages. 1981.

Vol. 858: E. A. Coddington, H. S. V. de Snoo: Regular Boundary Value Problems Associated with Pairs of Ordinary Differential Expressions. V, 225 pages. 1981.

Vol. 859: Logic Year 1979–80. Proceedings. Edited by M. Lerman, J. Schmerl and R. Soare. VIII, 326 pages. 1981.

Vol. 860: Probability in Banach Spaces III. Proceedings, 1980. Edited by A. Beck. VI, 329 pages. 1981.

Vol. 861: Analytical Methods in Probability Theory. Proceedings 1980. Edited by D. Dugué, E. Lukacs, V. K. Rohatgi. X, 183 pages. 1981.

Vol. 862: Algebraic Geometry. Proceedings 1980. Edited by A. Libgober and P. Wagreich. V, 281 pages. 1981.

Vol. 863: Processus Aléatoires à Deux Indices. Proceedings, 1980. Edited by H. Korezlioglu, G. Mazziotto and J. Szpirglas. V, 274 pages. 1981.

Vol. 864: Complex Analysis and Spectral Theory. Proceedings, 1979/80. Edited by V. P. Havin and N. K. Nikol'skii. VI, 480 pages. 1981.

Vol. 865: R. W. Bruggeman, Fourier Coefficients of Automorphic Forms. III, 201 pages. 1981.

Vol. 866: J.-M. Bismut, Mécanique Aléatoire. XVI, 563 pages. 1981.

Vol. 867: Séminaire d'Algèbre Paul Dubreil et Marie-Paule Malliavin. Proceedings, 1980. Edited by M.-P. Malliavin. V, 476 pages. 1981.

Vol. 868: Surfaces Algébriques. Proceedings 1976–78. Edited by J. Giraud, L. Illusie et M. Raynaud. V, 314 pages. 1981.

Vol. 869: A. V. Zelevinsky, Representations of Finite Classical Groups. IV, 184 pages. 1981.

Vol. 870: Shape Theory and Geometric Topology. Proceedings, 1981. Edited by S. Mardešić and J. Segal. V, 265 pages. 1981.

Vol. 871: Continuous Lattices. Proceedings, 1979. Edited by B. Banaschewski and R.-E. Hoffmann. X, 413 pages. 1981.

Vol. 872: Set Theory and Model Theory. Proceedings, 1979. Edited by R. B. Jensen and A. Prestel. V, 174 pages. 1981.